Metalloids in Biology

Metalloids belong to a class of elements that exhibit physiochemical characteristics intermediating between those of metals and non-metals. Some are quasi-essential for the overall growth and development of plants. Silicon, for instance, enhances plant structural integrity, while boron is crucial for cell wall formation, and selenium acts as an antioxidant, but some are toxic, like germanium (Ge) and arsenic (As), as they threaten the soil ecosystem and human health. Metalloid toxicity hinges on their cellular concentrations, where low levels aid plant development, whereas high levels cause harmful effects. Thus, it is crucial to encompass the underlying detoxification mechanisms behind metalloid uptake by the root system, their transport to other tissues, and their redistribution within and between cells. This book provides a comprehensive elucidation of the valuable insights of metalloids in green agriculture, emphasizing management strategies to mitigate their adverse effects through various detoxification pathways, including cell complexation, cell wall binding, efflux, vacuolar sequestration, and ultimately redistribution.

Key features:

- Explores databases of metalloid distribution in plants and other habitats
- Deliberates about metalloid transporters and detoxification strategies in plants
- Describes the interaction of metalloids with microbes and their impact on ecophysiology
- Unravels the mysteries of metalloid stress in plants by using multi-omics approaches
- Covers biological applications of metalloids in sustainable agricultural practices and in human health.

This book is aimed to giving updated and scientific insights to readers and researchers associated with plant stress physiology, agricultural sciences, and environmentalists working for the well-being of the environment. Apart from these, the present book will also be a boon for scientists, farmers, teachers, and undergraduate and postgraduate students as it provides a detailed account of the distribution, biochemistry, detoxification mechanisms, and biological applications of metalloids.

Metalloids in Biology

From Function to Toxicity

Edited by

Geetika Sirhindi, Renu Bhardwaj, Nitika Kapoor, and Chandra Shekhar Seth

CRC Press
Taylor & Francis Group
Boca Raton London New York

CRC Press is an imprint of the
Taylor & Francis Group, an informa business

Designed cover image: *Author's owned image*

First edition published in 2025
by CRC Press
2385 NW Executive Center Drive, Suite 320, Boca Raton, FL 33431

and by CRC Press
4 Park Square, Milton Park, Abingdon, Oxon, OX14 4RN

CRC Press is an imprint of Taylor & Francis Group, LLC

ISBN: 9781032673561 (hbk)
ISBN: 9781032673578 (pbk)
ISBN: 9781032673592 (ebk)

DOI: 10.1201/9781032673592

Typeset in Times
by codeMantra

Contents

Foreword ... viii
Acknowledgment ... ix
Editor Biographies ... x
List of Contributors ... xii

Chapter 1 Global Database of Metalloid Distribution and their Presence in Plants' Habitat ... 1

Manish Kumar, Gurvarinder Kaur, Shruti Kaushik, Rashmi Mishra, Palampreet Singh, Geetika Sirhindi, Isha Madaan, Sarbjeet Kaur, and Aprajita

Chapter 2 Metalloids: Functional, Deficiency, and Toxicity Aspects ... 15

Ashutosh Sharma, Priyanka Sharma, Pooja Sharma, Sonika Sharma, Ashish Sharma, Renu Bhardwaj, and Indu Sharma

Chapter 3 Molecular Mechanism Underlying Metalloid Transportation in Plants ... 32

Firdous Rasool Sheikh, Ram Naresh, and Rajesh Kumar Singh

Chapter 4 Metalloid Menace: Potential Accumulation and Detoxification Strategies for Plant Metalloid Tolerance ... 43

Rashmi Choudhary, Kamel A. Abd-Elsalam, Shraddha Singh, Haitao Liu, Faheem Ahmad, Praveen Kumar Srivastava, and Chandra Shekhar Seth

Chapter 5 Stress Ameliorative Role of Metalloids in Plants for Sustainable Agriculture ... 53

Gurvarinder Kaur, Shruti Kaushik, Isha Madaan, Anmol Sidhu, Rashmi Mishra, Aarti, Babita Dhiman, Palampreet Singh, and Geetika Sirhindi

Chapter 6 Role of Silicon in Green Agriculture ... 65

Manu Sharma, Roohi Sharma, Nahida Rehman Mir, Amandeep Singh, Harpreet Kaur, Sakshi Verma, Yogesh K Ahlawat, Kushi Yadav, Puja Ohri, Renu Bhardwaj, and Nitika Kapoor

Chapter 7 Biochemistry of Boron in Plants: Unraveling its Vital Role and Transportation ... 88

Arpit Jindal, Cengiz Kaya, Mohammad Wasim Siddiqui, Kamel A. Abd-Elsalam, Saurabh Chandra Saxena, Prabhakar Sharma, and Chandra Shekhar Seth

Chapter 8 Metalloid Toxicity in Plants: A Systemic Assessment on Recent Insights Concerning their Impact on Plants ... 97

Safoora Mariyam, Kamel A. Abd-Elsalam, Hrishikesh Upadhyaya, Muthu Thiruvengadam, Asish Kumar Parida, Cengiz Kaya, and Chandra Shekhar Seth

Chapter 9 Integrating Multi-Omics Approaches to Unravel the Mysteries Underlying Metalloid Stress in Plants ... 108

Nahida Rehman Mir, Roohi Sharma, Amandeep Singh, Manu Sharma, Bilal Ahmad Mir, Nitika Kapoor, Puja Ohri, Renu Bhardwaj, and Yogesh K. Ahlawat

Chapter 10 Impacts of Metalloids on Photosynthetic Apparatus in Plants ... 132

Twinkle, Ayushi Gautam, Dipanshu Ghosh, Rimpy Dhiman, and Vivek Dogra

Chapter 11 Cross-Talk between Metalloids and Secondary Metabolism in Plants ... 144

Anmol Sidhu, Shruti Kaushik, Gurvarinder Kaur, Isha Madaan, Rashmi Mishra, and Geetika Sirhindi

Chapter 12 Biofortification of Crops with Metalloids: Importance for Human Health and Sustainable Plant Growth ... 156

Nirupma Singh and Kajal Nagre

Chapter 13 Phytohormones and Metalloids: Crosstalk for Plant Growth and Development ... 177

Shruti Kaushik, Alok Ranjan, Anmol Sidhu, Gurvarinder Kaur, Isha Madaan, Rashmi Mishra, Anil Kumar Singh, and Geetika Sirhindi

Chapter 14 Phytoremediation of Metalloids ... 196

Alok Ranjan, Shruti Kaushik, Anil Kumar Singh, and Geetika Sirhindi

Chapter 15 Plant-Fungal Interaction in Remediation of Metalloids from Environment ... 211

Venkataramana Chapara and Rani Chapara

Chapter 16 Algal-Metalloid Dynamics: Unraveling the Impact on Plant Ecophysiology ... 221

Kushi Yadav, Swati Mangla, Gaurav Pathak, Pranshu Kalson, Veer Bhan, and Yogesh K. Ahlawat

Chapter 17 Siderophores and their Action against Metalloid Toxicity 246

Roohi Sharma, Deepak Kumar, Parkirti, Nahida Rehman Mir, Manu Sharma, Amandeep Singh, Bilal Ahmad Mir, Puja Ohri, Renu Bhardwaj, and Nitika Kapoor

Chapter 18 Perspective of Biotechnological Mechanisms of Metalloid Detoxification in Plants 270

Gagan Jain, Gurvarinder Kaur, Rashmi Mishra, Aarti, Babita Dhiman, Palampreet Singh, and Geetika Sirhindi

Chapter 19 Biological Applications of Metalloid Nanoparticles 279

Isha Madaan, Gurvarinder Kaur, Shruti Kaushik, Anmol Sidhu, Kuljinder Kaur, Shamsher Singh, Akanksha Sharma, Simranpreet Kaur, Manjot Kaur, Jahanvi, Jetina, Nivedita Sharma, Tarandeep Kaur, Renu Bhardwaj, and Geetika Sirhindi

Index 299

Foreword

As we set on a journey into the world of metalloids, deciphering their functionalities and implications on the biotic components of the ecosystem, this piece of scientific art comes to the forefront and treads the summit of scientific majesty. Metalloids, from being essential nutrient elements to potential toxins, have allured the scientific community for decades now. In *Metalloids in Biology: From Function to Toxicity*, the reader is put through the path that traverses through the varied realms of metalloids involving biochemistry, molecular biology, plant biology, and human health.

This literary work serves as evidence of the interdisciplinary collaboration of academicians and specialists who have devoted their professional lives to elucidating the enigmatic nature of metalloids. Employing thorough investigation and novel approaches, the chapters contained within this volume provide an all-encompassing comprehension of the dynamics of metalloids in the environment of plants, including their physiological operations, mechanisms of transport, and the delicate equilibrium between their advantageous and deleterious impacts. An intriguing facet of this book pertains to its investigation of metalloid stress and the exceptional adaptive mechanisms implemented by plants in order to alleviate the detrimental effects of such toxins. Through the exploration of omics approaches and the complex relationship between metalloids and phytohormones, every chapter sheds light on the intricate network of interactions that govern the way in which plants react to metalloid exposure.

Moreover, this book explores the promising genre of biotechnological interventions for metalloid detoxification, the potential of biofortification in enhancing human health, and the innovative applications of metalloid nanoparticles in various biological contexts. The exploration of phytoremediation, plant-fungal interactions, and the symbiotic relationships between algae and plants adds depth to our understanding of sustainable environmental practices.

As we peruse the contents of this manuscript, we are not only enriched with knowledge but also motivated to further push the limits of scientific investigation. I commend the authors, editors, and contributors for their invaluable contributions to this seminal work. I am confident that this book will serve as a fundamental resource for forthcoming scientific pursuits in the ever-evolving domain of metalloid biology.

Prof. Bashir Ahmad Ganai
Dean School of Unani Medicine,
Former Director CORD,
Former Director of North Campus,
University of Kashmir

Acknowledgment

Metalloids, a group of elements that exhibit properties intermediate between metals and non-metals, have garnered significant attention in recent years due to their multifaceted roles in biological systems. This book, *Metalloids in Biology: From Function to Toxicity*, aims to provide a comprehensive exploration of the various dimensions of metalloids, from their essential biological functions to their potential toxicological impacts. The chapters in this volume have been meticulously curated to cover a broad spectrum of topics related to metalloids. Starting with a global database of metalloid distribution and their presence in plant habitats, this book delves into the functional, deficiency, and toxicity aspects of metalloids. Detailed insights into the molecular mechanisms underlying metalloid transportation in plants and the strategies plants employ to accumulate and detoxify these elements are also provided.

Special emphasis is placed on the stress ameliorative role of metalloids in plants, highlighting their importance for sustainable agriculture. The role of silicon and boron, critical metalloids in green agriculture, is discussed in depth, along with the broader impacts of metalloid toxicity on plant growth and development. Omics approaches toward understanding metalloid stress further enrich the discourse. The book also examines the impact of metalloids on photosynthetic efficiency and the intricate crosstalk between metalloids and plant metabolism. Biofortification of crops with metalloids, and the interplay of phytohormones in plant growth and development, are explored to underscore the significance of these elements in enhancing plant health and productivity. The environmental aspects of metalloids are addressed through discussions on phytoremediation, plant-fungal interactions, and the role in algae. The perspective of biotechnological mechanisms in metalloid detoxification and the biological applications of metalloid nanoparticles provide a futuristic outlook on the subject.

This book is a collective effort of esteemed researchers and scientists, whose contributions have been instrumental in advancing our understanding of metalloids in biological systems. We hope that this compilation will serve as a valuable resource for researchers, students, and professionals in the fields of plant science, environmental science, and biotechnology.

We extend our deepest gratitude to all the authors for their invaluable contributions and to the reviewers for their critical insights. We are confident that this book will inspire further research and foster a deeper appreciation of the complex roles of metalloids in biological organisms.

Geetika Sirhindi

Renu Bhardwaj

Nitika Kapoor

Chandra Shekhar Seth

Editor Biographies

Dr. Geetika Sirhindi is a distinguished Professor of Plant Physiology in the Department of Botany at Punjabi University, Patiala. She holds an M.Sc. and a Ph.D. and has garnered several academic honors, including the Nature Publication Group Award in 2007, the A.C. Joshi Memorial Award, and the Hooker Memorial Botanical Society Award from the Department of Botany at Punjabi University for the years 1993–1995. With 17 years of extensive research and teaching experience, Dr. Sirhindi has guided numerous scholars, including 12 Ph.D. awardees, 3 registered Ph.D. candidates, 2 enrolled and 4 M.Phil. students, and 11 M.Sc. dissertation students. She has successfully completed three major research projects funded by SERB-DST and UGC, and she currently collaborates with Godrej Agrovet on an industrial project. Dr. Sirhindi has an impressive publication record with 50 research articles, multiple book chapters, and two edited books. Her academic contributions extend beyond research, as she has participated in 43 general orientation/refresher courses, workshops, women capacity-building programs, and conferences. She has also organized a workshop/conference. Her administrative roles include being a life member of several prestigious societies such as the Indian Society of Plant Physiology, Punjab Academy of Sciences, K. K. Nanda Foundation, Indian Botanical Society, Biotechnology Research Society of India, and the Indian Science Congress. Additionally, she serves as a reviewer for various scientific research journals published by Springer, Elsevier, Taylor & Francis, and MDPI. Dr. Sirhindi's research students have achieved placements at esteemed institutions. Her international academic footprint includes visits to Paris, Switzerland, Japan, the USA, and Russia, underscoring her global engagement in plant physiology research. She has been tirelessly working in the field of plant stress physiology, and her lab's recent work focuses on metal toxicity, which aligns with the objective of this book.

With over 40 years of research expertise and 33 years of teaching experience, **Dr. Renu Bhardwaj** is a renowned academic and researcher. Her outstanding list of credentials includes an M.Sc., an M.Phil., a Ph.D., and many fellowships (FLESA, FMSET, FICEE, FMSET, and FLS). Numerous awards have been given to her, including the Senior Scientist Award, Scientist of the Year Award, CSIR Visiting Associateship, National Environmental Science Academy Fellowship, Fellowship of the Linnean Society, INSA Visiting Fellowship, Nation Builder Award, and Pearl Foundation Lifetime Achievement Award. 42 Ph.D. students, 11 M.Phil. students, and 25 M.Sc. projects have been supervised by Dr. Bhardwaj. Major grants from esteemed institutions such as CSIR, UGC, DST, DBT, and MOEF, among others, have financed her research. She has finished twenty significant research projects and handled major departmental funds, such as UGC-DRS and DST-FIST. Her extensive body of work has yielded 500 papers, 7 patents, 8 edited volumes, an i10 index of 268 and over 15,427 citations, as well as a Scopus h-index of 61. She has coordinated and participated in a great deal of seminars, conferences, and capacity-building events. Dr. Bhardwaj has held administrative positions at GNDU as Director of IQAC, Dean of Life Sciences, and Director of Research. She has also had important positions in UGC and NAAC, such as coordinator of peer team visits and women's capacity-building programs. Several Indian universities and research institutes, Zhejiang University in China, the University of Doha in Qatar, and the National Institutes of Health in the United States have all hired her pupils. Due to her significant contributions to research and education, Dr. Bhardwaj is well known in her field.

Dr. Nitika Kapoor is an Assistant Professor at the Department of Botanical and Environmental Sciences, Guru Nanak Dev University, Amritsar, Punjab, India. Dr. Kapoor did M.Sc. (Botany) and Doctorate in Plant Physiology from Guru Nanak Dev University, Amritsar, Punjab. She has more than 15 years of teaching and research experience in the field of Plant Physiology. Her research interests include Plant Stress Physiology, Biochemistry, Ecophysiology, Phytoremediation, PGPR, and Biochar. She has received Junior and Senior research fellowships from UGC-CSIR, MHRD, the Govt. of India. She has also been Principle investigator of a Major research project funded by the UGC, Govt. of India. She has published more than 17 peer-reviewed research papers in journals of international repute and around 30 book chapters in edited books by Elsevier, Springer, John Wiley, Nova Science Publisher, Taylor and Francis group, etc. She has already edited 2 books. She has presented her research work in more than 50 international/national conferences. She has been a reviewer of many peer-reviewed research journals of international repute, including *Plant Stress* (Elsevier), *Vegetos* (Springer), *International Journal of Environmental Research* (Springer), *Heliyon* (Cell Press), *Frontiers in Plant Science*, and *Frontiers in Microbiology* (Frontiers Media). Dr. Kapoor is a life member of the Punjab Academy of Sciences, Bioclues Organization, and Asia Association of Plant Scientists.

Dr. Chandra Shekhar Seth is an Associate Professor at the Department of Botany, University of Delhi, New Delhi-110007, Delhi, India. He obtained his M.Sc. degree in Botany from Banaras Hindu University, Varanasi (U.P.) in 2003 and his Ph.D. degree in Botany from the University of Lucknow (U.P.) in 2008. He currently teaches the course papers naming Plant physiology and Biochemistry, Topics in Plant Physiology and Biochemistry, Algae, Environment and Human Welfare at the postgraduate level, and Phytohormone-assisted abiotic stress management in plants at the Ph.D. level. His major research interest areas include how various phytohormones, particularly salicylic acid (SA), jasmonic acid (JA), and nitric oxide (NO) mitigate the heavy metals (HMs) and salinity induced toxicity in plants, photosynthesis and nitrogen metabolism under abiotic stress, and the role of nanomaterials in mitigating the HMs and salt stresses in plants. Oxidative stress and tolerance mechanism via antioxidants and phytochelatins, phytoremediation of HMs, and potential use of algae as biofuels, nutrients, and pigments are other important interesting area of his work. He is the fellow of two prestigious societies, naming Fellow of the Academy of Environmental Biology (FAEB) Lucknow-226020 (U.P.) India and Fellow Society for Plant Research (FSPR) Meerut-250110, U.P., India. He has been the Life members of ISCA (The Indian Science Congress Association) Kolkata-700017, India, ISEB (International Society of Environmental Botanist) Lucknow-226001 (U.P.), AEB (The Academy of Environmental Biology) Lucknow-226020 (U.P.), India, ISPM (International Society of Plant Morphologists), New Delhi-110007, India, and DUBS (Delhi University Botanical Society) New Delhi-110007, India. He has already guided more than 25 M.Sc. dissertation, 04 M. Phil, 03 Ph.D. students, and is guiding 06 more Ph.D. students currently in his focused research areas. He has published many peer-reviewed excellent research papers, review papers, and several books and book chapters related to this field. He is an internationally recognized reviewer for many journals.

Contributors

Aarti
Department of Botany
Punjabi University
Patiala, Punjab, India

Kamel A. Abd-Elsalam
Agricultural Research Center
Plant Pathology Research Institute
Giza, Egypt

Yogesh K. Ahlawat
Department of Biotechnology
University Centre for Research and Development
Chandigarh University
Mohali, Punjab, India

Faheem Ahmad
Department of Botany
Aligarh Muslim University
Aligarh, Uttar Pradesh, India

Amritsar
Department of Microbiology
DAV University, Sarmastpur
Jalandhar, Punjab, India

Aprajita
Postgraduate Department of Botany
Dev Samaj College for Women
Ferozepur City, Punjab, India

Veer Bhan
Department of Biotechnology
UIET, MDU
Rohtak, Haryana, India

Renu Bhardwaj
Department of Botanical and Environmental Sciences
Guru Nanak Dev University
Amritsar, Punjab, India

Rani Chapara
Acharya N. G. Ranga Agricultural University,
Lam Farm, Guntur, Andhra Pradesh, India

Venkataramana Chapara
North Dakota Agricultural Experiment Station
North Dakota State University
Fargo, North Dakota

Rashmi Choudhary
Department of Botany
University of Delhi
New Delhi, India

Babita Dhiman
Department of Botany
Punjabi University
Patiala, Punjab, India

Rimpy Dhiman
Biotechnology Division
CSIR-Institute of Himalayan Bioresource Technology
Palampur, India

Vivek Dogra
Biotechnology Division
CSIR-Institute of Himalayan Bioresource Technology
Palampur, Himachal Pradesh, India
and
Academy of Scientific and Innovative Research (AcSIR)
Ghaziabad, India

Ayushi Gautam
Biotechnology Division
CSIR-Institute of Himalayan Bioresource Technology
Palampur, Himachal Pradesh, India

Dipanshu Ghosh
Biotechnology Division
CSIR-Institute of Himalayan Bioresource Technology
Palampur, Himachal Pradesh, India
and
Academy of Scientific and Innovative Research (AcSIR)
Ghaziabad, India

Jahanvi
Science Department
Government College of Education
Jalandhar, Punjab, India

Gagan Jain
Department of Land Management and Systems,
Lincoln University
Canterbury, New Zealand

Jetina
Science Department
Government College of Education
Jalandhar, Punjab, India

Arpit Jindal
Department of Botany
University of Delhi
New Delhi, India

Pranshu Kalson
Department of Biotechnology
UIET, MDU
Rohtak, Haryana, India

Nitika Kapoor
Department of Botanical and Environmental Sciences
Guru Nanak Dev University
Amritsar, Punjab, India

Gurvarinder Kaur
Department of Botany
Punjabi University
Patiala, Punjab, India

Harpreet Kaur
P.G. Department of Botany
Khalsa College
Amritsar, Punjab, India

Kuljinder Kaur
Department of Botany
Government College Hoshiarpur
Punjab, India.

Manjot Kaur
Science Department
Government College Hoshiarpur
Punjab, India

Sarbjeet Kaur
Postgraduate Department of Botany
Dev Samaj College for Women
Ferozepur City, Punjab, India

Simranpreet Kaur
Science Department
Government College of Education
Jalandhar, Punjab, India

Tarandeep Kaur
Science Department
Government College of Education
Jalandhar, Punjab, India

Shruti Kaushik
Young Professional II, Division of Soil and Crop Management
ICAR-Central Soil and Salinity Research Institute
Karnal, Haryana, India
and
Department of Botany
Punjabi University
Patiala, Punjab, India

Cengiz Kaya
Soil Science and Plant Nutrition Department
Harran University
Sanliurfa, Turkey

Deepak Kumar
Department of Zoology
Guru Nanak Dev University
Amritsar, Punjab, India

Manish Kumar
Postgraduate Department of Botany
Dev Samaj College for Women
Ferozepur City, Punjab, India

Haitao Liu
College of Resources and Environment
Henan Agricultural University
Zhengzhou, China

Isha Madaan
Department of Botany
Punjabi University
Patiala, Punjab, India
and
Science Department
Government College of Education
Jalandhar, Punjab, India

Swati Mangla
Department of Biotechnology
UIET, MDU
Rohtak, Haryana, India

Safoora Mariyam
Department of Botany
University of Delhi
New Delhi, India

Bilal Ahmad Mir
Department of Botany
North Campus, University of Kashmir
Delina, Baramulla, Jammu and Kashmir, India

Nahida Rehman Mir
Department of Botanical and Environmental Sciences
Guru Nanak Dev University
Amritsar, Punjab, India

Rashmi Mishra
Department of Botany
Punjabi University
Patiala, Punjab, India

Kajal Nagre
Division of Genetics
ICAR-Indian Agricultural Research Institute
New Delhi, India

Ram Naresh
Biotechnology Division
CSIR-Institute of Himalayan Bioresource Technology
Palampur, Himachal Pradesh, India

Puja Ohri
Department of Zoology
Guru Nanak Dev University
Amritsar, Punjab, India

Asish Kumar Parida
Plant Omics Division
CSIR-Central Salt and Marine Chemicals Research Institute (CSIR-CSMCRI)
Gijubhai Badheka Marg, Bhavnagar, Gujarat, India

Parkirti
Department of Zoology
Guru Nanak Dev University
Amritsar, Punjab, India

Gaurav Pathak
Department of Biotechnology
UIET, MDU
Rohtak, Haryana, India

Alok Ranjan
Department of Biotechnology
Patna Women's College
Bihar, India

Saurabh Chandra Saxena
Department of Biochemistry
Central University of Harayana
Jant-Pali, Mahendragarh, Haryana, India

Chandra Shekhar Seth
Department of Botany
University of Delhi
New Delhi, India

Akanksha Sharma
Science Department
Government College of Education
Jalandhar, Punjab, India

Ashish Sharma
Department of Botany and Environment Studies
DAV University, Sarmastpur
Jalandhar, Punjab, India

Ashutosh Sharma
Faculty of Agricultural Sciences
DAV University, Sarmastpur
Jalandhar, Punjab, India

Indu Sharma
Department of Life Sciences, University Institute of Sciences
Sant Baba Bhag Singh University Khiala
Jalandhar, Punjab, India

Manu Sharma
Department of Botanical and Environmental Sciences
Guru Nanak Dev University
Amritsar, Punjab, India

Nivedita Sharma
Science Department
Government College of Education
Jalandhar, Punjab, India

Pooja Sharma
Department of Botanical and Environmental Sciences
Guru Nanak Dev University
Amritsar, Punjab, India
and
Department of Microbiology
DAV University, Sarmastpur
Jalandhar, Punjab, India

Prabhakar Sharma
Department of Agricultural Engineering and Technology, School of Engineering and Technology
Nagaland University
Dimapur, Nagaland, India

Priyanka Sharma
School of Bioengineering Sciences & Research
MIT-Art, Design, and Technology University
Loni Kalbhor, Pune, Maharashtra, India

Roohi Sharma
Department of Zoology
Guru Nanak Dev University
Amritsar, Punjab, India

Sonika Sharma
Faculty of Agricultural Sciences
DAV University, Sarmastpur
Jalandhar, Punjab, India

Firdous Rasool Sheikh
Biotechnology Division
CSIR-Institute of Himalayan Bioresource Technology
Palampur, Himachal Pradesh, India
and
Academy of Scientific and Innovative Research (AcSIR)
Ghaziabad, India

Mohammad Wasim Siddiqui
Department of Food Science and Post-Harvest Technology
Bihar Agricultural University
Sabour, Bhagalpur, Bihar, India

Anmol Sidhu
Department of Botany
Punjabi University
Patiala, Punjab, India

Amandeep Singh
Department of Botanical and Environmental Sciences
Guru Nanak Dev University
Amritsar, Punjab, India

Anil Kumar Singh
ICAR-National Institute for Plant Biotechnology
New Delhi, India

Nirupma Singh
Division of Genetics
ICAR-Indian Agricultural Research Institute
New Delhi, India

Palampreet Singh
Department of Botany
Punjabi University
Patiala, Punjab, India

Rajesh Kumar Singh
Biotechnology Division
CSIR-Institute of Himalayan Bioresource Technology
Palampur, Himachal Pradesh, India

Shamsher Singh
Department of Botany
Government College Hoshiarpur
Punjab, India

Shraddha Singh
Nuclear Agriculture and Biotechnology Division
Bhabha Atomic Research Centre, Homi Bhabha National Institute
Mumbai, Maharashtra, India

Geetika Sirhindi
Department of Botany
Punjabi University
Patiala, Punjab, India

Praveen Kumar Srivastava
Rajiv Gandhi Institute of Petroleum Technology
Jais, Amethi, Uttar Pradesh, India

Muthu Thiruvengadam
Department of Applied Bioscience, College of Life and Environmental Sciences
Konkuk University
Seoul, South Korea

Twinkle
Biotechnology Division
CSIR-Institute of Himalayan Bioresource Technology
Palampur, Himachal Pradesh, India
and
Academy of Scientific and Innovative Research (AcSIR)
Ghaziabad, India

Hrishikesh Upadhyaya
Department of Botany
Cotton University
Pan Bazaar, Guwahati, Assam, India

Sakshi Verma
Department of Zoology
Hans Raj Mahila Maha Vidyalaya
Jalandhar, Punjab, India

Kushi Yadav
Department of Biotechnology
Amity University
Noida, India

1 Global Database of Metalloid Distribution and their Presence in Plants' Habitat

Manish Kumar, Gurvarinder Kaur, Shruti Kaushik, Rashmi Mishra, Palampreet Singh, Geetika Sirhindi, Isha Madaan, Sarbjeet Kaur, and Aprajita

1.1 INTRODUCTION

Metalloids consist of a group of chemical elements which are found commonly in nature, and most of their physical and chemical characteristics are intermediate between metals and nonmetals; hence, it is hard to classify them as either metals or nonmetals. Arsenic (As), antimony (Sb), boron (B), germanium (Ge), silicon (Si), and tellurium (Te) are usually recognized as metalloids (Vernon, 2013). Primary sources of metalloids are the weathering of parent material, dissociation of minerals, mining, atmospheric deposition, burning of fossil fuels, use of fertilizers and pesticides, industrial emissions, anthropogenic sources, sewage, agrochemicals, and municipal and industrial wastes (Nagajyoti et al., 2010; Alloway, 2013). They are important elements showing the effects ranging from the essential to highly toxic in nature. Generally, metalloids are not toxic; they only exhibit toxic effects when their internal concentration in the biological system exceeds a certain threshold. Few metalloids are essential in very small quantities for proper metabolism, growth, and development. In contrast, few metalloids show a toxic effect even at very low concentrations. The toxic effects of metalloids also depend on various factors, including the type of metalloid, its concentration, organisms, plant species, and the stage of growth (Appenroth, 2010). Their higher accumulation in soils, drinking water, and food chain affect plant, animal, and human health. The bioavailability of metalloids also depends on pH, soil organic matter, soil structure, and chemical composition. Here, we briefly discuss the distribution and presence of metalloids in plant habitats.

1.2 SOURCES OF METALLOID DISTRIBUTION

Metalloid contamination has emerged as a significant global concern. Soil contamination caused by metalloids is a matter of global significance, given its implications for the environment, biology, and human health (Karadaş & Kara, 2011). Mosa et al. (2016) reported that anthropogenic activities like mining, industrialization, release of industrial waste, as well as intensive agricultural practices, have resulted in a rise in the buildup of metalloids in agricultural soils and crops. This poses a significant threat to human health, particularly for those who rely on food from multiple contaminated regions for their survival. Yuan et al. (2019) demonstrated that annually, a substantial amount of metalloids are extracted from the lithosphere and introduced into the anthroposphere for various purposes, such as industrial, agricultural, chemical, and medical applications. Throughout the lifespan, metalloids are bound to be released into aquatic environments through various means such as mining waste, discharge from wastewater treatment plants, surface run-off, atmospheric deposition, and direct input (Fu et al., 2014). The discharge of substantial amounts of metalloids into natural aquatic environments leads to sediment enrichment due to hydrolysis, adsorption, and precipitation, while

DOI: 10.1201/9781032673592-1

only a minor fraction remains in a dissolved state within the water (Xiao et al., 2015). Moreover, due to their potential toxicity, persistence, inability to degrade, and subsequent potential for bioaccumulation, the metalloid contaminants can present significant hazards to invertebrates (Fu et al., 2013). As a result, the extent of metalloid pollution in the riverbed sediments of numerous vital natural rivers has been continuously capturing the attention of researchers (Dubovina et al., 2018; Islam et al., 2018). The migration of metalloids into their surroundings poses a significant threat to the environment and human health, thereby elevating environmental risks. Industrial and urban growth result in the introduction of substantial amounts of pollutants, such as heavy metals and metalloids, into surface soils (Liang et al., 2017). The buildup of metalloids in agricultural soil has garnered considerable interest due to their potential entry into the food chain and the associated risks they pose to human health (Deng et al., 2023). The sources of heavy metals in soils include the breakdown of soil minerals, the use of treated wastewater for land application, the application of sewage sludge and fertilizers, and various industrial activities.

1.2.1 Distribution of Metalloids in Soil System of Plants

Metalloids are present in the environment in association with various inorganic and organic substances, as well as other natural chemicals (Kroukamp et al., 2016). Some soils exhibit unusually high concentrations of metalloids due to extensive industrial exploitation, negatively impacting plant development and causing ecosystem degradation (Eiche et al., 2015). B, Si, and As are commonly found in soils, and their presence can be measured in most soil solutions as boric acid [$B(OH)_3$], silicic acid [$Si(OH)_4$], trivalent arsenious acid [$As(OH)_3$], or pentavalent arsenic acid [$AsO(OH)_3$]. These elements, As being poisonous, Si advantageous, and B essential for plants, have varying impacts on plant growth. Accordingly, plants have evolved physiological mechanisms to either utilize or detoxify these metalloids (Javaid et al., 2022).

The inorganic forms of As, specifically arsenite [As(III)] and arsenate [As(V)], are primarily absorbed by plant systems. Arsenic enters the environment through natural or human sources (Abbas et al., 2018; Dutta & Bandopadhyay, 2016). Terrestrial plants, such as *Pteris vittata*, exhibit a lower ability to absorb As from soils compared to wetland plants. The apoplastic pathway or a membrane pump facilitates As penetration into the root endodermal cells and entry into the root symplast (Dutta & Bandopadhyay, 2016). Aquaporins, also known as nodulin 26-like intrinsic proteins (NIPs), play a role in actively transporting arsenite inside plant cells (Deshmukh et al., 2017; Dutta & Bandopadhyay, 2016). NIPs are implicated in bidirectional transport, allowing both influx and efflux of As(III) (Pommerrenig et al., 2015). In contrast, arsenate is absorbed by plants via phosphate transporters Pht1;1 and Pht1;4 (Xu et al., 2015). The mechanisms governing the influx of organic arsenicals, particularly monomethyl arsenic acid (MMA) and dimethylarsinic acid (DMA), remain uncertain, and research on the general mechanism governing As acid translocation from roots to shoots is limited (Dutta & Bandopadhyay, 2016; Farooq et al., 2016).

Boron enters plant cells through bidirectional diffusion via lipid bilayer, selective or non-selective channels, and passive diffusion. It is exported into and out of cells in the form of uncharged boric acid. High natural B concentrations are found in arid region soils, but human activities such as coal mining, fly ash deposition, and irrigation with B-laden water can lead to increased soil B deposition. Plants absorb B through their roots, and transpiration stream moves it to the stems and leaves (Parveen et al., 2022).

Si, the second most common element in the Earth's crust, is absorbed by plants in the form of silicic acid, which is uncommon in most agricultural soils (Deshmukh & Bélanger, 2016; Deshmukh et al., 2017). Plants are categorized as accumulators, excluders, or intermediate forms based on the amount of Si that accumulates in their tissues. Si accumulation in the form of amorphous silica (SiO) in the roots and shoots of various plants varies, and this variation is connected to the uptake and transport of the metalloid in plants (Deshmukh et al., 2020). Rice, a well-known Si accumulator, has been extensively studied regarding its Si absorption processes (Zargar et al., 2019; Bhat et al., 2019). Generally, plant species with a higher capacity to absorb Si yield greater benefits, though some Solanacee species, like tomatoes, are considered excluders despite benefiting from Si supplementation (Hoffmann et al., 2020).

Silica (HaSiO) is identified as the plant-available form of silicon (PAS) in soil when it exists in its monomeric form (Tubana et al., 2016). Soybean plants growing in soil with <20 ppm Si accumulated approximately 0.4% Si in their leaves, while in soils with PAs concentration >60 ppm, Si reached approximately 1% (Deshmukh et al., 2020).

1.2.2 Distribution of Metalloids in Water System of Plants

The distribution of metalloids in plant water systems is a complex and critical phenomenon that significantly impacts the overall health of aquatic ecosystems. Metalloids, such as arsenic, antimony, and selenium, are present in varying concentrations in water sources, primarily due to natural geological processes and human activities. These metalloids exhibit different behaviors in water systems, influencing their availability for uptake by plants and subsequent effects on the water ecosystem (Andráš et al., 2018). The distribution of metalloids in plant water systems involves their movement from the environment into plants, their accumulation within plant tissues, and their subsequent impact on the plant and the surrounding ecosystem. Metalloids, such as arsenic, antimony, and selenium, can enter plant water systems through various pathways, including uptake from soil or water, atmospheric deposition, and direct absorption from water sources (Mykolenko et al., 2018).

The availability of metalloids for uptake by plants is influenced by several factors, including the physicochemical properties of the water, the presence of other ions, and the characteristics of the plant itself. In water systems, metalloids can exist in different forms, such as dissolved species, colloidal particles, or adsorbed onto solid surfaces. The speciation of metalloids can affect their mobility and bioavailability in plants (Pandey et al., 2019). One of the key factors influencing the availability of metalloids for uptake by plants is the physicochemical properties of the water. The pH of the water, for example, can greatly influence the solubility and mobility of metalloids. In general, metalloids are more soluble and mobile in acidic conditions, increasing their availability for plant uptake (Li et al., 2018). Conversely, in alkaline conditions, metalloid solubility decreases, reducing their uptake by plants. Redox potential, which is a measure of the tendency of a substance to gain or lose electrons, also plays a significant role in metalloid availability. The reduced forms of metalloids tend to form more soluble and mobile species, increasing their availability for uptake by plants. In contrast, under oxidizing conditions, metalloids may form less soluble and less mobile species, reducing their availability for plant uptake (Seraj et al., 2018).

Plants can take up metalloids through their roots or leaves, with root uptake being the primary pathway for most metalloids. The uptake of metalloids by plants is influenced by factors such as the concentration of metalloids in the water, the pH of the water, and the presence of competing ions. Once absorbed by the roots, metalloids can be transported within the plant through the xylem, reaching various plant tissues, including leaves, stems, and roots (Feng et al., 2021).

The distribution of metalloids within plant tissues can vary depending on the plant species and the specific metalloid involved. Some metalloids, such as arsenic and selenium, have been found to accumulate predominantly in the roots, while others, like antimony, can translocate to the aboveground parts of the plant. The accumulation of metalloids in plant tissues can have both beneficial and detrimental effects on plants. At low concentrations, some metalloids, such as selenium, can act as essential micronutrients, promoting plant growth and development. However, at higher concentrations, metalloids can become toxic to plants, leading to physiological disorders, reduced growth, and even plant death. The accumulation of metalloids in plant tissues can also have implications for the surrounding ecosystem (Deng et al., 2021; Clemens & Ma, 2016).

The presence of other ions in the water can also influence the availability of metalloids for uptake by plants. Competitive interactions between metalloids and other ions for binding sites on plant roots can affect the uptake of metalloids. Additionally, the presence of certain ions, such as phosphate, sulfate, and bicarbonate, can form complexes with metalloids, altering their speciation and mobility in water systems. Once absorbed by plants, metalloids can accumulate in different plant parts, such as roots, stems, and leaves, depending on the plant species and the specific metalloid involved (Kroukamp et al., 2016). This accumulation can have both beneficial and detrimental effects on plants. For example, some metalloids, like selenium, can be beneficial to plant growth at low concentrations, acting as essential micronutrients. However, at higher concentrations, they can become toxic, leading to physiological disorders and reduced plant growth.

Furthermore, the accumulation of metalloids in plants can have cascading effects on the entire aquatic ecosystem. When plants die or shed leaves containing accumulated metalloids, these metalloids can enter the water, affecting other organisms in the ecosystem. Metalloids can accumulate in the tissues of aquatic organisms, leading to bioaccumulation and biomagnifications along the food chain. This can pose risks to aquatic organisms, as high levels of metalloids can be toxic and disrupt normal physiological functions. Additionally, the presence of metalloids in plant tissues can alter their decomposition rates, affecting nutrient cycling and overall ecosystem dynamics. Metalloids can inhibit the activity of decomposers, leading to slower decomposition rates and the accumulation of organic matter in water bodies. This can alter nutrient cycling processes, affecting the availability of nutrients for other organisms in the ecosystem (Singh et al., 2022).

Overall, understanding the distribution of metalloids in plant water systems and their impact on the water ecosystem is crucial for effective water management and conservation efforts. Strategies to mitigate the impacts of metalloids on aquatic ecosystems may include reducing metalloid inputs from anthropogenic sources, implementing phytoremediation techniques to remove metalloids from water systems, and monitoring metalloid concentrations in water and biota to assess the health of aquatic ecosystems.

1.3 BIO-FUNCTIONAL ASPECTS AND TOXICITY OF METALLOIDS

Metalloids are concerned with various physiological processes in plants. Among them Si, B, selenium (Se), and As are required in small quantities for proper plant growth and development. The toxicity of metalloids largely depends on their concentration within the living cells. Some metalloids, at low concentrations, may be beneficial for plant growth and development; however, when present at high concentration, they often exert negative effects on plant growth, biomass, photosynthesis, and yield (Bienert et al., 2008). It has been demonstrated that the accumulation of metalloids in plants causes several biochemical, physiological, and molecular changes (Hodson, 2013).

Arsenic is highly toxic to plants in their ionic form; As is also toxic in its methylated forms, whereas methylated As, at least in the pentavalent state, is generally considered less toxic than As(III) and As(V). This element can damage different cellular structures and a variety of tissues and organs. One major cause of toxicity is the strong interaction with sulfhydryl groups; another is the interference with the homeostasis of essential elements.

Among the metalloids, Si is the second most abundant element on the earth's crust. Although Si accounts for up to 10% of the shoot dry weight in higher plants, it is not considered an essential element for plants. Several researchers demonstrated that Si is beneficial for plant growth and development (Ouellette et al., 2017; Deshmukh & Bélanger, 2016; Vivancos et al., 2015; Tripathi et al., 2017). It improves strength and structural integrity of the cell wall in rice plants (Asada & Tazaki, 2001). It alleviates diverse abiotic stresses like heat stress, low-temperature stress, drought stress, and metal toxicity and decreases lodging potential and salt stress (Řezanka & Sigler, 2008; Tripathi et al., 2016). Plants growing in soils with increased cadmium concentration, when supplemented with Si, displayed lower reactive oxygen species (ROS) compared to control plants (Hasanuzzaman et al., 2017), suggesting a role of Si in antioxidant defense mechanisms.

In addition, a negative correlation has been observed between external Si supply and uptake and concentration of As (III) in rice seedlings under hydroponic as well as soil culture conditions (Sanglard et al., 2016). Positive effects have been reported for Si application on chlorophyll content and photosynthesis under heavy metal toxicity. For instance, silicon accumulation in the exodermis and endodermis in the roots of rice plants decreases the transport of sodium which is strongly correlated with salt tolerance.

Boron is one of the micronutrients that is required by plants for their optimum growth and development. Decreased boron availability causes some of the most noted disorders in plants such as cracked stems of celery *Apium graveolens* L., brown heart of *Brassica napobrassica* Mill. and *Raphanus sativus* L. roots, brown heart of *Brassica oleracea* var. *botrytis* L., internal brown spots of *Ipomoea batatas* Lam., and heart rot of *Beta vulgaris* L. (Gupta & Gupta, 2016).

Boron is essential in regulating cell elongation and division, as a result of which it directly impacts root growth (Figure 1.1). In addition, several studies have reported the beneficial role of boron in plant growth and metabolism by regulating processes such as auxin and phenol metabolism (Camacho-Cristóbal et al., 2008), formation of flowers and subsequent seed production (Zohaib et al., 2018), and membrane function (González-Fontes et al., 2016).

Se is a widely distributed trace element with dual (beneficial or toxic) effects for plants. Anthropogenic activities affect the content of Se in the environment. The uptake and metabolism of Se are also greatly varied in different soil and plant systems due to the heterogeneity of physiological and biochemical nature. It has been reported that Se supplementation in a relatively low amount can alleviate the negative impact of a variety of abiotic stresses in plants by improving growth and development (Hasanuzzaman et al., 2020). For example, Se application via plant roots restricted the uptake and translocation of heavy metals/metalloids (Rizwan et al., 2020; Zwolak,

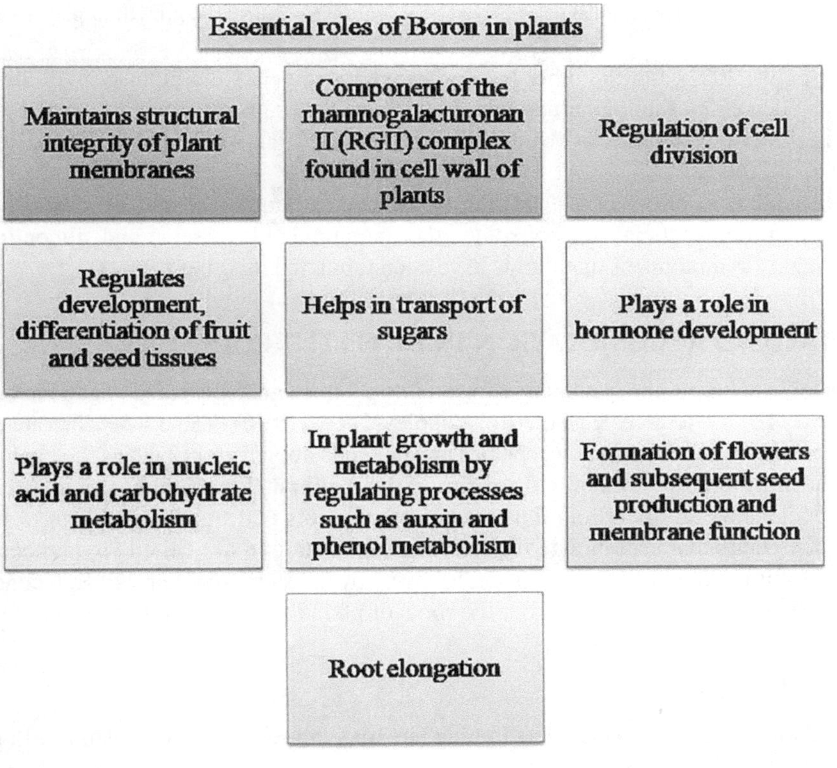

FIGURE 1.1 Beneficial roles of boron in plants.

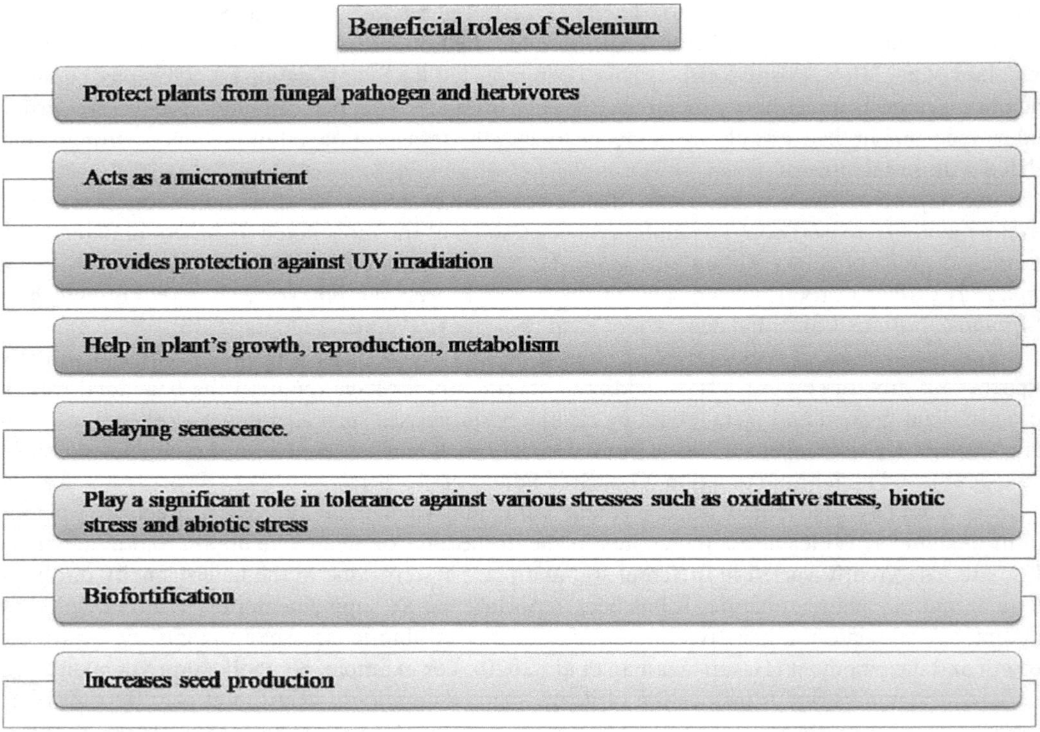

FIGURE 1.2 Essential physiological roles played by selenium to control the growth of plants.

2020; Feng et al., 2021). On the other hand, Se toxicity impedes the plant growth, development, and disturbs plant ecophysiology along with causing chlorosis and necrosis, restricted growth, and reduction in protein biosynthesis (Mroczek-Zdyrska et al., 2017; Molnár et al., 2018; Pyrzynska & Sentkowska, 2020).

The beneficial roles of boron and selenium are listed in Figures 1.1 and 1.2, respectively, along with elucidating the beneficial roles of some other metalloids like arsenic and silicon in Table 1.1. Also, the toxicity symptoms of metalloids have been tabulated in Table 1.2.

1.4 METALLOID ACCUMULATION IN DIFFERENT PLANT FAMILIES

High metalloid accumulation has been reported in wild trees and cultivated vegetable crop species (Schreck et al., 2012; Clemens & Ma, 2016; Vukojević et al., 2016) (Table 1.3). The rate of accumulation and translocation varies among species and depends upon factors such as the chemical nature and the concentration of the metalloid present in the medium (Finnegan & Chen, 2012). Most of the plant species take up metalloid ions through their roots and translocate them to harvestable parts. Various plants that accumulate high amounts of metalloid are called "hyperaccumulators." Most of the accumulation occurs in the root, followed by the stems and leaves. High concentrations (around 28–75 times) have been found in the roots of plants as compared to the shoots.

1.5 CONCLUSION

The accumulation of metalloid elements in high amounts induces toxicity in plants with great damage to root and leaf development. In recent years, marvelous progress has been made to recognize the molecular mechanisms of the uptake, transport, distribution, and detoxification of metalloids.

TABLE 1.1
Bio-Functional Aspects of Metalloids in Plants

Metalloids	Role in Plants
Arsenic	Increases biomass
	Enhances growth
	Enhances malondialdehyde (MDA) contents
Silicon	Protects plants from environmental stress by providing stable membrane lipids and triggers antioxidant defensive mechanisms
	Decreases the permeability of plasma membrane in leaf cells by reducing the lipid peroxidation levels
	Facilitates cell wall loosening and promotes cell wall expansion during salt stress
	Limits transpiration rate thereby reducing osmotic stress and increasing root activity
	By regulating the development of a waxy layer on the plant that significantly reduces the rate of transpiration by the plant
	To regulate the functioning of the stomatal valves, thereby affecting stomatal conductance
	While growing in greenhouses, the leaves transpired less with increased Si supplementation
	To alleviate heat stress in plants as it imparts thermal stability to lipids in cell membranes

TABLE 1.2
Toxicity of Metalloids in Plants

Metalloids	Plants	Concentration	Impacts	References
Arsenic	*Vigna mungo*	100 and 200 μm	Decreases in the length of shoot and root, as well as a decline in germination percentage and reduces photosynthetic pigment.	Srivastava and Sharma (2013)
	Vigna mungo	200 μm	Decreases catalase activity	Srivastava and Sharma (2013)
	Oryza sativa	>60 mg kg^{-1} total As.	Retardation in growth, brown spots, and scorching on leaves.	Khan et al. (2010)
	Lycopersicon esculentum	53.6 μmol L^{-1}	Reduction in length and dry weight of root and shoot	Singh and Upadhyay (2014)
	Wrightia arborea	2.0 mM	Decline in the growth of plant	Kumar et al. (2015)
Boron	*Spirodella polyrrhiza*	3.55 mg L^{-1}	Lessened number of frond	Davis et al. (2002)
	Lemna minor L. and Lemna gibba L.	16 mg L^{-1}	Negatively affects photosynthetic pigment. Chlorosis and necrosis occurred, thereby reducing plant growth.	Gür et al. (2016)
	Citrus sinensis	400 μM H_3BO_3	Exocytosis occurred in the companion cells and the parenchyma cells of the leaf phloem. Irregularity in cell wall thickening of cortex region and leaves phloem tissue Death of root epidermal cells	Huang et al. (2014)
	Citrus grandis	400 μM H_3BO_3	Irregular cell wall thickening in cortex cells and phloem tissue of leaves. Negatively affected photosynthesis, cell death of phloem tissue enhanced via autophagy in leaf veins, death of root epidermal cells	Huang et al. (2014)
	Vigna radiata	4 μg g^{-1}	Decline in total phenol and soluble leaf protein content	Seth and Aery (2014)

(Continued)

TABLE 1.2 (*Continued*)
Toxicity of Metalloids in Plants

Metalloids	Plants	Concentration	Impacts	References
Silicon	*Arachis hypogaea*	80 mg L^{-1}	Reduction in MDA content resulting in membrane damage and increase in root surface as well as root and shoot dry weight under Al stress	Shen et al. (2014)
Selenium	*Arabidopsis*	≥2 mg kg^{-1}	10% reduction in biomass	Kabata-Pendias (2010)
	Cucumis sativus	20 and 80 µM	Reduced biomass	Hawrylak-Nowak (2009)
	Cucumis sativus	50 or 100 µM	Decreased vegetative growth and reproductive development	Lehotai et al. (2016)
	Zea mays	0.1 and 0.5 µM	Decreased ALA content	Jain et al. (2017)

TABLE 1.3
Distribution of Metalloids in Different Plant Families

Metalloids	Families	Plants	References
Boron	Poaceae	*Zea mays*	Maze (1919)
	Fabaceae	*Vicia faba*	Warington (1923)
		Vigna unguiculata	
		Piscum sativa	
	Umbellieraceae	*Apium graveolens*	Gupta and Gupta (2016)
	Brassicaceae	*Brassica napobrassica*	Gupta and Gupta (2016)
		Raphanus sativa	Sinha et al. (2000)
		Brassica oleraceae	
		Brassica nigra	
	Solanaceae	*Lycopersicum esculentum*	Pommerrenig et al. (2015), Camacho-Cristóbal et al. (2008)
	Amaranthaceae	*Beta vulgaris*	Bonilla et al. (1997)
	Cucurbitaceae	*Cucurbita pepo*	Bohnsack and Albert (1977)
	Asteraceae	*Helianthus annus*	Josten and Kutschera (1999)
	Solanaceae	*Nicotiana tabacum*	Camacho-Cristóbal et al. (2008)
	Rutaceae	*Citrus sinensis*	Huang et al. (2014)
		Citrus grandis	
	Araceae	*Spirodella polyrrhiza*	Davis, Drake and Maier (2002)
	Fabaceae	*Vigna unguiculata*	Chatterjee et al. (1990)
		Glycine max	Yamagishi and Yamamoto (1994)
Silicon	Poaceae	*Oryza sativa*	Ma and Yamaji (2015)
		Triticum sativa	Zeyen et al. (2002)
	Cucurbitaceae	*Cucumis sativa*	Gonzalo et al. (2013)
	Malvaceae	*Gossypium hirsutum*	Anwaar et al. (2015)
	Solanaceae	*Lycopersion esculentum*	Li et al. (2015)
	Fabaceae	*Glycine max*	Gonzalo et al. (2013)
		Arachis hypogaea	Shen et al. (2014)
Selenium	Lecythidaceae	*Lecythis ollaria*	Pilon-Smits et al. (2017), Lindblom et al. (2018)

(*Continued*)

TABLE 1.3 (*Continued*)
Distribution of Metalloids in Different Plant Families

Metalloids	Families	Plants	References
	Brassicaceae	*Stanleya pinnata*	Pilon-Smits et al. (2017), Lindblom et al. (2018)
		Brassica rapa	Lyons et al. (2009)
		Brassica juncea	Hanson et al. (2004)
	Poaceae	*Tritium sativa*	Nawaz et al. (2015)
		Oryza sativa	Boldrin et al. (2013)
	Aceraceae	*Acer saccharinum*	Pukacka et al. (2011)
	Curcurbitaceae	*Cucumis sativa*	Hawrylak-Nowak (2009)
	Pharmidiaceae	*Spirulina platensis*	Belokobylsky et al. (2004)
	Asteraceae	*Lactuca sativa*	Xue et al. (2001)
	Fabaceae	*Astragalus praelongus*	Pilon-Smits et al. (2017), Lindblom et al. (2018)
Arsenic	Poaceae	*Oryza sativa*	Khan et al. (2010)
	Pteridaceae	*Pteris vittata*	Tu and Ma (2003)
	Fabaceae	*Vigna mungo*	Srivastva and Sharma (2013)
	Solanaceae	*Layopersicon esculentum*	Singh and Upadhyay (2014)
	Apocynaceae	*Wrightia arborea*	Kumar et al. (2015)
Antimony	Poaceae	*Zea mays*	Pan et al. (2011)
	Asteraceae	*Helianthus annus*	Ortega et al. (2017)
	Brassicaceae	*Brassica oleraceae*	Xi-Yuan et al. (2015)
	Urticaceae	*Boehmeria nivea*	Chai et al. (2016)

The identification and characterization of responsible genes could be exploited to develop better medicines, insecticides, antibiotics, and plant varieties which are resistant to metalloids. These can also reduce the uptake of metalloids and minimize the risk of toxicity through the food chain and enhance crop production in metalloid-contaminated soils and water. Further studies are required to explore possible ways of improving the potential of plants for removing/tolerating high levels of metalloids. A detailed understanding of the underlying molecular mechanisms of metalloid action may further improve tolerance against biotic and abiotic stresses and crop yield.

REFERENCES

Abbas, G., Murtaza, B., Bibi, I., Shahid, M., Niazi, N., Khan, M., Amjad, M., Hussain, M., & Natasha. (2018). Arsenic uptake, toxicity, detoxification, and speciation in plants: Physiological, biochemical, and molecular aspects. *International Journal of Environmental Research and Public Health, 15*(1), 59.

Alloway, B. J. (2013). Sources of heavy metals and metalloids in soils. In B. Alloway (Ed.), *Heavy Metals in Soils. Environmental Pollution* (vol. 22). Dordrecht: Springer. https://doi.org/10.1007/978-94-007-4470-7_2

Andráš, P., Matos, J. X., Turisová, I., Batista, M. J., Kanianska, R., & Kharbish, S. (2018). The interaction of heavy metals and metalloids in the soil–plant system in the São Domingos mining area (Iberian Pyrite Belt, Portugal). *Environmental Science and Pollution Research, 25*, 20615–20630.

Anwaar, S. A., Ali, S., Ali, S., Ishaque, W., Farid, M., Farooq, M. A., Najeeb, U., Abbas, F., & Sharif, M. (2015). Silicon (Si) alleviates cotton (*Gossypium hirsutum* L.) from zinc (Zn) toxicity stress by limiting Zn uptake and oxidative damage. *Environmental Science and Pollution Research International, 22*(5), 3441–3450. https://doi.org/10.1007/s11356-014-3938–9. Epub 2014 Dec 18. PMID: 25516248.

Appenroth, K.-J. (2010). What are “heavy metals” in plant sciences? *Acta Physiologiae Plantarum, 32*(4), 615–619.

Asada, R. & Tazaki, K. (2001). Silica biomineralization of unicellular microbes under strongly acidic conditions. *The Canadian Mineralogist, 39*(1), 1–16.

Bienert, G.P., Schüssler, M.D. and Jahn, T.P. (2008). Metalloids: essential, beneficial or toxic? Major intrinsic proteins sort it out. *Trends in Biochemical Sciences*, *33*(1), 20–26.

Belokobylsky, A. I., Ginturi, E. I., Kuchava, N. E., Kirkesali, E. I., Mosulishvili, L., Frontasyeva, M. V., Pavlov, S., & Aksenova, N. (2004). Accumulation of selenium and chromium in the growth dynamics of *spirulina platensis*. *Journal of Radioanalytical and Nuclear Chemistry*, *259*, 65–68.

Bhat, J. A., Shivaraj, S. M., Singh, P., Navadagi, D. B., Tripathi, D. K., Dash, P. K., Solanke, A. U., Sonah, H., & Deshmukh, R. (2019). Role of silicon in mitigation of heavy metal stresses in crop plants. *Plants, 8*(3), 71.

Bohnsack, C. W., & Albert, L. S. (1977). Early effects of boron deficiency on indole acetic acid oxidase levels of squash root tips. *Plant Physiology*, *59*, 1047–1050.

Boldrin, P. F., Faquin, V., Ramos, S. J., Boldrin, K. V. F., Ávila, F. W., & Guilherme, L. R. G. (2013). Soil and foliar application of selenium in rice biofortification. *Journal of Food Composition and Analysis*, *31*, 238–244.

Bonilla, I., Mergold-Villasenor, C., Campos, M. E., Sanchez, N., Perez, H., Lopez, L., Castrejón, L., Sánchez, F., & Cassab, G. I. (1997). The aberrant cell walls of boron-deficient bean root nodules have no covalently bound hydroxyproline-/proline-rich proteins. *Plant Physiology*, *115*, 1329.

Camacho-Cristóbal, J. J., Rexach, J., & González-Fontes, A. (2008). Boron in plants: Deficiency and toxicity. *Journal of Integrative Plant Biology*, *50*, 1247–1255.

Chai, L. Y., Mubarak, H., Yang, Z. H., Yong, W., Tang, C. J., & Mirza, N. (2016). Growth, photosynthesis, and defense mechanism of antimony (Sb)-contaminated Boehmeria nivea L. *Environmental Science and Pollution Research*, *23*(8), 7470–7481.

Chatterjee, C., Sinha, P., & Agarwala, S. C. (1990). Interactive effect of boron and phosphorus on growth and metabolism of maize grown in refined sand. *Canadian Journal of Plant Science*, *70*, 455–460.

Clemens, S., & Ma, J. F. (2016). Toxic heavy metal and metalloid accumulation in crop plants and foods. *Annual Review of Plant Biology,* 67, 489–512.

Davis, S. M., Drake, K. D., & Maier, K. J. (2002). Toxicity of boron to the duckweed, Spirodella polyrrhiza. *Chemosphere*, *48*(6), 615–620.

Deng, F., Zeng, F., Chen, G., Feng, X., Riaz, A., Wu, X., Gao, W., Wu, F., Holford, P., & Chen, Z.-H. (2021). Metalloid hazards: From plant molecular evolution to mitigation strategies. *Journal of Hazardous Materials, 409*, 124495.

Deng, W., Wang, F., & Liu, W. (2023). Identification of factors controlling heavy metals/metalloid distribution in agricultural soils using multi-source data. *Ecotoxicology and Environmental Safety, 253*, 114689.

Deshmukh, R., & Bélanger, R. R. (2016). Molecular evolution of aquaporins and silicon influx in plants. *Functional Ecology, 30*(8), 1277–1285.

Deshmukh, R., Tripathi, D. K., & Guerriero, G. (2020). *Metalloids in Plants: Advances and Future Prospects.* New York: John Wiley & Sons.

Deshmukh, R. K., Ma, J. F., & Bélanger, R. R. (2017). Role of silicon in plants. *Frontiers in Plant Science, 8*, 1858.

Dubovina, M., Krčmar, D., Grba, N., Watson, M. A., Rađenović, D., Tomašević-Pilipović, D., & Dalmacija, B. (2018). Distribution and ecological risk assessment of organic and inorganic pollutants in the sediments of the transnational Begej canal (Serbia-Romania). *Environmental Pollution, 236*, 773–784.

Dutta, P., & Bandopadhyay, P. (2016). Arsenic pollution in agriculture: Its uptake and metabolism in plant systems. *Agricultural Research and Technology, 1*(5), 113–117.

Eiche, E., Bardelli, F., Nothstein, A. K., Charlet, L., Göttlicher, J., Steininger, R., Dhillon, K. S., & Sadana, U. S. (2015). Selenium distribution and speciation in plant parts of wheat (*Triticum aestivum*) and Indian mustard (*Brassica juncea*) from a seleniferous area of Punjab, India. *Science of the Total Environment, 505*, 952–961.

Farooq, M. A., Islam, F., Ali, B., Najeeb, U., Mao, B., Gill, R. A., Yan, G., Siddique, K. H. M., & Zhou, W. (2016). Arsenic toxicity in plants: Cellular and molecular mechanisms of its transport and metabolism. *Environmental and Experimental Botany, 132*, 42–52.

Feng, R., Wang, L., Yang, J., Zhao, P., Zhu, Y., Li, Y., Yu, Y., Liu, H., Rensing, C., Wu, Z., Ni, R. X., & Zheng, S. A. (2021). Underlying mechanisms responsible for restriction of uptake and translocation of heavy metals (metalloids) by selenium via root application in plants. *Journal of Hazardous Materials, 402*, 123570.

Finnegan, P., & Chen, W. (2012). Arsenic toxicity: The effects on plant metabolism. *Frontiers in Physiology, 3*, 182.

Fu, J., Hu, X., Tao, X., Yu, H., & Zhang, X. (2013). Risk and toxicity assessments of heavy metals in sediments and fishes from the Yangtze River and Taihu Lake, China. *Chemosphere, 93*(9), 1887–1895.

Fu, J., Zhao, C., Luo, Y., Liu, C., Kyzas, G. Z., Luo, Y., Zhao, D., An, S., & Zhu, H. (2014). Heavy metals in surface sediments of the Jialu River, China: Their relations to environmental factors. *Journal of Hazardous Materials, 270*, 102–109.

González-Fontes, A., Navarro-Gochicoa, M. T., Camacho-Cristóbal, J. J., Herrera-Rodríguez, M. B., Gupta, R. D., & Gupta, S. K. (2016). Strategies for increasing the production of oilseed on a sustainable basis. In S. K. Gupta (Ed.), *Breeding Oilseed Crops for Sustainable Production* (pp. 1–18). Elsevier Inc.

Gür, N., Türker, O.C. & Böcük, H. (2016). Toxicity assessment of boron (B) by Lemna minor L. and Lemna gibba L. and their possible use as model plants for ecological risk assessment of aquatic ecosystems with boron pollution. *Chemosphere, 157*, 1–9.

Hanson, B., Lindblom, S. D., Loeffler, M. L., & Pilon-Smits, E. A. H. (2004). Selenium protects plants from phloem-feeding aphids due to both deterrence and toxicity. *New Phytologist, 162*, 655–662.

Hasanuzzaman, M., Bhuyan, M. B., Raza, A., Hawrylak-Nowak, B., Matraszek-Gawron, R., Al Mahmud, J., Nahar, K., & Fujita, M. (2020). Selenium in plants: Boon or bane? *Environmental and Experimental Botany, 178*, 104170.

Hasanuzzaman, M., Nahar, K., Anee, T.I. & Fujita, M., 2017. Exogenous silicon attenuates cadmium-induced oxidative stress in Brassica napus L. by modulating AsA-GSH pathway and glyoxalase system. *Frontiers in Plant Science, 8*, 1061.

Hawrylak-Nowak, B. (2009). Beneficial effects of exogenous selenium in cucumber seedlings subjected to salt stress. *Biological Trace Element Research, 132*, 259–269.

Hodson, Mark E. (2013). Effects of heavy metals and metalloids on soil organisms. In *Heavy Metals in Soils: Trace Metals and Metalloids in Soils and Their Bioavailability*, edited by Brian J. Alloway, Springer, Dordrecht, 2013, 141–160.

Hoffmann, J., Berni, R., Hausman, J. F., & Guerriero, G. (2020). A review on the beneficial role of silicon against salinity in non-accumulator crops: Tomato as a model. *Biomolecules, 10*(9), 1284.

Huang, J. H., Cai, Z. J., Wen, S. X., Guo, P., Ye, X., Lin, G. Z., & Chen, L. S. (2014). Effects of boron toxicity on root and leaf anatomy in two Citrus species differing in boron tolerance. *Trees, 28*(6), 1653–1666.

Islam, M. S., Hossain, M. B., Matin, A., & Sarker, M. S. I. (2018). Assessment of heavy metal pollution, distribution and source apportionment in the sediment from Feni River estuary, Bangladesh. *Chemosphere, 202*, 25–32.

Jain, M., Panwar, M., & Gadre, R. (2017). Influence of selenium supplementation on δ-aminolevulinic acid formation in greening maize leaf segments. *Research Journal of Phytochemistry, 11*, 111–117.

Javaid, S., Ashraf, K., Sultan, K., Siddiqui, M. H., Ali, H. M., Chen, Y., & Zaman, Q. U. (2022). Risk assessment of potentially toxic metals and metalloids in soil, water and plant continuum of fragrant rice. *Agronomy, 12*(10), 2480.

Josten, P., & Kutschera, U. (1999). The micronutrient boron causes the development of adventitious roots in sunflower cuttings. *Annals of Botany, 84*, 337–342.

Kabata-Pendias, A. (2010). *Trace Elements in Soils and Plants*. Boca Raton, FL: CRC Press/Taylor & Francis Group ICC.

Karadaş, C., & Kara, D. (2011). In vitro gastro-intestinal method for the assessment of heavy metal bioavailability in contaminated soils. *Environmental Science and Pollution Research, 18*, 620–628.

Khan, M. A., Stroud, J. L., Zhu, Y. G., Mcgrath, S. P., & Zhao, F. J. (2010). Arsenic bioavailability to rice is elevated in Bangladeshi paddy soils. *Environmental Science and Technology, 44*(22), 8515–8521.

Kroukamp, E. M., Wondimu, T., & Forbes, P. B. C. (2016). Metal and metalloid speciation in plants: Overview, instrumentation, approaches and commonly assessed elements. *TrAC Trends in Analytical Chemistry, 77*, 87–99.

Kumar, D., Singh, V. P., Tripathi, D. K., Prasad, S. M., & Chauhan, D. K. (2015). Effect of arsenic on growth, arsenic uptake, distribution of nutrient elements and thiols in seedlings of *Wrightia arborea* (Dennst.) Mabb. *International Journal of Phytoremediation, 17*(2), 128–134.

Lehotai, N., Lyubenova, L., Schröder, P., Feigl, G., Ördög, A., Szilágyi, K., Erdei, L., & Kolbert, Z. (2016). Nitro-oxidative stress contributes to selenite toxicity in pea (*Pisum sativum* L). *Plant Soil, 400*, 107–122.

Li, H., Zhu, Y., Hu, Y., Han, W., & Gong, H. (2015). Beneficial effects of silicon in alleviating salinity stress of tomato seedlings grown under sand culture. *Acta Physiologiae Plantarum, 37*(4), 71.

Li, L., Liu, H., & Li, H. (2018). Distribution and migration of antimony and other trace elements in a Karstic river system, Southwest China. *Environmental Science and Pollution Research, 25*, 28061–28074.

Liang, J., Feng, C., Zeng, G., Gao, X., Zhong, M., Li, X., Li, X., He, X., & Fang, Y. (2017). Spatial distribution and source identification of heavy metals in surface soils in a typical coal mine city, Lianyuan, China. *Environmental Pollution, 225*, 681–690.

Lindblom, S. D., Wangeline, A. L., Valdez Barillas, J. R., Devilbiss, B., Fakra, S. C., & Pilon-Smits, E. A. H. (2018). Fungal endophyte alternaria tenuissima can affect growth and selenium accumulation in its hyperaccumulator host *Astragalus bisulcatus*. *Frontiers in Plant Science*, *9*, 1213.

Lyons, G. H., Genc, Y., Soole, K., Stangoulis, J. C. R., Liu, F., & Graham, R. D. (2009). Selenium increases seed production in Brassica. *Plant and Soil, 318*, 73–80.

Gonzalo, M. J., Lucena, J. J., & Hernández-Apaolaza, L. (2013). Effect of silicon addition on soybean (*Glycine max*) and cucumber (*Cucumis sativus*) plants grown under iron deficiency. *Plant Physiology and Biochemistry*, 70, 455–461.

Ma, J.F. & Yamaji, N. (2015). A cooperative system of silicon transport in plants. *Trends in Plant Science*, *20*(7), 435–442.

Maze, P. (1919). Mineral nutrient solution for maize. *Annales de l'Institut Pasteur*, *33*, 139–173.

Molnár, Á., Kolbert, Z., Kéri, K., Feigl, G., Ördög, A., Szőllősi, R., & Erdei, L. (2018). Selenite-induced nitro-oxidative stress processes in Arabidopsis thaliana and *Brassica juncea*. *Ecotoxicology and Environmental Safety*, *148*, 664–674.

Mosa, K. A., Saadoun, I., Kumar, K., Helmy, M., & Dhankher, O. P. (2016). Potential biotechnological strategies for the cleanup of heavy metals and metalloids. *Frontiers in Plant Science, 7*, 303.

Mroczek-Zdyrska, M., Strubińska, J., & Hanaka, A. (2017). Selenium improves physiological parameters and alleviates oxidative stress in shoots of lead-exposed Vicia faba L. minor plants grown under phosphorus-deficient conditions. *Journal of Plant Growth Regulation, 36*, 186–199.

Mykolenko, S., Liedienov, V., Kharytonov, M., Makieieva, N., Kuliush, T., Queralt, I., Marguí, E., Hidalgo, M., Pardini, G., & Gispert, M. (2018). Presence, mobility and bioavailability of toxic metal (oids) in soil, vegetation and water around a Pb-Sb recycling factory (Barcelona, Spain). *Environmental Pollution, 237*, 569–580.

Nagajyoti, P., Lee, K., & Sreekanth, T. (2010). Heavy metals, occurrence and toxicity for plants: A review. *Environmental Chemistry Letters*, *8*(3), 199–216.

Nawaz, F., Ahmad, R., Ashraf, M. Y., Waraich, E. A., & Khan, S. Z. (2015). Effect of selenium foliar spray on physiological and biochemical processes and chemical constituents of wheat under drought stress. *Ecotoxicology and Environmental Safety, 113*, 191–200.

Ortega, A., Garrido, I., Casimiro, I., & Espinosa, F. (2017). Effects of antimony on redox activities and antioxidant defence systems in sunflower (*Helianthus annuus* L.) plants. *PLOS One, 12*(9), e0183991. https://doi.org/10.1371/journal.pone.0183991

Ouellette, S., Goyette, M.H., Labbé, C., Laur, J., Gaudreau, L., Gosselin, A., Dorais, M., Deshmukh, R.K. & Bélanger, R.R., 2017. Silicon transporters and effects of silicon amendments in strawberry under high tunnel and field conditions. *Frontiers in plant science*, *8*, 949.

Pan, X., Zhang, D., Chen, X., Bao, A., & Li, L. (2011). Antimony accumulation, growth performance, antioxidant defense system and photosynthesis of *Zea mays* in response to antimony pollution in soil. *Water, Air, & Soil Pollution*, *215*(1–4), 517–523.

Pandey, A. K., Gautam, A., & Dubey, R. S. (2019). Transport and detoxification of metalloids in plants in relation to plant-metalloid tolerance. *Plant Gene, 17*, 100171.

Parveen, N., Berni, R., Sudhakaran, S., Bhat, J. A., Shinde, S., Ramawat, N., Singh, V. P., Sahi, S. V., Deshmuk, R., Chauhan, D. K., & Tripathi, D. K. (2022). Metalloids in plants: A systematic discussion beyond description. *Annals of Applied Biology, 180*(1), 7–25.

Pilon-Smits, E. A. H., Winkel, L. H. E., & Lin, Z.-Q. (2017). *Selenium in Plants: Molecular, Physiological, Ecological and Evolutionary Aspects*. Cham: Springer International Publishing.

Pommerrenig, B., Diehn, T. A., & Bienert, G. P. (2015). Metalloido-porins: Essentiality of nodulin 26-like intrinsic proteins in metalloid transport. *Plant Science, 238*, 212–227.

Pukacka, S., Ratajczak, E., & Kalemba, E. (2011). The protective role of selenium in recalcitrant Acer saccharium L. seeds subjected to desiccation. *Journal of Plant Physiology*, *168*, 220–225.

Pyrzynska, K., & Sentkowska, A. (2020). Selenium in plant foods: Speciation analysis, bioavailability, and factors affecting composition. *Critical Reviews in Food Science and Nutrition, 61*, 1340–1352.

Řezanka, T. & Sigler, K. (2008). Biologically active compounds of semi-metals. *Studies in Natural Products Chemistry*, *35*, 835–921.

Rizwan, M., Ali, S., Rehman, M. Z., Rinklebe, J., Tsang, D. C. W., Tack, F. M. G., Abbasi, G. H., Hussain, A., Igalavithana, A. D., Lee, B. C., & Ok, Y. S. (2020). Effects of selenium on the uptake of toxic trace elements by crop plants: A review. *Critical Reviews in Environmental Science and Technology*, *51*, 2531–2566.

Sanglard, L. M., Detmann, K. C., Martins, S. C., Teixeira, R. A., Pereira, L. F., Sanglard, M. L., Fernie, A. R., Araújo, W. L. & DaMatta, F. M. (2016). The role of silicon in metabolic acclimation of rice plants challenged with arsenic. *Environmental and Experimental Botany*, *123*, 22–36.

Schreck, E., Foucault, Y., Sarret, Sobanska, S., Cécillon, L., Castrec-Rouelle, M., Uzu, G., & Dumat, C. (2012). Metal and metalloid foliar uptake by various plant species exposed to atmospheric industrial fallout: Mechanisms involved for lead. *Science of the Total Environment*, *427–428*, 253–262.

Seth, K. & Aery, N.C. (2014). Effect of boron on the contents of chlorophyll, carotenoid, phenol and soluble leaf protein in mung bean, Vigna radiata (L.) Wilczek. *Proceedings of the National Academy of Sciences, India Section B: Biological Sciences*, 84, 713–719.

Seraj, F., & Rahman, T. (2018). Heavy metals, metalloids, their toxic effect and living systems. *American Journal of Plant Sciences,* 9(13), 2626–2643.

Shen, X., Xiao, X., Dong, Z., & Chen, Y. (2014). Silicon effects on antioxidative enzymes and lipid peroxidation in leaves and roots of peanut under aluminum stress. *Acta Physiologiae Plantarum, 36*. https://doi.org/10.1007/s11738-014-1676–8

Singh, D., Pandey, H., Dhiman, V. K., Thakur, K., Sharma, I., Dhiman, V. K., Singh, V., & Pandey, D. (2022). Assessment of environmental impacts of metal/metalloid pollution on plants. In T. Aftab & K. Hakeem (Eds.), *Metals Metalloids Soil Plant Water Systems* (pp. 217–232). Academic Press.

Singh, V. K., & Upadhyay, R. S. (2014). Effects of arsenic on reactive oxygen species and antioxidant defense system in tomato plants. *Toxicological & Environmental Chemistry, 96*(9), 1374–1383.

Sinha, P., Jain, R., & Chatterjee, C. (2000). Interactive effect of boron and zinc on growth and metabolism of mustard. *Communications in Soil Science and Plant Analysis, 31*, 41–49.

Srivastava, S., & Sharma, Y. K. (2013). Arsenic phytotoxicity in black gram (Vigna Mungo L. Var. PU19) and its possible amelioration by phosphate application. *Journal of Plant Physiology & Pathology*, *1*, 3. https://doi.org/10.4172/2329-955X.1000107

Tripathi, D. K., Singh, S., Singh, V. P., Prasad, S. M., Chauhan, D. K. and Dubey, N. K. (2016). Silicon nanoparticles more efficiently alleviate arsenate toxicity than silicon in maize cultiver and hybrid differing in arsenate tolerance. *Frontiers in Environmental Science*, *4*, 46.

Tripathi, D. K., Singh, S., Singh, V. P., Prasad, S. M., Dubey, N. K. & Chauhan, D. K. (2017). Silicon nanoparticles more effectively alleviated UV-B stress than silicon in wheat (Triticum aestivum) seedlings. *Plant Physiology and Biochemistry*, *110*, 70–81.

Tu, C., & Ma, L. Q. (2003). Effects of arsenate and phosphate on their accumulation by an arsenic-hyperaccumulator *Pteris vittata* L. *Plant and Soil*, *249*(2), 373–382.

Tubana, B. S., Babu, T., & Datnoff, L. E. (2016). A review of silicon in soils and plants and its role in US agriculture: History and future perspectives. *Soil Science, 181*(9/10), 393–411.

Vernon, R. E. (2013). Which elements are metalloids? *Journal of Chemical Education*, *90*(12), 1703–1707.

Vivancos, J., Labbé, C., Menzies, J. G. & Bélanger, R. R. (2015). Silicon-mediated resistance of A rabidopsis against powdery mildew involves mechanisms other than the salicylic acid (SA)-dependent defence pathway. *Molecular Plant Pathology*, *16*(6), 572–582.

Vukojević, V., Trifković, J., Krgović, R., Milojković-Opsenica, D., Markovic, M., Ramadan, N., & Mutic, J. (2016). Uptake of metals and metalloids by *Conyza canadensis* L. from a thermoelectric power plant landfill. *Archives of Biological Sciences, 68*, 829–835.

Warington, K. (1923). The effect of boric acid and borax on the bean and certain other plants. *Annals of Botany, 37*, 629–672.

Xiao, R., Bai, J., Lu, Q., Zhao, Q., Gao, Z., Wen, X., & Liu, X. (2015). Fractionation, transfer, and ecological risks of heavy metals in riparian and ditch wetlands across a 100-year chronosequence of reclamation in an estuary of China. *Science of the Total Environment, 517*, 66–75.

Xi-Yuan, X. I. A. O., Zhao-Hui, G. U. O., Yue-Ping, L. U. O., Jun-Ping, B. I., Miao, Y. A. N. G., & Huang, D. Q. (2015). Effect of antimony on physiological responses of green Chinese cabbage and enzyme activities of allitic udic ferrisols. *Pedosphere*, *25*(1), 124–129.

Xu, W., Dai, W., Yan, H., Li, S., Shen, H., Chen, Y., Xu, H., Sun, Y., He, Z., & Ma, M. (2015). Arabidopsis NIP3; 1 plays an important role in arsenic uptake and root-to-shoot translocation under arsenite stress conditions. *Molecular Plant, 8*(5), 722–733.

Xue, T., Hartikainen, H., & Piironen, V. (2001). Antioxidative and growth-promoting effect of selenium on senescing lettuce. *Plant and Soil, 237*, 55–61.

Yamagishi, M., & Yamamoto, Y. (1994). Effects of boron on nodule development and symbiotic nitrogen fixation in soybean plants. *Soil Science and Plant Nutrition*, *40*, 265–274.

Yuan, Q., Wang, P., Wang, C., Chen, J., Wang, X., Liu, S., & Feng, T. (2019). Metals and metalloids distribution, source identification, and ecological risks in riverbed sediments of the Jinsha River, China. *Journal of Geochemical Exploration, 205*, 106334.

Zargar, S. M., Mahajan, R., Bhat, J. A., Nazir, M., & Deshmukh, R. (2019). Role of silicon in plant stress tolerance: Opportunities to achieve a sustainable cropping system. *Biotech,* 9(3), 73.

Zeyen, R. J., Kruger, W. M., Lyngkjær, M. F., & Carver, T. L. W. (2002). Differential effects of D –mannose and 2-deoxy-D-glucose on attempted powdery mildew fungal infection of inappropriate and appropriate Gramineae. *Physiological and Molecular Plant Pathology*, *61*, 315–323.

Zohaib, A., Tabassum, T., Jabbar, A., Anjum, S. A., Abbas, T., Mehmood, A., Irshad, S., Kashif, M., Nawaz, M., Farooq, N., Nasir, I. R., Rasool, T., Nadeem, M., & Ahmad, R. (2018). Author correction: Effect of plant density, boron nutrition and growth regulation on seed mass, emergence and offspring growth plasticity in cotton. *Scientific Reports, 8*, 10489.

Zwolak, I. (2020). The role of selenium in arsenic and cadmium toxicity: An updated review of scientific literature. *Biological Trace Element Research*, *193*, 44–63.

2 Metalloids
Functional, Deficiency, and Toxicity Aspects

Ashutosh Sharma, Priyanka Sharma, Pooja Sharma, Sonika Sharma, Ashish Sharma, Renu Bhardwaj, and Indu Sharma

2.1 INTRODUCTION

The presence of metalloids in the Earth's crust is a natural occurrence. However, due to human activity, the concentration of metalloids within the Earth's sphere has experienced a significant increase (Kour et al. 2022). Excessive hazardous metal concentrations in soils are mostly produced by global industrialization, the overuse of fertilizers, pesticides, biosolids and manure, poor wastewater management, metal mining, and industrial waste (Nguyen et al. 2021). Toxic metalloids from soils, sediments, and mining wastes enter streams. Gravitational movement and adsorptive components in aquatic sediments (like freshwater, estuarine, or marine) cause metalloids to accumulate, resulting in their increased concentrations. These toxic metalloids can come from both natural and man-made sources, and they can be released back into the water. Therefore, incessant anthropogenic activities contaminate the environment with high concentrations of metalloid ions, which pose serious health risks to plants, humans, and animals. This contamination can also have significant effects on food chains and global food security.

Metals and non-metals are separated by metalloids, which share characteristics of both. Metalloids are elements with intermediate physical and chemical characteristics, such as better heat and electricity conductivity, but not so well as metals, and form amphoteric oxides, grouped along a diagonal line in the periodic table (Lombi and Holm 2010). The eight most well-known metalloids are antimony (Sb), arsenic (As), astatine (At), boron (B), germanium (Ge), silicon (Si), tellurium (Te), and polonium (Po). Besides these, some researchers also considered selenium (Se) as an important metalloid (Lichtfouse et al. 2022). It is an essential element for humans and animals, but its unbalanced distribution in soils leads to deficiency and pollution worldwide (Feng et al. 2021). When given at low concentrations, Se is thought to be advantageous for plants, especially in mitigating the harmful effects of abiotic stressors and enhancing plant growth and development (Hasanuzzaman et al. 2020).

The abundance of metalloids varies, with Si being the most common and At being rare (Lombi and Holm 2010). Si is abundant in the Earth's crust, making up over a quarter of its mass. However, the rarest natural element, At, produced by uranium (U) and thorium (Th) decay, has a half-life of only 8.3 hours, containing only 25 g of the total amount on Earth. Other elements Te, Po, Ge, and At exhibit their presence in the soil environment in traces or even ultra-trace quantities (Dutta et al. 2020). Among all metalloids, As, B, Si, Se, and Sb have harmful environmental effects, whereas Ge, Si, Te, and Po have the least significance to environmental health. The impact of any metalloid varies with its oxidation state and species. According to Wan et al. (2013), Sb(III) levels were increased under waterlogged conditions, which further led to a breakdown of Fe and Mn oxides and a rise in Sb levels. As a result, in perennial ryegrass, Sb levels increased tenfold, while *Holcus lanatus* L. decreased by 80%. Thus, oxidation state and species are crucial in affecting Sb absorption.

DOI: 10.1201/9781032673592-2

Plants absorb both the metals and metalloids from the soil, some essential but toxic at high concentrations, and others non-essential and toxic even at low concentrations (Tang et al. 2023). Figure 2.1 depicts the beneficial roles and toxicity symptoms of major metalloids in plants. In plants, the excessive accumulation of metalloids can cause oxidative stress, genotoxicity, programmed cell death, and physiological disturbances (Tang et al. 2023). To overcome the negative impact of harmful metalloids, plants have developed strategies for detoxifying them. Besides genotype or species differences in plants, growth conditions or habitat can also have a significant impact on phytotoxicity to metalloids and their detoxifying mechanism. In soil, the phytotoxicity of metalloids is influenced by soil properties such as pH, the contents of clay and organic matter, competing ions, and redox potential, which affect the availability of metalloids for plant uptake. For example, inorganic As compounds are 100 times more toxic than methylated forms, with As (III) being more toxic biologically than As (V), and the As concentration varies based on geological environment redox conditions (Raju 2022). Humans can be exposed to As through various pathways, which include air, food, water, and soil. These pathways may differ depending on the geological composition and residential activities. The presence of As in natural water is associated with geochemical environments such as basin fill deposits, volcanic deposits, geothermal inputs, mining wastes, and landfills. Inorganic As compounds, used in pesticides, preservatives, and pharmaceuticals, pose global health risks due to As pollution (Su et al. 2023). Besides As, Sb is also reported as a hazardous pollutant that is geno- and cytotoxic and has the potential to cause human cancers (Bolan et al. 2022). It is also oncogenic and exposure to it can result in serious health effects, such as genotoxic, neurological, immunological, reproductive, developmental, and cancerous effects.

Increasing crop productivity is hampered by the shrinking amount of land under cultivation and the growing population (Sharma et al. 2023). More food production is required to meet the expanding demands of a growing population every day, emphasizing the need for more cultivated land (Feng et al. 2021). Therefore, it is not appropriate to just discard soils contaminated by metalloids. Crop loss results from both biotic and abiotic plant stressors (Sharma et al. 2023). Many strategies have been used to lessen stress however some raise environmental issues. Environmentally friendly and sustainable methods are required to safeguard crop plants from plant stressors. Through a variety of biological reactions, such as tissue growth, cytological, enzymatic, molecular,

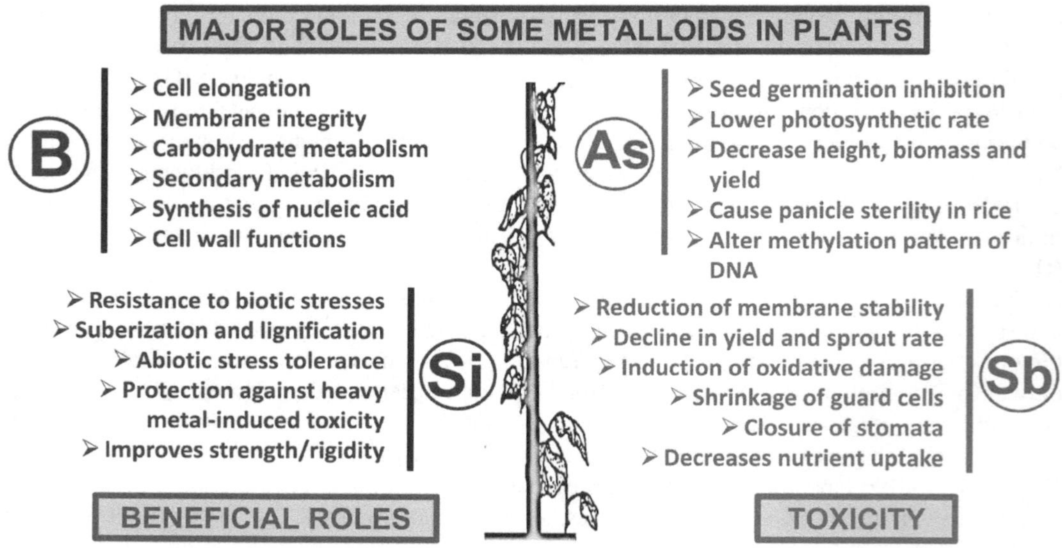

FIGURE 2.1 Beneficial roles and toxicity symptoms of major metalloids in plants.

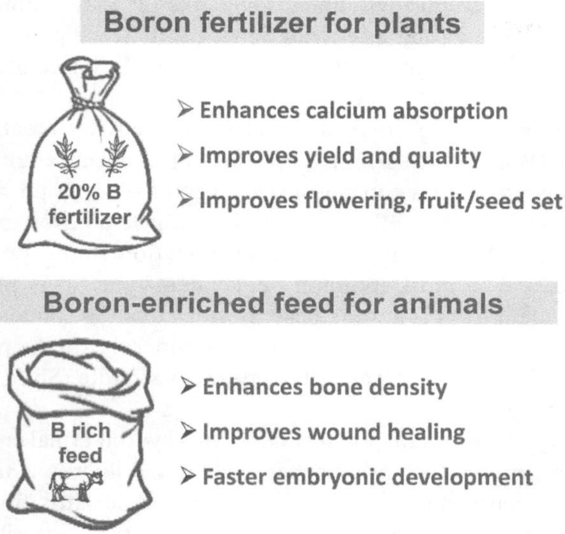

FIGURE 2.2 Role of B in plants and animals.

and genetic alterations, plants can withstand harmful levels of Sb (Maresca et al. 2020). These defense mechanisms shield genes, proteins, and other macromolecules from oxidative stress. It has been demonstrated that plant-growth-promoting rhizobacteria (PGPR) treatment in combination with other biological micro- and macro-organisms improves plant development and stress tolerance (Singh et al. 2023). Plants have adapted certain signaling pathways, including nitric oxide signaling, mitogen-activated protein kinase (MAPK), and the reactive oxygen species (ROS) phosphorylation cascade, to survive in the presence of metalloid toxicity (Kour et al. 2022). In addition to these, plant growth regulators (PGRs) such as gibberellins, cytokinin, auxin, ethylene, brassinosteroids, jasmonic acid, and salicylic acid also have a significant role in managing metalloid-induced phytotoxicities.

Furthermore, another metalloid, B is found to be a crucial micro-element that is essential for the growth of plants, animals, and humans (Figure 2.2). It has beneficial effects on physiological processes at lower concentrations, but it can be toxic at higher levels (Białek et al. 2019). B plays a vital role in the metabolism of essential life processes such as hormones, hormone action, and transmembrane signaling. Organic B compounds contribute to bone growth, metabolic functions, central nervous and immune systems, arthritis relief, hormone action, and reduced cancer risk. It is important to note that different metalloids can affect animals and plants differently. Thus, the primary aim of this book chapter is to summarize the most significant discoveries regarding the biological functions and toxicity aspects of metalloids in plants and animals.

2.2 METALLOIDS IN THE ENVIRONMENT

Metalloids are intermediate elements that exist between metals and non-metals (Dutta et al. 2020). They are abundant on Earth's crust and transfer from one environment matrix to another, making them toxic. The Earth's crust contains heavy metals and metalloids, which can be increased by natural and anthropogenic processes (Martínez-Alcalá and Bernal 2020). Soil factors like pH, redox potential, and organic matter composition affect the solubility and toxicity of these elements. The global average level of As in soil is 5 mg kg^{-1}, while in open seawater, it is 1–2 μg L^{-1} (Raju 2022).

It is found everywhere in the natural environment and ranks 12th in the human body. Soil concentrations of Se range from 0.01 to 2.0 mg kg^{-1}, with Se-rich soils reaching up to 1,200 mg kg^{-1} (Feng et al. 2021). Se-deficient soils are often defined as those with a Se concentration of less than 0.5 mg kg^{-1} (Mirlean et al. 2018).

Sb can be found in most living organisms and various environmental compartments, such as soil, water, and air (Bolan et al. 2022). This element is not as abundant as other elements present in the Earth's crust, with a concentration range of 0.2–0.3 μg g^{-1}. Its abundance typically ranges from 0.2 μg g^{-1} to several μg g^{-1}, but it can have an even greater abundance of up to 300 μg g^{-1}. Sb has two biologically significant oxidation states: pentavalent and trivalent (Periferakis et al. 2022). The most common pentavalent-type Sb is present in seawater and freshwater, but the trivalent form is a weak base. Besides this, Sb can also exist in both biological and inorganic forms, and pyrites and other sulfide minerals are the primary sources (Bolan et al. 2022). It is a rare element that frequently appears as stibnite (Sb_2S_3) and Sb bloom (Sb_2O_3) in natural environments.

Stibnite is the primary Sb ore, a sulfide mineral found in hydrothermal environments (Periferakis et al. 2022). It forms in deposits such as cinnabar, quartz, and fluorite. Other minerals containing Sb include boulangerite, bournonite, gudmundite, and polybasite. It is frequently connected with harmful substances such as Hg. Sb ores can be found in a variety of deposits, including epithermal, pegmatite, and hot spring replacement deposits. Sb deposits may be simple or complex polymetallic. Organic materials, fertilizers containing sulfur (S), and phosphate fertilizers are some of the options for acid soils, which absorb Sb less readily (Igdelou et al. 2020). Higher amounts of Sb can potentially result in soil oxidative stress. Also, natural water bodies that are free of pollution are reported to possess small amounts of Sb (Bolan et al. 2022). However, the amount of total dissolved Sb in surface and groundwater varies according to location and surrounding conditions. The mining operations enhance the levels of Sb in natural waterbodies and soils.

The rare element B is similar to carbon in that it can form stable molecules in covalently connected networks. It can be found in sedimentary rocks, shale, seas, coal, and soil (Białek et al. 2019). B is produced in trace amounts during the nucleosynthesis of cosmic rays during spallation, and it may be isolated in meteoroid materials and cosmic dust. About 3 billion years ago, a meteorite carrying RNA synthesis from Mars may have carried life to Earth through the catalysis of B and Mo (molybdenum). The USA and Turkey have the largest B reserves in the world. A hazardous metalloid and trace element, Se can be found in soil, water, and open spaces on Earth (Hasanuzzaman et al. 2020). However, Se is not usually evenly distributed in terrestrial habitats, and it can have low phytoavailability in soils (White 2018). Selenate, selenite, and organic selenides are consumed by plant roots as they take Se from the rhizosphere. Whereas selenite is plentiful in anaerobic soils with a pH of neutral to acidic, selenate is abundant in oxic soils. The soil solution contains trace amounts of organoselenium molecules.

Overuse of Se-containing fertilizers in subsurface water has resulted in elevated quantities of Se in Belgium, France, and Punjab (Gupta and Gupta 2017). Natural and anthropogenic activities contribute to the biogeochemical cycling of Se in the atmosphere, with SeO_2 becoming unstable and converting into selenious acid. Selenate is the most common type found in agricultural soils, and it is more water soluble than selenite. The second most common element, Si, is present in soils in a range of alumino-silicate forms at values of 0.5%–4% (Irfan et al. 2023). Although pure Si crystals are uncommon, plant roots can absorb Si as monosilicic acid (H_4SiO_4), which has a concentration in soil solution that is twice that of phosphorus.

2.3 BENEFICIAL ROLES OF METALLOIDS

The metalloids are generally present in the form of oxides, and some metalloids are essential for living organisms but can become toxic above a certain threshold level (Martínez-Alcalá and Bernal 2020). Metalloids like B are essential for the health of animal cell membranes, enzyme processes, and hormone metabolism (Białek et al. 2019). It has a variety of physiological effects and efficiently accumulates in bones (Figure 2.2). The following section of the chapter discusses the foremost beneficial roles of metalloids in animals and plants.

2.3.1 Beneficial Roles in Animals

Ayurveda mentions that As possesses medical uses such as treating syphilis and menstrual disorders (Raju 2022). Sb, used since ancient civilizations, is studied for lung tumor cell lines, with potential antibacterial and antifungal properties, and is considered an anticarcinogenic agent (Periferakis et al. 2022). Leishmaniasis is a disease caused by protozoans, transmitted by female phlebotome sandflies. It is endemic in Africa and Asia and affects HIV-infected and immunosuppressed patients. Sb-based drugs are used for treatment, but they can be toxic and ineffective in tropical climates. Sb was once considered a potential curative agent for tropical diseases, but its use is sparse due to its high toxicity. Antimonial medications, such as sodium antimony tartrate, were used to treat schistosomiasis, a tropical illness brought on by Schistosoma worms, up to the 1960s (Thétiot-Laurent et al. 2013; Periferakis et al. 2022). However, because of their toxicity and toxic Sb(III), they had to be abandoned because of possible negative effects on cardiograms and the myocardium (Periferakis et al. 2022).

Sb compounds may be useful in the treatment of specific illnesses (Lai et al. 2022). Certain Sb compounds have demonstrated selectivity against cancer; for example, they have been shown to control the signaling of mitogen-activated protein kinase 4 (SEK1)/c-jun kinase in acute promyelocytic leukemia cell lines and to degrade the promyelocytic leukemia/retinoic acid receptor alpha (PML-RARα) fusion protein. Synthetic Sb compounds exhibit comparable effects on both normal and tumoral cells. Additionally, arylstibonic acids containing Sb can inhibit DNA-binding activities. Sb compounds, including polyoxometalate SbW9, which stops the growth of non-small cell lung cancer cells and triggers their death, have been used to treat human malignant cells (Sun et al. 2019). A study reported that at least 26 distinct enzymes, which are crucial for energy status and metabolism, are affected by B, which has biological consequences (Abdelnour et al. 2018). B also modifies animal hormone or enzyme systems, which has an indirect effect on the metabolism of calcium (Ca), phosphorus (P), and magnesium (Mg). B helps to increase Mg levels in the body and insulin levels in the bloodstream, but at the same time, it reduces the absorption of Ca and P (Hunt 1996).

The amount of B in the human diet plays a significant role and helps in determining the metabolism of minerals under lower levels of vitamin D (Hunt 1996). B also increases the concentration of plasma triglycerides. Diets high in B enhance the structural integrity of cell membranes, bone calcification, and central nervous system performance (Białek et al. 2019). B reduces the severity of lipolysis and promotes glucose metabolism in perinatal diets. Animal hormone and mineral metabolism, enzymatic processes, and cell membrane function are all significantly impacted by this metalloid. B is a necessary metalloid for animals. When present at low concentrations, B compounds have a variety of physiological impacts. Although B builds up in bones, it does not readily accumulate in soft tissues. In animals, the urine excretes boric acid, which makes up about 90% of the excess borate. Because B influences skeletal metabolism, it also affects bone formation and the composition of soft tissues in both humans and animals.

Selenoproteins in animals have important roles in antioxidant defense, DNA synthesis, reproduction, immunological response, and thyroid hormone production (Gupta and Gupta 2017). Plants are crucial sources of dietary Se for humans and animals, with sulfate transporters (SULTR) facilitating its absorption and assimilation through the sulfur assimilating pathway. The World Health Organization (WHO) recommends a lower limit of 10 $\mu g\ L^{-1}$ for drinking water. Se has antioxidant properties that protect against liver, pancreatic, prostate, esophageal, and colon cancers. Selenoproteins such as glutathione (GSH) peroxidase and thioredoxin reductase maintain intracellular redox potential, whereas deiodinase regulates thyroid hormone metabolism.

2.3.2 Beneficial Roles in Plants

In plants, B is an anti-heat stress agent (Abdelnour et al. 2018); however, the distribution of B is uneven across the different species of flowering plants, with Dicotyledoneae having higher concentrations compared to Monocotyledoneae (Hunt 1996). B is an essential element for the growth and development of plants since it is primarily engaged in the metabolism of carbohydrates and cell division (Gupta and Solanki 2013). It has an impact on at least 14 plant processes, such as the synthesis of carbohydrates, cell division, and the preservation of cell wall integrity. Plants and soil should have between 10 and 50 mg of B per kg of dry weight; however, substantial amounts will differ among plants. The concentration of B in leaf margins is five to ten times higher than in the leaf blade as a whole. Figure 2.1 presents the beneficial roles of two important metalloids, namely, B and Si, in plants.

Si promotes plant growth and development under a variety of conditions, including drought (Irfan et al. 2023). It helps to create organic defense compounds, solidifies as silica, and polymerizes as silicic acid. In Si-hyperaccumulating plant species, Si is essential for primary metabolism, which enhances cell wall rigidity and strength. In drought-stressed plants, Si fertilization improves crop tolerance, lowers physiological abnormalities, and changes cellular metabolism to increase plant growth and biomass production. It influences the morphology, root structure, retention of leaves, photosynthesis, and physiological stressors in plants under stress. Si increases chlorophyll content, reduces malondialdehyde concentration, and improves iron translocation efficiency in sorghum plants, leading to greater dry weight accumulation under iron deficiency (Teixeira et al. 2020). Si application mitigated the harmful effects of nutritional deficiencies on forage crops like *Urochloa brizantha* and *Megathyrsus maximum* (Araújo et al. 2022). It reduces electrolyte extravasation, increases phenolic compounds, and promotes sustainable cultivation.

Se is an element that shares many chemical characteristics with S. It is found in various organic Se molecules such as selenoamino acids, selenocysteine (SeCys) and selenomethionine (SeMet), and selenoproteins (White 2018). Some species require Se proteins, while land plants and fungi do not. Selenoproteins are enzymes with redox activity that use selenocysteine as a catalytic residue. Excessive accumulation of Se can make most angiosperms poisonous. A small amount of Se has been found to enhance the activities of various antioxidants, which helps reduce plant oxidative stress (Hasanuzzaman et al. 2020).

The soil can be enriched with different forms of Se, such as selenate or selenite. Different plant species have varying abilities to accumulate Se in their tissues, and only a small amount of Se applied to the soil is utilized by plants (Hasanuzzaman et al. 2020). Certain organic forms of Se, like SeCys and SeMet, have better phytoavailability and are used as active components. SeCys and SeMet are taken up by root cells via transporters, which differ between plant species in terms of selectivity and regulation (White 2018). Selenate is the most abundant Se species in the xylem, with trace levels of SeMet and SeOMet being detected. Members of the aluminum-activated malate transporters (ALMT) transporter family carry selenate into the xylem sap. Thereafter, selenate enters leaf cells *via* SULTR and is stored in cell vacuoles. *Arabidopsis thaliana* sulfate transporter (AtSULTR4) homologs 1 and 2 catalyze selenate inflow from vacuoles. In most plants, Se deficiency or accumulation increases gene expression. Se is transferred *via* the phloem as SeMet and Se-methylselenocysteine. In *Arabidopsis*, AtSULTR1;3 and AtSULTR2;2 catalyze selenate absorption.

Plants can be protected from various abiotic stresses like cold, drought, desiccation, and metal stress when supplemented with low doses of Se (Gupta and Gupta 2017). Se-mediated dismutation can help reduce aluminum (Al) toxicity, accumulation of ROS, and heat stress in *Sorghum* and As stress in mung bean. However, higher doses of Se can be harmful and cause oxidative stress. Se can help maintain the chloroplast membrane structure, reduce electrolytic leakage, improve cell integrity, delay senescence, increase yield, and guard plants against pathogens and insects. Figure 2.3 illustrates the beneficial roles of Se in plants.

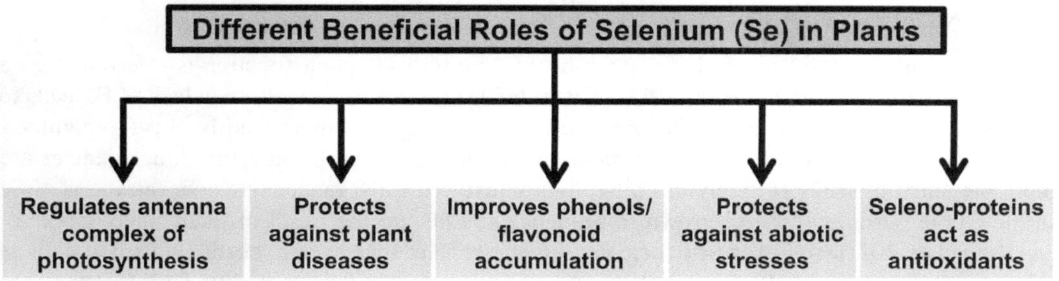

FIGURE 2.3 Beneficial roles of Se in plants.

2.4 METALLOID DEFICIENCY

Metalloid deficiencies impact plants and animals, causing physiological stress, growth decline, and potential death. Se deficiency significantly impacts human health, while micronutrient toxicities can lead to severe consequences. These metalloid deficits are economically more significant than soil-contaminated toxicities.

2.4.1 Metalloid Deficiency Symptoms in Animals

Low Se uptake by crop plants in Se-deficient soils results in low Se concentrations in edible parts, leading to Se deficiency in humans (Feng et al. 2021). Approximately 0.5–1 billion people worldwide are suspected to suffer from Se deficiency. The WHO suggests that 50–55 µg day^{-1} of Se should be included in human diets worldwide (Gupta and Gupta 2017). Se deficiency may happen when the dietary intake is less than 40 µg day^{-1}, while chronic toxicity can occur when levels of 400 µg day^{-1} are exceeded. Livestock require 0.05–0.10 mg kg^{-1} dry forage, and toxic concentrations in animal feed are between 2 and 5 mg kg^{-1}. Humans suffer from a shortage of Se as a result, since food is the primary supply of this vital component. Se deficiency occurs when the quantity of Se in regularly consumed meals is less than 0.1 µg g^{-1}, while selenosis is caused by concentrations greater than 1 µg Se g^{-1} (Feng et al. 2021).

As a component of the amino acid selenocysteine found in the active site of enzymes, Se is required in trace amounts by both humans and animals (Genchi et al. 2023). Se becomes poisonous in large quantities because it takes the place of S in enzymes. It is a part of numerous metabolic activities in both animal and human systems. Se increases the activity of immunological cells in the immune system, including natural killer, cytotoxic T, and helper T cells. Long-term Se deficiency in humans can lead to or exacerbate conditions like Keshan and Kashin–Beck illnesses, cancer, cardiovascular diseases, liver diseases, and cataracts (Fairweather-Tait et al. 2011; Feng et al. 2021; Genchi et al. 2023). Moreover, a lack of Se is linked to immune system enhancement, muscular necrosis, hypothyroidism, cardio-cerebrovascular illness, male infertility, and an elevated risk of many types of cancer (Genchi et al. 2023; Rayman 2012). However, these deficiency diseases might also be associated with the imbalance of multiple dietary nutrients (Feng et al. 2021).

B is a trace element that is essential for animal health. According to Abdelnour et al. (2018), it is essential to several physiological, metabolic, and biological processes. Low immune response, osteoporosis, and an elevated mortality risk can result from B deficiency. On the other hand, animals that receive B supplements in animal feed have improved their liver metabolism, bone density, wound healing, and embryonic development (Figure 2.2). B deficiency activates the MAPK pathway in animal cells grown in culture, leading to the development and proliferation of cell lines that have a deletion of the boron transporter namely Na^+-Coupled Borate Transporter (NaBC1).

2.4.2 Metalloid Deficiency Symptoms in Plants

A decrease in important minerals for plant health that impacts plant function is referred to as a mineral deficit (Gupta and Solanki 2013). A worldwide concern in agriculture, a lack of B, leads to problems with crop productivity. Because it percolates through soil more readily, B is more vulnerable to acidic soil conditions and heavy rains. In regions with significant rainfall, deficiencies are more prevalent in sandy or freely draining, lightly textured soils. Inadequate concentrations of Se in the edible portions of crops grown in Se-deficient soils are the result of inadequate uptake of Se (Feng et al. 2021). To lessen Se deficiency, numerous plant species have been biofortified with Se (Hasanuzzaman et al. 2020). Se can also increase crop output and food quality. Different types of Se, including selenite or seleniate, can be used to fertilize the soil.

According to Gupta and Solanki (2013), B deficiency affects 132 crops in over 80 countries and causes anatomical, physiological, and biochemical changes in plants, bacteria, and animals. B deficiency symptoms in plants include cell division in growing tips; death of terminal buds; distorted growth of lateral shoots, thick, curled, and brittle leaves; inhibited young leaves; and deformed flowers. B deprivation hinders flowering and fruiting processes, harms cell wall synthesis, and leads to cell death and tissue disintegration, resulting in abnormalities such as cracked or hollow stems. B deficiency causes male sterility which affects the yield of wheat, barley, and triticale (Lordkaew et al. 2011). It also decreased maize grain yield in sand culture trials and had little effect on the vegetative dry weight. B deficient plants displayed signs including narrow white streaks, malformed ears, and shriveled anthers, and they produced 0.4 grain ear^{-1} less than B-enriched plants.

Compared to herbaceous plants, there have been fewer studies conducted on B transporters in woody plants (Wang et al. 2015). This suggests that woody plants may be more susceptible to mild or chronic B deficiencies. Trees, in particular, have larger bodies, longer lifespans, and more B reserves. This may indicate that trees have more complex responses to and tolerances for B deficiency. When B is lacking in trees, they exhibit symptoms such as depressed growth points and distorted organs. This could be related to the way B functions in the membrane and cell wall. The efficiency of B uptake in trees is affected by three mechanisms: B uptake, B translocation and retranslocation, and B utilization.

According to Miyake and Takahashi's research (1978), Si deficiency has an impact on the reproduction of tomato plants. When grown in a Si-free solution during the initial bud blossoming stage, plants showed symptoms of Si deficiency. However, when grown in a solution containing 100 ppm SiO_2, normal growth was observed. To cause Si deficiency, the tomato plants were cultivated in ideal conditions with low planting density, perfect temperatures, light intensity, and aeration. Some symptoms of Si deficiency include malformed leaflets, delayed pollination, yellowed leaves, necrotic patches, and dryness. While Si-free plants can still bloom, pollination is frequently absent.

2.5 METALLOID TOXICITY

An important health concern is the global pollution of water bodies by metals, metalloids, and organometallic contaminants (Lichtfouse et al. 2022). Particularly concerning, especially in developing nations, is the presence of the Se metalloid in quite high concentrations in the environment, particularly in the water. These metalloids are absorbed by plants due to their increased concentration, which has an impact on plant growth and development (Kour et al. 2022). They change the physiological, metabolic, and biochemical factors of plants, which inhibits plant growth and development. One of the main causes of metalloid toxicity in plants and animals is As. Because of natural mobilization by minerals and human activities including mining, burning fossil fuels, using pesticides and herbicides, desiccating crops, adding As to animal feed, and preserving wood, living organisms are exposed to As (Raju 2022). The presence of As in groundwater used for agriculture irrigation can contaminate plants and soil, endangering the health of people and animals as well as harming food security (Su et al. 2023). The food safety of fruits and vegetables is at risk from As contamination. Animals can become poisoned by inadvertently ingesting As-contaminated food. Chronic As poisoning in humans can occur when it is mixed with food (Raju 2022). Although an essential metalloid, excessive concentrations of B may be hazardous (Białek et al. 2019). The following section discusses various harmful effects of metalloids on animals and plants.

2.5.1 Harmful Effects of Metalloids on Animals

The higher concentrations of B have been found to result in cell damage and toxicity in humans and other animal species (Abdelnour et al. 2018). When inorganic As enters the animal body, it produces oxidative stress, DNA damage, inflammation, immune system imbalances, and cell cycle abnormalities that lead to toxic damage to several organ systems (Su et al. 2023). Human exposure to As can lead to diseases like cognitive impairment, cardiovascular failure, and cancer. Long-term As intake through drinkable water and food consumption at specific concentrations is linked to several major health problems, including skin cancer, melanosis, hyperkeratosis, restricted lung disease, peripheral vascular disease, and gangrene in humans (Raju 2022). Prolonged As use in animals can result in chronic inflammation and hepatorenal toxicity (Su et al. 2023). Weak mutagen arsenite can intensify the mutagenic effects of other carcinogens, including UV and X-ray radiation. Besides this, As can also cause reproductive damage by hastening the aging of the testicles in rats. The WHO suggests minimum uptake of Sb should not exceed 0.6 μg kg^{-1} of body weight per day (Periferakis et al. 2022). Sb is a highly toxic metalloid that is just as toxic as As, if not more so. The trivalent inorganic form of Sb, Sb(III), is even more toxic than its organic counterpart, Sb(V). Generally, inorganic species are more potent in terms of toxicity. While Sb shares similarities with As, only the biochemical behaviors of the trivalent form are comparable. The pentavalent forms of Sb and As have different structures and may affect different physiological mechanisms of the human body.

Humans can get a wide range of illnesses, malignancies, and genetic problems as a result of exposure to Sb at different doses (Igdelou et al. 2020). It also results in gastritis, keratitis, dermatitis, and conjunctivitis. The threshold value of Sb in drinking water is 20 μg L^{-1} as per recommendations of the WHO (Lai et al. 2022). Sb can be found in drinking water, food, and dust, causing increased absorption. Chronic daily consumption can result from roadside dirt, polyethylene terephthalate (PET) plastic, and bottled water. Sb enters human erythrocytes through protein channels and metalloid transporters. It is harmful to human cells, reducing their viability and proliferation. Sb induces oxidative stress and apoptosis and increases the risk of cancer by activating related signaling pathways. In human cells, Sb also causes micronuclei and chromosome abnormalities. Its genotoxic effects on pulmonary cells could be attributed to indirect mechanisms of action such as increased ROS production, changed gene expression, or disruption of DNA repair mechanisms.

Sb poisoning is a result of Sb entering the human body through the respiratory system, skin contact, and food chain (Zhang et al. 2021). It is a mutagenic, carcinogenic, and deformed metalloid that can harm the liver, heart, and nervous system; excite the mucosa in the respiratory tract; decrease enzyme activity; and interfere with the metabolism of proteins and carbohydrates due to its great affinity for sulfhydryl groups in body proteins. There are two types of Sb toxicity: acute and chronic. Acute poisoning damages the visual nerves, but chronic poisoning might result in pulmonary dust disease. The toxicity of Sb results from its binding to thiol-containing enzymes, which produces thioantimonites (Periferakis et al. 2022). This depletes reduced GSH levels in cells like As exposure. Both Sb and As directly inhibit the pyruvate dehydrogenase enzyme, which affects the glucose oxidation pathway. Exposure to Sb can lead to a reduction in ATP levels, which may trigger anaerobic glycolysis. Se poisoning can cause a condition known as selenosis, which can result in symptoms such as garlic breath, gastrointestinal issues, hair loss, nail sloughing, and brain impairment (Gupta and Gupta 2017). Severe selenosis can even lead to liver cirrhosis, lung edema, or death.

2.5.2 Harmful Effects of Metalloids on Plants

The industrial revolution has resulted in severe environmental damage due to the overuse of metalloids, heavy metals, and hazardous substances. Being sessile creatures, plants have modified their anatomical structures among other characteristics to adapt to this environment. A thorough understanding of plant growth, development, defense, and production under metalloid/heavy metal stress is provided by the study of structural changes to plant tissues caused by metals and metalloids (Yadav et al. 2021). Different plant organs and tissues adapt anatomically to metalloids in different ways, influencing plant access to resources in the environment through structural alteration of plant tissues. Metalloids also act as environmental contaminants by negatively impacting crop productivity, especially when present in high concentrations in soil. Plants with the help of transporters maintain the concentrations of essential metal ions for normal metabolic activities (Alloway 2013). Non-essential trace metals are transported to plant tissues *via* membrane transporters for essential or beneficial elements (Tang et al. 2023). Excess metalloid ions damage cell biomolecules, affect metabolic processes, and inhibit plant growth, leading to decreased crop yield. As is one of the most common hazardous metalloids and its levels can go as high as 3,000 ppm.

Plants are known to absorb Sb, which can have negative effects on them, such as growth retardation, suppression of photosynthesis, and decreased synthesis of metabolites and hormones (Bolan et al. 2022). Sb species can also alter the shape of roots, obstructing cell division, elongation, and growth in plants. The metalloids modify the architectural structure of root diameter, epidermal cells, parenchymatous cells, central vein, pericycle, and conducting tissues (Yadav et al. 2021). They can also thicken the exo- and endodermis, suberizing and lignifying root tissues, and enhance cell thickness by retarding growth and reducing cell division. Sb has a deleterious impact on plants, resulting in slowed growth, decreased biomass, ROS production, and lipid peroxidation (Vidya et al. 2022). The toxicities of Sb to plants vary depending on the type, with inorganic Sb being more toxic than organic Sb (Zhu et al. 2020). Sb harms plants, causing physiology, anatomy, and ultrastructure variations (Maresca et al. 2020). Low hazardous storage accumulation takes place in certain tissues or within organisms. Sb suppresses soil enzymes, soil microbial populations, and their communities.

Contaminated soils are more mobile and bioavailable, making them more dangerous than natural sources (Nguyen et al. 2021). Plant biomass, germination, root length, plant height, and photosynthesis can all be negatively impacted by concentrations of Sb in the soil that are higher than the critical level (i.e., 35 mg kg^{-1}), leading to oxidative stress (Igdelou et al. 2020). The oxidative burst, a more intense buildup of ROS, results from a disruption in the delicate balance between the ROS-producing and ROS-scavenging processes (Sharma et al. 2024). Further degenerative changes in plants, such as membrane damage, lipid peroxidation, cell death, etc., are consequent to this rise in ROS accumulation. The toxic effects of Sb on plants are primarily caused by anti-oxidative stress and enhanced rate of lipid peroxidation; reduction in photosynthesis; nutrient imbalance; and inhibition of synthesis of soluble proteins, sugars, and starch in plants (Zhu et al. 2020). Phosphorylating substances such as arsenate are harmful to cells because they resemble phosphate in phosphorylation processes and produce unstable, rapidly hydrolyzing arsenate esters (Zhao et al. 2022). This cycle selects phosphate as a life-sustaining component by distancing photophosphorylation from oxidative phosphorylation. Conversely, arsenite binds to cysteine residues and inactivates enzymes. The detrimental effects of the two main metalloids, As and Sb, on plants are shown in Figure 2.1. In addition to these two metalloids, plants may have problems such as impaired protein biosynthesis, necrosis, chlorosis, and growth restriction due to high Se concentrations (Hasanuzzaman et al. 2020).

2.6 MITIGATION OF METALLOID TOXICITY

Plants absorb less Sb from acidic soils than from calcareous soils. The use of organic materials, phosphate fertilizers (phosphorus has an antagonistic impact on Sb), and sulfur-containing fertilizers can help in reducing Sb absorption in calcareous soils (Igdelou et al. 2020). Another metalloid, Se, is frequently found in complexation with other mixtures or compounds that pose problems with anions competing with one another (Lichtfouse et al. 2022). It is difficult to remove it from groundwater and industrial effluents Furthermore, water characteristics like pH and the concentration of competing anions, as well as Se speciation, affect how effective remediation techniques are. Se can be removed from wastewater, industrial effluents, and contaminated water through various techniques. Among them, zero-valent iron, iron-oxy-hydroxides, supporting materials, nanofiltration, reverse osmosis, improved ultrafiltration with chitosan, electrodialysis, and activated granular sludge are the major Se remediation technologies. Since metalloid uptake varies with plant species, therefore, effective mitigation techniques are required to safeguard plants and their associated organisms (Vidya et al. 2022). The remediation options for Sb include the use of Si and Se, charcoal, arbuscular mycorrhiza, salicylic acid, growth regulators, and nanoparticles. The next section reviews important strategies for metalloid detoxification in animals and plants.

2.6.1 Treatment Strategies for Detoxification of Metalloids in Animals

Metalloids can enter into the body of grazing animals through different modes, such as ingestion, inhalation, and consumption of contaminated plants, soil, and water (Roggeman et al. 2014). In cattle, the primary processes of regulating metalloid homeostasis involve changes in absorption rate, urine excretion, tissue deposition, secretion into milk, and endogenous excretion through feces. Chioneins (MTs), which are tiny, cysteine-rich metal-binding proteins, play a vital role in the binding, transport, and detoxification of metals/metalloids. Metalloid and MT concentrations were observed to be higher in the cattle's liver and kidney. Histopathological analysis of the tissue revealed that only the liver had moderate damage, which may have resulted from excessive metal buildup and inadequate detoxification by binding to MT.

Concerns over the presence of metalloids in the ecosystem have led to the development of new technologies to tackle the problem (Nishad and Bhaskarapillai 2021). Due to the complex nature of the pollutant, effective remediation methods are being developed. Researchers are exploring ways to limit Sb release from industrial activities and nuclear operations. Better remediation techniques are being developed for domestic water treatment, industrial effluent remediation, and safe radioactive waste isolation. The poisoning of inorganic As toxicity to animals primarily arises from the induction of oxidative stress, DNA damage, and cell death (Su et al. 2023). Therefore, As poisoning can be effectively treated by administrating drugs/medications that have antioxidant potential and can inhibit DNA damage and reverse the cell death effects. Figure 2.4 represents the toxic effects of important metalloids viz., As, Ge, Sb, and Te, in human beings.

2.6.2 Metalloid Detoxification Strategies in Plants

Plant roots absorb important nutrients and can reduce metalloid toxicity by changing the morphology of roots, the composition of root exudate, creating root iron plaques, and altering root cell wall components (Zhu et al. 2020). Root exudates (low molecular weight or high molecular weight) can regulate metalloid intake in plants by changing the kind and quantity of exudates, affecting rhizosphere pH, competing with heavy metals for adsorption sites, and producing chelates. Plants have to adopt various detoxification strategies to reduce the harmful effects of metalloids (Alloway 2013). Through the use of mutants, metalloid-binding peptides and proteins, such as phytochelatins (PCs) and MTs, can be overexpressed to sequester metalloids in the tissues and limit the uptake of numerous polluting metalloids by plants by decreasing the activities of transporters. Consequently,

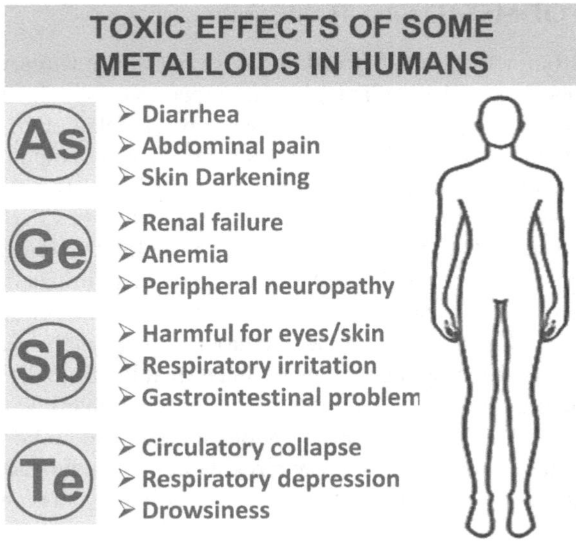

FIGURE 2.4 Toxic effects of some metalloids in human beings.

metalloid-tolerant plants would result from decreased metalloid movement in the tissues and higher sequestration of these particles within the cells. Some plants have naturally developed mechanisms to sequester and accumulate metalloids beyond normal tolerance limits within their above-ground plant parts as compared to other plants growing in similar conditions (Tang et al. 2023). These plants are natural metalloid hyperaccumulators and hypertolerant of the metals/metalloids they accumulate and can even flourish in severely contaminated soils (Krämer 2010). Some hyperaccumulator plants like *Houttuynia cordata*, *Pteris vittata*, *Ageratum houstonianum*, and *Potamogeton oxyphyllus* are good candidates for phytoremediation of As and multi-metal contaminated sites (Ha et al. 2011).

Many phytoremedial techniques, in which hyperaccumulating plants aid in the removal of these metals from the soil, have been employed to regulate the rising concentration of metal(loid)s (Kour et al. 2020). Rhizobacteria and endophytic bacteria are examples of helpful microbes that can be used to increase the efficiency of these approaches. These bacteria coexist in symbiotic relationships with the plants, facilitating the quick and simple uptake of metals through either stabilizing the metals in the soil or breaking them down into less dangerous forms. These bacteria employ a variety of strategies to aid in the process of removing metal(loid)s from the soil. In a hyperaccumulator plant, *Brassica juncea*, by regulating antioxidant enzyme activities, lowering the production of ROS, inhibiting lipid peroxidation, and enhancing cellular viability, the combination of PGPR and earthworms reduced the oxidative stress (Sharma et al. 2022). Increased production of ROS is one of the first reactions to abiotic stress including metalloid toxicity, and ROS play a critical role in these responses (Sharma et al. 2021). During stress-induced oxidative bursts, ROS levels are regulated by systems that generate ROS and systems that scavenge ROS.

The ROS are not the only species that cause cellular damage; due to their high reactivity and harmful lethality, reactive nitrogen species (RNS), reactive sulfur species (RSS), and reactive carbonyl species (RCS) are also produced (Dhiman et al. 2023). Nonetheless, RSS is also essential to signaling cascades. Low-molecular-weight RSS, such as thiol, disulfide, sulfenic acid, thiosulfate, and thiosulfonate, is generated as an intermediary in the disulfide synthesis pathway during oxidative stress. RSS may be created during the detoxifying processes between ROS and RNS as well as during the oxidation of disulfides by thiosulfinates, a kind of ROS that is eventually transformed into thiosulfonates. By decreasing ROS, boosting low-molecular-weight compounds, and upregulating antioxidant enzyme activity, Se stimulates the antioxidant defense system of plants (Feng et al. 2021). It also regulates various physiological metabolites that are involved in photosynthetic systems, chloroplast, and recovery of cell membrane integrity. The exogenous application of melatonin improves photosynthesis and preserves ion homeostasis, strengthening antioxidant mechanisms (Maurya et al. 2024). Various techniques including, the use of antioxidants and distinct PGRs, adjuvants, biopesticides, nano-engineered pesticides, integrated pest management, and others, can be employed to mitigate oxidative damage (Sharma et al. 2024).

Recently, Sb(III) has been found to increase the abscisic acid production in rice plants, hindering their cell division, elongation, and growth (Feng et al. 2020). However, rice crops have been found to enhance the activities of antioxidant enzymes including superoxide dismutase and ascorbate peroxidase as a response to Sb-induced oxidative stress. Antioxidants such as Se and others as well as proteins involved in DNA replication, stress response, energy, and carbohydrate metabolism are also protected against damage from excessive ROS formation (Feng et al. 2021). Se also promotes the absorption and redistribution of vital components for maintaining cell ion homeostasis. Application of Se has been found to lessen the toxicity of certain metalloids at low concentrations. The addition of Se to the soil may be a more effective way to lower the amounts of metalloids in plant tissues than foliar spraying. Se lessens the toxicity of metalloids in plants, including uptake and translocation from roots to shoots due to Se's capacity to prevent metalloid uptake, which lowers a variety of metalloids. Also, the co-application of both PGPR and earthworms may be useful in developing methods to improve plant development and yield in soils impacted by metals/metalloids (Sharma et al. 2022). Further, using PGPR treatment in conjunction with other soil-dwelling micro- and macro-organisms, such as earthworms, microalgae, and mycorrhizal fungus, can be a more effective stress-relieving treatment than using PGPR species alone (Singh et al. 2023).

Some microbes can eliminate Sb from affected environments through various mechanisms such as methylation, efflux, Sb(III) oxidation, and Sb(V) reduction (Kataoka et al. 2018). Certain microbes even use these mechanisms to thrive in Sb-rich environments. However, there has also been extensive research on the harmful effects of Sb on microbes, including methylation, reduction, oxidation, and efflux. Li et al. (2019) emphasized that Sb is harmful to microorganisms. The presence of Sb also affects As(III)-oxidizing bacterial communities in soil. However, some microbes can remove Sb from affected areas through chemical reactions like efflux, methylation, Sb(III) oxidation, and Sb(V) reduction (Bolan et al. 2022). Microorganisms also can decrease lethal Sb(III) accumulation in yeast and oxidize Sb(III) to a less toxic and immobilized form, making bioremediation easier.

In plants, the primary metabolism of Se and S leads to the accumulation of SeCys or SeMet, which can cause physiological malfunction by impairing protein functions (White 2018). However, this accumulation can be prevented by either preventing selenide entry into plastids, sequestration in vacuoles, or synthesizing Se-binding proteins. Alternatively, SeCys and SeMet can be converted into non-toxic or volatile Se metabolites, which is a more effective method of detoxification. The enzyme selenocysteinelyase can convert SeCys to elemental Se and alanine, which is relatively non-toxic. Other compounds produced to detoxify Se include Se-methylselenomethionine (SeMSeMet) and γ-GluSeMSeMet.

Plants can detoxify arsenite, arsenate, and toxic metals by binding them with PCs and storing them in vacuoles (Zhao et al. 2022). By enhancing PC biosynthesis through metabolic alterations, rice can increase its tolerance toward As and cadmium (Cd). In *Arabidopsis*, the zinc-finger transcription factor 6 (ZAT6) positively regulates reduced GSH and PC biosynthesis, leading to an increase in Cd tolerance (Chen et al. 2016). Signaling can regulate GSH-PC biosynthesis and Cd tolerance through MYB4 (Myeloblastosis Viral Oncogene Homolog 4) and MNB1(Mannose-Binding-Lectin 1), while cytokinin acts as a negative regulator (Chen et al. 2016; Zhao et al. 2022). The PC pathway is ineffective at detoxifying dimethylarsenate, possibly because PCs do not form stable complexes with dimethylarsenate. Presently, numerous studies on *Arabidopsis* and rice have helped understand the molecular mechanisms of As and Cd uptake, translocation, and detoxification. Gene editing and transgenic methods have identified targets for reducing their uptake and translocation. Research is still ongoing to understand how plants like wheat, barley, maize, and vegetables detoxify other metalloids due to existing knowledge gaps.

2.7 CONCLUSION AND FUTURE PROSPECTS

The ingestion or uptake of metalloids can be harmful to both plants and animals, affecting their overall health. The contamination of metalloids in the soil can also impact food safety and security, making it crucial to understand how plants absorb, transport, and detoxify these harmful substances. One potential solution could be through genetic modification or the use of hyperaccumulating plants for phytoremediation. It's important to note that metalloids can enter food chains and have an impact on the entire ecosystem. Further research is needed to understand the microstructure, morphology, and reaction mechanisms of metalloids at natural mineral interfaces. It's also essential to compare metalloids with other elements and assess their potential health risks and phytotoxic effects. Generally, the biogeochemical behavior of metalloids is influenced by their chemical speciation. Immobilization, mobilization, and microbial remediation are techniques to remove the harmful metalloids, but there are still knowledge gaps. Future research should focus on collecting data on distribution and speciation, ecotoxicology, chemical and microbiological interactions, developing biosensors for remediation of different types of metalloids, and conducting integrated field studies. Moreover, understanding the underlying mechanisms of metalloid detoxification and how plants resist metalloids is crucial for improving plant growth and ensuring food safety.

REFERENCES

Abdelnour, Samir A., Mohammed E. Abd El-Hack, Ahmed A. Swelum, Alessio Perillo, and Cristina Losacco. 2018. The vital roles of boron in animal health and production: A comprehensive review. *Journal of Trace Elements in Medicine and Biology* 50: 296–304.

Alloway, Brian J., (Ed.). 2013. Heavy metals and metalloids as micronutrients for plants and animals. In *Heavy Metals in Soils: Trace Metals and Metalloids in Soils and Their Bioavailability*, 195–209. Springer, Dordrecht.

Araújo, Wellington B.S., Guilherme C.M. Teixeira, Rômulo de Mello Prado, and Ana Maria S. Rocha. 2022. Silicon mitigates nutritional stress of nitrogen, phosphorus, and calcium deficiency in two forage plants. *Scientific Reports* 12, no. 1: 6611.

Białek, Marta, Michał Czauderna, Katarzyna Krajewska, and Witold Przybylski. 2019. Selected physiological effects of boron compounds for animals and humans: A review. *Journal of Animal and Feed Sciences* 28, no. 4: 307.

Bolan, Nanthi, Mukesh Kumar, Etiene Singh, Arvind Kumar, Lakhveer Singh, Suraj Kumar, Santhosh Keerthanan, Ahmed El-Naggar, Mithun Vithanage, and Banwari Sarkar. 2022. Antimony contamination and its risk management in complex environmental settings: A review. *Environment International* 158: 106908.

Chen, Jun, Lin Yang, Xiujuan Yan, Yan Liu, Rui Wang, Tao Fan, Yan Ren, Xugang Tang, Fasong Xiao, Yongbiao Liu, and Shuqing Cao. 2016. Zinc-finger transcription factor ZAT6 positively regulates cadmium tolerance through the glutathione-dependent pathway in Arabidopsis. *Plant Physiology* 171, no. 1: 707–719.

Dhiman, Sunil, Mohammed Ibrahim, Khushboo Khanna, Tanu Bhardwaj, Kiran Devi, Isha Sharma, Upasna Arora, Bilal Ahmad Mir, and Renu Bhardwaj. 2023. Cross-talk between ROS, RNS, RCS, and RSS in plants under abiotic stresses. In M. Iqbal, R. Khan, Noushina Iqbal, Peter Poor, Antonio Ferrante (Eds.), *Nitric Oxide in Developing Plant Stress Resilience*, 305–326. Academic Press.

Dutta, Sutapa, Abhishek Datta, Akhtar H. Zaid, and Javaid A. Bhat. 2020. Metalloids and their impact on the environment. In Rupesh Deshmukh, Gea Guerriero, Durgesh Kumar Tripathi (Eds.), *Metalloids in Plants: Advances and Future Prospects*, 19–26. John Wiley & Sons, Incorporated, Newark.

Fairweather-Tait, Susan J., Yu Bao, Malcolm R. Broadley, Rachel Collings, David Ford, John E. Hesketh, and Ray Hurst. 2011. Selenium in human health and disease. *Antioxidants & Rredox Signaling* 14, no. 7: 1337–1383.

Feng, Ruili, Ling Lei, Jiazhen Su, Rui Zhang, Yun Zhu, Wei Chen, Lina Wang, Rui Wang, Jin Dai, Zhiwei Lin, and Yunxia Li. 2020. Toxicity of different forms of antimony to rice plant: Effects on root exudates, cell wall components, endogenous hormones and antioxidant system. *Science of the Total Environment* 711: 134589.

Feng, Ruili, Ping Zhao, Yun Zhu, Jiaqi Yang, Xin Wei, Lijun Yang, Hongbin Liu, Christopher Rensing, and Yongchao Ding. 2021. Application of inorganic selenium to reduce accumulation and toxicity of heavy metals (metalloids) in plants: The main mechanisms, concerns, and risks. *Science of the Total Environment* 771: 144776.

Genchi, Giuseppe, Giuseppe Lauria, Alessandra Catalano, Maria Stefania Sinicropi, and Antonella Carocci. 2023. Biological activity of selenium and its impact on human health. *International Journal of Molecular Sciences* 24, no. 3: 2633.

Gupta, Manoj, and Sunita Gupta. 2017. An overview of selenium uptake, metabolism, and toxicity in plants. *Frontiers in Plant Science* 7: 234638.

Gupta, Umesh, and Himani Solanki. 2013. Impact of boron deficiency on plant growth. *International Journal of Bioassays* 2, no. 7: 1048–1050.

Ha, Nguyen Thi Hong, Masahiko Sakakibara, Shuhei Sano, and Mai Trong Nhuan. 2011. Uptake of metals and metalloids by plants growing in a lead–zinc mine area, Northern Vietnam. *Journal of Hazardous Materials* 186, no. 2–3: 1384–1391.

Hasanuzzaman, Mohammed, M. B. H. Bhuyan, Atiqur Raza, Barbara Hawrylak-Nowak, Renata Matraszek-Gawron, Jubayer Al Mahmud, Kawsar Nahar, and Masayuki Fujita. 2020. Selenium in plants: Boon or bane? *Environmental and Experimental Botany* 178: 104170.

Hunt, D. C. 1996. Dietary boron deficiency and supplementation. In Ronald R. Watson (Ed.), *Methods in Nutrition Research: A New CRC Series on Research in Nutrition*, 229–253. CRC Press.

Igdelou, Nana Koua Makita, Ahmad Golchin, and Tahereh Rahmani Kelarlou. 2020. Antimony and its effects on the components of environment. *Iranian Journal of Soil and Water Research* 50, no. 9: 2373–2399. https://doi.org/10.22059/ijswr.2019.280131.668180

Irfan, Muhammad, Muhammad Asif Maqsood, Hafiz U. Rehman, Waqas Mahboob, Nasir Sarwar, Omer B. A. Hafeez, Samina Hussain, Serkan Ercisli, Muhammad Akhtar, and Tehseen Aziz. 2023. Silicon nutrition in plants under water-deficit conditions: Overview and prospects. *Water* 15, no. 4: 739.

Kataoka, Takayuki, Shinya Mitsunobu, and Naoki Hamamura. 2018. Influence of the chemical form of antimony on soil microbial community structure and arsenite oxidation activity. *Microbes and Environments* 33, no. 2: 214–221.

Kour, Jaskirat, Khushboo Khanna, Poonam Bakshi, Isha Sharma, Puneet Ohri, Bilal Ahmad Mir, Neelam Kapoor, and Renu Bhardwaj. 2020. Role of beneficial microbes in the molecular phytotoxicity of heavy metals. In Mohammad Faisal, Quaiser Saquib, Abdulrahman A. Alatar, Abdulaziz A. Al-Khedhairy (Eds.), *Cellular and Molecular Phytotoxicity of Heavy Metals*, 227–262. Springer, Cham.

Kour, Jaskirat, Khushboo Khanna, Parul Sharma, Ashish Deep Singh, Tanu Bhardwaj, Nisha Sharma, Anshika Sharma, Renu Kaur, and Renu Bhardwaj. 2022. Salicylic acid-mediated physiological and molecular mechanism in plants under metal (loid) stress. In Anket Sharma, Renu Bhardwaj, Vinod Kumar, Bingsong Zheng, Durgesh Kumar Tripathi (Eds.), *Managing Plant Stress Using Salicylic Acid: Physiological and Molecular Aspects*, 138–162. Wiley, Hoboken, NJ.

Krämer, Ute. 2010. Metal hyperaccumulation in plants. *Annual Review of Plant Biology* 61: 517–534.

Lai, Zhigang, Ming He, Chunxiao Lin, Weiping Ouyang, and Xueqing Liu. 2022. Interactions of antimony with biomolecules and its effects on human health. *Ecotoxicology and Environmental Safety* 233: 113317.

Li, Jun, Yan Zhang, Shuping Zheng, Feng Liu, and Guodong Wang. 2019. Anaerobic bacterial immobilization and removal of toxic Sb (III) coupled with Fe (II)/Sb (III) oxidation and denitrification. *Frontiers in Microbiology* 10: 421021.

Lichtfouse, Eric, Naceur Morin-Crini, Cristian Bradu, Younès Achour Boussouga, Mehrdad Aliaskari, Andrew I. Schäfer, Sagar Das, Lindsey D. Wilson, Masaru Ike, Daisuke Inoue, and Mika Kuroda. 2022. Methods for selenium removal from contaminated waters: A review. *Environmental Chemistry Letters* 20, no. 3: 2019–2041.

Lombi, Enzo, and Per H. Holm. 2010. Metalloids, soil chemistry and the environment. In Thomas P. Jahn, Gerd P. Bienert (Eds.), *MIPs and Their Role in the Exchange of Metalloids*, 33–44. Springer Science+Business Media, New York.

Lordkaew, Sakhan, Barry Dell, Suchada Jamjod, and Benjavan Rerkasem. 2011. Boron deficiency in maize. *Plant and Soil* 342: 207–220.

Maresca, Viviana, Milad Heydari, and Adriana Basile. 2020. Antimony and plants: Accumulation, toxic effects, and plants' defense systems. In Rupesh Deshmukh, Gea Guerriero, Durgesh Kumar Tripathi (Eds.), *Metalloids in Plants: Advances and Future Prospects*, 275–299. John Wiley & Sons, Incorporated, Newark.

Martínez-Alcalá, Iván, and María del P. Bernal. 2020. Environmental impact of metals, metalloids, and their toxicity. In Rupesh Deshmukh, Gea Guerriero, Durgesh Kumar Tripathi (Eds.), *Metalloids in Plants: Advances and Future Prospects*, 451–488. John Wiley & Sons, Incorporated, Newark.

Maurya, Vishnu, Neeraj Singh, Inderdeep Sharma, Anket Sharma, and Ritesh Kumar. 2024. Effect of melatonin in regulating salt stress responses in plants. In Anket Sharma, Golam Jalal Ahammed (Eds.), *Melatonin in Plants: Role in Plant Growth, Development, and Stress Response*, 109–139. Springer Nature Singapore, Singapore.

Mirlean, Nicolae, Eduardo R. Seus-Arrache, and Olga Vlasova. 2018. Selenium deficiency in subtropical littoral pampas: Environmental and dietary aspects. *Environmental Geochemistry and Health* 40: 543–556.

Miyake, Yasuto, and Eiichi Takahashi. 1978. Silicon deficiency of tomato plant. *Soil Science and Plant Nutrition* 24, no. 2: 175–189.

Nguyen, Trung Hieu, Sunghyun Won, Myung Gyu Ha, Duong Dinh Nguyen, and Hyun Young Kang. 2021. Bioleaching for environmental remediation of toxic metals and metalloids: A review on soils, sediments, and mine tailings. *Chemosphere* 282: 13118.

Nishad, Pratibha A., and Arumugam Bhaskarapillai. 2021. Antimony, a pollutant of emerging concern: A review on industrial sources and remediation technologies. *Chemosphere* 277: 130252.

Periferakis, Athanasios, Ana Caruntu, Alexandru T. Periferakis, Anca Elena Scheau, Ioana Andreea Badarau, Cristian Caruntu, and Constantin Scheau. 2022. Availability, toxicology and medical significance of antimony. *International Journal of Environmental Research and Public Health* 19, no. 8: 4669.

Raju, Nandimandalam Janardhana. 2022. Arsenic in the geo-environment: A review of sources, geochemical processes, toxicity and removal technologies. *Environmental Research* 203: 111782.

Rayman, Margaret P. 2012. Selenium and human health. *The Lancet* 379, no. 9822: 1256–1268.

Roggeman, Stijn, Gert De Boeck, Hendrik De Cock, Ronny Blust, and Lieven Bervoets. 2014. Accumulation and detoxification of metals and arsenic in tissues of cattle (Bos taurus), and the risks for human consumption. *Science of the Total Environment* 466: 175–184.

Sharma, Anket, Puneet Sharma, Ritesh Kumar, Vineet Sharma, Renu Bhardwaj, and Inderdeep Sharma. 2021. Role of reactive oxygen species in the regulation of abiotic stress tolerance in legumes. In Vijay Pratap Singh (Ed.), *Abiotic Stress and Legumes*, 217–243. Academic Press.

Sharma, Anket, Sanjay K. Sharma, Neeraj Singh, Vishnu Maurya, Sukhdeep Kaur, Ritesh Kumar, and Inderdeep Sharma. 2024. Pesticides-mediated ROS generation in plants. In Anket Sharma, Vinod Kumar and Bingsong Zheng (Eds.), *Pesticides in a Changing Environment*, 179–202. Elsevier.

Sharma, Inderdeep, Anket Sharma, Renu Bhardwaj, and Gaurav Sirhindi. 2023. *PGPR (Plant Growth Promoting Rhizobacteria)* for Plant Stress Management. Nova Science Publishers, Inc.

Sharma, Parul, Renu Bhardwaj, Rajni Chouhan, Poonam Bakshi, Satveer Gandhi, Rupinder Kaur, Anket Sharma, and Renu Bhardwaj. 2022. Amelioration of chromium-induced oxidative stress by combined treatment of selected plant-growth-promoting rhizobacteria and earthworms via modulating the expression of genes related to reactive oxygen species metabolism in Brassica juncea. *Frontiers in Microbiology* 13: 802512.

Singh, Neeraj, Vishnu Maurya, Inderdeep Sharma, Yuvraj Kumar, Ritesh Kumar, and Anket Sharma. 2023. Combinations of PGPR with other soil micro-and macro-organisms for improved stress tolerance in plants. In Indu Sharma (Ed.), *PGPR (Plant Growth Promoting Rhizobacteria) for Plant Stress Management*, 83–107. Nova Science Publishers, Inc.

Su, Qian, Ying He, Hang Pan, Haiyan Liu, Khalid Mehmood, Zhaoxin Tang, Lianmei Hu. 2023. Toxicity of inorganic arsenic to animals and its treatment strategies. *Comparative Biochemistry and Physiology Part C: Toxicology & Pharmacology* 271: 109654.

Sun, Hai-Bo, Li Xu, Zhi-Xin Wang, Yi Zheng, Yan Zhao, Ying-Ying Yin, Xiao-Li Han, and Zhi-Nan Xu. 2019. Polyoxometalate SbW9 regulates proliferation and apoptosis of NSCLC cells via PTEN-dependent AKT signaling pathway. *European Review for Medical & Pharmacological Sciences* 23, no. 18: 7959.

Tang, Zhen-Yu, Hai-Qiang Wang, Jun Chen, Ji-Dong Chang, and Fang-Jie Zhao. 2023. Molecular mechanisms underlying the toxicity and detoxification of trace metals and metalloids in plants. *Journal of Integrative Plant Biology* 65, no. 2: 570–593.

Teixeira, Guilherme C. M., Rômulo de Mello Prado, Kelen Soraya Oliveira, Vinícius D'Amico-Damião, and Gilberto da Silveira Sousa Junior. 2020. Silicon increases leaf chlorophyll content and iron nutritional efficiency and reduces iron deficiency in sorghum plants. *Journal of Soil Science and Plant Nutrition* 20: 1311–1320.

Thétiot-Laurent, Stéphanie A. L., Julien Boissier, Alain Robert, and Bernard Meunier. 2013. Schistosomiasis chemotherapy. *Angewandte Chemie International Edition* 52, no. 31: 7936–7956.

Vidya, C. S. N., R. Shetty, Milada Vaculíková, and Martin Vaculík. 2022. Antimony toxicity in soils and plants, and mechanisms of its alleviation. *Environmental and Experimental Botany* 202: 104996.

Wan, Xin Mei, Samuel Tandy, Kathrin Hockmann, and Rainer Schulin. 2013. Changes in Sb speciation with waterlogging of shooting range soils and impacts on plant uptake. *Environmental Pollution* 172: 53–60.

Wang, Na, Chong Yang, Zhen Pan, Yan Liu, and Shao An Peng. 2015. Boron deficiency in woody plants: Various responses and tolerance mechanisms. *Frontiers in Plant Science* 6: 128665.

White, Philip J. 2018. Selenium metabolism in plants. *Biochimica et Biophysica Acta (BBA) General Subjects* 1862, no. 11: 2333–2342.

Yadav, Vinay, Nadeem Arif, Jozef Kováč, Vinay Pratap Singh, Devendra Kumar Tripathi, Dharmendra Kumar Chauhan, and Martin Vaculík. 2021. Structural modifications of plant organs and tissues by metals and metalloids in the environment: A review. *Plant Physiology and Biochemistry* 159: 100–112.

Zhang, Yu, Chunyan Ding, Dengfeng Gong, Yanting Huang, Jiayu Zheng, Suyun Xiong, Renfang Tang, Yijun Wang, and Liang Su. 2021. A review of the environmental chemical behavior, detection and treatment of antimony. *Environmental Technology & Innovation* 24: 102026.

Zhao, Fang-Jie, Zhen Tang, Ji-Jiang Song, Xiao-Yan Huang, and Ping Wang. 2022. Toxic metals and metalloids: Uptake, transport, detoxification, phytoremediation, and crop improvement for safer food. *Molecular Plant* 15, no. 1: 27–44.

Zhu, Yun, Jia Yang, Lina Wang, Zhiwei Lin, Jin Dai, Rui Wang, Yuan Yu, Hongbin Liu, Christopher Rensing, and Ruili Feng. 2020. Factors influencing the uptake and speciation transformation of antimony in the soil-plant system, and the redistribution and toxicity of antimony in plants. *Science of the Total Environment* 738: 140232.

3 Molecular Mechanism Underlying Metalloid Transportation in Plants

Firdous Rasool Sheikh, Ram Naresh, and Rajesh Kumar Singh

3.1 INTRODUCTION

The elements falling in between metals and non-metals in terms of their physical as well as chemical characteristics are known as semimetals or metalloids. This group of elements forms a diagonal line in the periodic table and consists of boron (B), silicon (Si), germanium (Ge), arsenic (As), antimony, and tellurium (Zangi & Filella, 2012). Some metalloids including B, Si, and selenium are essential for plants in trace amounts but are toxic in higher quantities. However, metalloids namely As, Ge, antimony, and tellurium are non-essential and toxic to plants at lower concentrations. When plants are grown in such a contaminated soil, these elements move from soil to plant roots and other plant parts through different channels and transporters (Zangi & Filella, 2012; Pandey et al., 2019; Mariyam et al., 2023a,b). These transporters comprise a diverse array of genome-encoded transporter proteins that govern translocation through varying degrees of selectivity, distinct expression patterns, and distinct cellular localizations (Zangi & Filella, 2012). Numerous transporter protein families, such as phosphate transporters (Pi), sulfate transporters, aquaglyceroporins, nodulin 26-like intrinsic proteins, Si influx transporters, hexose transporters, etc., play a crucial role in metalloids transport (Farooq et al., 2009). Under normal growth conditions, plants express these transporters, and their expression upregulates and downregulates in response to increasing metalloid concentration in the soil (Singh et al., 2017). After being absorbed by the root cells, metalloids are eliminated from the plant system by sequestration or active efflux (Fu et al., 2000). Consequently, potentially fatal circumstances may be caused by an increase in the availability of metalloids that might damage vital cellular functions (Agnihotri & Seth, 2020). The organism's simplest adaptation would be to regulate how these metalloids enter into the cell (influx) and exit it (efflux). Nonetheless, the creation of metalloid complexes by compartmentalization and subsequent sequestration of metalloids through complex formation with proteins or other peptide legends is a crucial step in the plant tolerance mechanism (Dhankher, 2005). Global worry over metalloid contamination of soil has resulted in decline in agricultural production (Detmann et al., 2012). This demands a thorough understanding of how plants absorb metalloids and how they are transported, sequestered, and tolerated. In this chapter, we are summarizing the biotechnological and molecular approaches to understand the molecular mechanisms underlying metalloid transportation in plants.

3.2 METALLOID TRANSPORT

Plants use two major processes i.e., active and passive transport to uptake metalloids from the soil into their cells and then move them to aerial tissues (Zangi & Filella, 2012; Kroukamp et al., 2016; Kumar et al., 2023). Transport through these membrane-spanning proteins can be driven by the concentration gradient, which is the difference in solute concentration when the diffusion route is from high to low concentration or by energy-fueled pumps where transport happens in the opposite

DOI: 10.1201/9781032673592-3

direction of the concentration gradient (Zangi & Filella, 2012). The energy source for active transport is adenosine triphosphate (ATP), which is directly used by major active transporters or other ion-pumping proteins that move Na^+ or H^+. This results in an electrochemical potential difference across the cell membrane (Choudhary et al., 2024). The force for transport is provided by this difference in electrochemical potential, and this sort of transporter, known as a secondary active transporter, is typically involved in the movement of ions (Zangi & Filella, 2012).

The cell membrane is predominantly hydrophobic at the molecular level, which prevents the passage of hydrophilic molecules (Mariyam et al., 2023a). Nonetheless, water and other tiny, uncharged hydrophilic molecules can permeate the lipid bilayer directly and larger molecules like sugars cannot cross the membrane (Deamer & Bramhall, 1986; Jansen & Blume, 1995). For the larger molecules to be able to move from one side of the cell to the other, a channel with a sufficient hydrophilic environment must be formed using particular transmembrane proteins.

Metalloids like Si, B, As, and antimony are stable in aqueous state and exist in multivalent poly-hydroxylated species. Being intermediate and hydrophilic in size, these metalloids can cross the cell membrane and sometimes via channel proteins also. Initially, it was thought that metalloids enter the cell only by passive unaided diffusion. However, several pieces of evidence suggest the involvement of integral membrane proteins in the uptake and/or exclusion of metalloids from the cell. These channels can be driven by energy-fueled pumps that can actively transport the solute against the concentration gradient, or by differences in the solute's concentration, where the direction of the diffusion is from high to low concentration (facilitated diffusion). The channels known as aquaporins and aquaglyceroporins are important for the passive transport of metalloids. These channels are members of the superfamily of channel proteins known as major intrinsic protein (MIP), which is present in all spheres of life.

3.3 BORON (B)

The metalloid that is most likely to exhibit unassisted passive diffusion is B, since it is the smallest of the metalloids. B is carried into and out of cells via uncharged boric acid, which is also its most stable form. In fact, it has been reported that B diffuses passively throughout a variety of plants (Raven, 1980; Day & Aasim, 2020). However, other research has also suggested the active transport of B (Dannel et al., 2000). It has been found that B can move passively both with and without the assistance of a solute channel facilitator (Perica et al., 2001). There aren't many aquaporins and aquaglyceroporins that have been shown to help Xenopus oocytes and yeast to absorb B (Dordas & Brown, 2000; Fitzpatrick & Reid, 2009). It should be noted that protein channels also aid in the efflux of metalloids out of the cell in addition to facilitating their uptake into the cell. It has been shown that this enhanced efflux gives plants resistance against excessive B concentrations, which can be hazardous (Hayes & Reid, 2004).

3.4 MOLECULAR MECHANISMS OF BORON (B) TRANSPORT

Boric acid is known to be transported across membranes via three different mechanisms including passive diffusion across the lipid bilayer, active transport by borate transporter (BOR), and facilitated transport by nodulin-like intrinsic protein (NIP) channel. Passive diffusion was determined more than 80 years ago. B is essential for plant growth and development at lower concentrations (Warington, 1923). There is evidence that insufficient B lowers plant productivity and decreases the quality and amount of agricultural output in more than 70 nations worldwide (Shorrocks, 1997). Boric acid has a relatively high permeability coefficient for lipid bilayers because it is an uncharged molecule (Tanaka et al., 2008). According to Raven's estimation, boric acid's theoretical lipid permeability coefficient was 8×10^{-6} cm s^{-1} (Raven, 1980). When there is a relatively higher boric acid supply, transport across the lipid bilayer should be the major portion of transmembrane boric acid transport. When the permeability coefficient was measured experimentally by Dordas and Brown, the result fell within a comparable range to Raven's estimate (Dordas & Brown, 2000). Nevertheless, they also measured the permeability coefficients of boric acid in the large internodal cells of the charophyte alga (*Chara corallina*) and the plasma membrane of squash (*Cucurbita pepo*) roots and discovered that the values were one or two orders of magnitude lower than those calculated by Raven (Dordas & Brown, 2001). The findings suggest that passive diffusion through lipid bilayers becomes noteworthy solely in situations where there is a strong concentration gradient or a substantial supply of B.

Before Dannel et al.'s research, it was widely thought that B transport in plants occurs passively (2000). However, it was demonstrated that the transport of boric acid in sunflowers occurs against concentration gradient (Dannel et al., 2000). Forward genetics was used to identify BOR1, the first B transporter in biological systems from Arabidopsis (Noguchi et al., 1997). It was shown that under low B concentrations, BOR1 is involved in the xylem loading required for normal shoot growth in *Arabidopsis* (Noguchi et al., 1997). BOR1 is a member of the solute carrier family 4 (SLC4) bicarbonate transporter superfamily. It is localized in the plasma membrane, and when produced in yeast, the amount of boric acid in the cells is reduced. It suggests that BOR1 acts as a borate exporter. The root stele's pericycle cells also express BOR1. When B is deficient, *Arabidopsis* plant with a point mutation in BOR1 exhibits a significant drop in plant growth. When considered collectively, it was determined that BOR1 serves in the xylem loading of boric acid in low B concentrations (Noguchi et al., 1997; Takano et al., 2002). Ibrahim Ilker Ozyigit et al. 2020 have identified 80 BOR1 and 34 NIP5;1 homolog from 18 different plant species. *NIP5;1, 6;1*, and *BOR1* share distinct expression patterns. *NIP5;1, 6;1* is expressed in the root and petiole, while BOR1 is expressed in the primary root, radical, and flower. The schematic presentation of B transport is depicted in Figure 3.1.

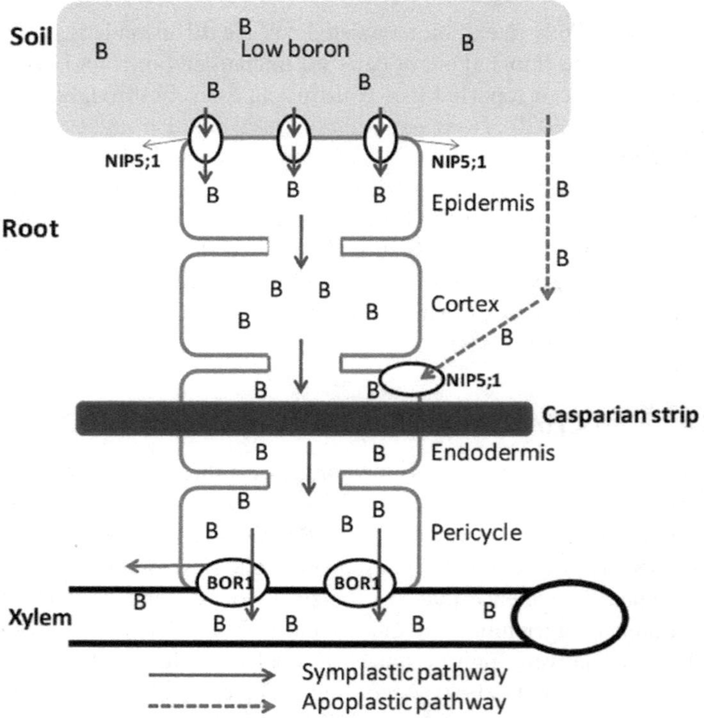

FIGURE 3.1 Diagrammatic representation of how BOR1 and NIP5;1 transport B in plant cell.

3.5 SILICON (SI)

Si is the most abundant essential element in soil after oxygen. Si always exists in silica or silicate form due to its strong affinity with oxygen. Si is an essential element for plant growth and development. Also, Si has an important role in biotic and abiotic stress in plants. For example, Si decreases sheath blight in rice, ring spot in sugarcane, powdery mildew in cucumber (Vasanthi et al., 2014). It also suppresses insect and pest attacks like leaf spiders, plant hoppers, etc. (Feng et al., 2011). All living organisms are capable of absorbing Si in its uncharged hydroxylated form Amo and Brzezinski (1999). The majority of plants, especially dicots, are unable to build up significant Si concentrations in their shoots. Variations in the roots capacity to absorb Si have been linked to the variation in Si build-up among species (Ma & Takahashi, 2002). Si is taken by the roots as silicic acid [$Si(OH)_4$], an uncharged monomeric molecule when the pH of the solution is less than 9. Si is transported to the shoot through the xylem after being absorbed by the roots. The process of transporting Si from the extracellular medium into the cortical cells and then into the xylem is called "xylem loading." Through transpiration, silicic acid in the shoot becomes even more concentrated and undergoes polymerization (Ma & Yamaji, 2006).

3.6 MOLECULAR MECHANISM OF SI TRANSPORT

Every plant has certain transporters that allow Si to be absorbed and accumulated in various plant species, such as rice (OsLsi), *Arabidopsis* (AtLsi), and soybean (ZmLsi), etc. In addition, Si is transferred from soil to root, from root to cortical cells, and finally into xylem vessels (Mitani & Ma, 2005). The energy-dependent process of Si absorption has also been shown by Mitani and Ma (2005) to function at low temperatures and in the presence of metabolic inhibitors. Furthermore, they discovered that the transport of Si in xylem sap in *O. sativa* is mediated by the transporter genes, whereas in *C. sativus* and *L. esculentum*, diffusion is responsible for the transport of Si (Ma & Yamaji, 2006).

The gene *Lsi1* has been found to be responsible for active Si absorption in rice. It is a bidirectional transporter that carries out both influx and efflux types of transport, but operates only as an influx transporter in the roots of *O. sativa*. The low silicon rice 1 (Lsi1) transporter in roots carries silicic acid from the soil, whereas the Lsi2 transporter carries silicic acid from one cell to another (Gaur et al., 2020). *Lsi1* gene is responsible for silicon uptake in rice, which encodes a plasma membrane protein that is a member of the NIP subfamily of aquaporins (Ma et al., 2006). The concentration of Si outside of cells controls the constitutive expression level of this gene. *Lsi1* is localized mainly in roots. It is expressed in the lateral and main roots, but not in the root hair (Ma & Yamaji, 2006). This confirms that only roots are involved in Si transport and not root hairs. Additional research revealed that the transport protein is located on the distal side of the exodermis and endodermis cells of plasma membranes where Casparian strips exist (Ma & Yamaji, 2006). It has also been proven that lateral roots considerably contribute to Si uptake, but root hair do not participate in the process (Ma et al., 2001). Si accumulated in these root cells need to be transported to other locations (Zangi & Filella, 2012; Maryam & Seth, 2023). It was also shown that the homologous protein HvLsi1 (*Hordeum vulgare* silicon influx transporter), which is a member of the same class of aquaporins, helps the influx of silicic acid in barley (Chiba et al., 2009). Three more homologs (*Zeamays* zinc-regulated transporter protein 2 (ZmNIP2-1), ZmNIP2-2, and ZmNIP2-3) were discovered in maize, responsible for the uptake of Si (Ma et al., 2006). Similarly, *Zea mays* likewise had a high concentration of Si build-up, and *Zeamays* silicon influx transporter (*ZmLsi1*) and *ZmLsi6* transporter genes are responsible for this. They also demonstrated that these genes were OsLsi1 homologs, Si influx transporters of *Oryza sativa*. In *Trirticum aistivum*, it is reported that TaLsi1 and OsLsi1 forming a heterologous system possess intake of low Si uptake. The schematic presentation of B transport is depicted in Figure 3.2.

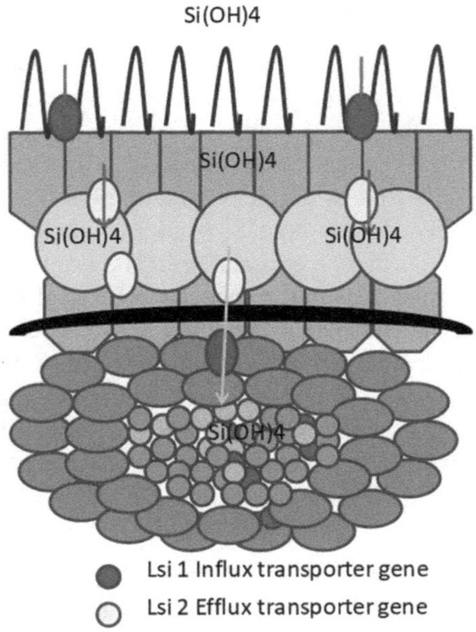

FIGURE 3.2 Schematic representation of Si transport in the plant through transporter proteins LSi1 and LSi2.

3.7 ARSENIC (AS)

As is a non-essential metalloid having potent carcinogenic and mutagenic properties. As has been put in group 1 human carcinogen among all the metalloids by the US environmental protection agency (EPA) and agency for Research on Cancer (IARC) (Abbas et al., 2018). Inorganic As can be found in soil and water in four oxidation forms: arsenite [As(iii)), arsenate (As(v), As(III), and As(0)], and organic forms [arsenobetain (AsB), trimethyl arsine oxide (TMAO), monomethylarsonic acid (MMA), etc.) (Patel et al., 2013)]. Inorganic As is more toxic and mobile in nature than organic forms. Out of these forms, As(V) and As (III) are highly toxic and mobile in nature. As(V) is dominating under oxidized conditions, while As(III) is dominating under reduced conditions (Khalid et al., 2017).

As is considered a non-essential element for plants. In very minute concentrations, it has positive effects in plant growth and development, while at higher concentrations, it disturbs the normal functioning of the plant like seed germination, plant growth, and yield (Gupta & Seth, 2019; Beniwal & Yadav, 2023; Yadav et al., 2022). For example, in *Brassica napus* and *Brassica juncea*, As toxicity causes a reduction in plant growth parameters like leaf size, leaf number, branch number, and dry weight (Niazi et al., 2018). In the roots of soybean (*Glycine max*), As deposits in root cortex area and xylem vessels, and reduction in chlorophyll content (Armendariz et al., 2016). In *Oryza sativa*, a high concentration of As (III) (50 µM) affects carbohydrate metabolism, photosynthesis and also affects root and shoot growth (Sanglard et al., 2016; Upadhyay et al., 2016; Zemanová et al., 2021). High concentration of As also affects the grain quality. For example, in *Oryza sativa* and *Cicer arietinum* L, As accumulation causes a decrease in essential and non-essential amino acids and starch and sugar accumulation (Malik et al., 2011; Dwivedi et al., 2012; Tripathi et al., 2017). As accumulation in plants also increases H_2O_2, ROS accumulation, and lipid peroxidation (Gunes et al, 2009; Jin et al., 2010; Hu et al., 2020; Saleem et al., 2023).

3.8 MOLECULAR MECHANISM OF AS TRANSPORT

As is taken up by the plant root from the As-contaminated soil and also transported from the different parts of the plants through various transporters. There is no transporter reported for As in the plant system. They share the transporters of the essential elements because of its structural analogy. For example, the analogy of phosphate transporters (Pi), with arsenate (AsV) enabling it to transport through Pi, while the analogy of arsenite [As(III)] with silicic acid enables them to transport them through aquaporins (Thounaojam et al., 2021). There are various transporters for transport of As in plant system which leads to accumulation and transit of As to various plant parts. The different Pi transporter proteins (PHT) are reported that involved in the uptake of As(V). For example, PTH1 proteins are known to be involved in the uptake of (AsV). LeBlanc et al. (2013) reported that *Arabidopsis* plants over-expressing *AtPht1;1* or *AtPht1;7* are hypersensitive to arsenate indicating that arsenate uptake increases upon the expression of high-affinity Pi transporters. These transporters are unidirectional. However, pht1 and pht4 double mutants *Arabidopsis* plants are significantly tolerant to As. Rice OSTP4 (*Oryza sativa* Tonoplast Intrinsic Protein 4) mutants when grown in flooded soil with AsV exhibit AsV significantly lower AsV accumulation in grains (Cao et al., 2017) compared to control. The adenosine triphosphate-binding cassette (ABC) transporter yeast cadmium factor (YCF1) is localized to the vacuolar membrane. Over-expression of YCF1 increases the As accumulation and tolerance in *Arabidopsis* (LeBlanc et al., 2013). Once within the plant cell, AsV can probably migrate readily from one cellular compartment to another, crossing internal membranes via the numerous Pi transporters. Das et al. (2018) reported that secondary transporter family protein multidrug and toxic compound extrusion (MATE) transports universally present in the living organism and transports metabolite and xenotoxic compounds. He reported that *Oryza sativa* Multidrug and Toxic Compound Extrusion 2 (*OsMATE2*) up-regulated during seed development under As(V) stress. The endosperm-specific silencing of *OsMATE2* reduces grain As content by approx. 50%.

Furthermore, AsIII transport in plants occur majorly through aquaporin channels. Several aquaporins like *Oryza sativa* Nodulin 26-like Intrinsic Protein (OsNIP1;1), OsNIP2;1, OsNIP2;2, OsNIP3;1, OsNIP3;2, OsNIP3;3, *Oryza sativa* Plasma Membrane Intrinsic Protein 2;4 (OsPIP2;4), OsPIP2;6, and OsPIP2;7 are reported actively transporting As(III) (Seth et al., 2007). Bienert et al. (2008) reported that over-expression of *OsNIP2*;1(Lsi1) and *OsNIP3;2* (Lsi2) from *Oriza sativa* and *LjNIP5;1* and *LjNIP6;1* from *Lotus japonicas* increases yeast cell to uptake AsIII. The transporter Lsi1 is responsible for the uptake of AsIII, while Lsi2 drives efflux from root to xylem cells and translocated to shoot (Ma et al., 2008). Lsi1 and Lsi2 are located on the distal and proximal side of plasma membrane, respectively. Loss of function of OsLsi1 and OsLsi2 reduces the As uptake and its transportation to shoot. The schematic presentation of As transport is depicted in Figure 3.3

As methylation is mediated by the soil microorganisms and algae in various methylated forms like MMAs(III), dimethylarsenite [DMAs(III))>As(III)>As(V)>monomethylarsenate (MMAs(V)], dimethylarsenate [DMAs(V)]>trimethylarsine [TMAs(III)], and trimethylarsine oxide (TMAs(V) (Akter et al., 2005). Methylation causes As to volatilize, resulting in toxic effects on the environment. Other methylated As forms dimethylmonothioarsenate (DMMTA) and monomethylmonothioarsenate (MMMTA) are also reported in rice grain, and these forms of As are highly toxic to human (Di et al., 2019). The complete information of As uptake is not elucidated yet. OsLsi1 is the first reported protein that is involved in the translocation of MMA and DMA. Rahman et al., 2011 reported that OsLsi1 mutant lowers 80% and 49% MMA and DMA accumulation, respectively, in rice. As transport to the grain is mainly carried out through phloem (Carey et al., 2010). Duan et al. (2015) reported that the loss of function of *Arabidopsis thaliana* Inositol Transporter 2 (AtINT2) and AtINT4 in *Arabidopsis* reduced As content in phloem, silique, and seeds. Another report by Tang et al. (2019) indicated that *Oryza sativa* C-type ATP-binding cassette transporter (OsABCC1) is localized at the tonoplast membrane of the phloem that restricts the transport of As(III) to the grain by promoting the As(III) conjugate to the vacuole. Total As content in the *Oryza sativa* Acetyl-CoA Carboxylase 1 (Osacc1) knockout mutant rice seed is increased but decreased in the stem nodes.

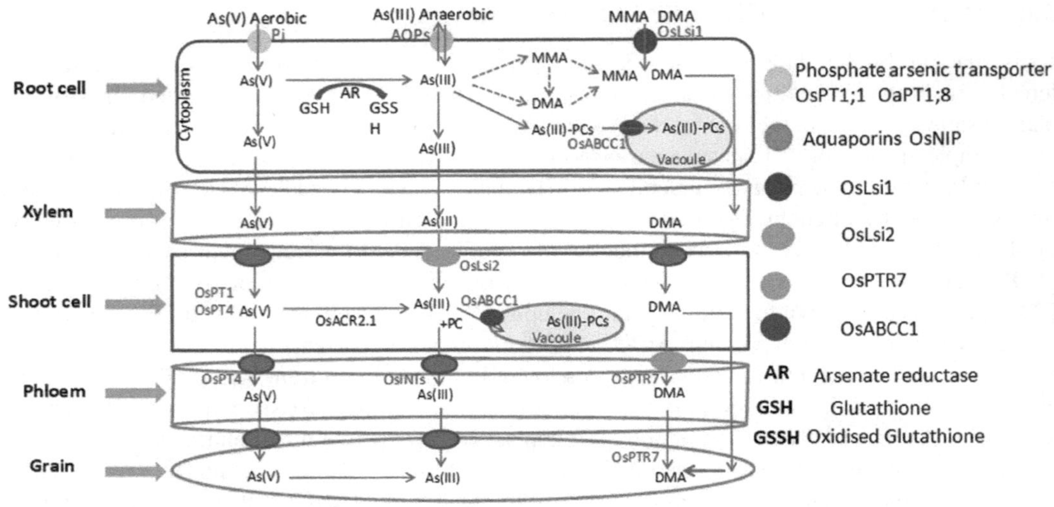

FIGURE 3.3 Schematic diagram for the uptake, transport, redistribution, and detoxification mechanisms of As.

3.9 ANTIMONY (SB) AND THEIR TRANSPORT

Antimony is a highly toxic metalloid that deteriorates plant growth and development. The main source of the antimony toxicity in the environment is the weathering of the sulfide ores. It is present in the organic and inorganic forms in the soil (Zhuang et al., 2018). Crops grown in such contaminated places introduced Sb toxin into the food web. The Sb contamination causes leaf chlorosis, damages photosynthetic machinery, reduces membrane stability, and increases the oxidative stress, thereby affecting the growth and development of the plant (Bolan et al., 2022). Sb(III) exists mainly in the trihydroxylated form Sb(III) and Sb(V). Sb(III) and Sb(V) mainly enters in the cell via aquaglyceroporins channel and hexose transporters (Maciaszczyk-Dziubinska et al., 2012).

3.10 GERMANIUM (GE)

The Earth's crust is rich in Ge. Ge, an element cognate to Si, exhibits remarkably similar chemical properties and has a significant impact on its behavior in soil–plant systems. In contrast to Si, Ge is more reactive (forms extremely stable compounds with a large number of functional groups at ecophysiologically relevant conditions) and at moderate-to-high concentrations becomes harmful to plants (Pokrovski, 1998; Delvigne, 2009;). Plant species with high Si values contained a high concentration of Ge. In hydroponic cultures, Ge concentrations greater than 20 μmol L^{-1} in barley result in necrosis and enhanced peroxidase activity (Halperin, 1995). Ge also causes harmful symptoms in Si accumulators like rice, which have been used to screen rice mutants with impaired Si uptake quite successfully (Ma & Takahashi, 2002).

3.11 MECHANISM OF GERMANIUM TRANSPORT

The uptake of Ge in plants occurs in different ways:

1. The capacity of certain plant species to actively absorb Ge from the soil and incorporate it into their aboveground organs.
2. Passive Ge uptake by mass flow.
3. The existence of Ge in pools that might be accessible to plants.

There is no doubt that the presence or absence of active transport mechanisms or a fluctuating density of the transporters in the root cell membranes accounts for the majority of the diversity in the ability of various species to accumulate Ge and Si (Ma & Yamaji, 2006; Nikolic, 2007). It is reported that plants take up Ge in the form of $Ge(OH)_4$ similar to silicic acid $Si(OH)_4$ from soil solutions (Ma & Yamaji 2006; Rains et al., 2006). As proven for $Si(OH)_4$ by Raven (2001) and Nikolic et al. (2007), a concentration-dependent passive absorption of $Ge(OH)_4$ by diffusion across the lipid component of the plasma membrane and/or by assisted diffusion is most likely widespread in all plants. Some plant species, especially grasses in the Cyperaceae and Gramineae groups, have high amounts of Si and Ge, which have been related to metabolically active concentration-independent assimilation (Nikolic, 2007; Wiche, 2018). Following uptake, Ge is carried by the transpiration stream into stems and leaves, where Si oversaturation causes it to precipitate as amorphous silica bodies, also referred to as plant opals or phytoliths (Blecker et al., 2007). The main organelles where the phytoliths are kept are in the intercellular spaces, cell walls, and cell lamina which are close to the plant's evaporating surfaces (Wiche, 2018). It is assumed that both Ge and Si accumulate at these locations equally, with no distinction between Ge and Si (Wiche, 2018). However, as of now, very little information is available regarding the molecular mechanism of Ge accumulation and transport in plants.

3.12 CONCLUSIONS

World industrialization and urbanization are the main causes of heavy metal and metalloid toxicity in the environment. Although some of these elements are essential for growth and development, they become toxic when they are present abundantly in the soil. The intricate process of metalloid transportation in plants reveals a fascinating interplay of molecular mechanisms, environmental cues, and plant adaption strategies. The selective uptake, translocation, and sequestering of metalloids by the plant not only influence their health but also have implications for the ecosystem. The knowledge gained from the above studies not only aids in the development of resilient and metalloid-tolerant crop varieties but also enlightens on strategies for phytoremediation, contributing to the remediation of metal-contaminated soils. Furthermore, the integration of advanced technologies, such as genomics, proteomics, and imaging techniques, will undoubtedly contribute to a more comprehensive understanding of metalloid transport dynamics in diverse plant species.

REFERENCES

Abbas G, Murtaza B, Bibi I, Shahid M, Niazi NK, Khan MI, Amjad M, Hussain M, & Natasha. Arsenic uptake, toxicity, detoxification, and speciation in plants: physiological, biochemical, and molecular aspects. *International Journal of Environmental Research and Public Health*. 2018;15(1):59.

Agnihotri A, Seth CS. Does jasmonic acid regulate photosynthesis, clastogenecity, and phytochelatins in *Brassica juncea* L. in response to Pb-subcellular distribution? *Chemosphere*. 2020;243:125361.

Akter KF, Owens G, Davey DE, Naidu R. Arsenic speciation and toxicity in biological systems. *Reviews of Environmental Contamination and Toxicology*. 2005:97–149.

Amo YD, Brzezinski MA. The chemical form of dissolved Si taken up by marine diatoms. *Journal of Phycology*. 1999 Dec;35(6):1162–1170.

Armendariz AL, Talano MA, Villasuso AL, Travaglia C, Racagni GE, Reinoso H, Agostini E. Arsenic stress induces changes in lipid signalling and evokes the stomata closure in soybean. *Plant Physiology and Biochemistry*. 2016;103: 45–52.

Beniwal R, Yadav R, Ramakrishna W. Multifarious effects of arsenic on plants and strategies for mitigation. *Agriculture*. 2023;13(2):401.

Bienert GP, Schüssler MD, Jahn TP. Metalloids: essential, beneficial or toxic? Major intrinsic proteins sort it out. *Trends in Biochemical Sciences*. 2008;33(1):20–26.

Blecker SW, King SL, Derry LA, Chadwick OA, Ippolito JA, Kelly EF. The ratio of germanium to silicon in plant phytoliths: quantification of biological discrimination under controlled experimental conditions. *Biogeochemistry*. 2007;86:189-199.

Bolan N, Kumar M, Singh E, Kumar A, Singh L, Kumar S, Keerthanan S, Hoang SA, El-Naggar A, Vithanage M, Sarkar B. Antimony contamination and its risk management in complex environmental settings: a review. *Environment International*. 2022;158:106908.

Cao Y, Sun D, Ai H, Mei H, Liu X, Sun S, Xu G, Liu Y, Chen Y, Ma LQ. Knocking out OsPT4 gene decreases arsenate uptake by rice plants and inorganic arsenic accumulation in rice grains. *Environmental Science & Technology*. 2017;51(21):12131–12138.

Carey AM, Scheckel KG, Lombi E, Newville M, Choi Y, Norton GJ, Charnock JM, Feldmann J, Price AH, Meharg AA. Grain unloading of arsenic species in rice. *Plant Physiology*. 2010;152(1):309–319.

Chiba Y, Mitani N, Yamaji N, Ma JF. HvLsi1 is a silicon influx transporter in barley. *The Plant Journal*. 2009;57(5):810–818.

Choudhary R, Rajput VD, Ghodake G, Ahmad F, Meena M, Rehman RU, Seth CS. Comprehensive journey from past to present to future about seed priming with hydrogen peroxide and hydrogen sulfide concerning drought, temperature, UV and ozone stresses-a review. *Plant and Soil*. 2024;500:1–23.

Dannel F, Pfeffer H, Römheld V. Characterization of root boron pools, boron uptake and boron translocation in sunflower using the stable isotopes 10 B and 11 B. *Functional Plant Biology*. 2000;27(5):397–405.

Das N, Bhattacharya S, Bhattacharyya S, Maiti MK. Expression of rice MATE family transporter OsMATE2 modulates arsenic accumulation in tobacco and rice. *Plant Molecular Biology*. 2018;98:101–120.

Day S, Aasim M. Role of boron in growth and development of plant: deficiency and toxicity perspective. *Plant Micronutrients: Deficiency and Toxicity Management*. 2020:435–453.

Deamer DW, Bramhall J. Permeability of lipid bilayers to water and ionic solutes. *Chemistry and Physics of Lipids*. 1986;40(2–4):167–188.

Delvigne C, Opfergelt S, Cardinal D, Delvaux B, André L. Distinct silicon and germanium pathways in the soil-plant system: Evidence from banana and horsetail. *Journal of Geophysical Research: Biogeosciences*. 2009;114(G2).

Detmann KC, Araújo WL, Martins SC, Sanglard LM, Reis JV, Detmann E, Rodrigues FÁ, Nunes-Nesi A, Fernie AR, DaMatta FM. Silicon nutrition increases grain yield, which, in turn, exerts a feed-forward stimulation of photosynthetic rates via enhanced mesophyll conductance and alters primary metabolism in rice. *New Phytologist*. 2012;196(3):752–762.

Dhankher OP. Arsenic metabolism in plants: an inside story. *New Phytologist*. 2005:503–505.

Di X, Beesley L, Zhang Z, Zhi S, Jia Y, Ding Y. Microbial arsenic methylation in soil and uptake and metabolism of methylated arsenic in plants: a review. *International Journal of Environmental Research and Public Health*. 2019;16(24):5012.

Dordas C, Brown PH. Permeability of boric acid across lipid bilayers and factors affecting it. *The Journal of Membrane Biology*. 2000;175:95–105.

Dordas C, Brown PH. Evidence for channel mediated transport of boric acid in squash (Cucurbita pepo). *Plant and Soil*. 2001;235:95–103.

Duan GL, Hu Y, Schneider S, McDermott J, Chen J, Sauer N, Rosen BP, Daus B, Liu Z, Zhu YG. Inositol transporters AtINT2 and AtINT4 regulate arsenic accumulation in Arabidopsis seeds. *Nature Plants*. 2015;2(1):1–6.

Dwivedi S, Mishra A, Tripathi P, Dave R, Kumar A, Srivastava S, Chakrabarty D, Trivedi PK, Adhikari B, Norton GJ, Tripathi RD. Arsenic affects essential and non-essential amino acids differentially in rice grains: inadequacy of amino acids in rice based diet. *Environment International*. 2012;46:16–22.

Farooq M, Wahid A, Kobayashi NS, Fujita DB, Basra SM. Plant drought stress: effects, mechanisms and management. *Sustainable Agriculture*. 2009:153–188.

Feng J, Yamaji N, Mitani-Ueno N. Transport of silicon from roots to panicles in plants. *Proceedings of the Japan Academy, Series B*. 2011;87(7):377–385.

Fitzpatrick KL, Reid RJ. The involvement of aquaglyceroporins in transport of boron in barley roots. *Plant, Cell & Environment*. 2009;32(10):1357–1365.

Fu D, Libson A, Miercke LJ, Weitzman C, Nollert P, Krucinski J, Stroud RM. Structure of a glycerol-conducting channel and the basis for its selectivity. Science. 2000;290(5491):481–486.

Gaur S, Kumar J, Kumar D, Chauhan DK, Prasad SM, Srivastava PK. Fascinating impact of silicon and silicon transporters in plants: a review. *Ecotoxicology and Environmental Safety*. 2020;202:110885.

Gunes A, Pilbeam DJ, Inal A. Effect of arsenic–phosphorus interaction on arsenic-induced oxidative stress in chickpea plants. *Plant and Soil*. 2009;314:211–220.

Gupta P, Seth CS. Nitrate supplementation attenuates As (V) toxicity in *Solanum lycopersicum* L. cv Pusa Rohini: insights into As (V) sub-cellular distribution, photosynthesis, nitrogen assimilation, and DNA damage. *Plant Physiology and Biochemistry*. 2019;139:44–55.

Halperin SJ, Barzilay A, Carson M, Roberts C, Lynch J, Komarneni S. Germanium accumulation and toxicity in barley. *Journal of Plant Nutrition*. 1995;18(7):1417–1426.

Hayes JE, Reid RJ. Boron tolerance in barley is mediated by efflux of boron from the roots. *Plant Physiology*. 2004;136(2):3376–3382.

Hu Y, Li J, Lou B, Wu R, Wang G, Lu C, Wang H, Pi J, Xu Y. The role of reactive oxygen species in arsenic toxicity. *Biomolecules*. 2020;10(2):240.

Jansen M, Blume A. A comparative study of diffusive and osmotic water permeation across bilayers composed of phospholipids with different head groups and fatty acyl chains. *Biophysical Journal*. 1995;68(3):997–1008.

Jin JW, Xu YF, Huang YF. Protective effect of nitric oxide against arsenic-induced oxidative damage in tall fescue leaves. *African Journal of Biotechnology*. 2010;9(11):1619–1627.

Khalid S, Shahid M, Niazi NK, Rafiq M, Bakhat HF, Imran M, Abbas T, Bibi I, Dumat C. Arsenic behaviour in soil-plant system: biogeochemical reactions and chemical speciation influences. In Anjum NA, Gill SS, Tuteja N (Eds.), *Enhancing Cleanup of Environmental Pollutants: Volume 2: Non-Biological Approaches*. 2017:97–140. Springer, Cham.

Kroukamp EM, Wondimu T, Forbes PB. Metal and metalloid speciation in plants: overview, instrumentation, approaches and commonly assessed elements. *TrAC Trends in Analytical Chemistry*. 2016;77:87–99.

Kumar D, Dhankher OP, Tripathi RD, Seth CS. Titanium dioxide nanoparticles potentially regulate the mechanism (s) for photosynthetic attributes, genotoxicity, antioxidants defense machinery, and phytochelatins synthesis in relation to hexavalent chromium toxicity in Helianthus annuus L. *Journal of Hazardous Materials*. 2023;454:131418.

LeBlanc MS, McKinney EC, Meagher RB, Smith AP. Hijacking membrane transporters for arsenic phytoextraction. *Journal of Biotechnology*. 2013;163(1):1–9.

Ma JF, Takahashi E. *Soil, Fertilizer, and Plant Silicon Research in Japan*. Amsterdam: Elsevier; 2002.

Ma JF, Tamai K, Yamaji N, Mitani N, Konishi S, Katsuhara M, Ishiguro M, Murata Y, Yano M. A silicon transporter in rice. *Nature*. 2006;440(7084):688–691.

Ma JF, Yamaji N. Silicon uptake and accumulation in higher plants. *Trends in Plant Science*. 2006;11(8):392–397.

Ma JF, Yamaji N, Mitani N, Xu XY, Su YH, McGrath SP, Zhao FJ. Transporters of arsenite in rice and their role in arsenic accumulation in rice grain. *Proceedings of the National Academy of Sciences*. 2008;105(29):9931–9935.

Maciaszczyk-Dziubinska E, Wawrzycka D, Wysocki R. Arsenic and antimony transporters in eukaryotes. *International Journal of Molecular Sciences*. 2012;13(3):3527–3548.

Malik JA, Goel S, Sandhir R, Nayyar H. Uptake and distribution of arsenic in chickpea: effects on seed yield and seed composition. *Communications in Soil Science and Plant Analysis*. 2011;42(14):1728–1738.

Mariyam S, Bhardwaj R, Khan NA, Sahi SV, Seth CS. Review on nitric oxide at the forefront of rapid systemic signaling in mitigation of salinity stress in plants: crosstalk with calcium and hydrogen peroxide. *Plant Science*. 2023a;336:111835.

Mariyam S, Seth CS. Nitric oxide: a key player in reinforcement of photosynthetic efficiency under abiotic stress. In Khan MIR, Iqbal N, Poor P, Ferrante A (Eds.), *Nitric Oxide in Developing Plant Stress Resilience*. 2023:157–171. Academic Press, Elsevier, UK. ISBN: 978-0-323–91209-9

Mariyam S, Upadhyay SK, Chakraborty K, Verma KK, Duhan JS, Muneer S, Seth CS. Nanotechnology, a frontier in agricultural science, a novel approach in abiotic stress management and convergence with new age medicine-a review. *Science of the Total Environment*. 2023b;912:169097.

Mitani N, Ma JF. Uptake system of silicon in different plant species. *Journal of Experimental Botany*. 2005;56(414):1255–1261.

Niazi NK, Bibi I, Shahid M, Ok YS, Shaheen SM, Rinklebe J, Wang H, Murtaza B, Islam E, Nawaz MF, Lüttge A. Arsenic removal by Japanese oak wood biochar in aqueous solutions and well water: investigating arsenic fate using integrated spectroscopic and microscopic techniques. *Science of the Total Environment*. 2018;621:1642–1651.

Nikolic M, Nikolic N, Liang Y, Kirkby EA, Romheld V. Germanium-68 as an adequate tracer for silicon transport in plants. Characterization of silicon uptake in different crop species. *Plant Physiology*. 2007;143(1):495–503.

Noguchi K, Yasumori M, Imai T, Naito S, Matsunaga T, Oda H, Hayashi H, Chino M, Fujiwara T. bor1-1, an Arabidopsis thaliana mutant that requires a high level of boron. *Plant Physiology*. 1997;115(3):901–906.

Ozyigit II, Filiz E, Saracoglu IA, Karadeniz S. Exploration of two major boron transport genes BOR1 and NIP5; 1 in the genomes of different plants. *Biotechnology & Biotechnological Equipment*. 2020;34(1):455–468.

Pandey AK, Gautam A, Dubey RS. Transport and detoxification of metalloids in plants in relation to plant-metalloid tolerance. *Plant Gene*. 2019;17:100171.

Patel KS, Pandey PK, Martín-Ramos P, Corns WT, Varol S, Bhattacharya P, Zhu Y. A review on arsenic in the environment: contamination, mobility, sources, and exposure. *RSC Advances*. 2023;13(13):8803–8821.

Perica S, Brown PH, Connell JH, Nyomora AM, Dordas C, Hu H, Stangoulis J. Foliar boron application improves flower fertility and fruit set of olive. *HortScience*. 2001;36(4):714–716.

Pokrovski GS, Schott J. Experimental study of the complexation of silicon and germanium with aqueous organic species: implications for germanium and silicon transport and Ge/Si ratio in natural waters. *Geochimica et Cosmochimica Acta*. 1998;62(21-22):3413–3428.

Rahman MA, Kadohashi K, Maki T, Hasegawa H. Transport of DMAA and MMAA into rice (Oryza sativa L.) roots. *Environmental and Experimental Botany*. 2011;72(1):41–46.

Rains DW, Epstein E, Zasoski RJ Aslam M. Active silicon uptake by wheat. *Plant and Soil*. 2006;280:223–228.

Raven JA. Short-and long-distance transport of boric acid in plants. *New Phytologist*. 1980;84(2):231–249.

Raven JA. Silicon transport at the cell and tissue level. In *Studies in plant science*, edited by L. E. Datnoff, G. H. Snyder, and G. H. Korndörfer, 2001; Vol. 8, pp. 41–55, Elsevier, Amsterdam.

Saleem K, Asghar MA, Javed HH, Raza A, Seleiman MF, Ullah A, Rahman A, Iqbal S, Hanif A, Imran S, Nadeem SM. Alleviation of arsenic toxicity-induced oxidative stress in lemon grass by methyl jasmonate. *South African Journal of Botany*. 2023;160:547–559.

Sanglard LM, Detmann KC, Martins SC, Teixeira RA, Pereira LF, Sanglard ML, Fernie AR, Araújo WL, DaMatta FM. The role of silicon in metabolic acclimation of rice plants challenged with arsenic. *Environmental and Experimental Botany*. 2016;123:22–36.

Seth CS, Chaturvedi PK, Misra V. Toxic effect of arsenate and cadmium alone and in combination on giant duckweed (*Spirodela polyrrhiza* L.) in response to its accumulation. *Environmental Toxicology: An International Journal*. 2007;22(6):539–549.

Shorrocks VM. The occurrence and correction of boron deficiency. *Plant and Soil*. 1997;193(1–2):121–148.

Singh D, Agnihotri A, Seth CS. Interactive effects of EDTA and oxalic acid on chromium uptake, translocation and photosynthetic attributes in Indian mustard (Brassica juncea L. var. Varuna). *Current Science*. 2017;112:2034–2042.

Takano J, Noguchi K, Yasumori M, Kobayashi M, Gajdos Z, Miwa K, Hayashi H, Yoneyama T, Fujiwara T. Arabidopsis boron transporter for xylem loading. *Nature*. 2002;420(6913):337–340.

Tanaka M, Fujiwara T. Physiological roles and transport mechanisms of boron: perspectives from plants. *Pflügers Archiv-European Journal of Physiology*. 2008 Jul;456:671–677.

Tanaka M, Wallace IS, Takano J, Roberts DM, Fujiwara T. NIP6; 1 is a boric acid channel for preferential transport of boron to growing shoot tissues in Arabidopsis. *The Plant Cell*. 2008 Oct 1;20(10):2860–2875.

Tang Z, Chen Y, Miller AJ, Zhao FJ. The C-type ATP-binding cassette transporter OsABCC7 is involved in the root-to-shoot translocation of arsenic in rice. *Plant and Cell Physiology*. 2019;60(7):1525–1535.

Thounaojam TC, Khan Z, Meetei TT, Srivastava S, Panda SK, Upadhyaya H. Transporters: the molecular drivers of arsenic stress tolerance in plants. *Journal of Plant Biochemistry and Biotechnology*. 2021;30:730–743.

Tripathi P, Singh PC, Mishra A, Srivastava S, Chauhan R, Awasthi S, Mishra S, Dwivedi S, Tripathi P, Kalra A, Tripathi RD. Arsenic tolerant *Trichoderma* sp. reduces arsenic induced stress in chickpea (Cicer arietinum). *Environmental Pollution*. 2017;223:137–145.

Upadhyay, A. K., Singh, N. K., Singh, R., & Rai, U. N. (). Amelioration of arsenic toxicity in rice: comparative effect of inoculation of Chlorella vulgaris and Nannochloropsis sp. on growth, biochemical changes and arsenic uptake. *Ecotoxicology and Environmental Safety*. 2016;124:68–73.

Vasanthi N, Saleena LM, Raj SA. Silicon in crop production and crop protection-A review. *Agricultural Reviews*. 2014;35(1):14–23.

Warington K. The effect of boric acid and borax on the broad bean and certain other plants. *Annals of Botany*. 1923;37(148):629–672.

Wiche O, Székely B, Moschner C, Heilmeier H. Germanium in the soil-plant system—a review. *Environmental Science and Pollution Research*. 2018;25:31938–31956.

Yadav M, Gupta P, Seth CS. Foliar application of α-lipoic acid attenuates cadmium toxicity on photosynthetic pigments and nitrogen metabolism in *Solanum lycopersicum* L. *Acta Physiologiae Plantarum*. 2022;44(11):112.

Zangi R, Filella M. Transport routes of metalloids into and out of the cell: a review of the current knowledge. *Chemico-Biological Interactions*. 2012;197(1):47–57.

Zemanová V, Pavlíková D, Hnilička F, Pavlík M. Arsenic toxicity-induced physiological and metabolic changes in the shoots of Pteris cretica and Spinacia oleracea. *Plants*. 2021;10(10):2009.

Zhuang W, Lai X, Wang Q, Liu Y, Chen Q, Liu C. Distribution characteristics, sources and ecological risk of antimony in the surface sediments of Changjiang Estuary and the adjacent sea, East China. *Marine Pollution Bulletin*. 2018;137:474–480.

4 Metalloid Menace
Potential Accumulation and Detoxification Strategies for Plant Metalloid Tolerance

Rashmi Choudhary, Kamel A. Abd-Elsalam, Shraddha Singh, Haitao Liu, Faheem Ahmad, Praveen Kumar Srivastava, and Chandra Shekhar Seth

4.1 INTRODUCTION

Due to their high reactivity and likelihood of toxic effects, metalloids are a significant ecological contaminant that reduces global crop yields. But the extent of plant toxicity is decided by an array of parameters, including the type, nature, and quantity of metalloids to which plants are exposed. Excessive metalloid exposure can harm plant growth and development, resulting in lower crop yields and environmental degradation (Abd-Elsalam et al., 2023; Xalxo et al., 2021). Excessive levels of harmful metalloid ions in plant tissues can destroy cell macromolecules, disrupt essential metabolic processes, and impede plant growth, eventually resulting in lower crop yields. Previously present coping strategies or the deployment of specific machinery for managing the risk will be stimulated in response to the plants' conditions. The last barrier regarding metalloid absorption through root systems is the leaf cells. Plants attempt to restrict the entry or redistribution of these metals into tissues of photosynthetic apparatus. However, in some plant systems, avoiding this seems impossible, and unfavorable outcomes include a decrease in leaf dimension, modifications in the organization of the protective upper layer of epidermal cells, and a decrease in interstitial gaps of mesophyll cells (Angulo-Bejarano et al., 2021; Seth et al., 2012, 2007; Gupta et al., 2019). Plants can absorb metalloids extracted from soil using transporter proteins present in their roots. These transporter proteins are encoded in the plant's genome and have different specificities, expression patterns, and cellular localization. They are critical in regulating metalloid transport to distinct portions of the plant system. Protein families responsible for the transport of silicon (Si); phosphate; hexose; sulfate, including, nodulin-26-like intrinsic proteins, aquaglyceroporins are proteins that transport metalloids (Kumar et al., 2024b; Pandey et al., 2019). Plants have developed mechanisms to limit the harmful effects of metalloids and reduce their accumulation. When metalloids enter plant parts, they are removed from the plant either by efflux or by sequestration of the metalloid complex into the vacuole (Fu et al., 2000; Gupta et al., 2017). Formation of metalloid complexes by compartmentalization and sequestration by peptides or proteins, on the other hand, is a key step connected with plant tolerance mechanisms. Plants have a natural detoxification mechanism that incorporates compartmentalization and chelation by phytochelatins (PCs), glutathione (GSH), and metallothioneins (MTs), among other chemicals like amino acids, organic acids (OAs), etc. (Agnihotri and Seth, 2020; Agnihotri et al., 2018). In addition, they alter membrane-bound transporters to decrease metalloid absorption. These treatments aid in the reduction of metalloid toxicity. As a consequence, it is critical to have an in-depth knowledge of metalloid absorption, movement, sequestration, and resistance strategies in plants, especially given the rising urbanism and manufacturing trends, to devise stress-resistance strategies and develop metalloid-resistant crops (Kumar et al., 2023; Singh et al., 2017).

DOI: 10.1201/9781032673592-4

This chapter explores plant accumulation and tolerance to extremely high amounts of heavy metalloids and various beneficial detoxification strategies to reduce metalloid toxicity and stress in plants. By enhancing plant capacity to survive in environmental extremes, metalloid-resilient crops can be developed.

4.2 METALLOID METABOLISM IN PLANTS AND ACCUMULATION

Metalloids are hydrophilic anions, which means despite the process of diffusion, they are unable to cross plasma membranes. This is because of the membrane's hydrophobic lipid bilayer. Transporter protein families regulate metalloid influx and efflux via active and passive transport pathways (Kim et al., 2024; Pandey et al., 2019). Metalloid absorption, transport, translocation, and elimination are all aided by transport proteins. Each metalloid, like other mineral elements, requires transporters that are expressed in the roots for uptake. In the case of metalloids, influx transporters are required for the initial radial transfer of minerals from the soil to the root stele. Efflux channels then move the material from the cells into the xylem (Yamaji and Ma, 2021; Tang and Zhao, 2021; Pommerrenig et al., 2015).

It has been reported that metalloid buildup is higher in trees and various cultivated vegetable varieties found in nature (Pathak et al., 2024; Clemens and Ma 2016). Different species experience accumulation and translocation at different rates, which are influenced by the organic structure and levels of metalloids in the media (Finnegan and Chen, 2012). Many metalloid ions are taken up by various species of plants through their roots and moved to areas that can be harvested. Plants that accumulate large metalloid ion concentrations are referred to as "hyperaccumulators." It has been reported that fern species like *Pteris vittata* and *Pteris cretica* can accumulate significant amounts of arsenic (As) in their aerial portions, with concentrations exceeding 1 g kg^{-1} dry weight (DW) (Chaturvedi et al., 2007). Antimony (Sb) hyperaccumulation has also been noticed ranging from 0.2 to 50 g kg^{-1} in ferns and terrestrial vascular plants. Hyperaccumulator species, such as *Sinapis arvensis* and *Brassica napus*, members of the Brassicaceae family, have been reported to accumulate 1,000–15,000 mg kg^{-1} DW of selenium (Babula et al., 2010). For members of Poaceae, such as wheat, barley, Si uptake has been reported to be high. Barley (*H. vulgare*) has reportedly increased germanium (Ge) accretion in both roots and shoots at its seedlings stage. Plants treated with GeO_2 showed modest levels of Ge accumulation in the shoots (above 25 mg L^{-1}) and in the roots (below 10 mg L^{-1}). Boron (B) levels in submerged and emergent water plants have been observed to range from 11.3 to 57 mg kg^{-1} and 26 to 382 mg kg^{-1}, respectively (Yadav et al., 2022; Chen et al., 2021). Table 4.1 illustrates the major accumulation sites of different metalloids in a plant.

TABLE 4.1
The Major Plant Parts Accumulating Different Metalloids in the Plant System

S. No.	Metalloid	Accumulation Part	References
1.	B	Move up through the xylem and accumulate at the end of transpiration streams primarily in leaf tips and margins. Accumulate in older leaves before relocating to young tissues	Brdar-Jokanović (2020); Shireen et al. (2018)
2.	As	As was mostly accumulated in the roots. A little As translocates to the aerial part of the plant that includes entire shoots, leaves, and grains too.	Moulick et al. (2023); Carbonell-Barrachina et al. (1997)
3.	Si	Si accumulates in shoots but not in roots mostly observed as deposition on stem, leaves, and hull or as amorphous hydrated silica for future deposition	Shanmugaiah et al. (2023); Mitani-Ueno and Ma (2021)
4.	Sb	Accumulates majorly in root tissues and seeds following a general pattern of roots > stems > leaves	Lu et al. (2023); Feng et al. (2013)
5.	Ge	Accumulates in shoots along with other parts in the lower concentration	Wiche et al. (2018)

Metalloid uptake and accumulation by roots involve a complex network of mechanisms of physiological and molecular processes. These processes include the attachment of metals to cell-wall constituents and extracellular discharges; the removal of dangerous metals from the inside of cells and their partitioning into vacuoles; the production of amino acids, OAs, MTs, or PC; the combination of metal complexes and production of antioxidant enzymes triggered by heavy metals (HM); and altering plant metabolism to promote restoration and recovery damaged cellular structures (Al Majeed et al., 2019; Deng et al., 2021). Important physiological and biochemical processes, such as modifications to gene expression, differences in proteins, and changes in the composition of metabolites, are brought about by these activities. All of these changes are in charge of creating the proper signals that, in the presence of HM toxicity, activate the defense and tolerance mechanisms in plants. Plants may have distinct detoxification processes depending on the metals or metalloids they are exposed to (Angulo-Bejarano et al., 2021).

4.3 METALLOID DETOX: UNVEILING STRATEGIC PATHWAYS FOR PLANT WELLNESS

Plants are impacted by metalloids in a number of ways, such as ion compartmentalization, chelation, and immobilization shown in Figure 4.1. Thiols that bond to metals like GSH, PCs, and proteins like MTs all play important roles in metalloid chelation and tolerance. Metalloids are hazardous substances that can significantly affect the growth of plants and development (Tang et al., 2023). Metals and metalloids have been identified as the most dangerous elements by the Agency for Dangerous Substances and Disease Registry. Metalloid-containing plants and goods can impair human and animal health by passing through the food chain. Toxic metalloids have been shown in tests to impair plant growth, development, and overall yield (Sharma et al., 2016; Ali et al., 2019). Metalloids influence plant productivity by increasing reactive oxygen species (ROS) generation *via* several processes. ROS are signaling molecules that participate in a variety of biological processes

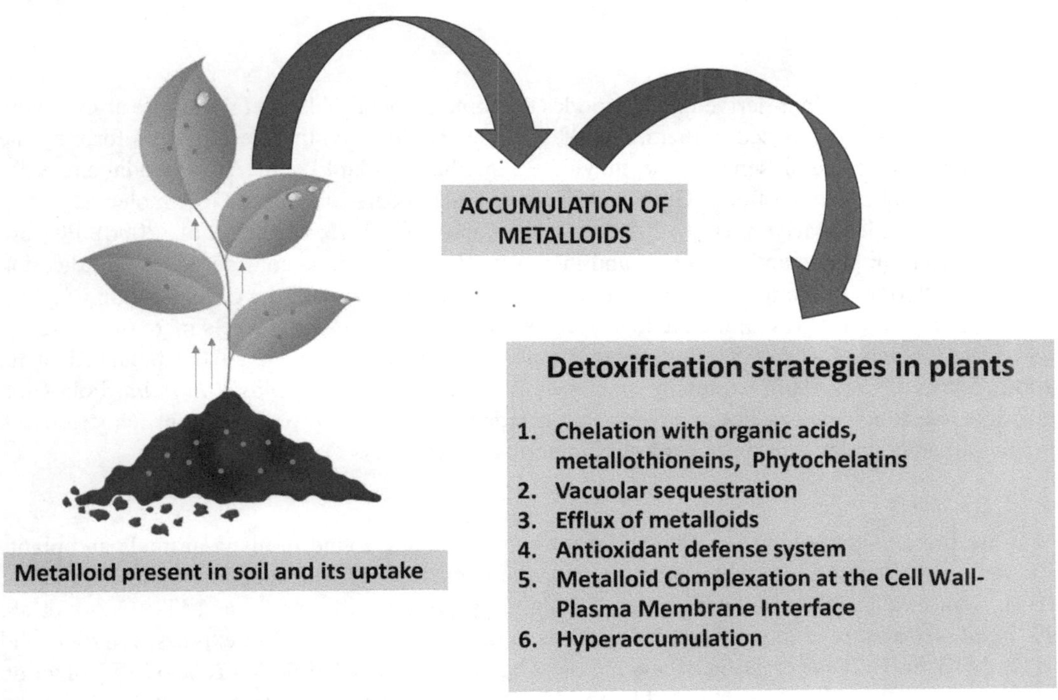

FIGURE 4.1 The internal defense system of plants for mitigation of metalloid toxicity.

such cell division, gene regulation, stress sensing, hypersensitive response, and senescence. ROS can also initiate plant reactions in response to several external stimuli (Gupta et al., 2013). As is known as the deadly metalloid; however, because of the increased internal phosphate content, it does not exhibit direct toxicity under typical ambient conditions. As plays no role in the growth and maturation of plants. As disrupts the redox state, causing membrane and protein damage, resulting in oxidative stress and finally cell death (Parveen et al., 2022).

4.3.1 Complexation/Chelation

Metal binding is a primary detoxifying method in plants that uses strong ligands for detoxification. Thiols, notably, PCs, MTs, and GSH, are the most well-known ligands. Furthermore, there are various non-thiol ligands involved; notably, non-thiol ligands play pivot roles during hyperaccumulation. Recently, nicotinamide has been implicated in metal buildup. Non-hyperaccumulated metals undergo processes such as temporary binding during transit to storage sites, storage, and remobilization when needed for enzyme active centers. PCs, on the other hand, appear to have been synthesized expressly for metal(loid) detoxification (Kumar et al., 2024c; Leitenmaier and Küpper, 2013).

4.3.1.1 Thiol Ligands

4.3.1.1.1 GSH

GSH is a tripeptide consisting of the amino acids -Glu-Cys-Gly. It is a thiol-containing molecule that acts as an antioxidant (Kumar et al., 2024b; Anjum et al., 2015). GSH, or reduced GSH, includes a thiol group that permits it to behave as a metal and metalloid chelator. This indicates that it can aid in the detoxification of these chemicals by plants. GSH generates non-toxic compounds with metalloids that can be stored safely in cells. When As binds to GSH, the resultant vacuoles retain the As-GSH complex (Choudhary et al., 2024; Seth et al., 2012). When As III is present in the medium, the GSH concentration and glutamylcysteine synthetase activity of an aquatic fern, *Salvinia molesta*, increased. This suggests that GSH plays a crucial role against the toxicity of As III (Samanta et al., 2023; Da Silva et al., 2018).

4.3.1.1.2 PCs

PCs are examples of plant-derived thiol peptides that bind to metalloids. Metalloids are absorbed by plants when they are exposed to them, and their tissues begin to synthesize PCs. PCs form by the reaction of GSH in the presence of the enzyme PC synthase (Solanki et al., 2024; Jindal and Seth, 2023) Transpeptidylation of one -Glu-Cys moiety of GSH occurs to another GSH molecule, forming a PC molecule (Mariyam et al., 2023; Agnihotri and Seth, 2016; Mishra et al., 2006). PCs are widely present in plants and microbes, and increasing PC synthesis is considered as an indication of metal pollution. In the presence of excess metalloids in the medium, the concentration of glutathione (GSH) in plant growth media decreases, while phytochelatin (PC) levels increase; however, this effect is reversed when GSH is added to the medium (Chandrakar et al., 2020; Chaturvedi et al., 2006). Excess PC synthase expression enhances As tolerance in *Arabidopsis thaliana.* Following transit across tonoplast vesicles, a process that is independent of the proton gradient sequesters the PC-metalloid complex into the vacuoles (Kumar and Seth, 2022).

4.3.1.1.3 MTs

MTs are proteins that bind to metals and metalloids. They are found in many animals and plants and act as chelators. Some metalloids increase the expression of MT-encoding genes (Jin et al., 2024; Seth et al., 2008). Metal and metalloid-binding proteins are known as MTs (Gupta et al., 2023). MTs are expressed in different plant parts in the plant system as MT1 expresses in roots and MT2, MT3, and MT4 expression occurs in seeds, but not in leaves. While MT2a and MT3 function as chaperones, these MTs, such as MT1a and MT2b, aid in the redistribution of metalloids via the phloem (Verma et al., 2023; Zhao et al., 2022). Studies on the confined MTs have revealed their involvement in the metalloid detoxification process. *A. thaliana* immunoblot studies demonstrated that MT production is in correlation with metalloid-induced mRNA expression (Gupta et al., 2021; Yadav et al., 2022). The location, arrangement, and expression of Cys residues dictate the capacity of MTs to bind and sequester metalloids.

4.3.2 Non-Thiol Ligands

Histidine and nicotianamine are two metal ligands that are gaining more recognition for their importance in plants. While additional non-thiol ligands have been proven to participate in certain stress situations, their true significance may not be fully appreciated. For instance, anthocyanins, oxalate, malic or citric acid are examples that have been associated with accumulation (Chauhan et al., 2023; Leitenmaier and Küpper, 2013).

4.4 COMPARTMENTALIZATION

Plant homeostasis is maintained by a variety of protein transporters during HM stress. Root hair influx or efflux and HM detoxification sustain tolerance as a result of their specific absorption and compartmentalization. Numerous transporter proteins were recently found in higher plants that aid in the transit of metalloids and serve as a key role in metalloid tolerance and seclusion. The most widely accepted paradigm for metalloid detoxification is the conception of a PC-metalloid complex, which further reaches vacuoles for compartmentalization (Kim et al., 2024; Khalid et al., 2021). The sequestration methods involve the production of metalloid formation of complexes with ligands and the removal of these complexes from the cytoplasm by relocating them into dormant organelles, primarily cell walls and vacuoles. The sequestration of HMs/metalloids largely depends upon the constitutive overexpression of genetic material that encodes proteins that regulate HMs/metalloids transport across the tonoplast or otherwise plasma membrane that are accountable for HMs/metalloids exclusion from the cytoplasm, according to transcriptome analyses comparing hyperaccumulator and related non-hyperaccumulator species (Xalxo et al., 2021; Thakur et al., 2016).

4.5 EFFLUX OF METALLOIDS

Tolerance could have an impact on the plasma membrane, for example, by limiting HM intake or by increasing the efflux pumping of metals that have reached the cytosol. HM efflux from root cells is expected to be an upcoming way to dropping metalloid buildup in plants. Cellular metalloid exclusion is an essential adaptation strategy for the resistance of plants to phytotoxicity. After absorbing metalloids, they are carried to shoot by apoplastic movement, accumulating a huge number of HMs/metalloids in apoplastic space. Metalloid transporter proteins are believed to be concerned with the exclusion of fatal metalloids from the symplastic to apoplastic region (Mariyam et al., 2023; Xalxo et al., 2021). Several studies also suggested that efflux transporters are critical in metal purification by excreting metalloids outside the cell. Low silicon rice transporter 1 (Lsi1) is involved in As(III) uptake and also has been verified to be involved in metalloid efflux (Zhao et al., 2022; Misra et al., 2009). Other plant aquaporins that have been identified in *Arabidopsis thaliana* and *Pteris vittate* are *Arabidopsis thaliana* Nodulin-26-like Intrinsic Protein (AtNIP3;1, AtNIP6;1), and *Pteris vittata* Tonoplast Intrinsic Protein 4;1 (PvTIP4;1).

4.6 SYSTEM OF ANTIOXIDANT DEFENSE

Plants are capable of providing resistance to metalloid toxicity. Antioxidant defense systems include antioxidant enzymes and antioxidants such as GSH, ascorbate, and carotenoids play a crucial part in stress resilience and contribute largely to metalloid detoxification (Singh et al., 2021; Khan et al.,

2017). Numerous reactions inside plant cell components, such as the apoplast, chloroplasts, cytosol, endoplasmic reticulum, mitochondria, and nucleus, were catalyzed by these enzymes that were beneficial during metal and metalloid stress conditions.

4.7 METALLOID BINDING AT THE WALL–PLASMA INTERFACE

Because the interface between the plasma membrane and plant cell wall accumulates a high number of metalloids, it is assumed that this is the area for metalloid tolerance. An increased number of metalloids is considered to be present at this site in *Lolium multiflorum*. However, more research is needed to understand how this space functions in HMs/metalloid tolerance (Vinogradova et al., 2023; Xalxo et al., 2021).

4.8 HYPERACCUMULATION

Plant species can acquire a substantial metalloid through improved translocation from roots to the apical region of the plant, significantly limiting metal presence in soil. When grown in excess metal concentrations, they perform normal metabolic and physiological functions without displaying any obvious stress symptoms. These are known as hyperaccumulator plants. Metals are frequently stored in plant arial parts rather than the roots. To further refine the original definition, only plant leaves should be considered when evaluating above-ground tissue to determine its hyperaccumulator status (Upadhyay et al., 2024; Pollard et al., 2014). Thus, rather than at the level of certain tissues or organelles (like the epidermis), or at the level of latex, hyperaccumulation is set at the organ level (leaves). Moreover, the word "accumulation" should only be used to describe active accumulation that occurs via the roots in the plant's leaf tissue (Gupta et al., 2022; Van der Ent et al., 2013).

4.9 CONCLUSIONS AND FUTURE PROSPECTS

Excess metalloids can be absorbed by plants *via* various transporters such as phosphate, aquaglyceroporins, hexose, and sulfate transporters. Some metalloids are absorbed by important nutrient transporters such as phosphate and sulfate. The phosphate transporter absorbs As(V); the hexose transporter absorbs As(III) and Sb(III); and the sulfate transporters absorb selenium. Different subtypes of aquaglyceroporin proteins, including As (III), B (III), Si (III), and $Sb(OH)_3$, mediate the intake of metalloids. Furthermore, PCs and MTs play important roles in metalloid sequestration inside vacuoles and metalloid tolerance. Biotechnological techniques are being developed to improve plant resistance to metalloids by altering the genes and uptake mechanisms of metalloid transporters to reduce hazardous metalloid uptake. The approaches of increasing metalloid uptake through carriers as well as raising the ability of crop phytoremediation and cleaning up metal-contaminated places utilizing plants may be greatly enhanced by the development of metalloid chelating agents like GSH, PCs, and MTs. Compartmentalization of these chelated compounds along with efflux and activation of antioxidant defense system can mitigate the toxic effects of metalloids. Effective chelation of metalloids with PCs and MTs in the cytoplasm, as well as their sequestration and storage in vacuoles, antioxidant defense system, efflux of metalloids, hyperaccumulation would result in improved plant tolerance to metalloids.

REFERENCES

Abd-Elsalam, Kamel A., Chandra Shekhar Seth, and Mousa A. Alghuthaymi. Nanofertilizers: Challenges and future trends. In Abd-Elsalam, Kamel A., and Mousa A. Alghuthaymi (Eds.), *Nanofertilizers for Sustainable Agroecosystems: Recent Advances and Future Trends*, pp. 561–570. Cham: Springer Nature Switzerland, 2023.

Agnihotri, Ashish, and Chandra Shekhar Seth. Exogenously applied nitrate improves the photosynthetic performance and nitrogen metabolism in tomato (Solanum lycopersicum L. cv Pusa Rohini) under arsenic (V) toxicity. *Physiology and Molecular Biology of Plants* 22 (2016): 341–349.

Agnihotri, Ashish, and Chandra Shekhar Seth. Does jasmonic acid regulate photosynthesis, clastogenecity, and phytochelatins in Brassica juncea L. in response to Pb-subcellular distribution? *Chemosphere* 243 (2020): 125361.

Agnihotri, Ashish, Praveen Gupta, Anuj Dwivedi, and Chandra Shekhar Seth. Counteractive mechanism (s) of salicylic acid in response to lead toxicity in Brassica juncea (L.) Czern. cv. Varuna. *Planta* 248 (2018): 49–68.

Ali, Muhammad Arif, Shah Fahad, Idrees Haider, Niaz Ahmed, Shakeel Ahmad, Sajjad Hussain, and Muhammad Arshad. Oxidative stress and antioxidant defense in plants exposed to metal/metalloid toxicity. In Hasanuzzaman, Mirza, Vasileios Fotopoulos, Kamrun Nahar, Masayuku Fujita (Eds.), *Reactive Oxygen, Nitrogen and Sulfur Species in Plants: Production, Metabolism, Signaling and Defense Mechanisms*, pp. 353–370. Newark: John Wiley & Sons, Incorporated, 2019.

Angulo-Bejarano, Paola I., Jonathan Puente-Rivera, and Rocío Cruz-Ortega. Metal and metalloid toxicity in plants: An overview on molecular aspects. *Plants* 10, no. 4 (2021): 635.

Anjum, Naser A., Mirza Hasanuzzaman, Mohammad A. Hossain, Palaniswamy Thangavel, Aryadeep Roychoudhury, Sarvajeet S. Gill, Miguel A. Merlos Rodrigo, Masayuki Fujita, Rene Kizek, and Armando C. Duarte. Jacks of metal/metalloid chelation trade in plants—An overview. *Frontiers in Plant Science* 6 (2015): 124587.

Babula, Petr, Vojtech Adam, Radka Opatrilova, Josef Zehnalek, Ladislav Havel, and Rene Kizek. Uncommon heavy metals, metalloids and their plant toxicity: A review. In Lichtfouse, Eric (Ed.), *Organic Farming, Pest Control and Remediation of Soil Pollutants*, pp. 275–317. Dordrecht: Springer, 2010.

Brdar-Jokanović, Milka. Boron toxicity and deficiency in agricultural plants. *International Journal of Molecular Sciences* 21, no. 4 (2020): 1424.

Carbonell-Barrachina, A. A., Fernando Burló, Alba Burgos-Hernandez, Emilio López, and J. Mataix. The influence of arsenite concentration on arsenic accumulation in tomato and bean plants. *Scientia Horticulturae* 71, no. 3–4 (1997): 167–176.

Chandrakar, Vibhuti, Bhumika Yadu, Roseline Xalxo, Meetul Kumar, and S. Keshavkant. Mechanisms of plant adaptation and tolerance to metal/metalloid toxicity. In Hasanuzzaman, Mirza (Ed.), *Plant Ecophysiology and Adaptation under Climate Change: Mechanisms and Perspectives II: Mechanisms of Adaptation and Stress Amelioration*, pp. 107–135. Singapore: Springer, 2020.

Chaturvedi, Pranav Kumar, Chandra Shekhar Seth, and Virendra Misra. Sorption kinetics and leachability of heavy metal from the contaminated soil amended with immobilizing agent (humus soil and hydroxyapatite). *Chemosphere* 64, no. 7 (2006): 1109–1114.

Chaturvedi, Pranav Kumar, Chandra Shekhar Seth, and Virendra Misra. Selectivity sequences and sorption capacities of phosphatic clay and humus rich soil towards the heavy metals present in zinc mine tailing. *Journal of Hazardous Materials* 147, no. 3 (2007): 698–705.

Chauhan, Poonam, Neha Sharma, Ashwani Tapwal, Ajay Kumar, Gaurav Swaroop Verma, Mukesh Meena, Chandra Shekhar Seth, and Prashant Swapnil. Soil microbiome: Diversity, benefits and interactions with plants. *Sustainability* 15, no. 19 (2023): 14643.

Chen, Xuan, Wei Jiang, Tao Tong, Guang Chen, Fanrong Zeng, Sunghoon Jang, Wei Gao, Zhen Li, Michelle Mak, Fenglin Deng, and Zhong-Hua Chen. Molecular interaction and evolution of jasmonate signaling with transport and detoxification of heavy metals and metalloids in plants. *Frontiers in Plant Science* 12 (2021): 665842.

Choudhary Rashmi, Vishnu Dayal Rajput, Gajanan Ghodake, Faheem Ahmad, Mukesh Meena, Reiaz ul Rehman, Ram Prasad, Rajesh Kumar Sharma, Rachana Singh, and Chandra Shekhar Seth. Comprehensive journey from past to present to future about seed priming with hydrogen peroxide and hydrogen sulfide concerning drought, temperature, UV and ozone stresses-A review. *Plant and Soil* 500, (2024): 1–23.

Clemens, Stephan and Ma, Jian Feng. Toxic heavy metal and metalloid accumulation in crop plants and foods. *Annual Review of Plant Biology*, 67, no. 1 (2016): 489–512.

da Silva, Adinan Alves, Juraci Alves de Oliveira, Fernanda Vidal de Campos, Cleberson Ribeiro, Fernanda dos Santos Farnese, and Alan Carlos Costa. Phytoremediation potential of Salvinia molesta for arsenite contaminated water: Role of antioxidant enzymes. *Theoretical and Experimental Plant Physiology* 30, no. 4 (2018): 275–286.

Deng, Fenglin, Fanrong Zeng, Guang Chen, Xue Feng, Adeel Riaz, Xiaojian Wu, Wei Gao, Feibo Wu, Paul Holford, and Zhong-Hua Chen. Metalloid hazards: from plant molecular evolution to mitigation strategies. *Journal of Hazardous Materials*, 409 (2021): 124495.

Feng, Renwei, Chaoyang Wei, Shuxin Tu, Yongzhen Ding, Ruigang Wang, and Junkang Guo. The uptake and detoxification of antimony by plants: A review. *Environmental and Experimental Botany* 96 (2013): 28–34.

Finnegan, Patrick, and Weihua Chen. Arsenic toxicity: The effects on plant metabolism. *Frontiers in Physiology* 3 (2012): 24498.

Fu, Daxiong, Andrew Libson, Larry JW Miercke, Cindy Weitzman, Peter Nollert, Jolanta Krucinski, and Robert M. Stroud. Structure of a glycerol-conducting channel and the basis for its selectivity. *Science* 290, no. 5491 (2000): 481–486.

Gupta, Dharmendra Kumar, Hildegarde Vandenhove, and Masahiro Inouhe. *Role of Phytochelatins in Heavy Metal Stress and Detoxification Mechanisms in Plants*. Berlin, Heidelberg: Springer, 2013.

Gupta, Praveen, and Chandra Shekhar Seth. 24-Epibrassinolide regulates functional components of nitric oxide signalling and antioxidant defense pathways to alleviate salinity stress in Brassica juncea L. cv. Varuna. *Journal of Plant Growth Regulation* 42, no. 7 (2023): 4207–4222.

Gupta, Praveen, and Chandra Shekhar Seth. Nitrate supplementation attenuates As (V) toxicity in *Solanum lycopersicum* L. cv Pusa Rohini: Insights into As (V) sub-cellular distribution, photosynthesis, nitrogen assimilation, and DNA damage. *Plant Physiology and Biochemistry* 139 (2019): 44–55.

Gupta, Praveen, Sudhakar Srivastava, and Chandra Shekhar Seth. 24-Epibrassinolide and sodium nitroprusside alleviate the salinity stress in Brassica juncea L. cv. Varuna through cross talk among proline, nitrogen metabolism and abscisic acid. *Plant and Soil* 411 (2017): 483–498.

Gupta, Samta, and Chandra Shekhar Seth. Salicylic acid alleviates chromium (VI) toxicity by restricting its uptake, improving photosynthesis and augmenting antioxidant defense in Solanum lycopersicum L. *Physiology and Molecular Biology of Plants* 27 (2021): 2651–2664.

Gupta, Samta, Sarda Devi Thokchom, Monika Koul, and Rupam Kapoor. Arbuscular Mycorrhiza mediated mineral biofortification and arsenic toxicity mitigation in Triticum aestivum L. *Plant Stress*, 5 (2022): 100086.

Jin, Weihuan, Lan Cheng, Chunyan Liu, Haitao Liu, Qiujuan Jiao, Haoyang Wang, Zhaolong Deng, Chandra Shekhar Seth, Hongxiang Guo, and Yong Shi. Cadmium negatively affects the growth and physiological status and the alleviation effects by exogenous selenium in silage maize (Zea mays L.). *Environmental Science and Pollution Research* 31 (2024): 1–13.

Jindal, A. and Seth, C.S. Nitric oxide mediated post-translational modifications and its significance in plants under abiotic stress. In M. Iqbal R Khan, Noushina Iqbal, Peter Poor, Antonio Ferrante (Eds.), *Nitric Oxide in Developing Plant Stress Resilience*. Academic Press, 2023. 233–250.

Khalid, Hafiza Sania, Muhammad Farhan Akhtar, Munazza Ijaz, Muhammad Iqbal, Shazia Anwer Bukhari, Ghulam Mustafa, and Kanval Shaukat. Role of metal-binding proteins and peptides in bioremediation of toxic metals. In Hasanuzzaman, Mirza, and Majeti Narasimha Vara Prasad (Eds.), *Handbook of Bioremediation*, pp. 437–444. Academic Press, 2021.

Khan, M. Nasir, M. Mobin, Zahid Khorshid Abbas, and Manzer H. Siddiqui. Nitric oxide-induced synthesis of hydrogen sulfide alleviates osmotic stress in wheat seedlings through sustaining antioxidant enzymes, osmolyte accumulation and cysteine homeostasis. *Nitric Oxide*, 68 (2017): 91–102.

Kim, Dae-Young, Min Kim, Jung-Suk Sung, Janardhan Reddy Koduru, Shivraj Hariram Nile, Asad Syed, Ali H. Bahkali, Chandra Shekhar Seth, and Gajanan Sampatrao Ghodake. Extracellular synthesis of silver nanoparticle using yeast extracts: Antibacterial and seed priming applicationss. *Applied Microbiology and Biotechnology* 108, no. 1 (2024): 150.

Kumar, Amit, Vinod Kumar, Andrés Rodríguez-Seijo, Raj Setia, Sumit Singh, Ashok Kumar, Chandra Shekhar Seth, and Renato Somma. Appraisal of heavy metal(loid)s contamination in rice grain and associated health risks. *Journal of Food Composition and Analysis* 131 (2024a): 106215.

Kumar, Dharmendra, and Chandra Shekhar Seth. Photosynthesis, lipid peroxidation, and antioxidative responses of Helianthus annuus L. against chromium (VI) accumulation. *International Journal of Phytoremediation* 24, no. 6 (2022): 590–599.

Kumar, Dharmendra, Om Parkash Dhankher, Rudra Deo Tripathi, and Chandra Shekhar Seth. Titanium dioxide nanoparticles potentially regulate the mechanism(s) for photosynthetic attributes, genotoxicity, antioxidants defense machinery, and phytochelatins synthesis in relation to hexavalent chromium toxicity in Helianthus annuus L. *Journal of Hazardous Materials* 454 (2023): 131418.

Kumar, Gokul Anil, Sumit Kumar, Rupesh Bhardwaj, Prashant Swapnil, Mukesh Meena, Chandra Shekhar Seth, and Ankush Yadav. Recent advancements in multifaceted roles of flavonoids in plant–rhizomicrobiome interactions. *Frontiers in Plant Science* 14 (2024c): 1297706.

Kumar, Vinod, Monika Thakur, and Chandra Shekhar Seth. In situ remediation techniques for removal of arsenic in the environment. *Current Opinion in Environmental Science & Health* 38 (2024b): 100538.

Leitenmaier, Barbara, and Hendrik Küpper. Compartmentation and complexation of metals in hyperaccumulator plants. *Frontiers in Plant Science* 4 (2013): 55114.

Lu, Yi, Zhaoxue Zhang, Yingyang Wang, Fangyuan Peng, Zhaoguang Yang, and Haipu Li. Uptake, tolerance, and detoxification mechanisms of antimonite and antimonate in Boehmeria nivea L. *Journal of Environmental Management* 334 (2023): 117504.

Majeed, A., Muhammad, Z., Hasanuzzaman, M., Hakeem, K. R., Nahar, K., & Alharby, H. F. *Plant Abiotic Stress Tolerance: Agronomic, Molecular and Biotechnological Approaches 2019*, pp. 221–257. Cham: Springer, 2019.

Mariyam, Safoora, Renu Bhardwaj, Nafees A. Khan, Shivendra V. Sahi, and Chandra Shekhar Seth. Review on nitric oxide at the forefront of rapid systemic signaling in mitigation of salinity stress in plants: Crosstalk with calcium and hydrogen peroxide. *Plant Science* 336 (2023): 111835.

Mishra, Seema, Shikha Srivastava, Rama Dwivedi Tripathi, Ravindra Kumar, Chandra Shekhar Seth, and Dinesh Kumar Gupta. Lead detoxification by coontail (Ceratophyllum demersum L.) involves induction of phytochelatins and antioxidant system in response to its accumulation. *Chemosphere* 65, no. 6 (2006): 1027–1039.

Misra, Virendra, Anjana Tiwari, Bhaskar Shukla, and Chandra Shekhar Seth. Effects of soil amendments on the bioavailability of heavy metals from zinc mine tailings. *Environmental Monitoring and Assessment* 155 (2009): 467–475.

Mitani-Ueno, Namiki, and Jian Feng Ma. Linking transport system of silicon with its accumulation in different plant species. *Soil Science and Plant Nutrition* 67, no. 1 (2021): 10–17.

Moulick, Debojyoti, Dibakar Ghosh, Jajati Mandal, Subhamoy Bhowmick, Debapriya Mondal, Shuvasish Choudhury, Subhas Chandra Santra, Meththika Vithanage, and Jayanta Kumar Biswas. A cumulative assessment of plant growth stages and selenium supplementation on arsenic and micronutrients accumulation in rice grains. *Journal of Cleaner Production* 386 (2023): 135764.

Pandey, Akhilesh Kumar, Arti Gautam, and Rama Shanker Dubey. Transport and detoxification of metalloids in plants in relation to plant-metalloid tolerance. *Plant Gene* 17 (2019): 100171.

Parveen, Nishat, Roberto Berni, Sreeja Sudhakaran, Javaid A. Bhat, Suhas Shinde, Naleeni Ramawat, Vijay P. Singh, Shivendra V. Sahi, Rupesh Deshmukh, Devendra Kumar Chauhan, and Durgesh K. Tripathi. Metalloids in plants: A systematic discussion beyond description. *Annals of Applied Biology* 180, no. 1 (2022): 7–25.

Pathak, Himanshu K., Prabhat K. Chauhan, Chandra Shekhar Seth, Gopal Dubey, and Sudhir K. Upadhyay. Mechanistic and future prospects in rhizospheric engineering for agricultural contaminants removal, soil health restoration, and management of climate change stress. *Science of the Total Environment* 927 (2024): 172116.

Pollard, A. Joseph, Roger D. Reeves, and Alan J. M. Baker. Facultative hyperaccumulation of heavy metals and metalloids. *Plant Science* 217 (2014): 8–17.

Pommerrenig, Benjamin, Till Arvid Diehn, and Gerd Patrick Bienert. Metalloido-porins: Essentiality of Nodulin 26-like intrinsic proteins in metalloid transport. *Plant Science* 238 (2015): 212–227.

Samanta, Santanu, Chandra Shekhar Seth, and Aryadeep Roychoudhury. The molecular paradigm of ROS and RNS with different phytohormone signaling during drought stress in plants. *Plant Physiology and Biochemistry* 206 (2023): 108259.

Seth, Chandra Shekhar, Pranav Kumar Chaturvedi, and Virendra Misra. The role of phytochelatins and antioxidants in tolerance to Cd accumulation in Brassica juncea L. *Ecotoxicology and Environmental Safety* 71, no. 1 (2008): 76–85.

Seth, Chandra Shekhar, Pranav Kumar Chaturvedi, and Virendra Misra. Toxic effect of arsenate and cadmium alone and in combination on giant duckweed (Spirodela polyrrhiza L.) in response to its accumulation. *Environmental Toxicology: An International Journal* 22, no. 6 (2007): 539–549.

Seth, Chandra Shekhar, V. Misra, and L. K. S. Chauhan. Accumulation, detoxification, and genotoxicity of heavy metals in Indian mustard (Brassica juncea L.). *International Journal of Phytoremediation* 14, no. 1 (2012): 1–13.

Shanmugaiah, Vellasamy, Ankita Gauba, Sharath Kumar Hari, Ram Prasad, Vellaisamy Ramamoorthy, and Mahaveer P. Sharma. Effect of silicon micronutrient on plant's cellular signaling cascades in stimulating plant growth by mitigating the environmental stressors. *Plant Growth Regulation* 100, no. 2 (2023): 391–408.

Sharma, Shanti S., Karl-Josef Dietz, and Tetsuro Mimura. Vacuolar compartmentalization as indispensable component of heavy metal detoxification in plants. Plant, *Cell & Environment* 39, no. 5 (2016): 1112–1126.

Shireen, Fareeha, Muhammad Azher Nawaz, Chen, Qikai Zhang, Zuhua Zheng, Hamza Sohail, Jingyu Sun, Haishun Cao, Yuan Huang, and Zhilong Bie. Boron: Functions and approaches to enhance its availability in plants for sustainable agriculture. *International Journal of Molecular Sciences* 19, no. 7 (2018): 1856.

Singh, Durga, Ashish Agnihotri, and Chandra Shekhar Seth. Interactive effects of EDTA and oxalic acid on chromium uptake, translocation and photosynthetic attributes in Indian mustard (Brassica juncea L. var. Varuna). *Current Science* 112 (2017): 2034–2042.

Singh, Simranjeet, Vijay Kumar, Shivika Datta, Daljeet Singh Dhanjal, Satyender Singh, Sanjay Kumar, Dhriti Kapoor, Ram Prasad, and Joginder Singh. Physiological responses, tolerance, and remediation strategies in plants exposed to metalloids. *Environmental Science and Pollution Research* 28, no. 30 (2021): 40233–40248.

Solanki, Swati, Surbhi Sinha, Chandra Shekhar Seth, Shivanshi Tyagi, Aarushi Goyal, and Rachana Singh. Enhanced adsorption of Bismark Brown R dye by chitosan conjugated magnetic pectin loaded filter mud: A comprehensive study on modeling and mechanisms. *International Journal of Biological Macromolecules* 270 (2024): 131987.

Tang, Zhong, and Fang-Jie Zhao. The roles of membrane transporters in arsenic uptake, translocation and detoxification in plants. *Critical Reviews in Environmental Science and Technology* 51, no. 21 (2021): 2449–2484.

Tang, Zhong, Han-Qing Wang, Jie Chen, Jia-Dong Chang, and Fang-Jie Zhao. Molecular mechanisms underlying the toxicity and detoxification of trace metals and metalloids in plants. *Journal of Integrative Plant Biology* 65, no. 2 (2023): 570–593.

Thakur, Sveta, Lakhveer Singh, Zularisam Ab Wahid, Muhammad Faisal Siddiqui, Samson Mekbib Atnaw, and Mohd Fadhil Md Din. Plant-driven removal of heavy metals from soil: Uptake, translocation, tolerance mechanism, challenges, and future perspectives. *Environmental Monitoring and Assessment* 188 (2016): 1–11.

Upadhyay, Sudhir K., Garima Singh, Nitu Rani, Vishnu D. Rajput, Chandra Shekhar Seth, Padmanabh Dwivedi, Tatiana Minkina, Ming Hung Wong, Pau Loke Show, and Kuan Shiong Khoo. Transforming bio-waste into value-added products mediated microbes for enhancing soil health and crop production: Perspective views on circular economy. *Environmental Technology & Innovation* 34 (2024): 103573.

Vander Ent, Antony, Alan J. M. Baker, Roger D. Reeves, A. Joseph Pollard, and Henk Schat. Hyperaccumulators of metal and metalloid trace elements: Facts and fiction. *Plant and Soil* 362 (2013): 319–334.

Verma, Krishan K., Xiu-Peng Song, Lin Xu, Hai-Rong Huang, Qiang Liang, Chandra Shekhar Seth, and Yang-Rui Li. Nano-microplastic and agro-ecosystems: A mini-review. *Frontiers in Plant Science* 14 (2023): 1283852.

Vinogradova, Natalya, Elena Vinogradova, Victor Chaplygin, Saglara Mandzhieva, Pradeep Kumar, Vishnu Dayal Rajput, Tatiana Minkina et al. Phenolic compounds of the medicinal plants in an anthropogenically transformed environment. *Molecules* 28, no. 17 (2023): 6322.

Wiche, Oliver, Balázs Székely, Christin Moschner, and Hermann Heilmeier. Germanium in the soil-plant system—A review. *Environmental Science and Pollution Research* 25 (2018): 31938–31956.

Xalxo, Roseline, Jipsi Chandra, Neha Pandey, and Keshavkant Sahu. Physiological and molecular mechanism of metalloid tolerance in plants. In Hasanuzzaman, Mirza (Ed.), *Approaches to the Remediation of Inorganic Pollutants*, pp. 197–221. Springer, 2021.

Yadav, Manisha, Praveen Gupta, and Chandra Shekhar Seth. Foliar application of α-lipoic acid attenuates cadmium toxicity on photosynthetic pigments and nitrogen metabolism in Solanum lycopersicum L. *Acta Physiologiae Plantarum* 44, no. 11 (2022): 112.

Yamaji, Naoki, and Jian Feng Ma. Metalloid transporters and their regulation in plants. *Plant Physiology* 187, no. 4 (2021): 1929–1939.

Zhao, Fang-Jie, Zhong Tang, Jia-Jun Song, Xin-Yuan Huang, and Peng Wang. Toxic metals and metalloids: Uptake, transport, detoxification, phytoremediation, and crop improvement for safer food. *Molecular Plant* 15, no. 1 (2022): 27–44.

5 Stress Ameliorative Role of Metalloids in Plants for Sustainable Agriculture

Gurvarinder Kaur, Shruti Kaushik, Isha Madaan, Anmol Sidhu, Rashmi Mishra, Aarti, Babita Dhiman, Palampreet Singh, and Geetika Sirhindi

5.1 INTRODUCTION: ELUCIDATING BENEFICIAL ROLE OF METALLOIDS IN PLANTS

In the era of burgeoning environmental changes and agricultural threats, it has become imperative to develop innovative strategies that enhance crop productivity sustainably and hold tendency to make plants resilient under conditions of environmental dynamism (Sarraf et al., 2022). Different types of stressors like water scarcity, contaminants in the environment, changing global temperatures, increased pathogen distribution, disturbed growth pattens and population, and increased unprecedented disease outbreaks pose formidable threats to crop productivity (Jalil et al., 2020; Zahra et al., 2022). Conditions of stress affect the overall metabolism and physiology of the plants and deteriorate the growth and yield attributes (Raza, 2021). Therefore, research needs to target unconventional approaches to crop protection and productivity enhancement and should be targeted toward understanding and exploring an intriguing role of metalloids and their potential in defending plants against conditions of stress.

Metalloids form a unique class of elements which have recently been explored for their stress amelioration potential in plants (Sarraf et al., 2022). Even though, some of the metalloids like arsenic and antimony pose toxicity effects on plants (Mathur et al., 2022), but there are some of the metalloids that help in enhancing the growth of plants under normal and stress conditions (Parveen et al., 2022). These beneficial metalloids like boron (B), silicon (Si), selenium hold some of the essential physiological roles in plants like secondary metabolism, cell elongation, maintenance of cell wall integrity, and many more (Parveen et al., 2022). This chapter aims to discuss the beneficial role of metalloids like selenium, Si, and B in reducing the negative impacts of stress in plants and their role in maintaining the crop productivity under conditions of environmental cues. The numerous pathways that metalloids take to improve the growth of plants include uplifting antioxidant defense system which helps in minimizing oxidative stress, synthesizing osmoprotectants like proline to help plants to combat water stress, improving secondary metabolism and physiology which leads to improved biomass and productivity (Seliem et al., 2020; Wang et al., 2021; Cao et al., 2022). The insights into the role of selenium, Si, and B in plants against stress will be discussed in later sections suggesting their utilization in promoting growth of plants as a sustainable approach.

5.2 SHIELDING CAPACITY OF METALLOIDS IN PLANTS

5.2.1 Drought Stress Tolerance

Plants in their natural habitat face stress related to water and drought stress is one of the major stress conditions that affect crop yield and productivity (Iqbal et al., 2020). Plants are versatile sessile organisms that are capable of withstanding conditions of stress; however, prolonged conditions of

DOI: 10.1201/9781032673592-5

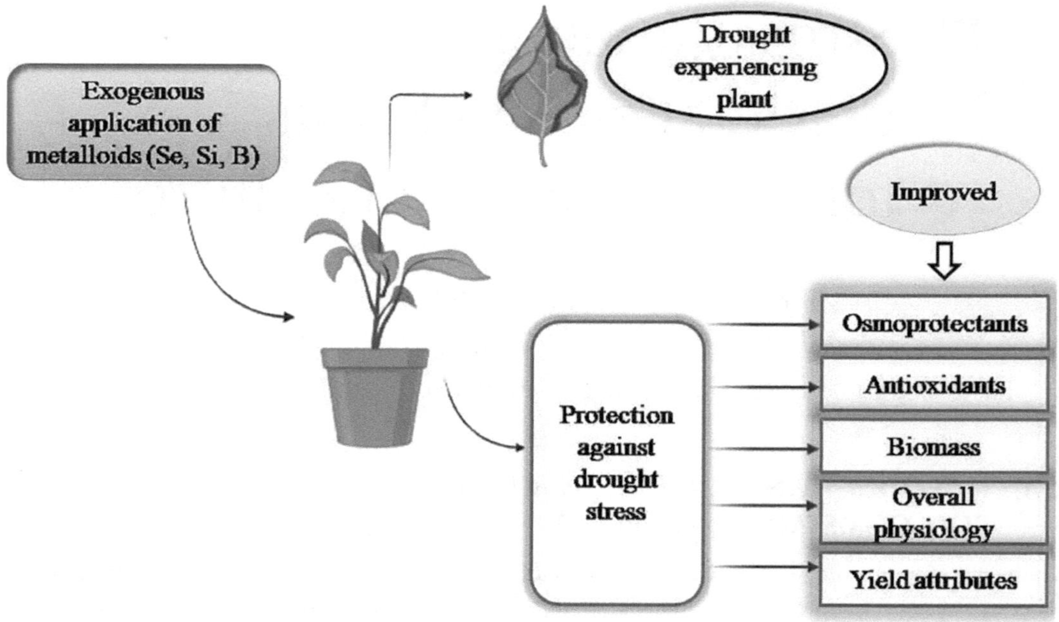

FIGURE 5.1 Mitigation of drought stress by metalloids (Se, Si, and B).

stress require plants to develop effective endurance with the help of biostimulants (Lamaoui et al., 2018). Three major metalloids have been known to play a positive role in mitigating drought stress in plants (Figure 5.1).

Exogenous application of Si is known to be helpful for plants to overcome conditions of drought and enable them to withstand such environmental cues while maintaining productivity (Amin et al., 2014; Ming et al., 2012). Si can be applied in different forms like K_2SiO_3, Na_2SiO_3, H_2SiO_3, and others, and it has been reported that Si application in the form of seed treatment not only helps to inculcate drought stress tolerance in plants but also prepares plants against any future stress conditions (Bokor et al., 2021; Biju et al. 2017). Si application enables plants to strengthen their physiology in terms of water uptake and regulate water loss through stomata thereby modulating stomatal movements, accumulating different osmolytes like proline, and strengthening antioxidant defense system to curb oxidative species to protect plants against oxidative stress (Wang et al., 2021). Si application is known to be helpful against salt and drought stress as it helps the plants to maintain their productivity, grain yield, and biomass by improving ion homeostasis which is achieved by improving nutrient uptake from the soil-water system, maintenance of cellular homeostasis by regulating osmolytes, and improving photosynthetic efficiency of plants (Rizwan et al., 2015).

A research study has suggested that the application of Si can help in mitigating drought stress in rice. This is achieved by maintaining various parameters of productivity such as plant height, root health, rice straw, and grain yield. These parameters are greatly impacted by drought stress, which can lead to a significant reduction in yield. Furthermore, Si application improved N, P, and K status of the rice plants exposed to conditions of low soil moisture content (Kumar et al., 2023; Ibrahim et al., 2018). Similar results have been reported in wheat exposed to drought stress where Si application helped in reducing levels of reactive oxygen species (ROS) by uplifting antioxidant defense systems including both enzymatic and non-enzymatic antioxidants like phenols, flavonoids, ascorbate, and others (Ma et al., 2016). Another study of drought-stressed wheat reported the ameliorative potential of SiO_2 Nanoparticles (NPs) which helped in maintaining yield attributes in wheat plants exposed to drought-stressed conditions (Behboudi et al., 2018). Enhanced efficiency of antioxidant defense system has also been reported in Pearl and Malka varieties of maize exposed to conditions of drought where an increase in activity of different antioxidants like superoxide dismutase (SOD), peroxidase (POD), and catalase (CAT) was observed. Furthermore, Si enhanced levels of osmolytes like sugars and proline and maintained the morphological attributes of the plants thereby demonstrating its promising potential against drought conditions in plants (Choudhary et al., 2024; Parveen et al., 2019).

Application of B along with drought stress led to the development of stress-related responses and uplifted antioxidant defense system in tomato plants thereby suggesting that drought stress tolerance was well supported by B application (Aydin et al., 2019). Foliar spray application of B in wheat plants growing under conditions of calcareous soils leads to the improvement of the yield attributes like improved plant height, plant biomass, number of spikelets, number of grains, and weight of grains along with maintaining levels of proline and other stress markers (Abd-Elsalam et al., 2024; Abdel-Motagally et al., 2018).

Application of Se foliar spray helps in mitigating drought stress in wheat plants (Nawaz et al., 2015). Se treatment improved yield attributes by improving grain yield which can be attributed to improved parameters like antioxidants, cell turgor, transpiration rate, and biochemical attributes like total sugars and free amino acids (Kim et al., 2023; Nawaz et al., 2015). Se nanoparticles (SeNPs) application of 30 mg L^{-1} concentration helped in mitigating drought stress in wheat plants by improving plant biomass in terms of fresh weight, dry weight, root and shoot length (Ikram et al., 2020). Co-application of Si and Se helped in mitigating drought stress in rice plants and improved yield of rice plants by reducing the water loss during drought stress conditions and enhancing photosynthetic pigments (Deshwal et al., 2023; Ghouri et al., 2022). The application of Se helped in mitigating drought stress by uplifting antioxidants and improving growth parameters (Rady et al., 2020).

5.2.2 Endurance against Temperature Extremes

5.2.2.1 High-Temperature Stress Tolerance

High temperature directly causes oxidative stress, which, in turn, causes an excessive production of a significant amount of reactive species. This leads to further changes in physiological and biochemical activity, including the production of different osmolytes and heat shock proteins (HSPs), which can protect proteins and cell structures and strengthen antioxidant mechanisms to restore redox potential and homeostasis.

It has been discovered that NPs like Se are useful in reducing the effects of high temperatures. Applying SeNPs to sorghum enhanced its antioxidant system's ability to scavenge ROS generated by heat, thereby reducing high-temperature stress (Verma et al., 2023; Djanaguiraman et al., 2018). Se inhibits ribulose-1,5-bisphosphate carboxylase/oxygenase (RuBisCO) activity, raises photorespiration, and tampers with the electron transport chain of plants' photosynthetic system (Mathur et al., 2018). Se can shield chlorophyll and lessen the destruction caused by photosynthesis due to temperature stress. SeNPs enhance photosynthetic processes and raise the total chlorophyll content of various varieties of *Chrysanthemum morifolium* when exposed to high-temperature stress of 40°C (Seliem et al., 2020).

Si could lower the stress level, particularly that from the source K_2SiO_3. Two significant photosynthetic proteins, Photosystem I subunit A (*PsaA*) and Photosystem II protein D1 (*PsbA*), exhibited lower expression levels in without Si-supplied plants under temperature stress of 41°C, according to immunoblot. However, the expression levels were maintained in the plants supplied with Si, especially when K_2SiO_3 was used as the Si source. No variations in *PsaA* and *PsbA* expressions were seen at 25°C or 33°C. It is interesting to note that under high-temperature stress (41°C), the expression of three key stress-responsive proteins—superoxide SOD, ascorbate peroxidase (APX), and CAT—was significantly increased in the Si-treated plants and decreased in without Si-treated plants, except for those grown at 25°C or 33°C. Si, especially in the form of K_2SiO_3, has a major role in limiting the harmful effects of high-temperature stress by maintaining the photosynthetic proteins and stress-responsive proteins of the ascorbate glutathione defense mechanism, according to the observed responses to Si supply in high-temperature stressed plants. Furthermore, these findings show that Si application is an effective means of preserving plant health during the propagation stage, even in greenhouses with high temperatures (Gupta and Seth, 2021; Muneer et al., 2017).

Spraying foliar B compounds resulted in reduced oxidative damage (represented by the production of proline and malondialdehyde in the leaves) and increased photochemical efficiency of photosystem II (PSII) in seedlings under heat stress (0.76 with B vs. 0.72 without B). When high daytime stress periods are anticipated in rice areas, the use of B (boric acid or sodium borate) can be taken into consideration as an agronomic strategy to reduce the harmful effects of heat stress on rice (Calderón-Páez et al., 2021). The most damaging effect of high temperatures on rice production in tropical regions is spikelet sterility. B plays a crucial role in sugar translocation, cell wall formation, and reproduction and may also be able to lessen the harmful effect of high temperature on three cultivars of rice crops. For all three cultivars, the application of B increases grain yield over control under both ambient and high-temperature conditions; however, the susceptible cultivar showed a greater increase than the tolerant one. The findings suggest that the yield was significantly affected by the exogenous application of B, which also had a significant impact on sugar mobilization, pollen viability, spikelet fertility, and cell membrane stability. Different cultivars behave differently under high-temperature stress. Those that were more susceptible to the high temperature showed more response to the application of B (Shahid et al., 2018).

Crop plants are highly prone to abiotic stresses such as drought, cold. These stresses can damage numerous biochemical, physiological, and molecular processes throughout the plant's lifecycle leading to affecting crop production both qualitatively and quantitatively (Kordrostami et al., 2023). Low temperature is a major factor that affects plant growth and has an impact on plant reproduction, particularly in temperate and subtropical regions (Kapoor et al., 2023). Se has been shown to enhance photosynthesis and glucose metabolism in wheat and sorghum by reducing chlorophyll loss at suboptimal temperatures (Liu et al., 2022). High levels of free proline in cold-stressed plants may increase their resistance to chilling, but they can also be a sign of damage. Low Se doses increased proline accumulation in cold-stressed cucumbers and sorghums (Rasool et al., 2023). Cold temperatures can indirectly affect plant physiology by causing lipid peroxidation and secondary oxidative stress, which lead to an excess of ROS. Ascorbate, antioxidant enzyme activity (APX, POX, and CAT), anthocyanins, carotenoids, flavonoids, phenolic compounds, and ascorbate are all increased by applying Se. Both non-enzymatic and enzyme-induced antioxidants may decrease the effects of Se-induced oxidative damage induced by cooling. Applying Se to tomato plants under short-term suboptimal as well as high-temperature conditions proves to be advantageous in building plants' ability to survive in cold temperatures and this is positively correlated with free endogenous proline accumulation (Haghighi and Abolghasemi, 2018). Similarly, unless proline is provided exogenously, most plants that are sensitive to cold cannot build up free proline and develop more resistance to this stressful event. Se-biofortified plants show less lipid peroxidation in their roots as compared to non-enriched plants. Applying Se to cucumber tissue resulted in an altered physiological response to cold stress by increasing proline concentrations in leaf tissues and blocking lipid peroxidation. According to a recent study, Se may improve the plants' ability to produce antioxidants and grow more rapidly in wheat seedlings (Cao et al., 2022). Application of 1.0 mg of Se per kilogram of soil reduced both lipid peroxidation and the formation of ROS. It has been shown that the application of Si significantly improves plant growth and development in cold-stress situations. It has been reported that the application of Si can improve various aspects of plant growth such as biomass, relative water content, and plant height. Additionally, Si has been found to increase the levels of important substances like total protein, amino acids, carotenoids, anthocyanins, and chlorophylls. When plants are exposed to cold stress, it can negatively affect their performance by deactivating certain characteristics, inhibiting electron transport, and trapping excitation energy. However, by applying Si, these negative effects can be reduced since it helps to enhance the very characteristics that cold stress can deactivate, thus allowing plants to better withstand the stress. Moreover, Si application can improve mineral ion balance, lower ROS levels, and promote the synthesis of phytohormones, all of which can increase a plant's resistance to cold stress (Habibi, 2016).

5.2.3 Protection against Environmental Pollutants

The importance of Si in increasing plant resistance to heavy metal pollutants in plants is well established, and the beneficial role of Si in detoxifying can be attributed to both external (growth media) and internal plant processes (Sahebi et al., 2015). Si enhances soil physical and chemical properties, decreases soil acidity (i.e., raises soil pH), and coprecipitates with heavy metals in the soil, generating metal-silicate complexes, which inhibit heavy metal uptake by plants (Khan et al., 2021; Bhardwaj et al., 2023). Plants differ in their ability to absorb Si, with monocots being better accumulators. Si can reduce metal toxicity in both monocots and dicots (Shi et al., 2018; Greger et al., 2018). Si-rich amendments reduced heavy metal accumulation and improved growth in multi-metal [cadmium (Cd), zinc (Zn), copper (Cu), and lead (Pb)] contaminated acidic soil in rice (Hou et al., 2023). Si has also been shown to improve seedling biomass while decreasing Zn concentration in rice seedling roots and shoots, as well as the xylem sap flow. Furthermore, Si supply considerably decreased Zn buildup in several plant sections such as cotton roots and maize leaves (Anwaar et al., 2015). According to Shi et al. (2018), Si reduces Cd metal toxicity in seedlings of rice by lowering ion absorption and translocation from root to shoot. Si amendments have been reported in sorghum, tomato, soybean, and maize as an alternate detoxifying approach for aluminum (Al) toxicity (Sahebi et al., 2015).

There are several widely accepted Si-mediated metal detoxification mechanisms, including toxic metal immobilization in soil (at the soil level), stimulation of enzymatic and non-enzymatic antioxidants, metal co-precipitation, metal ion chelation, compartmentation, and structural alterations of plant tissues, and changes in molecular responses (at the plant level). Si aids in transforming the speciation of metals, converting them from toxic to non-toxic forms through the creation of silicate complexes in the soil solution. When soluble silicates hydrolyze in the soil solution, they generate gel-like metasilicic acid. This acid binds heavy metals, shifting them from toxic to non-toxic forms, as demonstrated in research by Bhat et al. (2019). Cd toxicity promotes oxidative stress in plants through elevated ROS production. Under Cd toxic effects, the application of Si increases the activities of OD), POD, CAT, and APX in various wheat cultivars, with the degree of enhancement in antioxidant enzyme activity being greater in Cd-tolerant cultivars than in Cd-sensitive cultivars (Naeem et al., 2018; Hussain et al., 2019).

Se also plays a vital role in enhancing resistance to oxidative stress, boosting immunity, and mitigating heavy metal toxicity, as highlighted in studies by Ma et al. (2016) and Sonah et al. (2017). In *Lactuca sativa*, the addition of selenite has been shown to significantly reduce the aggregation of Pb and Cd while simultaneously increasing the uptake of essential elements, including Se (Zargar et al., 2019). Se not only promotes plant growth but also acts as a countermeasure against heavy metals, serving as a necessary micronutrient with antioxidative and physiological properties, as discussed by Gui et al. (2022). The presence of Se can impact the distribution of elements crucial for plant growth and metabolism, reducing metal toxicity and regulating micronutrient levels through modifications in soil conditions such as pH, adsorption, and organic matter (Gui et al., 2022). Optimal concentrations of Se application have been found to improve the growth and biomass of *Cucumis sativus* plants under heavy metal stress (Cd, Pb, Ni) by detoxifying damages induced by ROS through the regulation of various antioxidant enzymes (Aroiee et al., 2019). Se application has also been shown to reduce oxidative stress caused by combined treatments of Cd and Pb in *Brassica napus* by minimizing the uptake and accumulation of both metals (Wu et al., 2016). Overall, Se plays a pivotal role in neutralizing heavy metal stresses in plants.

B is essential for a plant cell wall, where a significant portion (70%–90%) of total Cu, Zn, and Cd is located in certain plants (Riaz et al., 2021). It increases the likelihood of adjusting Cd accumulation and toxicity. Therefore, B alleviates plant Cd toxicity by reducing Cd aggregation and relieving oxidative stress induced by Cd exposure (Chen et al., 2019). *Brassica napus* improves Cd adsorption by acting as an integral part of the cell wall, preventing Cd exit and facilitating Cd chelation (Zhang et al., 2020). Additionally, B is involved in the formation and cross-linking of different cell wall components (Wu et al., 2017). Hence, it can be inferred that managing Cd aggregation and toxicity may contribute to successful crop growth. Hussain et al. (2019) reported that soil-applied B and biochar improved wheat grain yield by triggering the activities of enzymatic antioxidants. Individual or combined B and biochar applications improved proline contents and reduced H_2O_2 and malondialdehyde (MDA) contents in plants. The combined application of B and biochar enhanced soluble sugars and total phenolics contents as compared to the control and Cd-contaminated plants.

5.2.4 Shielding against Biotic Attacks

Climate change is a significant global issue that impacts all forms of life, including plants. The rise in average temperatures has contributed to the proliferation of different plant diseases. Pests and diseases pose a constant threat to agriculture, resulting in significant crop yield losses (Bajwa et al. 2020). Significant progress has been made in recent decades toward understanding the potential of Si to enhance plant resistance against biotic stress while maintaining crop productivity. Numerous studies have provided evidence that the application of Si can effectively suppress various plant diseases across a wide range of crops. Notable examples include the mitigation of rice blast, brown spot, and leaf scald (Van Bockhaven et al., 2015; Domiciano et al., 2015; Tatagiba et al., 2016), as well as the control of barley and wheat powdery mildew (Moldes et al., 2016). The accumulation of Si in the leaves of *Magnaporthe grisea* has been found to reduce lesion formation (Sakr, 2016) and enhance tolerance capacity in the stem, stalk, and shoot of various plant species including *Triticum aestivum, Oryza sativa, Zea mays,* and *Saccharum spp.* against biotic stress (Debona et al., 2017; Bakat et al., 2018). Additionally, Si may hinder digestibility by reducing the availability of nitrogen and carbohydrates during the digestion process (Dos Santos et al., 2015). The presence of Si in plant tissue triggers metabolic changes that protect against insect pests (Kim et al., 2024a, b; Liang et al., 2015). Insecticides typically target the midgut cells of insects initially, but the presence of inert amorphous silica in plant tissues can make it challenging for insects to bite and digest the plant material (Agnihotri and Seth, 2020; Coskun et al., 2019). Bakhat et al. (2018) demonstrated that Si application has the potential to decrease pest infestations and could offer a sustainable, environmentally friendly integrated approach as a substitute for excessive pesticide utilization. The role of Si in shielding plants against biotic attacks has been summarized in Figure 5.2.

Se is considered to be a valuable trace element for plants, as it stimulates various physiological processes in plants. Plants primarily absorb Se in the form of selenate, utilizing root sulfate transporters. Additionally, Se impacts the uptake of sulfur (S) by interfering with self-regulatory mechanisms. Se affects the pathway of sulfate uptake and the synthesis of S organic compounds, which are crucial for the plant's response to biotic stress within plant cells (Hashem et al., 2022). Se in its nano form exhibits enhanced antibacterial and antioxidant effects compared to its other forms (Abbas et al., 2021). The antimicrobial activity of SeNPs against various phytopathogenic bacteria and fungi was evaluated using different concentrations (100, 200, and 250 µg mL^{-1}). Notably, the concentration of 100 µg mL^{-1} demonstrated the most effective control, inhibiting 99% of different bacteria, including *Pseudomonas aeruginosa*, *Staphylococcus aureus*, and *Escherichia coli*. Upon employing SeNPs as a novel method of plant fertilization (Srivastava et al., 2015), it was found that these NPs are less toxic, possess higher bioavailability, and exhibit greater biological activity. In the case of *Vicia faba* plants under biotic stress caused by *Rhizoctonia solani* (Hashem et al., 2021), SeNPs enhanced the phenolic compounds, total chlorophyll contents, antioxidant defense systems,

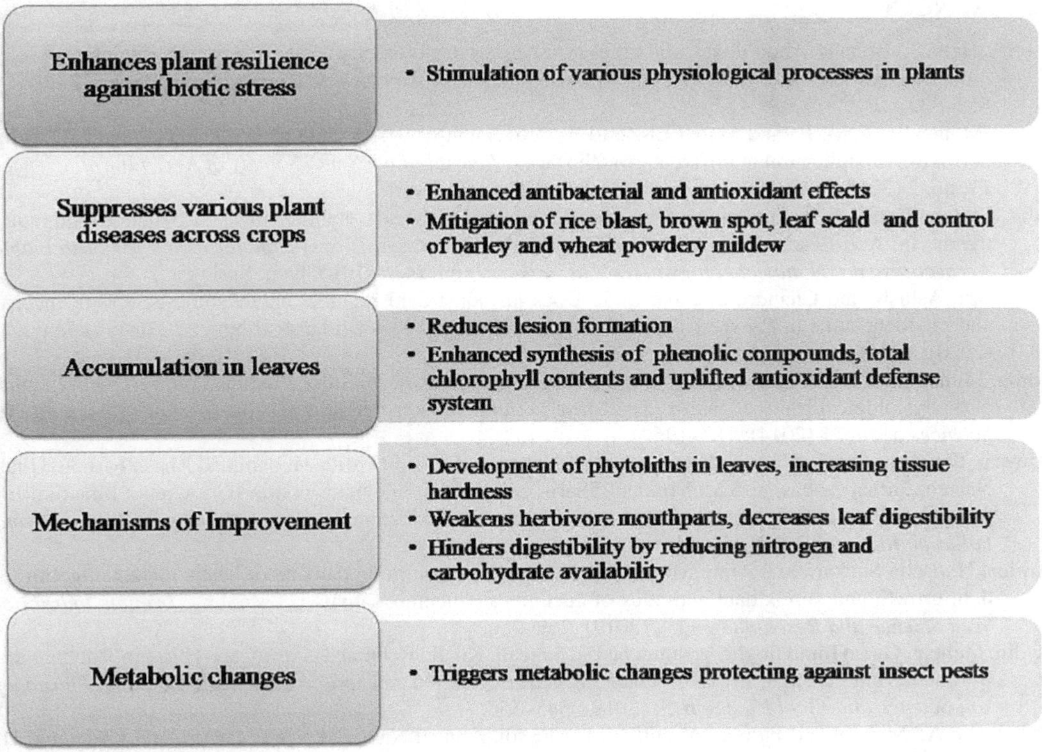

FIGURE 5.2 Mechanism adopted by Si in shielding plants against biotic attacks.

as well as the morphological and genetical attributes (El-Gazzar et al., 2020). Mosallam et al. (2018) elucidated that the SeNPs that were synthesized exhibited activity against *Acinetobacter calcoaceticus* (15.0 mm zone of inhibition (ZOI)) and *Staphylococcus aurus* (16.6 mm ZOI). Furthermore, they demonstrated inhibitory effects on *Candida albicans* (15.3 mm ZOI) and *Aspergillus flavus*, a producer of mycotoxins (29.6 mm ZOI). Due to their unique characteristics and the innovative biosynthesis process, these SeNPs have the potential to be utilized in various fields such as biomedicine, prevention of food spoilage, cosmetics, and pharmaceuticals as environmentally friendly antimicrobial agents.

5.3 CONCLUSION: FUTURE PROSPECTS AND PROMOTING SUSTAINABLE AGRICULTURE USING METALLOIDS

Se, Si, and B form a major group of metalloids that help plants to inculcate stress tolerance against different conditions of stress like drought, high temperature, cold stress, heavy metal contamination, and biological attacks. Plants are shielded by enhancing their antioxidant defense system, promoting the synthesis of osmolytes and other protectants, and promoting growth attributes such as improved biomass, chlorophyll content, photosynthetic efficiency, soil-water relations, and other physiological processes. These metalloids hold the tendency to strengthen the plants in very minute quantities which suggests their utilization as a paramount cost-effective tool in sustainable agriculture. The exact mechanisms behind the positive role of some of the metalloids still need to be explored to get further insights into the shielding capacity of metalloids for their future large-scale utilization in sustainable agriculture to protect plants against a broad spectrum of biotic and abiotic stress conditions.

REFERENCES

Abbas, Heba S., Doha H. Abou Baker, and Entesar A. Ahmed. Cytotoxicity and antimicrobial efficiency of selenium nanoparticles biosynthesized by Spirulina platensis. *Archives of Microbiology* 203, no. 2 (2021): 523–532.

Abdel-Motagally, F. M. F., and Manal El-Zohri. Improvement of wheat yield grown under drought stress by boron foliar application at different growth stages. *Journal of the Saudi Society of Agricultural Sciences* 17, no. 2 (2018): 178–185.

Abd-Elsalam, Kamel A., Seth, Chandra Shekhar, Alghuthaymi, Mousa A. Nanofertilizers: Challenges and future trends. In: Abd-Elsalam, Kamel A., and Mousa A. Alghuthaymi (eds) *Nanofertilizers for Sustainable Agroecosystems. Nanotechnology in the Life Sciences* (pp. 561–570). Cham: Springer, 2024.

Agnihotri, Ashish, and Chandra Shekhar Seth. Does jasmonic acid regulate photosynthesis, clastogenecity, and phytochelatins in *Brassica juncea* L. in response to Pb-subcellular distribution? *Chemosphere* 243 (2020): 125361.

Amin, Muhammad, Riaz Ahmad, Shahzad Basra, and Ghulam Murtaza. Silicon induced improvement in morpho-physiological traits of maize (Zea Mays L.) under water deficit. *Pakistan Journal of Agricultural Sciences* 51, no. 1 (2014): 187–196.

Anwaar, Shad Ali, Shafaqat Ali, Skhawat Ali, Wajid Ishaque, Mujahid Farid, Muhammad Ahsan Farooq, Ullah Najeeb, Farhat Abbas, and Muhammad Sharif. Silicon (Si) alleviates cotton (Gossypium hirsutum L.) from zinc (Zn) toxicity stress by limiting Zn uptake and oxidative damage. *Environmental Science and Pollution Research* 22 (2015): 3441–3450.

Aroiee, H., Leila Shekari, and Amin Mirshekari. Effects of selenium on damage of heavy metals in germination, growth and antioxidant activities of cucumber (Cucumis sativus L.) seedling. *Iranian Journal of Seed Science and Research* 6, no. 2 (2019): 269–286.

Aydin, Mehtap, GuzinTombuloglu, Mehmet Serdal Sakcali, Khalid Rehman Hakeem, and Huseyin Tombuloglu. Boron alleviates drought stress by enhancing gene expression and antioxidant enzyme activity. *Journal of Soil Science and Plant Nutrition* 19 (2019): 545–555.

Bajwa, Ali A., Muhammad Farooq, Abdullah M. Al-Sadi, Ahmad Nawaz, Khawar Jabran, and Kadambot H. M. Siddique. Impact of climate change on biology and management of wheat pests. *Crop Protection* 137 (2020): 105304.

Bakhat, Hafiz Faiq, Najma Bibi, Zahida Zia, Sunaina Abbas, Hafiz Mohkum Hammad, Shah Fahad, Muhammad Rizwan Ashraf, Ghulam Mustafa Shah, Faiz Rabbani, and Shafqat Saeed. Silicon mitigates biotic stresses in crop plants: A review. *Crop Protection* 104 (2018): 21–34.

Behboudi, Fatemeh, Zahra Tahmasebi Sarvestani, Mohammad Zaman Kassaee, Seyed Ali Mohammad Modares Sanavi, and Ali Sorooshzadeh. Improving growth and yield of wheat under drought stress via application of SiO2 nanoparticles. *Journal of Agricultural Science and Technology* 20, no. 7 (2018): 1479–1492.

Bhardwaj, Savita, Dhriti Sharma, Simranjeet Singh, Praveen C. Ramamurthy, Tunisha Verma, Mamta Pujari, Joginder Singh, Dhriti Kapoor, and Ram Prasad. Physiological and molecular insights into the role of silicon in improving plant performance under abiotic stresses. *Plant and Soil* 486, no. 1–2 (2023): 25–43.

Bhat, Javaid Akhter, S. M. Shivaraj, Pritam Singh, Devanna B. Navadagi, Durgesh Kumar Tripathi, Prasanta K. Dash, Amolkumar U. Solanke, Humira Sonah, and Rupesh Deshmukh. Role of silicon in mitigation of heavy metal stresses in crop plants. *Plants* 8, no. 3 (2019): 71.

Biju, Sajitha, Sigfredo Fuentes, and Dorin Gupta. Silicon improves seed germination and alleviates drought stress in lentil crops by regulating osmolytes, hydrolytic enzymes and antioxidant defense system. *Plant Physiology and Biochemistry* 119 (2017): 250–264.

Bokor, Boris, Carla S. Santos, Dominik Kostolani, Joana Machado, Marta Nunes da Silva, Susana MP Carvalho, Marek Vaculik, and Marta W. Vasconcelos. Mitigation of climate change and environmental hazards in plants: Potential role of the beneficial metalloid silicon. *Journal of Hazardous Materials* 416 (2021): 126193.

Calderón-Páez, Sara Eliana, Yeraldin Andrea Cueto-Niño, Alefsi David Sánchez-Reinoso, Gabriel Garces-Varon, CristhianCamilo Chávez-Arias, and Hermann Restrepo-Díaz. Foliar boron compounds applications mitigate heat stress caused by high daytime temperatures in rice (Oryza sativa L.) boron mitigates heat stress in rice. *Journal of Plant Nutrition* 44, no. 17 (2021): 2514–2527.

Cao, Qingjun, Wangtian Wang, Saqib Saleem Akhtar, Fentuan Yang, Fanli Kong, Zhengguo Cui, Xiaoli Jiang, Enping Zhang, and Gang Li. Exogenous foliar spray of selenium (Se) alleviates cold stress by promoting photosynthesis and antioxidant defense in waxy maize. *Research Square* (2022). https://doi.org/10.21203/rs.3.rs-1719103/v1

Chen, Dongmei, Daoqian Chen, Rongrong Xue, Jun Long, Xianhui Lin, Yibin Lin, Lianghai Jia, Rensen Zeng, and Yuanyuan Song. Effects of boron, silicon and their interactions on cadmium accumulation and toxicity in rice plants. *Journal of Hazardous Materials* 367 (2019): 447–455.

Choudhary, Rashmi, Vishnu Dayal Rajput, Gajanan Ghodake, Faheem Ahmad, Mukesh Meena, Reiaz ul Rehman, Ram Prasad, Rajesh Kumar Sharma, Rachana Singh, and Chandra Shekhar Seth. Comprehensive journey from past to present to future about seed priming with hydrogen peroxide and hydrogen sulfide concerning drought, temperature, UV and ozone stresses-A review. *Plant Soil* 2024 (2024): 351–373.

Coskun, Devrim, Rupesh Deshmukh, Humira Sonah, James G. Menzies, Olivia Reynolds, Jian Feng Ma, Herbert J. Kronzucker, and Richard R. Bélanger. The controversies of silicon's role in plant biology. *New Phytologist* 221, no. 1 (2019): 67–85.

Debona, Daniel, Fabrício A. Rodrigues, and Lawrence E. Datnoff. Silicon's role in abiotic and biotic plant stresses. *Annual Review of Phytopathology* 55 (2017): 85–107.

Deshwal, Akanksha, Kirti Saxena, Garima Sharma, Rajesh, Faheem A. Sheikh, Chandra Shekhar Seth, and Ravi Mani Tripathi. Nanozymes: A comprehensive review on emerging applications in cancer diagnosis and therapeutics. *International Journal of Biological Macromolecules* 256 (2023): 128272.

Djanaguiraman, Maduraimuthu, Remya Nair, Juan Pablo Giraldo, and PagadalaVenkataVara Prasad. Cerium oxide nanoparticles decrease drought-induced oxidative damage in sorghum leading to higher photosynthesis and grain yield. *ACS Omega* 3, no. 10 (2018): 14406–14416.

Domiciano, Gisele Pereira, Isaías Severino Cacique, Cecília Chagas Freitas, Marta Cristina Corsi Filippi, Fábio Murilo DaMatta, Francisco Xavier Ribeiro do Vale, and Fabrício Ávila Rodrigues. Alterations in gas exchange and oxidative metabolism in rice leaves infected by Pyricularia oryzae are attenuated by silicon. *Phytopathology* 105, no. 6 (2015): 738–747.

Dos Santos, M. C., Ana Maria Resende Junqueira, V. G. Mendes de Sá, J. C. Zanúncio, and J. E. Serrão. Effect of silicon on the morphology of the midgut and mandible of tomato leafminer Tuta absoluta (Lepidoptera: Gelechiidae) larvae. *Invertebrate Survival Journal* 12, no. 1 (2015): 158–165.

El-Gazzar, Nashwa, and Ahmed Mahmoud Ismail. The potential use of titanium, silver and selenium nanoparticles in controlling leaf blight of tomato caused by Alternaria alternata. *Biocatalysis and Agricultural Biotechnology* 27 (2020): 101708.

Ghouri, Fozia, Zeeshan Ali, Muhammad Naeem, Sami Ul-Allah, Muhammad Babar, Faheem Shehzad Baloch, Waqas Shafqat Chattah, and Muhammad Qasim Shahid. Effects of silicon and selenium in alleviation of drought stress in rice. *Silicon* 14, no. 10 (2022): 5453–5461.

Greger, Maria, Tommy Landberg, and Marek Vaculík. Silicon influences soil availability and accumulation of mineral nutrients in various plant species. *Plants* 7, no. 2 (2018): 41.

Gui, Jia-Ying, Shen Rao, Xinru Huang, Xiaomeng Liu, Shuiyuan Cheng, and Feng Xu. Interaction between selenium and essential micronutrient elements in plants: A systematic review. *Science of the Total Environment* 853 (2022): 158673.

Gupta, Samta, and Chandra Shekhar Seth. Salicylic acid alleviates chromium (VI) toxicity by restricting its uptake, improving photosynthesis and augmenting antioxidant defense in *Solanum lycopersicum* L. *Physiology and Molecular Biology of Plants* 27 (2021): 2651–2664.

Habibi, Ghader. Effect of foliar-applied silicon on photochemistry, antioxidant capacity and growth in maize plants subjected to chilling stress. *Acta Agriculturae Slovenica* 107, no. 1 (2016): 33–43.

Haghighi, Maryam, and Reza Abolghasemi. The effect of metallic selenium and nano on the physiological characteristics of tomato plants. *Journal of Soil and Plant Interactions-Isfahan University of Technology* 9, no. 1 (2018): 63–78.

Hashem, Amr H., Amer M. Abdelaziz, Ahmed A. Askar, Hossam M. Fouda, Ahmed M. A. Khalil, Kamel A. Abd-Elsalam, and Mona M. Khaleil. Bacillus megaterium-mediated synthesis of selenium nanoparticles and their antifungal activity against Rhizoctonia solani in faba bean plants. *Journal of Fungi* 7, no. 3 (2021): 195.

Hashem, Amr H., Amer M. Abdelaziz, Mohamed S. Attia, and Salem S. Salem. Selenium and nano-selenium-mediated biotic stress tolerance in plants. In Hossain, Mohammad Anwar, Golam Jalal Ahammed, Zsuzsanna Kolbert, Hassan Ragab El-Ramady, Tofazzal Islam, and Michela Schiavon (eds) *Selenium and Nano-Selenium in Environmental Stress Management and Crop Quality Improvement*, pp. 209–226. Cham: Springer International Publishing, 2022.

Hou, Lei, Shengzhe Ji, Yao Zhang, Xiuzhe Wu, Li Zhang, and Peng Liu. The mechanism of silicon on alleviating cadmium toxicity in plants: A review. *Frontiers in Plant Science* 14 (2023): 1141138.

Hussain, Afzal, Muhammad Rizwan, Qasim Ali, and Shafaqat Ali. Seed priming with silicon nanoparticles improved the biomass and yield while reduced the oxidative stress and cadmium concentration in wheat grains. *Environmental Science and Pollution Research* 26 (2019): 7579–7588.

Ibrahim, Mostafa A., Abdel-Rahman M. Merwad, and Elsayed A. Elnaka. Rice (Oryza sativa L.) tolerance to drought can be improved by silicon application. *Communications in Soil Science and Plant Analysis* 49, no. 8 (2018): 945–957.

Ikram, Muhammad, Naveed Iqbal Raja, Bilal Javed, Zia-ur-Rehman Mashwani, Mubashir Hussain, Mujahid Hussain, Maria Ehsan, Noman Rafique, Khafsa Malik, Tahira Sultana, and Abida Akram. Foliar applications of bio-fabricated selenium nanoparticles to improve the growth of wheat plants under drought stress. *Green Processing and Synthesis* 9, no. 1 (2020): 706–714.

Iqbal, Mohammed Shariq, Akhilesh Kumar Singh, and Mohammad Israil Ansari. Effect of drought stress on crop production. *New Frontiers in Stress Management for Durable Agriculture* (2020): 35–47.

Jalil, Syed Uzma, and Mohammad Israil Ansari. Stress implications and crop productivity. In Hasanuzzaman, Mirza (ed) *Plant Ecophysiology and Adaptation under Climate Change: Mechanisms and Perspectives I: General Consequences and Plant Responses*, pp. 73–86. Singapore: Springer, 2020.

Kapoor, Bhuvnesh, Pankaj Kumar, Navjot Singh Gill, Rajnish Sharma, Naresh Thakur, and Mohammad Irfan. Molecular mechanisms underpinning the silicon-selenium (Si-Se) interactome and cross-talk in stress-induced plant responses. *Plant and Soil* 486, no. 1 (2023): 45–68.

Khan, Imran, Samrah Afzal Awan, Muhammad Rizwan, Shafaqat Ali, Muhammad Jawad Hassan, Marian Brestic, Xinquan Zhang, and Linkai Huang. Effects of silicon on heavy metal uptake at the soil-plant interphase: A review. *Ecotoxicology and Environmental Safety* 222 (2021): 112510.

Kim, Dae-Young, Min Kim, Jung-Suk Sung, Janardhan Reddy Koduru, Shivraj Hariram Nile, Asad Syed, Ali H. Bahkali, Chandra Shekhar Seth, and Gajanan Sampatrao Ghodake. Extracellular synthesis of silver nanoparticle using yeast extracts: Antibacterial and seed priming applications. *Applied Microbiology and Biotechnology* 108, no. 1 (2024a): 150.

Kim, Dae-Young, Sanjay Kumar Singh Patel, Kashif Rasool, Nasreena Lone, Shashi Kant Bhatia, Chandra Shekhar Seth, and Gajanan Sampatrao Ghodake. Bioinspired silver nanoparticle-based nanocomposites for effective control of plant pathogens: A review. *Science of the Total Environment* 908 (2023): 168318.

Kim, Dae-Young, Tianxi Yang, Priyanka Srivastava, Shivraj Hariram Nile, Chandra Shekhar Seth, Umesh Jadhav, Asad Syed, Ali H. Bahkali, Gajanan Sampatrao Ghodake (2024). Alginic acid-functionalized silver nanoparticles: A rapid monitoring tool for detecting the technology-critical element tellurium. *Journal of Hazardous Materials* 465 (2024b): 133161.

Kordrostami, Mojtaba, Ali Akbar Ghasemi-Soloklui, Mohammad Anwar Hossain, and Mohammad Golam Mostofa. Breaking barriers: Selenium and silicon-mediated strategies for mitigating abiotic stress in plants. *Phyton* 92, no. 9 (2023): 2713–2736.

Kumar, Dharmendra, Ritu Singh, Sudhir K. Upadhyay, Krishan K. Verma, Ravi Mani Tripathi, Haitao Liu, Om Parkash Dhankher, Rudra Deo Tripathi, Shivendra V. Sahi, Chandra Shekhar Seth. Review on interactions between nanomaterials and phytohormones: Novel perspectives and opportunities for mitigating environmental challenges. *Plant Science* (2023): 111964.

Lamaoui, Mouna, Martin Jemo, Raju Datla, and Faouzi Bekkaoui. Heat and drought stresses in crops and approaches for their mitigation. *Frontiers in Chemistry* 6 (2018): 26.

Liang, Yongchao, Miroslav Nikolic, Richard Bélanger, Haijun Gong, and Alin Song. Silicon in agriculture. In *Silicon-Mediated Tolerance to Salt Stress*, pp. 123–142. Springer Science, 2015.

Liu, Haodong, Chunmei Xiao, Tianci Qiu, Jie Deng, Hua Cheng, Xin Cong, Shuiyuan Cheng, Shen Rao, and Yue Zhang. Selenium regulates antioxidant, photosynthesis, and cell permeability in plants under various abiotic stresses: A review. *Plants* 12, no. 1 (2022): 44.

Ma, Dongyun, Dexiang Sun, Chenyang Wang, Haixia Qin, Huina Ding, Yaoguang Li, and Tiancai Guo. Silicon application alleviates drought stress in wheat through transcriptional regulation of multiple antioxidant defense pathways. *Journal of Plant Growth Regulation* 35 (2016): 1–10.

Mariyam, Safoora, Renu Bhardwaj, Nafees A. Khan, Shivendra V. Sahi, and Chandra Shekhar Seth. Review on nitric oxide at the forefront of rapid systemic signaling in mitigation of salinity stress in plants: Crosstalk with calcium and hydrogen peroxide. *Plant Science* 336, (2023a): 111835.

Mariyam, Safoora, Sudhir K. Upadhyay, Koushik Chakraborty, Krishan K. Verma, Joginder Singh Duhan, Sowbiya Muneer, Mukesh Meena, Rajesh Kumar Sharma, Gajanan Ghodake, and Chandra Shekhar Seth. Nanotechnology, a frontier in agricultural science, a novel approach in abiotic stress management and convergence with new age medicine-A review. *Science of the Total Environment* 912 (2023b): 169097.

Mathur, Piyush, Durgesh Kumar Tripathi, František Baluška, and Soumya Mukherjee. Auxin-mediated molecular mechanisms of heavy metal and metalloid stress regulation in plants. *Environmental and Experimental Botany* 196 (2022): 104796.

Mathur, Sonal, Mahaveer P. Sharma, and AnjanaJajoo. Improved photosynthetic efficacy of maize (Zea mays) plants with arbuscularmycorrhizal fungi (AMF) under high temperature stress. *Journal of Photochemistry and Photobiology B: Biology* 180 (2018): 149–154.

Ming, De-Feng, Pei, Zhi-Feng, Naeem, Muhammad S., Gong, Hai-Jun, and Zhou, Wei-Jun. Silicon alleviates PEG-induced water-deficit stress in upland rice seedlings by enhancing osmotic adjustment. *Journal of Agronomy and Crop Science* 198, no. 1 (2012): 14–26.

Moldes, Carlos A., Oscar Fontão de Lima Filho, Luciano J. Merini, Siu M. Tsai, and José M. Camiña. Occurrence of powdery mildew disease in wheat fertilized with increasing silicon doses: A chemometric analysis of antioxidant response. *Acta Physiologiae Plantarum* 38 (2016): 1–9.

Mosallam, Farag M., Gharieb S. El-Sayyad, Rasha M. Fathy, and Ahmed I. El-Batal. Biomolecules-mediated synthesis of selenium nanoparticles using Aspergillus oryzae fermented Lupin extract and gamma radiation for hindering the growth of some multidrug-resistant bacteria and pathogenic fungi. *Microbial Pathogenesis* 122 (2018): 108–116.

Muneer, Sowbiya, YooGyeong Park, Soohoon Kim, and Byoung Ryong Jeong. Foliar or subirrigation silicon supply mitigates high temperature stress in strawberry by maintaining photosynthetic and stress-responsive proteins. *Journal of Plant Growth Regulation* 36 (2017): 836–845.

Naeem, Asif, Muhammad Zia-ur-Rehman, Tasneem Akhtar, Munir Hussain Zia, and Muhammad Aslam. Silicon nutrition lowers cadmium content of wheat cultivars by regulating transpiration rate and activity of antioxidant enzymes. *Environmental Pollution* 242 (2018): 126–135.

Nawaz, Fahim, Rashid Ahmad, Muhammad Yasin Ashraf, Ejaz Ahmad Waraich, and Syed Zulfiqar Khan. Effect of selenium foliar spray on physiological and biochemical processes and chemical constituents of wheat under drought stress. *Ecotoxicology and Environmental Safety* 113 (2015): 191–200.

Parveen, Abida, Wei Liu, Saddam Hussain, Jaleel Asghar, Shagufta Perveen, and Yousheng Xiong. Silicon priming regulates morpho-physiological growth and oxidative metabolism in maize under drought stress. *Plants* 8, no. 10 (2019): 431.

Parveen, Nishat, Roberto Berni, Sreeja Sudhakaran, Javaid A. Bhat, Suhas Shinde, NaleeniRamawat, Vijay P. Singh, Shivendra V. Sahi, Rupesh Deshmukh, Devendra Kumar Chauhan, and Durgesh K Tripathi. Metalloids in plants: A systematic discussion beyond description. *Annals of Applied Biology* 180, no. 1 (2022): 7–25.

Rady, Mohamed O. A., Wael M. Semida, Taia A. Abd El-Mageed, Saad M. Howladar, and Ahmed Shaaban. Foliage applied selenium improves photosynthetic efficiency, antioxidant potential and wheat productivity under drought stress. *International Journal of Agriculture and Biology* 24 (2020): 1293–1300.

Rasool, Aadil, Wasifa Hafiz Shah, Sajad Ahmad Padder, Inayatullah Tahir, Hesham F. Alharby, Khalid Rehman Hakeem, and Reiazul Rehman. Exogenous selenium treatment alleviates salinity stress in Proso Millet (Panicum miliaceum L.) by enhancing the antioxidant defence system and regulation of ionic channels. *Plant Growth Regulation* 100, no. 2 (2023): 479–494.

Raza, Ali. Eco-physiological and biochemical responses of rapeseed (Brassica napus L.) to abiotic stresses: Consequences and mitigation strategies. *Journal of Plant Growth Regulation* 40, no. 4 (2021): 1368–1388.

Riaz, Muhammad, Muhammad Kamran, Yizeng Fang, Guoling Yang, Muhammad Rizwan, Shafaqat Ali, Yaoyu Zhou, Qianqian Wang, Lulu Deng, Youjuan Wang, and Xiurong Wang. Boron supply alleviates cadmium toxicity in rice (Oryza sativa L.) by enhancing cadmium adsorption on cell wall and triggering antioxidant defense system in roots. *Chemosphere* 266 (2021): 128938.

Rizwan, Muhammad, Shafaqat Ali, Muhammad Ibrahim, Mujahid Farid, Muhammad Adrees, Saima Aslam Bharwana, Muhammad Zia-ur-Rehman, Muhammad Farooq Qayyum, and Farhat Abbas. Mechanisms of silicon-mediated alleviation of drought and salt stress in plants: A review. *Environmental Science and Pollution Research* 22 (2015): 15416–15431.

Sahebi, Mahbod, Mohamed M. Hanafi, Abdullah Siti Nor Akmar, Mohd Y. Rafii, Parisa Azizi, F. F. Tengoua, Jamaludin Nurul MayzaitulAzwa, and M. Shabanimofrad. Importance of silicon and mechanisms of biosilica formation in plants. *BioMed Research International* 2015 (2015): 396010.

Sakr, N. The role of silicon (Si) in increasing plant resistance against fungal diseases. *Hellenic Plant Protection Journal* 9, no. 1 (2016): 1–15.

Sarraf, Mohammad, Kanchan Vishwakarma, Vinod Kumar, Namira Arif, Susmita Das, Riya Johnson, Edappayil Janeeshma, Jos T. Puthur, Sasan Aliniaeifard, Devendra Kumar Chauhan, Masayuki Fujita, and Mirza Hasanuzzaman. Metal/metalloid-based nanomaterials for plant abiotic stress tolerance: An overview of the mechanisms. *Plants* 11, no. 3 (2022): 316.

Seliem, Mayada K., Yaser Hafez, and Hassan El-Ramady. Using nano-selenium in reducing the negative effects of high temperature stress on Chrysanthemum morifolium Ramat. *Journal of Sustainable Agricultural Sciences* 46, no. 3 (2020): 47–60.

Shahid, Mohammad, Amaresh Kumar Nayak, Rahul Tripathi, JawaharLalKatara, PriyankaBihari, BanwariLal, and PriyankaGautam. Boron application improves yield of rice cultivars under high temperature stress during vegetative and reproductive stages. *International Journal of Biometeorology* 62 (2018): 1375–1387.

Shi, Zhenya, Suqin Yang, Dan Han, Zhen Zhou, Xuanzhen Li, Ye Liu, and Biao Zhang. Silicon alleviates cadmium toxicity in wheat seedlings (Triticum aestivum L.) by reducing cadmium ion uptake and enhancing antioxidative capacity. *Environmental Science and Pollution Research* 25 (2018): 7638–7646.

Sonah, Humira, Rupesh K. Deshmukh, Caroline Labbé, and Richard R. Bélanger. Analysis of aquaporins in Brassicaceae species reveals high-level of conservation and dynamic role against biotic and abiotic stress in canola. *Scientific Reports* 7, no. 1 (2017): 2771.

Srivastava, Nishant, and Mausumi Mukhopadhyay. Green synthesis and structural characterization of selenium nanoparticles and assessment of their antimicrobial property. *Bioprocess and Biosystems Engineering* 38 (2015): 1723–1730.

Tatagiba, Sandro Dan, Fábio Murilo DaMatta, and Fabrício Ávila Rodrigues. Silicon partially preserves the photosynthetic performance of rice plants infected by Monographellaalbescens. *Annals of Applied Biology* 168, no. 1 (2016): 111–121.

Van Bockhaven, Jonas, Lukáš Spíchal, Ondřej Novák, Miroslav Strnad, Takayuki Asano, Shoshi Kikuchi, Monica Höfte, and David De Vleesschauwer. Silicon induces resistance to the brown spot fungus Cochliobolusmiyabeanus by preventing the pathogen from hijacking the rice ethylene pathway. *New Phytologist* 206, no. 2 (2015): 761–773.

Verma, Krishan K., Song, Xiu-Peng, Xu, Lin, Huang, Hai-Rong, Seth, Chandra Shekhar, and Li, Yang-Rui. Nano-microplastic and agro-ecosystems: A mini-review. *Frontiers in Plant Science* 14 (2023): 1283852.

Wang, Min, Ruirui Wang, Luis Alejandro Jose Mur, Jianyun Ruan, Qirong Shen, and Shiwei Guo. Functions of silicon in plant drought stress responses. *Horticulture Research* 8 (2021): 254.

Wu, Songwei, Chengxiao Hu, Qiling Tan, Shoujun Xu, and Xuecheng Sun. Nitric oxide mediates molybdenum-induced antioxidant defense in wheat under drought stress. *Frontiers in Plant Science* 8 (2017): 1085.

Wu, Zhilin, Xuebin Yin, Gary S. Bañuelos, Zhi-Qing Lin, Ying Liu, Miao Li, and Linxi Yuan. Indications of selenium protection against cadmium and lead toxicity in oilseed rape (Brassica napus L.). *Frontiers in Plant Science* 7 (2016): 1875.

Zahra, Noreen, Marwa Sulaiman Al Hinai, Muhammad Bilal Hafeez, Abdul Rehman, Abdul Wahid, Kadambot H. M. Siddique, and Muhammad Farooq. Regulation of photosynthesis under salt stress and associated tolerance mechanisms. *Plant Physiology and Biochemistry* 178 (2022): 55–69.

Zargar, Safa Mohammed, Mahajan, Rakesh, Bhat, Javaid Ahmad, Nazir, Mushtaq, and Deshmukh, Ramesh. Role of silicon in plant stress tolerance: opportunities to achieve a sustainable cropping system. *3 Biotech*, 9 (3) (2019): 73.

Zhang, Fang, Xiao, Xiaowei, and Wu, Xuelian. Physiological and molecular mechanism of cadmium (Cd) tolerance at initial growth stage in rapeseed (Brassica napus L.). Ecotoxicology and Environmental Safety, 197 (2020): 110613.

6 Role of Silicon in Green Agriculture

Manu Sharma, Roohi Sharma, Nahida Rehman Mir, Amandeep Singh, Harpreet Kaur, Sakshi Verma, Yogesh K Ahlawat, Kushi Yadav, Puja Ohri, Renu Bhardwaj, and Nitika Kapoor

6.1 INTRODUCTION

Silicon (Si) ranks as the second most prominent element on earth, exhibiting concentrations varying between 50 and 400 g Si kg^{-1} of soil samples that makes up approximately 25% of the crust of the earth (Adrees et al., 2015; Zaid et al., 2018). Carbonites, limestones, and carbonaceous rocks are the primary sources of Si (Monger and Kelly, 2002). Si commonly exists in the form of monosilic acid (H_4SiO_4) which gets assimilated by plants and subsequently undergoes polymerization, resulting in silica gel or biogenic-based opal forms within the intercellular spaces of roots, wall of the cells, cells of the leaves, and bracts (Paye et al., 2018). Si is a metalloid that belongs to group 14 of the periodic table with 14 as its atomic number and an atomic weight of 28.085. Si has two allotrophic forms: amorphous and crystal and has three oxidation states of −4, +2, and +4. The metalloid has a specific gravity of 2.42 and a melting point of 1,420°C. It also has three naturally occurring isotopes, namely ^{28}Si, ^{29}Si, and ^{30}Si. Of these isotopes, the most prevalent is ^{28}Si, accounting for 92% of the whole mass (Tubaña and Heckman, 2015). Chemically, silicates are salts of silicic acid, formed from the two common elements, Si and O, and they also carry accessory elements that give them different characteristics (Dematte et al., 2011).

Si is regarded as neither an essential nor a non-essential nutrient for plants. Although most plants can survive without Si, but the presence of Si provides additional benefits by making the plants more tolerant to various environmental stress circumstances (Luyckx et al., 2017). Si has played a key role in the alleviation of several abiotic stresses, such as drought (Verma et al., 2019), radiation (Mavrič et al., 2021), and heavy-metal-induced toxicity (Bhat et al., 2019). It can alleviate salt stress by enhancing relative water content (RWC), membrane stability index (MSI), and antioxidant enzyme activity (Ali et al., 2021). Si treatment in drought-stressed plants enhances nutrient absorption, RWC and maintains water balance by reducing the rate of transpiration and loss of water (Alam et al., 2021; Patel et al., 2021). Further, Si enhances tolerance in plants against unfavorable growth conditions by upregulating activities of antioxidative enzymes, modulating gene expression, which are related to stress, by controlling the biosynthesis of plant hormones and by forming conjugates with toxic heavy metal ions (Mostofa et al., 2021; Khan et al., 2023).

Si also positively affects the development and growth of plants and can be used in soilless settings like hydroponics and aeroponics for boosting plant productivity (Luyckx et al., 2017; López-Pérez et al. 2018). It functions as a beneficial element as its application as fertilizer improves the quality of fruits and increases the photosynthetic activity in plants (Li et al., 2022).

Green agriculture, embodying a sustainable farming ethos, integrates novel, eco-friendly technologies to meet food demands while safeguarding the environment. Si's utilization in this paradigm amplifies food crop biomass, augments mineral uptake, fosters sugar phosphorylation, and bolsters carbohydrate synthesis (Neu et al., 2017). Recognized increasingly for its multifaceted

DOI: 10.1201/9781032673592-6

advantages, Si's role as a metalloid in green agriculture is pivotal. Its integration enhances plant resilience to various stressors like drought, salinity, and pests, thereby nurturing sustainable crop production. Notably, Si's efficacy in bolstering nutrient absorption and activating innate defense mechanisms reduces reliance on chemical fertilizers and pesticides, aligning with eco-conscious agricultural practices (Yan et al., 2018). The current chapter discussed Si's positive influence on soil structure, and microbial vitality augments long-term soil health and fertility. Embracing Si in agricultural strategies holds significant promise for cultivating resilient, environmentally mindful farming systems imperative for global food security and ecological sustainability (Li et al., 2022).

6.2 SI AS AN ESSENTIAL METALLOID FOR PLANTS

Si is taken up by plant roots as silicic acid, which subsequently moves via the xylem to various plant tissues, where it gets accumulated in the cell wall as silica, providing structural support and enhancing resistance to biotic and abiotic stresses (Mandlik et al., 2020).

6.2.1 Significance of Si in Plant Growth and Development

In kingdom Plantae, Si is acknowledged as a vital nutrient for only two specific groups, namely diatoms and some members of the algae family (Souri et al., 2021). Si is generally considered a non-essential element for plant species; however, it has a beneficial effect on the health of the plants (Luyckx et al., 2017; López-Pérez et al., 2018). Si plays a pivotal role in various processes, including the synthesis of phenolic compounds, lignin, and phytoalexins, which in turn enhance plant defense mechanisms (Islam et al., 2020; Souri et al., 2021). Furthermore, Si provides significant benefits to plant growth by improving the rate of photosynthesis, minimizing transpiration loss, and increasing plant resilience to various abiotic and biotic stresses (as shown in Figure 6.1). This is primarily accomplished through the regulation of the electron transport chain (ETC), cellular homeostasis, oxidative phosphorylation, and the entire photosynthetic complex, as supported by studies (Yan et al., 2018). For a wide variety of agricultural plants, Si has been shown to reduce the effects of salt stress by

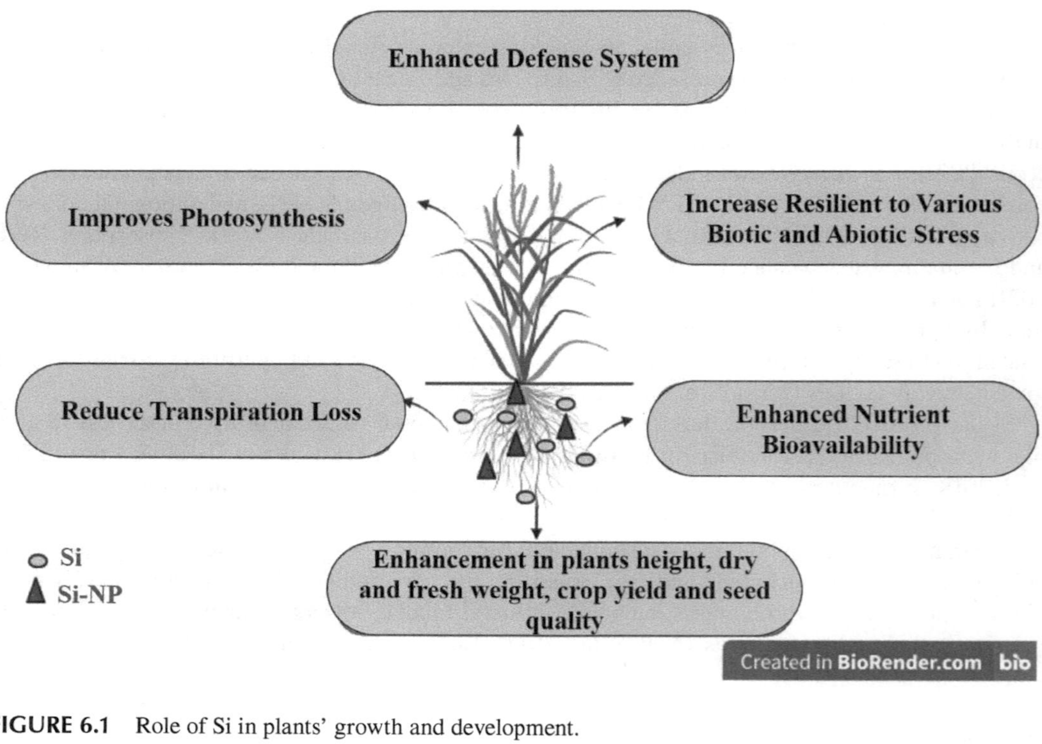

FIGURE 6.1 Role of Si in plants' growth and development.

lowering the sodium (Na) ions concentration (Luyckx et al., 2017; Dhiman et al., 2021). Recently, nanotechnology is regarded as an advanced method that has demonstrated its ability to enhance crop yield by affecting nutrient availability in the soil surrounding plant roots (rhizospheric soil) and facilitating nutrient absorption by plants. The enhanced nutrient bioavailability in plants due to silicon-nanoparticles (Si-NPs) application is a key factor contributing to increased plant growth (Liu et al., 2015a, b). Thus, given Si's pivotal role in plant nutrition akin to other essential macronutrients, researchers have turned their focus to employing Si-NPs in the soil to augment plant growth, recognizing their potential to bolster agricultural productivity. Numerous studies also favor Si-NPs function as fertilizer (Verma et al., 2021). Additionally, Si-NPs have the capacity to decrease the plant's transpiration rate, making it more resilient to conditions of low water availability as well as high temperatures and humidity (Rajput et al., 2021a). Si-NPs also serve as reinforcing agent, enhancing the plant's ability to resist diseases by guarding against fungal, bacterial, and nematode infections. Under various stress conditions, plants naturally produce mineralized nanoparticles to support their growth and development (Rajput et al., 2021b). Also, the application of Si-NPs has been found to lessen the negative effects of salt stress on plant growth and soil moisture levels, thereby significantly improving the plant length, fresh weight, dry weight, overall produce of the crops, and the quality of the seeds (Rajput et al., 2021a). Thus, the application of Si-NPs contributes to enhanced plant performance and increased yield even in adverse environmental situations. The substantial Si-NPs' ratio of surface area to volume is known to heighten their metabolic processes and reactive ability, which plays a key role in delivering beneficial impacts to plants (Adrees et al., 2015; Khan et al., 2021 a).

6.2.1.1 Uptake and Transport of Si in Plants

The largest Si reservoir (50%–70%) exists in the solid phase of soil as silicon dioxide, but its accessibility to plants is limited (Wang et al., 2019 a). Plants absorb Si in the form of silicic acid [$Si(OH)_4$], which is a neutral molecule at pH levels below 9. The concentration of Si in soil solutions typically ranges between 0.1 and 0.6 mM (Ma and Takahashi, 2002). For the intake of Si, three distinct water-based Si absorption mechanisms have been recommended, considering the amount of Si in plants and how rapidly they absorb it compared to water (Kaur et al., 2019). Plants with Si uptake rates higher than that of water are categorized as active (e.g., rice, wheat, and barley), while those with comparable rates are designated as passive (e.g., oat), and those with lower uptake rates are designated as rejective. Si transporters are a class of transmembrane proteins that are essential to the entire method of Si uptake (Marron et al., 2016).

6.2.1.1.1 Active Transport System

The uptake system for Si in plants is orchestrated by specific transporters, such as low silicon rice transporter 1 (*Lsi1*) and low silicon rice transporter 2 (*Lsi2*), which play a crucial role in Si accumulation. In rice, Si is taken up in a polar manner, with Lsi1 at the distal side of the roots importing Si into the exodermis and *Lsi2* at the proximal side exporting Si to the aerenchyma (Mitani-Ueno and Ma, 2021). Si is also moved both within and outside the endodermis by *Lsi1* and *Lsi2*, ultimately reaching the xylem. *Lsi1* allows the passive transport of Si, while *Lsi2* is an energy-dependent efflux transporter (Sakurai et al., 2015; Sharma et al., 2023). This polar localization of *Lsi1* and *Lsi2* in the same cell, along with the availability of Casparian strips, is critical for efficient Si uptake (Sakurai et al., 2015).

In barley and maize, Si uptake occurs through *Hordeum vulgare* silicon transporter 1 (*HvLsi1*) and *Zea mays* silicon transporter 1 (*ZmLsi*1), which are localized in epidermal, hypodermal, and cortical cells. Since these plants lack developed aerenchyma, Si is transported to the endodermis via the symplastic pathway, and *Hordeum vulgare* silicon transporter 2 (*HvLsi2*) or *Zea mays* silicon transporter 2 (*ZmLsi2*) releases Si to the stele. While this system also leads to Si accumulation in the shoots, it is less efficient than in rice due to differences in the levels of expression and polar localization of *Lsi1* and *Lsi2*, the presence of Casparian strips, as well as root arrangement. The expression levels of *Oryza sativa* silicon transporter 1 (*OsLsi1*) and *Oryza sativa* silicon transporter 2 (*OsLsi2*) in rice are likely higher than those of *HvLsi1/ZmLsi1* and *HvLsi2/ZmLsi2* in barley and maize, contributing to the superior Si accumulation in rice shoots. Additionally, *HvLsi2* and *ZmLsi2* do not show polar localization in the same cells, resulting in lower Si uptake compared to rice (Mitani-Ueno and Ma, 2021).

For wheat, an Si transporter called *Triticum aestivum* silicon transporter 1 (*TaLsi1*) was identified, but information on its expression unique to cell types is absent. An *Lsi2*-like transporter is absent from wheat, but its Si uptake system is expected to be like that of barley and maize based on Si accumulation patterns (Montpetit et al., 2012).

6.2.1.1.2 Passive Transport System

Plants like cucumber and pumpkin represent passive uptake systems (intermediate Si accumulation) that differ from active uptake systems. In these plants, *Cucurbita moschata* influx silicon transporter 1 (*CmLsi1*) and *Cucumis sativus* silicon transporter 1 (*CsLsi1*) are expressed in various root cells and *Cucumis sativus* silicon transporter 2 (*CsLsi2*) in the endodermal cells (Sun et al., 2018), but unlike in the high Si-accumulating species, most of them don't exhibit polar location in the cells of the cortex, except for *CsLsi1* which is polarly localized at the endodermis (Sun et al., 2017). Additionally, the Lsi1 and Lsi2 in these intermediate Si-accumulating plants are not localized in the same cell, unlike high Si-accumulating plants where polar localization of both transporters is critical (Sun et al., 2018). Since *Lsi1* is a dual-directional transporter, the lack of polar localization and the presence of *Lsi1* and *Lsi2* in these species together produce a less efficient Si uptake system. This explains the lower Si accumulation in these plants compared to Si-accumulating species.

6.2.1.1.3 Rejective Transport System

The Si uptake system used by non-Si accumulators like tomatoes is different from high Si-accumulating plants. Previous research suggested that *Lsi1* in tomatoes was inoperative due to greater distances between asparagine-proline-alanine (NPA) domains (Deshmukh et al., 2015). But the newest investigation has identified a functional *Lsi1* in tomato, SlLsi1, which is present in every root cell but lacks polar localization (Mitani-Ueno et al., 2021). The little buildup of Si in tomato appears to be due to the dearth of an *Lsi2*-type transporter. This was confirmed when a functional *Lsi2* from cucumber was introduced into tomato, resulting in elevated Si accumulation in tomato plants that are transgenic. The presence of *Lsi2* is crucial for actively pumping out Si toward the xylem (Mitani-Ueno and Ma, 2021) (Figure 6.2).

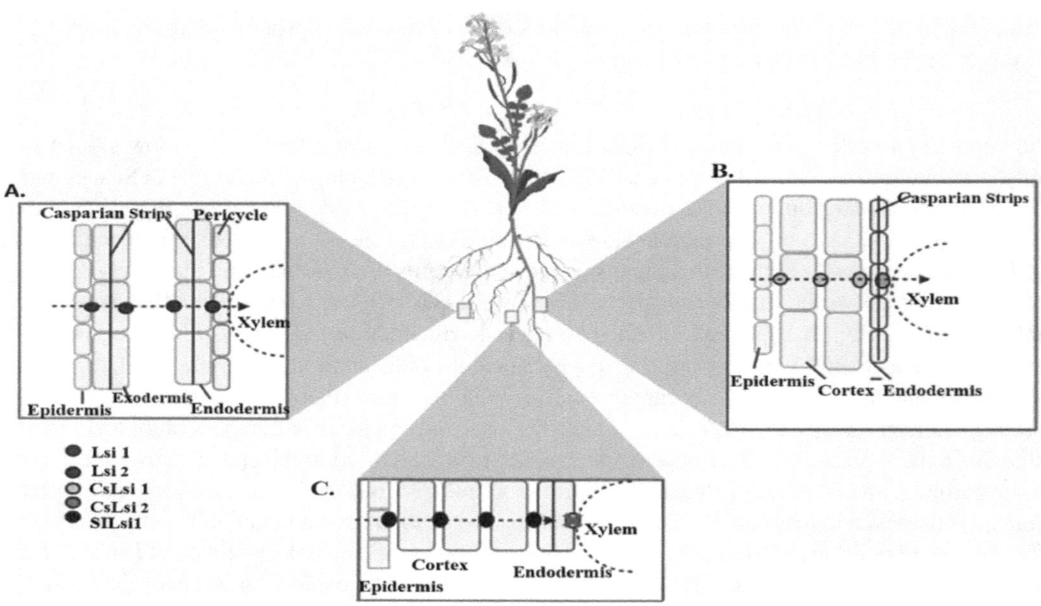

FIGURE 6.2 Different Si uptake systems (a) active transport system (b) passive transport system (c) rejective transport system.

6.2.2 Si Accumulation in Plants

Si accumulation and deposition in plants have been the subject of thorough research which is referred to as silicification. Silicification occurs in various locations within plants, such as the cell wall, cell lumens, spaces between the cells of roots and shoots, and specialized silica cells. The deposition pattern, quantity, and function of Si exhibit significant variations across different tissue types (Mandlik et al., 2020). Si has been detected in the roots of multiple Poaceae members where the endodermis is the most frequently found location of silicification. It is discovered that in rice, it is confined to the proximal and distal regions of the tissue as a ring in the endodermal cell walls (Wang et al., 2019 b). The inner tangential wall and occasionally the radial walls of the endodermis have the thickest deposits (Mandlik et al., 2020). For instance, Si deposition in date palm (*Phoenix dactylifera*) is observed in specialized cells called stegmata cells, positioned in rows to surround the root's sclerenchyma bundles (Bokor et al., 2019). In shoot tissues, the leaves serve as the primary location for Si deposition. In the case of rice leaf blades, a 2.5-μm thick double layer of Si is deposited beneath the cuticle (Ma and Yamaji, 2006). Silicification takes place in nearly all cells within the leaf blade, with over 90% of the leaf's total Si content getting stored in the epidermis as a consequence of the transpiration stream. Si is only present in specific silica cells and bulliform, or 'motor,' cells in immature leaves. Nevertheless, in aging leaves, Si is distributed across practically all cell types. Notably, older leaves exhibit more extensive Si deposition, with a gradual decrease from the leaf apex to its base, indicating that Si deposition intensifies as the plant matures. Si deposition is also evident in the cells all around the trichome hair base as well as inside the trichomes themselves (Mandlik et al., 2020). In a study, Si deposition was examined in the trichomes of six distinct Cucurbitaceae species, including *Cucumus sativus*, *Cucurbita pepo*, *Cucumis melo*, Citrullus lanatus, etc. (Abe, 2019). The research revealed that in watermelon leaves, Si is found in the trichomes as well as the cells that encircle their bases. In contrast, in cucumber, pumpkin, and melon, Si is exclusively placed in the cells at the trichome bases.

6.3 SOURCES OF SI IN GREEN AGRICULTURE

Si is available to the soil as amendments, conditioners, or as fertilizers. Different forms of calcium silicate from natural sources like wollastonite and slag, a byproduct of the steel industry, are examples of Si. In the initial years of the 20th century, there were 550,000–600,000 metric tons of wollastonite ore produced worldwide (Moore, 2003). Si-based chemicals are widely used in agriculture to produce crops, i.e., wollastonite; $CaSiO_3$ has been used as the vast source in many places with low concentrations of soluble Si in soils. Si fluid is applied to crops as an agrochemical dispersing agent (Vasanthi et al., 2012). Si is deemed safe for utilization and possesses favorable wetting qualities. Silicate materials are practical bioproduct transporters. To address the rising issue of remnants of pesticides on food items and environment pollution, organic agricultural practices use natural insecticides composed of Si such as talc and phytosil (Vasanthi et al., 2012). It has been established that each of these products works well to increase crop output and growth. Plants naturally absorb Si through the weathering and erosion of soil minerals like silicates and clay minerals, which occur over time as rocks break down. There are several Si sources in the market, which function as soil additives or fertilizers, such as diatomite, calcium silicate, sodium metasilicate, potassium silicate, magnesium silicate, orthosilicic acid, hydrated silicon dioxide, and calcium metasilicate (Ferreira et al., 2017). Sodium silicate and silica gel have also been utilized in scientific studies and for growing high-value crops. Although sodium silicates are entirely soluble in water and may be used topically, but they are often too costly for in-soil treatment. Applications involving horticulture or greenhouse crops are more frequently utilized for potassium and sodium silicates. These compounds are soluble and can be sprayed directly on leaves or added to nutrient solutions. Soil contains more Si than foliar treatments do for plants. This is due to the fact that Si cannot effectively control disease if it is not continuously supplied to plant roots. Thermophosphate is a commercial fertilizer used in Brazil and other countries to supply P, Ca, and Mg silicate ingredients (Table 6.1).

TABLE 6.1
Chemical and Natural Sources of Si in Green Agriculture

Source of Silicon	Composition	References
	Chemical Sources of Si	
Wollastonite	Calcium silicate	Panasenko et al. (2023)
Fly ash	SiO_2	Mustakim et al. (2021)
Ground granulated blast furnace slag		Mustakim et al. (2021)
Potassium silicate	K_2SiO_3	Ali et al. (2021)
Silicic acid	$Si(OH)_4$	Zhang et al. (2020)
Blast furnace slag	Aluminosilicate	Jiang et al. (2019)
Fly ash	Aluminum silicate, calcium/magnesium silicate	Aughenbaugh et al. (2016)
Iron/steel slag	Calcium silicate	Haynes et al. (2013)
Wollastonite	Calcium silicate	Sebastian et al. (2013)
$MgSiO_3$	Magnesium silicate	Sebastian et al. (2013)
$CaSiO_3$/$MgSiO_3$ blend	Calcium/magnesium silicate	Sebastian et al. (2013)
$NaSiO_3$ (aqueous)	Sodium silicate	Abed-Ashtiani et al. (2012)
	Natural Sources of Si	
Swine manure compost	Silica	Zhang et al. (2022)
Sugarcane bagasse ash	Silica	Chindaprasirt and Rattanasak (2020)
Palm oil ash	Silica	Pa et al. (2016)
Rice hull ash	Silica	Mejía et al. (2016)
Miscanthus biochar	Silica	Houben et al. (2014)
Cattle manure compost	Silica	Komiyama et al. (2013)
Rice hull fresh	Silica	Júnior et al. (2010)
Compost from poultry manure	Silica	Kobayashi et al. (2008)

6.4 ROLE OF SI IN MITIGATING BIOTIC AND BIOTIC STRESSES IN PLANT

Si serves as an essential metalloid for plants, enhancing their resilience to abiotic stresses like water stress; salinity stress; and biotic stress like diseases, etc. It improves nutrient uptake, boosts natural defense mechanisms, and fosters soil health, reducing the need for chemical inputs and promoting sustainable agricultural practices (Ranjan et al., 2021; Song et al., 2021).

6.4.1 Significance of Si in Abiotic Stress Tolerance in Plants

Abiotic stresses like heavy metal stress, salinity, heat, drought stress, etc., can negatively impact a plant's overall well-being, resulting in decreased agricultural output and productivity. Being a beneficial element, Si can help mitigate these stresses through various mechanisms, as illustrated in Table 6.2.

6.4.1.1 Heavy Metal Stress

Si assumes a significant function in mitigating stress caused by heavy metals in plants, providing an intricate mechanism to augment the adaptive resilience of plants (Vasilachi et al., 2023). Si's crucial role includes forming complexes with harmful heavy metals, a pivotal function for environmental remediation and industrial processes. This symbiotic affiliation curtails the inherent

TABLE 6.2
Si's Role in the Mitigation of Abiotic Stress in Different Species of Plants

Abiotic Stress	Plant	Observations	References
Chromium	*Oryza sativa*	Ameliorated rhizosphere acidification and yielded favorable outcomes by alleviating the burden of heavy metal stress.	Pang et al. (2023)
Zinc	*Oryza sativa*	Induced and controlled genes linked to photosynthesis in response to elevated zinc stress, hence boosting photosynthesis.	Arif et al. (2021)
Nickel	*Zea mays*	Ameliorated the stress induced by nickel (Ni) by maintaining membrane stability and having an effect on defense systems that are non-enzymatic (Pro and AsA) as well as enzymatic (SOD, POX, and CAT).	Fiala et al. (2021)
Copper	*Salvia officinalis*	Ameliorated oxidative damage by elevating proline content, bolstered the functioning of antioxidant enzymes like CAT and SOD; SOD gene expression was upregulated.	Pirooz et al. (2021)
Cd	*Brassica napus*	Treatment leads to decreased H_2O_2 and malondialdehyde contents, while enhancing antioxidant defense systems via the augmentation of AsA and GSH levels, as well as the activities of the AsA-GSH cycle, glyoxalase system enzymes, and CAT.	Hasanuzzaman et al. (2017)
Cd	*Poa annua*	Ameliorated the toxic effects of Cd reinstated the activity and G6PDH expression and decreased the oxidative damage brought on by exposure to Cd.	Li et al. (2017)
Salt	*Medicago sativa*	Reduced oxidative damage, regulated SOD and polyphenol oxidase activities, and increased the levels of flavonoids, total polyphenols, and carotenoids.	El Moukhtari et al. (2021)
Salt	*Solanum tuberosum*	Increased the amount of proline and soluble carbohydrates in leaves as compared to the control, while Si treatment also decreased the ratio of Na^+ to K^+ and enhanced antioxidant enzyme activity.	Kafi et al. (2021)
Salt	*Oryza sativa*	Elevated levels of endogenous polyamines through upregulating synthesis enzymes and decreased levels of gamma-aminobutyric acid (GABA) accumulation through downregulating catabolism, thereby maintaining the GABA shunt and lessening oxidative damage.	Das et al. (2021)
Salt	*Cucumis sativus*	Enhanced various growth-related parameters and mitigated the detrimental effects of salinity, concurrently decreasing shoot chloride (Cl^-) concentrations.	Kaloterakis et al. (2021)
Drought	*Zea mays*	Reduced EL from 0.64% to 0.52% increasing stability of membrane structure by 12% and boosted Chl 'a' by 35%, Chl 'b' by 31%, and carotenoids contents by 51%.	Bijanzadeh et al. (2022)
Drought	*Cucumis melo*	Fortified the inherent resistance to lodging, reinforcing cell walls, curtailing fungal disease infestations by insects, reducing reciprocal shading, improving water equilibrium, and reducing loss of water and transpiration.	Alam et al. (2021)
Drought	*Triticum aestivum*	Mitigated oxidative stress and also alleviated the harmful impact of drought by elevating the antioxidant enzymes' activity.	Naz et al. (2021)

(*Continued*)

TABLE 6.2 (*Continued*)
Si's Role in the Mitigation of Abiotic Stress in Different Species of Plants

Abiotic Stress	Plant	Observations	References
Drought	*Arachis hypogaea*	Enhanced the nutrient absorption and elevated the RWC, augmentation of leaf chlorophyll content, and a notable increase in biomass.	Patel et al. (2021)
Heat	*Phoenix dactylifera*	Served as a facilitator for plant growth and metabolic regulation which imparted resilience against environmental stressors.	Khan et al. (2020)
Heat	*Oryza sativa*	Enhanced the composition and arrangement of cellular membrane components.	Ashraf (2010)
Heat	*Oryza sativa*	Diminished the transpirational loss and mitigated the occurrence of sunburn.	Ashraf (2010)

mobility of these toxic entities, thereby diminishing their absorption by plant roots. The consequential impediment to their ingress impedes the accrual of hazardous elements in plant tissues, erecting a formidable defense against the adversities synonymous with heavy metal toxicity (Mostofa et al., 2021). Furthermore, Si undertakes a sophisticated duty in fortifying the structural and physiological robustness of plant cells through the strategic reinforcement of cell walls via the deposition of silica. This fortified architecture serves as a tangible bulwark, circumscribing the infiltration of heavy metals into the cellular matrix of plants, thereby ameliorating the repercussions of metal-induced oxidative stress (Akter et al., 2022).

The ubiquity of prevalent heavy metals namely aluminum (Al), cadmium (Cd), arsenic (As), lead, mercury, copper, and zinc within domesticated agricultural crops, predominantly attributed to watering techniques in agriculture, exerts a profound influence on soil dynamics. This pervasive presence not only jeopardizes the intricacies of soil composition but also poses substantial hazards to crop production and human well-being (Haseeb et al., 2022; Raza et al., 2022). These heavy metals have several deleterious impacts on crops, including the disruption of vital metabolic pathways, including lipid peroxidation, photosynthesis, and the ETC, by binding to sulfhydryl groups of proteins (Raza, 2021). The repercussions are evident in the form of compromised crop tissues, characterized by necrosis, leaf chlorosis, impaired chlorophyll synthesis, and membrane lipid breakage (Vasilachi et al., 2023). This intricate interplay underscores the intricate challenges posed by heavy metal contamination in agricultural ecosystems, necessitating a comprehensive understanding of effective mitigation strategies. In the rhizosphere, Si promotes the exudation of chelated metal ions, reducing metal ion uptake, and creates a physical barrier in the endodermis, lowering the amount of metal ions that are absorbed through the xylem to lessen the toxicity of metal ions in wheat seedlings (Khan et al., 2021b). Plants designated as Si accumulators including *Oryza sativa, Hordeum vulgare, Vicia faba, Cucurbita pepo, Cucumis sativus,* and *Vigna unguiculata* demonstrate an augmented resilience to manganese (Mn) toxicity, primarily achieved through the reduction of Mn within cell walls (Abou et al., 2022). The adaptive capacity of Si is further exemplified in its efficacy in ameliorating Cd toxicity across diverse crops. Si supplementation not only diminishes Cd uptake from shoots to grains but also orchestrates the movement of Cd-associated metal-binding proteins, thereby mitigating its toxic effects (Tang et al., 2023). Beyond its prowess in combating Mn and Cd stresses, Si emerges as a versatile guardian against heavy metal adversities. Si supplementation enhances plant tolerance to other noxious heavy metals such as Al and chromium (Cr). Concurrently, Si imparts positive modulation to chlorophyll biosynthesis and the photosynthetic machinery under conditions of heavy-metal-induced stress, as elucidated by studies (Singhal et al., 2023).

This multifaceted role of Si in mitigating heavy metal stresses not only accentuates its adaptability across diverse metal types but also underscores its potential as an invaluable tool in fostering sustainable agricultural practices. Si has proven to be a formidable mitigator against the stress induced by well-known heavy metals on plants. For instance, Cd represents a significant environmental hazard as a mobile and non-biodegradable heavy metal pollutant with global persistence. It disrupts metabolic and cellular functions in plants, inducing oxidative stress and mineral nutrient imbalance. Si emerges as a mitigating factor, enhancing plant height; biomass; photosynthetic apparatus; ion homeostasis; and antioxidant defense systems while reducing Cd content through various mechanisms, including blocking the apoplastic pathway and improving biological accumulation and translocation factors (Khan et al., 2019). Si also enhances heat shock proteins, reducing degraded protein content and ameliorating the toxic effects of Cd stress (Rehman et al., 2019).

Similarly, Cr, a common pollutant from industrial activities, also adversely affects plant physiobiochemical attributes. Si application diminishes Cr toxicity by inducing antioxidant enzymatic machinery, improving photosynthesis attributes and gas exchange parameters, and modulating membrane structures. Si serves as a Cr scavenger that enhances plant growth and nutrition during Cr stress (Singhal et al., 2023). Additionally, supplementing with Si improves the photosynthetic rate by increasing the intake and transfer of vital minerals, gaseous exchange, and chloroplast ultrastructure, taking on a vital part in mitigating the harmful effects of Cr (Ashraf et al., 2016). Stress due to heavy metal Pb also negatively impacts plant cells and their genetic material, limiting germination of seeds, development, plant matter, and photosynthetic characteristics. By improving photosynthetic and morphological characteristics, Si-NPs reduce Pb stress. Si promotes antioxidant activity, reduces oxidative stress, and decreases electrolyte leakage (EL) in coriander, ultimately improving growth and developmental processes (Fatemi et al., 2021). In *Bambusa vulgaris*, SiO_2 NPs alleviate Pb stress by boosting the activity of enzymes such as phenylalanine ammonia lyase (PAL), catalase (CAT), superoxide dismutase (SOD), and glutathione reductase (GR) and thereby enhancing bamboo growth, total biomass, pigment content, and photosynthetic apparatus (Emamverdian et al., 2020).

Besides all this, Si is essential in reducing the toxicity of metalloid As by enhancing germination, stature, biomass, photosynthesis, metabolism of carbohydrates, and nutritional efficacy. Si enhances antioxidant and thiolic systems, increasing plant yield and aiding in remediation against As stress. It promoted the synthesis and activities of antioxidants and antioxidative enzymes like SOD, CAT, and glutathione (GSH), which help to neutralize reactive oxygen species (ROS) generated under heavy metal stress (Ahmad et al., 2021). In wheat, Si improves morpho-physiological characteristics, total protein content, mineral uptake, and tensile strength of the membrane to detoxify As. Si decreases As translocation and accumulation while increasing the expression of genes linked to phytochelatin synthase and metallothionein synthase to improve As tension (Liu et al., 2023). Moreover, alterations in plant metabolism induced by Si contribute significantly to the amelioration of heavy metal stresses. Si has also demonstrated the capacity to promote the activity of pivotal antioxidant enzymes, such as CAT and SOD, crucial in detoxifying ROS generated amid conditions of metal stress (Mostofa et al., 2021). The heightened expression of these antioxidant defenses serves as a pivotal mechanism for plants to counteract oxidative damage induced by heavy metals, thereby fostering comprehensive stress tolerance (Wei et al., 2020). Si's role in alleviating heavy metal stress in plants is multifaceted, involving chemical, physical, and physiological mechanisms. Beyond mere physical reinforcement, it fosters biochemical adaptations, enhancing plant resilience in contaminated environments (Asgher et al., 2023). Its intricate interplay aids plants in thriving amid toxic metal elements, offering comprehensive support.

6.4.1.2 Drought Stress

A serious challenge to global food production is posed by drought stress, impacting morphological traits, yield, and various physiobiochemical processes such as photosynthesis, EL, and water use efficiency. Plants employ avoidance or tolerance strategies to cope with drought conditions, involving morpho-physiological and osmotic modulation (Khan et al., 2019). Si application emerges as an effective solution to mitigate water scarcity in crops like strawberries. Si application promotes stomatal conductance and proline concentration by improving plant water status, increases chlorophyll pigment fluorescence which enhances the plant's photosynthetic rate, and other quantitative and qualitative characteristics of the plant amid various stress situations (Kumar et al., 2023, 2024).

Additionally, by decreasing transpiration and electrolyte loss, it increases the efficiency of water usage (Xu et al., 2022). In *Glycine max*, drought induced by polyethylene glycol (PEG) causes increased oxidative stress and a reduction in plant turgor and osmotic potential. Si fortification is quite effective at reducing these negative effects by enhancing the growth, turgor, photosynthetic attributes, and antioxidant enzymes of the plant. Moreover, Si application upregulates endogenous plant hormones like gibberellic acid (GA), salicylic acid (SA), and jasmonic acid (JA) (Kosakivska et al., 2022). The use of Si-NPs further improves drought resistance in plants by improving growth characteristics, yield, photosynthetic process, stomatal conductance, water potential of leaves and xylem, and RWC of leaves. Si also enhances the integrity and stability of cell membranes while reducing the effects of water-limited conditions on lipid peroxidation, damage from oxidative stress, EL, and membrane selectivity alterations. Additionally, Si supplementation enhances the amount of Si, Na, and phosphorus (P), improving physical characteristics and effectively mitigating drought stress (Etesam et al., 2022a).

Drought has a deleterious impact on rice's fluorescence characteristics, photosynthetic properties, and chlorophyll concentration. Si application reverses these effects by increasing the expressions of proteins associated with chlorophyll a and b, ETC, photosystem II (PSII) protein, and ATP synthase subunit. Si also modulates thylakoid structure, boosting growth traits and ultimately improving morphological and photosynthetic traits, thereby attenuating the impact of drought conditions (Zahedi et al., 2023). By mitigating oxidative damage, Si contributes to the overall stress tolerance of plants. Understanding and harnessing these mechanisms hold great potential for sustainable agriculture and environmental remediation strategies.

6.4.1.3 Thermal Stress

Si plays a crucial role in mitigating thermal stress in plants through several mechanisms *viz.* enhanced heat tolerance; cellular stability; regulation of stomatal function and activities of antioxidative enzymes; improved nutrient uptake; etc.

Si is known to enhance the heat tolerance of plants by acting as a physical barrier, reducing water loss through transpiration. It forms a layer on the leaf surface, creating a shield against excessive solar radiation and preventing the plant from overheating (Mitani and Ma, 2005). Si deposition in plant cell walls also contributes to increased structural integrity. This reinforcement helps cells withstand thermal stress by maintaining cell turgor pressure and preventing cell collapse during extreme temperatures (Emam et al., 2014). Furthermore, Si influences stomatal conductance, reducing water loss through transpiration. By regulating stomatal openings, plants can conserve water during hot conditions, improving their ability to cope with thermal stress (Debona et al., 2017). Additionally, Si has shown its ability to increase antioxidant activity in plants. This is particularly important during periods of dry spells when the generation of ROS tends to rise. Si helps in the activation of antioxidant enzymes, scavenging ROS and minimizing oxidative damage (He et al., 2023). Also, Si improves the availability of essential nutrients to plants particularly those involved in stress response. This improved nutrient status contributes to the overall resilience of plants facing thermal stress (Debona et al., 2017).

6.4.2 Role of Si in Biotic Stress Management

Apart from abiotic stress management, Si also plays an important role in the management of biotic stress. It also promotes plant resistance against plant pathogens such as microbes like bacteria, fungi, viruses, nematodes and insect pests, etc. Si has a profound effect on plant–pathogen interactions by various types of mechanisms including physical, physiological, and biochemical.

However, a multitude of studies indicate Si deposition in the form of SiO_2 plays a crucial role in the physical defense mechanism of plants by the formation of phytoliths which promote Si accumulation in plant parts (Katz, 2015). Furthermore, phytolith deposition enhanced plant resistance against fungal penetration. Insect feeding mouth parts or mandibles may also be worn away because of Si deposition (Jeer et al., 2017). Furthermore, a thick layer of Si deposited mediated long and small epidermal cells beneath the cell wall, and a thickened Si–cellulose membrane formed by a cuticle–Si double layer, all contribute to Si-enhanced cell wall rigidity and reinforcement.

The physical barrier prevents pathogen penetration by making plants less susceptible to the denaturation of enzymes caused by pathogen invasion.

Many researchers also studied that Si application upregulates plant resistance mechanism by reducing fungi colonization and delaying pathogen ingression by inhibiting fungal entry into the epidermal cells and by papilla formation that keeps pathogens out of plants. Lignin, cellulose, pectin, suberin, chitin, lipids, Si, and proteins such as hydroxyproline-rich glycoproteins (HRGPs) and peroxidases are all components of papilla formation (Deb and Dutta, 2022). Si application in wheat improved physiological mechanisms by limiting the entry of fungal pathogens into first epidermal cells of wheat plants (Ahammed et al., 2021). Plant cells respond to pathogen attack by undergoing a variety of morphological changes as well as rapid and extensive metabolic changes. Silica causes changes in trichome morphology that contribute toward pathogen resilience (Hall et al., 2020).

Si treatment increases biochemical resistance in plants by altering the gene expression of pest-penetration-resistance enzymes involving peroxidase, chitinase, and PAL which increase the production of anti-microbial compounds like total phenolics and polyphenols and defense metabolites such as phytoalexins, flavonoids in plants and further alleviate systemic signal including biosynthesis of SA, JA, and ethylene (Wang et al., 2017; Deb and Dutta, 2022). Also, jasmonate signaling activation increases Si accumulation in leaves of *Oryza sativa*, demonstrating a clear association between Si and jasmonate in rice against insect herbivores (Reynolds et al., 2016). Si increases the activity of chitinase in bitter gourd leaves after the infection of the pathogen and upregulates antioxidant enzyme activities (Ratnayake et al., 2016).

The utilization of Si on plants afflicted with pathogens has been shown to effectively bolster immunity, thereby aiding in the management of insect herbivory and other arthropod-related challenges within agricultural systems (Wang et al., 2017). ROS levels typically increase with pathogen infection, and Si-mediated processes play a pivotal role in activating the antioxidant defense mechanism by stimulating enzymes such as SOD, CAT, GPX, and APOX (Deb and Dutta, 2022). Additionally, Si contributes to the activation of chitinase enzymes and facilitates cell wall modification, thereby enhancing cell wall rigidity. Moreover, the presence of silicified long and small epidermal cells, along with a dense layer of cells, is associated with the augmented stiffness and reinforcement of cell walls due to Si supplementation. Furthermore, Si deposition in the cuticle acts as a physical barrier, potentially limiting pathogen invasion (Islam et al., 2020) (Figure 6.3).

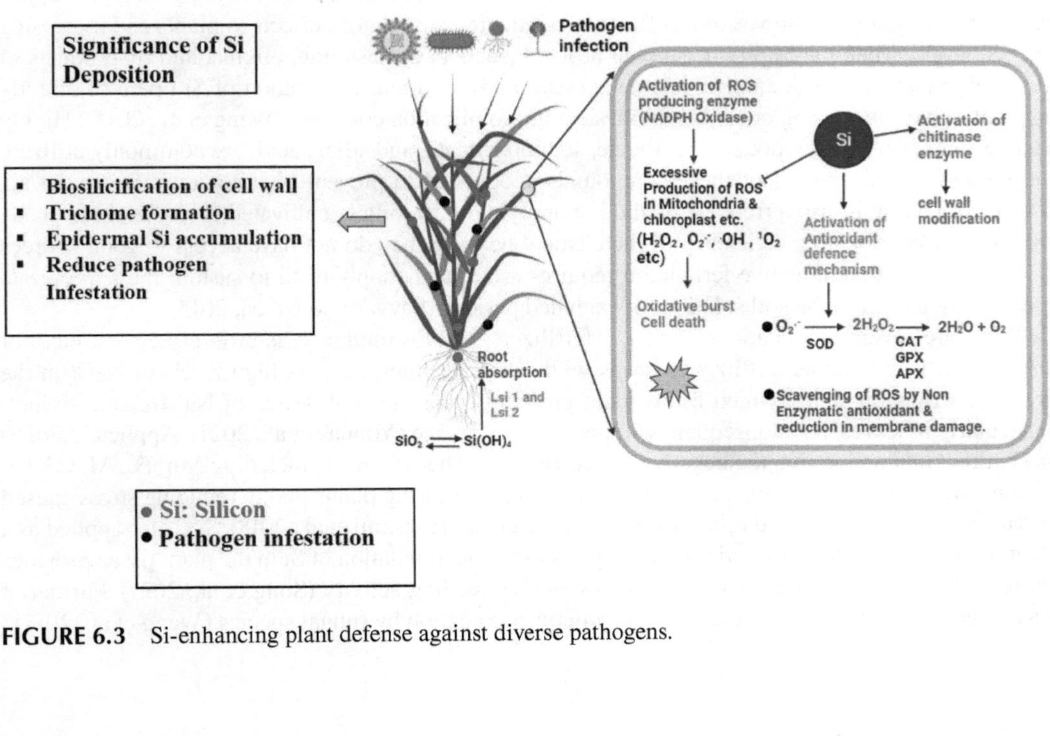

FIGURE 6.3 Si-enhancing plant defense against diverse pathogens.

Si plays a crucial role in the world's food safety and security by fostering the creation of cutting-edge, high-yielding cultivars. In common beans, *Phaseolus vulgaris* Si promotes SOD, ascorbate peroxidase (APX), and GR enzymes activities and reduces H_2O_2 content against penetration of *Colletotrichum lindemuthianum* which leads to anthracnose disease (Polanco et al., 2014). Similarly, Si decreases the primary infection in the root and stem and reduces disease incidence in oil palm (*Elaeis guinnensis)* against basal stem rot disease (Najihah et al., 2015). Si-induced thickening of the leaf cuticle in stem borer-infested sugarcane plants thereby facilitating resistance in sugarcane against the pathogen (Vilela et al., 2014). Table 6.3 summarize various studies related to the role of Si in plant pathogen management.

6.5 USE OF SI IN AN INTEGRATED MANNER FOR SUSTAINABLE AGRICULTURE

Si being an essential plant nutrient and the substantial body of research demonstrating its potential to enhance crop productivity, silica has been widely applied as fertilizer in agricultural production systems across the globe (Tayade et al., 2022). Many workers have advocated the potential of Si to strengthen the cell walls of plants (Sadeghi et al., 2016) and enhance resistance in the field due to diseases and pests (Han et al., 2018) and in storage (Ahsan et al., 2023). The mechanism underlying the physiological resistance in plants provided by Si is considered to be its catalytic role in generating phenolic and tannic compounds, among other substances (Swain and Rout, 2017). Additionally, rice hulls and diatomaceous earth are natural sources of Si, which are used both in field and greenhouse production as a drench or foliage spray (Mejía et al., 2016). Besides, it leaves no traces of pesticides in food or the surroundings and is readily used with additional pest control methods, such as biological control; therefore, Si treatment in plants offers a feasible option for integrated management of insect pests and diseases (Nagaratna et al., 2023).

Due to the potential of Si to alleviate biotic and abiotic stresses such as osmotic and nutritional imbalance and its potential to accumulate at either macro- or micro-concentrations in all plants, Si acts as a biostimulant and has been widely used in rice and sugarcane crops (Bocharnikova et al., 2023). Further, the application of Si as a biostimulant to various plants such as cucumber, zucchini squash, bean, tomato, roses, etc., has also been shown to enhance growth and productivity (Toresano-Sánchez et al., 2012). The various methods of Si application as a biostimulant in horticultural crops include foliar spraying through incorporation into the soil (Savvas and Ntatsi, 2015). Among these, the application of Si through soil proves more effective in inducing stimulatory effects in plants and increasing their Si levels, when the Si products being utilized (such as wollastonite, olivine, and by-products of the steel industry) release enough soluble Si. Alternatively, foliar application of Si operates directly on plant tissues, bypassing chemical or physical immobilization concerns (Wang et al., 2015). Highly soluble sources of Si, like potassium silicate, sodium silicate, and silicic acid, are commonly utilized. Before application, the pH of these compounds is adjusted to prevent phytotoxicity (Savvas et al., 2009). Fertigation is also a frequent method of applying Si to soilless cultivated crops, as these plants are more susceptible to a deficiency of the element because they do not have access to good sources of soil. Furthermore, effective fertigation requires a frequent supply of Si to sustain the appropriate concentration of Si in the soil solution for extended periods (Savvas and Ntatsi, 2015).

There are several applications of Si *viz.* fertilizers, as biostimulants, as crop protectors, etc. For instance, Si is applied as fertilizer in crops with salt difficulties, such as high levels of NaCl in the available irrigation water, which limits plant growth because of high levels of Na^+ in plant tissues, particularly in leaves, and consequently impacts gas exchange (Younas et al., 2021). Application of Si has befitted in ameliorating the negative effects of several heavy metals including Mn, Fe, Al, Cd, Cr, Ni, Zn, and many more in vegetables, fruits, as well as flowering plants and in reducing stress caused by salinity, heat, drought, and mineral nutrient deficiencies (Etesami et al., 2018). Si is also applied as a plant protector against various diseases and pests as the accumulation of Si in the plant tissue provides hardness that prevents insect larval boring as well as feeding activity (Song et al., 2021). Further, it makes the plant cells less vulnerable to enzymatic degradation by fungal species (Wang et al., 2017).

TABLE 6.3
Positive Function of Si in Biotic Stress Management against Different Pathogen Species

Biotic Stress/Disease	Pathogen Species	Plant Species	Observations	References
Powdery mildew	*Pythium* spp.	*Vigna mungo*	Si elevated the expression of some disease-resistant genes.	Parthasarathy and Jaiganesh (2016)
Powdery mildew	*Eryisphe* sp.	*Momordica charantia* L.	Si increased the activity of chitinase in bitter gourd leaves after the infection of the pathogen and upregulation of antioxidant enzymes.	Ratnayake et al. (2016)
Anthracnose	*Colletotrichum dematium*	*Solanum lycopersicum*	Si served to enhance cuticle thickness and fruit firmness.	Somapala et al. (2016)
Banana Xanthomonas wilt (BXW)	*Xanthomonas campestris* pv. Musacearum	*Musa* spp.	Si-induced POD activity, Si accumulation in leaves which inhibited bacterial infection.	Mburu et al. (2016)
Basal stem rot	*Ganoderma boninense*	*Elaeis guinnensis jacq.*	Si decreased the primary infection in the root and stem and reduced the disease incidence.	Najihah et al. (2015)
Anthracnose	*Colletotrichum lindemuthianum*	*Phaseolus vulgaris*	Si promoted SOD, APX, GR activities and reduced H_2O_2 content.	Polanco et al. (2014)
Sugarcane borer	*Diatraea saccharalis* (Lepidoptera)	*Saccharum officinarum*	Si-induced leaf cuticle thickening.	Vilela et al. (2014)
Bacterial wilt and blight	*Ralstonia solanacearum*	*Solanum lycopersicum*	Si induced the production of antioxidant enzymes.	Kurabachew et al. (2013)
Wheat blast	*Pyricularia grisea*	*Triticum aestivum L.*	Si helped in upregulation of antioxidant enzymes and chitinase.	Filha et al. (2011)
Bacterial wilt	*Ralstonia solanacearum*	*Solanum lycopersicum*	Si increased the response factor of JA and ethylene.	Ghareeb et al. (2011)
Rice blast	*Magnaporthe oryzae*	*Oryza sativa*	At the transcriptional level, Si influences the rice response to rice blast infection.	Brunings et al. (2009)
Post harvest decay	*Fusarium semitectum*	*Cucumis melo*	Si inhibited the mycelial growth of the pathogen.	Bi et al. (2006)
Fruit decay (blue mold and brown rot)	*Penicillium expansum* and *Monilinia fructicola*	*Prunus avivum*	Si reduced the decay incidence and lesion diameter.	Qin et al. (2005)
Powdery mildew	*Blumeria graminis*	*Hordeum vulgare*	Si increased the osmotic stress-pathogen resistance in plants.	Wiese et al. (2005)
Brown spot disease	*Mycosphaerella pinodes*	*Pisum sativum*	Si stimulated the activities of beta-1,3-glucanase and chitinase in leaves.	Dann et al. (2002)

In recent years, compared to bulk Si, nano-silicon has proved more efficient in reducing environmental stress because of their tiny particle size, large surface-area-to-volume ratio, higher absorption rates, solubility, reactivity, and adherence which allow them to permeate easily and communicate effectively within plant tissues (Wang et al., 2015; Tripathi et al., 2017). There are various chemical-based methods of synthesis of Si-NPs, but research is now inclined toward the green synthesis of Si nanoparticles using plant products such as rice husk, sugarcane, and bamboo as these methods are cost-effective and eco-friendly (Babu et al., 2018). Many works have also demonstrated the function of nano-Si in controlling the detrimental effects of abiotic and biotic stresses in plants which may be attributed to the unique properties of Si-NPs and their interaction with the plants thus affecting their morphology and physiology (Mathur et al., 2020). Furthermore, Si-NPs are also used as nanofertilizers; nanopesticides; nanoherbicides; and as delivery systems for the slow and continuous release of nucleotides, proteins, and other chemicals at the targeted sites in plants (Wanyika et al., 2012; Rastogi et al., 2019). Therefore, the application of nano-silicon possesses the potential to be used as sole fertilizer, as a facilitator for herbicides, as a plant protector against various diseases, and may be integrated into disease and pest management programs (Etesami et al., 2022b). Although nanotechnology is an emerging field in Si research, enough literature isn't available that compares the uptake and transport of bulk silicate fertilizer in plants with that of Si-NP. However, Si-NPs' absorption rate hinges on factors like Si bioavailability, size, concentration, plant morphology, environmental conditions, and application methods, warranting scrutiny for ecological implications and safety concerns (Bhat et al., 2021; Etesami et al., 2022c).

Numerous investigations have documented the synergistic effect of the application of Si and microorganisms in promoting plant development and crop output (Verma et al., 2020; Etesami et al., 2022b). For example, strawberry plant under drought stress when treated with Si and inoculated with arbuscular mycorrhizal fungus (AMF) showed significant enhancement in plant biomass production, rate of photosynthesis, water content, antioxidative enzyme activities, and the nutritional status of particularly Zn (Moradtalab et al., 2019). The study showed that AMF and Si work together synergistically to improve strawberry development under both stress and control circumstances, which declined the stress metabolite (malondialdehyde (MDA) and H_2O_2) levels and elevated the capacity of the antioxidant defense system (Etesami et al., 2022c). In a similar trend, PGPR strains of *Bacillus drentensis* and *Enterobacter cloacae* combined with an aerial spray of Si to mung bean under salt stress; improved mineral uptake (Mg, Zn, Fe, Cu) and antioxidative enzyme functioning (CAT, APOX, Peroxidase); and lowered levels of ROS thereby increasing the biomass, numbers of pods per plant, and seed weights (Mahmood et al., 2022). Thus, the combinatorial application of PGPRs and Si in cultivating crops may prove to be a promising approach to alleviating various stresses (Etesami and Jeong 2018; Verma et al., 2020). However, the mechanisms underlying the regulating effects of Si upon the interaction of plants and microorganisms are yet to be clearly understood. Therefore, well-designed field trials should be carried out on a large scale to evaluate the feasibility of integration of plant growth-promoting microorganisms and Si while applying them to the crops under various environmental conditions to achieve the goal of sustainability in agriculture.

6.6 CONCLUSION

Si appears as an indispensable element that plays a pivotal role in fostering sustainable and resilient agricultural practices. It contributes significantly to plant growth and development by improving plant process functionalities. Si's active role in plant abiotic and biotic stress management through physical, physiological, and biochemical means also positions it as a key player in enhancing plant resilience against diverse challenges. The significant role of Si in sustainable crop production is evident from its broad range of applications in agriculture from soil amendments to foliar sprays. The essentiality of Si needs to be further proven and the extent to which it can be utilized in the field of agriculture is still an area of research. Thus, future research should be encouraged to unveil the full potential of Si gearing toward green agriculture.

REFERENCES

Abe, Jun. Silicon deposition in leaf trichomes of cucurbitaceae horticultural plants: a short report. *American Journal of Plant Sciences* 10, no. 3 (2019): 486–490.

Abed-Ashtiani, Farnaz; Kadir, Jugah Bin; Selamat, Abdul B.; Hanif, Abdul H. B. M.; and Nasehi, Abbas. "Effect of foliar and root application of silicon against rice blast fungus in MR219 rice variety." *Journal of Plant Pathology & Microbiology* 3, no. 5 (2012): 1–5.

Abou Seeda, Mohamed, El-Zanaty Abou El-Nour, Maha Abdallah, and Hala El-Bassiouny. Impacts of metal, metalloid and their effects in plant physiology: a review. *Middle East Journal of Agriculture Research* 11, no. 03 (2022): 838–931.

Adrees, Muhammad, Shafaqat Ali, Muhammad Rizwan, Muhammad Zia-ur-Rehman, Muhammad Ibrahim, Farhat Abbas, Mujahid Farid, Muhammad Farooq Qayyum, and Muhammad Kashif Irshad. Mechanisms of silicon-mediated alleviation of heavy metal toxicity in plants: a review. *Ecotoxicology and Environmental Safety* 119 (2015): 186–197.

Ahammed, Golam Jalal, and Youxin Yang. Mechanisms of silicon-induced fungal disease resistance in plants. *Plant Physiology and Biochemistry* 165 (2021): 200–206.

Ahmad, Aqeel, Waheed Ullah Khan, Anis Ali Shah, Nasim Ahmad Yasin, Shagufta Naz, Aamir Ali, Arifa Tahir, and Aima Iram Batool. Synergistic effects of nitric oxide and silicon on promoting plant growth, oxidative stress tolerance and reduction of arsenic uptake in Brassica juncea. *Chemosphere* 262 (2021): 128384.

Ahsan, Muhammad, Mohammad Valipour, Fahim Nawaz, Muhammad Raheel, Hafiz Tassawar Abbas, Mateen Sajid, Abdul Manan, Shamsa Kanwal, Eman A. Mahmoud, Ryan Casini, Hosam O. Elansary, Emanuele Radicetti, and Hira Zulfiqar. Evaluation of silicon supplementation for drought stress under water-deficit conditions: an application of sustainable agriculture. *Agronomy* 13, no. 2 (2023): 599.

Akter, Rokeya, Md Habibur Rahman, Md Arifur Rahman Chowdhury, Manirujjaman, and Shimaa E. Elshenawy. Advances of nanotechnology in plant development and crop protection. In: Elngar, Ahmed A., Chowdhury, Rajdeep, Elhoseny, Mohamed, Balas, Valentia Emilia (Eds.), *Applications of Computational Intelligence in Multi-Disciplinary Research*, pp. 143–157. Academic Press, 2022.

Alam, Ashraful, Bambang Hariyanto, Hayat Ullah, Krishna R. Salin, and Avishek Datta. Effects of silicon on growth, yield and fruit quality of cantaloupe under drought stress. *Silicon* 13 (2021): 3153–3162.

Ali, Muhammad, Sobia Afzal, Aasma Parveen, Muhammad Kamran, Muhammad Rizwan Javed, Ghulam Hassan Abbasi, Zaffar Malik, Muhammad Riaz, Salman Ahmad, Muhammad Sohaib Chattha, Mohsin Ali, Qurban Ali, Muhammad Zahir Uddin, Muhammad Rizwan, and Shafaqat Ali. Silicon mediated improvement in the growth and ion homeostasis by decreasing Na+ uptake in maize (Zea mays L.) cultivars exposed to salinity stress. *Plant Physiology and Biochemistry* 158 (2021): 208–218.

Arif, Yamshi, Priyanka Singh, Andrzej Bajguz, Pravej Alam, and Shamsul Hayat. Silicon mediated abiotic stress tolerance in plants using physio-biochemical, omic approach and cross-talk with phytohormones. *Plant Physiology and Biochemistry* 166 (2021): 278–289.

Asgher, Mohd, Abdul Rehaman, Syed Nazar ul Islam, and Nafees A. Khan. Multifaceted roles of silicon nano particles in heavy metals-stressed plants. *Environmental Pollution* 341 (2023): 122886.

Ashraf, M., Muhammad Afzal, Rashid Ahmad, Muhammad A. Maqsood, Sher Muhammad Shahzad, Ahsan Aziz, and Naeem Akhtar. Silicon management for mitigating abiotic stress effects in plants. *Plant Stress* 4, no. 2 (2010): 104–114.

Ashraf, Muhammad, Shahzad, Syed Muhammad, Yousaf, Basharat, Shafiq, Muhammad, and Ahmad, Muhammad. Silicon occurrence, uptake, transport and mechanisms of heavy metals, minerals and salinity enhanced tolerance in plants with future prospects: A review. *Journal of Environmental Management* 183, no. 3 (2016): 521–529.

Aughenbaugh, Katherine L., Paul Stutzman, and Maria C. G. Juenger. Identifying glass compositions in fly ash. *Frontiers in Materials* 3 (2016): 1.

Babu, Reddla Hari, Pulicherla Yugandhar, and Nataru Savithramma. Synthesis, characterization and antimicrobial studies of bio silica nanoparticles prepared from *Cynodon dactylon* L.: a green approach. *Bulletin of Materials Science* 41 (2018): 1–8.

Bhat, Javaid Akhter, Nitika Rajora, Gaurav Raturi, Shivani Sharma, Pallavi Dhiman, Sandhya Sanand, S. M. Shivaraj, Humira Sonah, and Rupesh Deshmukh. Silicon nanoparticles (SiNPs) in sustainable agriculture: major emphasis on the practicality, efficacy and concerns. *Nanoscale Advances* 3, no. 14 (2021): 4019–4028.

Bhat, Javaid Akhter, S. M. Shivaraj, Pritam Singh, Devanna B. Navadagi, Durgesh Kumar Tripathi, Prasanta K. Dash, Amolkumar U. Solanke, Humira Sonah, and Rupesh Deshmukh. Role of silicon in mitigation of heavy metal stresses in crop plants. *Plants* 8, no. 3 (2019): 71.

Bi, Yang, S. P. Tian, Y. R. Guo, Y. H. Ge, and G. Z. Qin. Sodium silicate reduces postharvest decay on Hami melons: induced resistance and fungistatic effects. *Plant Disease* 90, no. 3 (2006): 279–283.

Bijanzadeh, Ehsan, Vahid Barati, and Todd P. Egan. Foliar application of sodium silicate mitigates drought stressed leaf structure in corn (Zea mays L.). *South African Journal of Botany* 147 (2022): 8–17.

Bocharnikova, Elena A., Amin Nikpay, Sabyasachi Majumdar, Masumeh Ziaee, and Vladimir V. Matichenkov. Bioactive silicon: approach to enhance sugarcane yield under stress environment. In: Verma, Krishan V., Song, Xiu-Peng, Rajput, Vishnu D., Solomon, Sushil, Li, Yang-Rui, Rao, Govind P. (Eds.), *Agro-industrial Perspectives on Sugarcane Production under Environmental Stress*, pp. 85–105. Singapore: Springer Nature Singapore, 2023.

Bokor, Boris, Milan Soukup, Marek Vaculík, Peter Vd'ačný, Marieluise Weidinger, Irene Lichtscheidl, Silvia Vávrová, Katarína Šoltys, Humira Sonah, Rupesh Deshmukh, Richard R. Bélanger, Philip J. White, Hamed A. El-Serehy, and Alexander Lux. Silicon uptake and localisation in date palm (Phoenix dactylifera)–a unique association with sclerenchyma. *Frontiers in Plant Science* 10 (2019): 988.

Brunings, A. M., Lawrence E. Datnoff, Jian Feng Ma, N. Mitani, Y. Nagamura, Bala Rathinasabapathi, and Matias Kirst. Differential gene expression of rice in response to silicon and rice blast fungus Magnaporthe oryzae. *Annals of Applied Biology* 155, no. 2 (2009): 161–170.

Chindaprasirt, Prinya, and Rattanasak, Ubolluk. Eco-production of silica from sugarcane bagasse ash for use as a photochromic pigment. *Scientific Reports,* 10, (2020): 9890.

Dann, Elizabeth K., and Sally Muir. Peas grown in media with elevated plant-available silicon levels have higher activities of chitinase and β-1,3-glucanase, are less susceptible to a fungal leaf spot pathogen and accumulate more foliar silicon. *Australasian Plant Pathology* 31 (2002): 9–13.

Das, Prabal, Indrani Manna, Palin Sil, Maumita Bandyopadhyay, and Asok K. Biswas. Silicon augments salt tolerance through modulation of polyamine and GABA metabolism in two indica rice (Oryza sativa L.) cultivars. *Plant Physiology and Biochemistry* 166 (2021): 41–52.

Debona, Daniel, Fabrício A. Rodrigues, and Lawrence E. Datnoff. Silicon's role in abiotic and biotic plant stresses. *Annual Review of Phytopathology* 55 (2017): 85–107.

Demattê, José Alexandre Melo, Paggiaro, Carlos Marcelo; Beltrame, José Antonio; and Ribeiro, Sérgio. Uso de silicatos em canade-açucar. *Informações Agronômicas* 133, (2011): 7–12.

Deshmukh, Rupesh Kailasrao, Julien Vivancos, Gowsica Ramakrishnan, Valérie Guérin, Gabriel Carpentier, Humira Sonah, Caroline Labbé, Paul Isenring, Francois J. Belzile, and Richard R. Bélanger. A precise spacing between the NPA domains of aquaporins is essential for silicon permeability in plants. *The Plant Journal* 83, no. 3 (2015): 489–500.

Dhiman, Pallavi, Nitika Rajora, Shubham Bhardwaj, Sreeja S. Sudhakaran, Amit Kumar, Gaurav Raturi, Koushik Chakraborty, Om Prakash Gupta, B. N. Devanna, Durgesh Kumar Tripathi, and Rupesh Deshmukh. Fascinating role of silicon to combat salinity stress in plants: an updated overview. *Plant Physiology and Biochemistry* 162 (2021): 110–123.

El Moukhtari, Ahmed, Pierre Carol, Mohammed Mouradi, Arnould Savoure, and Mohamed Farissi. Silicon improves physiological, biochemical, and morphological adaptations of alfalfa (Medicago sativa L.) during salinity stress. *Symbiosis* 85, no. 3 (2021): 305–324.

Emam, Manal M., Hemmat E. Khattab, Nesma M. Helal, and Abdelsalam E. Deraz. Effect of selenium and silicon on yield quality of rice plant grown under drought stress. *Australian Journal of Crop Science* 8, no. 4 (2014): 596–605.

Emamverdian, Abolghassem, Yulong Ding, Farzad Mokhberdoran, Yinfeng Xie, Xiao Zheng, and Yujun Wang. Silicon dioxide nanoparticles improve plant growth by enhancing antioxidant enzyme capacity in bamboo (Pleioblastus pygmaeus) under lead toxicity. *Trees* 34 (2020): 469–481.

Etesami, Hassan, and Byoung Ryong Jeong. Silicon (Si): review and future prospects on the action mechanisms in alleviating biotic and abiotic stresses in plants. *Ecotoxicology and Environmental Safety* 147 (2018): 881–896.

Etesami, Hassan, Ehsan Shokri, and Byoung Ryong Jeong. The combined use of silicon/nanosilicon and arbuscular mycorrhiza for effective management of stressed agriculture: action mechanisms and future prospects. In: Etesami, Hassan, Al Saeedi, Abdullah H., El-Ramady, Hassan, Fujita, Masayuki, Pessarakli, Mohammad, Anwar Hossain, Mohammad (Eds.), *Silicon and Nano-silicon in Environmental Stress Management and Crop Quality Improvement*, pp. 241–264. Academic Press, 2022a.

Etesami, Hassan, Fatemeh Noori, and Byoung Ryong Jeong. Directions for future research to use silicon and silicon nanoparticles to increase crops tolerance to stresses and improve their quality. In: Etesami, Hassan, Al Saeedi, Abdullah H., El-Ramady, Hassan, Fujita, Masayuki, Pessarakli, Mohammad, Anwar Hossain, Mohammad (Eds.) *Silicon and Nano-silicon in Environmental Stress Management and Crop Quality Improvement*, pp. 349–367. Academic Press, 2022b.

Etesami, Hassan, Zimin Li, Frans J. M. Maathuis, and Julia Cooke. The combined use of silicon and arbuscular mycorrhizas to mitigate salinity and drought stress in rice. *Environmental and Experimental Botany* 201 (2022c): 104955.

Fatemi, Hamideh, Behrooz Esmaiel Pour, and Muhammad Rizwan. Foliar application of silicon nanoparticles affected the growth, vitamin C, flavonoid, and antioxidant enzyme activities of coriander (Coriandrum sativum L.) plants grown in lead (Pb)-spiked soil. *Environmental Science and Pollution Research* 28 (2021): 1417–1425.

Ferreira, Thais H. S., Max S. Tsunada, Denis Bassi, Pedro Araújo, Lucia Mattiello, Giovanna V. Guidelli, Germanna L. Righetto, Vanessa R. Gonçalves, Prakash Lakshmanan, and Marcelo Menossi. Sugarcane water stress tolerance mechanisms and its implications on developing biotechnology solutions. *Frontiers in Plant Science* 8, (2017): 1–18.

Fiala, Roderik, Ivana Fialová, Marek Vaculik, and Miroslava Luxova. Effect of silicon on the young maize plants exposed to nickel stress. *Plant Physiology and Biochemistry* 166 (2021): 645–656.

Filha, Maria Santina Xavier, Fabrício Aparecido, Domiciano, Gisele P., Oliveira, Hélio V., Silveira, Paulo R, and W. R. Moreira. Wheat resistance to leaf blast mediated by silicon. *Australasian Plant Pathology* 40, no. 1 (2011): 28–38.

Ghareeb, Hassan, Zoltán Bozsó, Peter G. Ott, Cornelia Repenning, Frank Stahl, and Kerstin Wydra. Transcriptome of silicon-induced resistance against Ralstonia solanacearum in the silicon non-accumulator tomato implicates priming effect. *Physiological and Molecular Plant Pathology* 75, no. 3 (2011): 83–89.

Hall, Casey R., Vaibhav Dagg, Jamie M. Waterman, and Scott N. Johnson. Silicon alters leaf surface morphology and suppresses insect herbivory in a model grass species. *Plants* 9, no. 5 (2020): 643.

Han, Yong-Qiang, Ji-Hui Wen, Zhao-Pu Peng, De-Yong Zhang, and Mao-Lin Hou. Effects of silicon amendment on the occurrence of rice insect pests and diseases in a field test. *Journal of Integrative Agriculture* 17, no. 10 (2018): 2172–2181.

Hasanuzzaman, Mirza, Kamrun Nahar, Taufika Islam Anee, and Masayuki Fujita. Exogenous silicon attenuates cadmium-induced oxidative stress in Brassica napus L. by modulating AsA-GSH pathway and glyoxalase system. *Frontiers in Plant Science* 8 (2017): 1061.

Haseeb, Muhammad, Shahid Iqbal, Muhammad Bilal Hafeez, Muhammad Sohail Saddiq, Noreen Zahra, Ali Raza, Muhammad Usman Lbrahim, Javaid Iqbal, Muhammad Kamran, Qasim Ali, Talha Javed, Hayssam M. Ali, and Manzer H. Siddiqui. Phytoremediation of nickel by quinoa: morphological and physiological response. *Plos One* 17, no. 1 (2022): e0262309.

Haynes, Richard J., Oxana N. Belyaeva, and Graham Me Kingston. Evaluation of industrial wastes as sources of fertilizer silicon using chemical extractions and plant uptake.x *Journal of Plant Nutrition and Soil Science* 176, no. 2 (2013): 238–248.

He, Yingdui, Yang, Qinghuo Lin, Tao Jin, Xiaoping Zang, Tianyan Yun, Zheli Ding, Saudi A. Rekaby, Zengxian Zhao, and Mamdouh A. Eissa. Physio-biochemical evaluation of Si-rich biochar amendment to improve the salt stress tolerance of Grand Nain and Williams banana genotypes. *Industrial Crops and Products* 204 (2023): 117333.

Houben, David, Philippe Sonnet, and Jean-Thomas Cornelis. Biochar from Miscanthus: a potential silicon fertilizer. *Plant and Soil* 374 (2014): 871–882.

Islam, Waqar, Muhammad Tayyab, Farghama Khalil, Zhang Hua, Zhiqun Huang, and Han Y. H. Chen. Silicon-mediated plant defense against pathogens and insect pests. *Pesticide Biochemistry and Physiology* 168 (2020): 104641.

Jeer, Mallikarjuna, Maheswari Telugu, S. R. Voleti, and A. P. Padmakumari. Soil application of silicon reduces yellow stem borer, Scirpophaga incertulas (Walker) damage in rice. *Journal of Applied Entomology* 141, no. 3 (2017): 189–201.

Jiang, Haojie, Hongwei Guo, Peng Li, Yang Li, and Bingji Yan. Preparation of CaMgAl-LDHs and mesoporous silica sorbents derived from blast furnace slag for CO_2 capture. *RSC Advances* 9, no. 11 (2019): 6054–6063.

Kafi, Mohammad, Jafar Nabati, Mohammad Javad Ahmadi-Lahijani, and Armin Oskoueian. Silicon compounds and potassium sulfate improve salinity tolerance of potato plants through instigating the defense mechanisms, cell membrane stability, and accumulation of osmolytes. *Communications in Soil Science and Plant Analysis* 52, no. 8 (2021): 843–858.

Kaloterakis, Nikolaos, Sander H. van Delden, Sue Hartley, and Gerlinde B. De Deyn. Silicon application and plant growth promoting rhizobacteria consisting of six pure Bacillus species alleviate salinity stress in cucumber (Cucumis sativus L). *Scientia Horticulturae* 288 (2021): 110383.

Katz, Ofir. Silica phytoliths in angiosperms: phylogeny and early evolutionary history. *New Phytologist* 208, no. 3 (2015): 642–646.

Kaur, Harmanjit, and Maria Greger. A review on Si uptake and transport system. *Plants* 8, no. 4 (2019): 81.

Khan, Adil, Abdul Latif Khan, Sowbiya Muneer, Yoon-Ha Kim, Ahmed Al-Rawahi, and Ahmed Al-Harrasi. Silicon and salinity: crosstalk in crop-mediated stress tolerance mechanisms. *Frontiers in Plant Science* 10 (2019): 1429.

Khan, Adil, Saqib Bilal, Abdul Latif Khan, Muhammad Imran, Raheem Shahzad, Ahmed Al-Harrasi, Ahmed Al-Rawahi, Masood Al-Azhri, Tapan Kumar Mohanta, and In-Jung Lee. Silicon and gibberellins: synergistic function in harnessing ABA signaling and heat stress tolerance in date palm (Phoenix dactylifera L.). *Plants* 9, no. 5 (2020): 620.

Khan, Imran, Samrah Afzal Awan, Muhammad Rizwan, Marian Brestic, and Wengang Xie. Silicon: an essential element for plant nutrition and phytohormones signaling mechanism under stressful conditions. *Plant Growth Regulation* 100, no. 2 (2023): 301–319.

Khan, Imran, Samrah Afzal Awan, Muhammad Rizwan, Shafaqat Ali, Muhammad Jawad Hassan, Marian Brestic, Xinquan Zhang, and Linkai Huang. Effects of silicon on heavy metal uptake at the soil-plant interphase: a review. *Ecotoxicology and Environmental Safety* 222 (2021a): 112510.

Khan, Masudulla, Azhar U. Khan, Mohd Abul Hasan, Krishna Kumar Yadav, Marina M. C. Pinto, Nazia Malik, Virendra Kumar Yadav, Afzal Husain Khan, Saiful Islam, and Gulshan Kumar Sharma. Agro-nanotechnology as an emerging field: a novel sustainable approach for improving plant growth by reducing biotic stress. *Applied Sciences* 11, no. 5 (2021b): 2282.

Kobayashi, Noriko, Mikio Morioka, Teppei Komiyama, Toyoaki Ito, and Masahiko Saigusa. Silicon content in livestock manure compost and a simple estimation method for it (in Japanese with English summary). *Journal of the Japan Society of Waste Management Experts*, 19, (2008): 150–154.

Komiyama, Teppei, Arata Kobayashi, and Manabu Yahagi. The chemical characteristics of ashes from cattle, swine and poultry manure. *Journal of Material Cycles and Waste Management* 15 (2013): 106–110.

Kosakivska, Iryna V., Nina P. Vedenicheva, Lidiya M. Babenko, Lesya V. Voytenko, Kateryna O. Romanenko, and Valentyna A. Vasyuk. Exogenous phytohormones in the regulation of growth and development of cereals under abiotic stresses. *Molecular Biology Reports* 49, no. 1 (2022): 617–628.

Kurabachew, Henok, Frank Stahl, and Kerstin Wydra. Global gene expression of rhizobacteria-silicon mediated induced systemic resistance in tomato (Solanum lycopersicum) against Ralstonia solanacearum. *Physiological and Molecular Plant Pathology* 84 (2013): 44–52.

Kumar, Dharmendra, Om Parkash Dhankher, Rudra Deo Tripathi, and Chandra Shekhar Seth. Titanium dioxide nanoparticles potentially regulate the mechanism (s) for photosynthetic attributes, genotoxicity, antioxidants defense machinery, and phytochelatins synthesis in relation to hexavalent chromium toxicity in Helianthus annuus L. *Journal of Hazardous Materials* 454 (2023): 131418.

Kumar, Dharmendra, Safoora Mariyam, Kapuganti Jagadis Gupta, Muthu Thiruvengadam, Gajanan Ghodake, Baoshan Xing, and Chandra Shekhar Seth. Comparative investigation on chemical and green synthesized titanium dioxide nanoparticles against chromium (VI) stress eliciting differential physiological, biochemical, and cellular attributes in Helianthus annuus L. *Science of the Total Environment* 930 (2024): 172413.

Li, Peng, Zhao, Chao-Zhong, Zhang, Yu-Qing, Wang, Xiao-Min, Wang, Jian-Feng, Wang, Fang, and Bi, Yan-Rong. Silicon enhances the tolerance of Poa annua to cadmium by inhibiting its absorption and oxidative stress. *Biologia Plantarum* 61, no. 4 (2017): 741–750.

Li, Zhaozhuang, Zeci Liu, Zhibin Yue, Jie Wang, Li Jin, Zhiqi Xu, Ning Jin, Bo Zhang, Jian Lyu, and Jihua Yu. Application of exogenous silicon for alleviating photosynthetic inhibition in tomato seedlings under low–calcium stress. *International Journal of Molecular Sciences* 23, no. 21 (2022): 13526.

Lipa, Deb, and Dutta Pranab. Biogenic role of silicon in abiotic and biotic stress management in plants. *Journal of Mycology and Plant Pathology* 52, no. 2 (2022): 103–122.

Liu, Haitao, Qiujuan Jiao, Lina Fan, Ying Jiang, Mohammed Nasser Alyemeni, Parvaiz Ahmad, Yinglong Chen, Mo Zhu, Haiping Liu, Ying Zhao, Fang Liu, Shiliang Liu, and Gezi Li. Integrated physio-biochemical and transcriptomic analysis revealed mechanism underlying of Si-mediated alleviation to cadmium toxicity in wheat. *Journal of Hazardous Materials* 452 (2023): 131366.

Liu, Peng, Lina Yin, Shiwen Wang, Meijuan Zhang, Xiping Deng, Suiqi Zhang, and Kiyoshi Tanaka. Enhanced root hydraulic conductance by aquaporin regulation accounts for silicon alleviated salt-induced osmotic stress in Sorghum bicolor L. *Environmental and Experimental Botany* 111 (2015a): 42–51.

Liu, Ruiqiang, and Rattan Lal. Potentials of engineered nanoparticles as fertilizers for increasing agronomic productions. *Science of the Total Environment* 514 (2015b): 131–139.

López-Pérez, Mari C., Fabián Pérez-Labrada, Lino J. Ramírez-Pérez, Antonio Juárez-Maldonado, América B. Morales-Díaz, Susana González-Morales, Luis R. García-Dávila, Jesús García-Mata, and Adalberto Benavides-Mendoza. Dynamic modeling of silicon bioavailability, uptake, transport, and accumulation: applicability in improving the nutritional quality of tomato. *Frontiers in Plant Science* 9 (2018): 647.

Luyckx, Marie, Jean-Francois Hausman, Stanley Lutts, and Gea Guerriero. Silicon and plants: current knowledge and technological perspectives. *Frontiers in Plant Science* 8 (2017): 411.

Ma, Jian Feng, and Eiichi Takahashi. *Soil, Fertilizer, and Plant Silicon Research in Japan*. Elsevier, 2002.

Ma, Jian Feng, and Naoki Yamaji. Silicon uptake and accumulation in higher plants. *Trends in Plant Science* 11, no. 8 (2006): 392–397.

Mahmood, Sajid, Ihsanullah Daur, Muhammad Yasir, Muhammad Waqas, and Heribert Hirt. Synergistic practicing of rhizobacteria and silicon improve salt tolerance: implications from boosted oxidative metabolism, nutrient uptake, growth and grain yield in mung bean. *Plants* 11, no. 15 (2022): 1980.

Mandlik, Rushil, Vandana Thakral, Gaurav Raturi, Suhas Shinde, Miroslav Nikolić, Durgesh K. Tripathi, Humira Sonah, and Rupesh Deshmukh. Significance of silicon uptake, transport, and deposition in plants. *Journal of Experimental Botany* 71, no. 21 (2020): 6703–6718.

Marron, Alan O., Sarah Ratcliffe, Glen L. Wheeler, Raymond E. Goldstein, Nicole King, Fabrice Not, Colomban de Vargas, and Daniel J. Richter. The evolution of silicon transport in eukaryotes. *Molecular Biology and Evolution* 33, no. 12 (2016): 3226–3248.

Mathur, Piyush, and Swarnendu Roy. Nanosilica facilitates silica uptake, growth and stress tolerance in plants. *Plant Physiology and Biochemistry* 157 (2020): 114–127.

Mavrič Čermelj, Anja, Aleksandra Golob, Katarina Vogel-Mikuš, and Mateja Germ. Silicon mitigates negative impacts of drought and UV-b radiation in plants. *Plants* 11, no. 1 (2021): 91.

Mburu, Kellen, Oduor, Richard, Mgutu, Anthony, and Tripathi, Leena. Silicon application enhances resistance to xanthomonas wilt disease in banana. *Plant Pathology* 65, no. 5 (2016): 807–818.

Mejía, Johanna M., Ruby Mejía de Gutiérrez, and Carlos Montes. Rice husk ash and spent diatomaceous earth as a source of silica to fabricate a geopolymeric binary binder. *Journal of Cleaner Production* 118 (2016): 133–139.

Mitani, Namiki, and Jian Feng Ma. Uptake system of silicon in different plant species. *Journal of Experimental Botany* 56, no. 414 (2005): 1255–1261.

Mitani-Ueno, Namiki, and Jian Feng Ma. Linking transport system of silicon with its accumulation in different plant species. *Soil Science and Plant Nutrition* 67, no. 1 (2021): 10–17.

Monger, H. Curtis, and Eugene F. Kelly. Silica minerals. *Soil Mineralogy with Environmental Applications* 7 (2002): 611–636.

Montpetit, Jonatan, Julien Vivancos, Namiki Mitani-Ueno, Naoki Yamaji, Wilfried Rémus-Borel, François Belzile, Jian Feng Ma, and Richard R. Bélanger. Cloning, functional characterization and heterologous expression of TaLsi1, a wheat silicon transporter gene. *Plant Molecular Biology* 79 (2012): 35–46.

Moore, Paul. Changing tracks. *Industrial Minerals* 432 (2003): 69–73.

Moradtalab, Narges, Roghieh Hajiboland, Nasser Aliasgharzad, Tobias E. Hartmann, and Günter Neumann. Silicon and the association with an arbuscular-mycorrhizal fungus (Rhizophagus clarus) mitigate the adverse effects of drought stress on strawberry. *Agronomy* 9, no. 1 (2019): 41.

Mostofa, Mohammad Golam, Md Mezanur Rahman, Md Mesbah Uddin Ansary, Sanjida Sultana Keya, Mostafa Abdelrahman, Md Giashuddin Miah, and Lam-Son Phan Tran. Silicon in mitigation of abiotic stress-induced oxidative damage in plants. *Critical Reviews in Biotechnology* 41, no. 6 (2021): 918–934.

Mustakim, Syed Mohammed, Shaswat Kumar Das, Jyotirmoy Mishra, Asif Aftab, Thamer Salman Alomayri, Hasan Suliman Assaedi, and Cyriaque Rodrigue Kaze. Improvement in fresh, mechanical and microstructural properties of fly ash-blast furnace slag based geopolymer concrete by addition of nano and micro silica. *Silicon* 13 (2021): 2415–2428.

Nagaratna, Wangi, Chicknayakanahalli Marulasiddappa Kalleshwaraswamy, Bhakthanakatte Chandrappa Dhananjaya, Nagabovanalli B. Prakash, Sharanabasappa S. Deshmukh, Chandrashekar Sunil, Mohammad Anwar Hossain, and Hosamane Basvarajappa Mallikarjuna. Silicon accumulation in leaves reduces the herbivory by invasive fall armyworm *Spodoptera frugiperda* and enhances the yield of maize. *International Journal of Plant Biology* 14, no. 3 (2023): 701–713.

Najihah, Nor Ismail, Mohamed Musa Hanafi, Abu Seman Idris, and Md Abdul Hakim. Silicon treatment in oil palms confers resistance to basal stem rot disease caused by Ganoderma boninense. *Crop Protection* 67 (2015): 151–159.

Naz, Rabia, Sana Batool, Muhammad Shahid, Rumana Keyani, Humaira Yasmin, Asia Nosheen, Muhammad Nadeem Hassan, Saqib Mumtaz, and Manzer Hussain Siddiqui. Exogenous silicon and hydrogen sulfide alleviates the simultaneously occurring drought stress and leaf rust infection in wheat. *Plant Physiology and Biochemistry* 166 (2021): 558–571.

Neu, Silke, Jörg Schaller, and E. Gert Dudel. Silicon availability modifies nutrient use efficiency and content, C: N: P stoichiometry, and productivity of winter wheat (Triticum aestivum L.). *Scientific Reports* 7, no. 1 (2017): 40829.

Pa, Faizul Che, Abdullah Chik, and Md Fazlul Bari. Palm ash as an alternative source for silica production. *MATEC Web of Conferences* 78 (2016): 01062. EDP Sciences.

Panasenko, A. E., Sofya B. Yarusova, and Pavel S. Gordienko. Wollastonite-containing composite material from rice straw. *Protection of Metals and Physical Chemistry of Surfaces* 59, no. 3 (2023): 434–439.

Pang, Zhihao, Weisong Yin, Yuxiao Wang, Wen Zeng, Hongyun Peng, and Yongchao Liang. Silicon-phosphorus pathway mitigates heavy metal stress by buffering rhizosphere acidification. *Science of the Total Environment* 904 (2023): 166887.

Parthasarathy, Seethapathy, and Vaigundaperumal Jaiganesh. Plant activator and silicon nutrient mediated resistance against powdery mildew of Black gram (Vigna mungo L. Hepper). *Global Journal of Research Analysis* 4, no. 6 (2016): 48–49.

Patel, Monika, Dhara Fatnani, and Asish Kumar Parida. Silicon-induced mitigation of drought stress in peanut genotypes (Arachis hypogaea L.) through ion homeostasis, modulations of antioxidative defense system, and metabolic regulations. *Plant Physiology and Biochemistry* 166 (2021): 290–313.

Paye, Wooiklee, Brenda Tubana, Dustin Harrell, Tapasya Babu, Yumiko Kanke, and Lawrence Datnoff. Determination of critical soil silicon levels for rice production in Louisiana using different extraction procedures. *Communications in Soil Science and Plant Analysis* 49, no. 17 (2018): 2091–2102.

Pirooz, Pariya, Rayhaneh Amooaghaie, Alimohammad Ahadi, and Fariba Sharififar. Silicon-induced nitric oxide burst modulates systemic defensive responses of Salvia officinalis under copper toxicity. *Plant Physiology and Biochemistry* 162 (2021): 752–761.

Polanco, Leonora R., Fabrício A. Rodrigues, Kelly J. T. Nascimento, Maria F. A. Cruz, Carmen R. S. Curvelo, Fábio M. DaMatta, and Francisco X. R. Vale. Photosynthetic gas exchange and antioxidative system in common bean plants infected by Colletotrichum lindemuthianum and supplied with silicon. *Tropical Plant Pathology* 39 (2014): 35–42.

Qin, Guo Z., and Shi P. Tian. Enhancement of biocontrol activity of Cryptococcus laurentii by silicon and the possible mechanisms involved. *Phytopathology* 95, no. 1 (2005): 69–75.

Rajput, Vishnu D., Tatiana Minkina, Arpna Kumari, Harish, Vipin Kumar Singh, Krishan K. Verma, Saglara Mandzhieva, Svetlana Sushkova, Sudhakar Srivastava, and Chetan Keswani. Coping with the challenges of abiotic stress in plants: new dimensions in the field application of nanoparticles. *Plants* 10, no. 6 (2021b): 1221.

Rajput, Vishnu D., Tatiana Minkina, Morteza Feizi, Arpna Kumari, Masudulla Khan, Saglara Mandzhieva, Svetlana Sushkova, Hassan El-Ramady, Krishan K. Verma, Abhishek Singh, Eric D. van Hullebusch, Rupesh Kumar Singh, Hanuman Singh Jatav, and Ravish Choudhary. Effects of silicon and silicon-based nanoparticles on rhizosphere microbiome, plant stress and growth. *Biology* 10, no. 8 (2021a): 791.

Ranjan, Alok, Ragini Sinha, Meenu Bala, Ashwani Pareek, Sneh L. Singla-Pareek, and Anil Kumar Singh. Silicon-mediated abiotic and biotic stress mitigation in plants: underlying mechanisms and potential for stress resilient agriculture. *Plant Physiology and Biochemistry* 163 (2021): 15–25.

Rastogi, Anshu, Durgesh Kumar Tripathi, Saurabh Yadav, Devendra Kumar Chauhan, Marek Živčák, Mansour Ghorbanpour, Nabil Ibrahim El-Sheery, and Marian Brestic. Application of silicon nanoparticles in agriculture. *3 Biotech* 9 (2019): 1–11.

Ratnayake, R. M. R. N. K., W. A. M. Daundasekera, H. M. Ariyarathne, and M. Y. U. Ganehenege. Some biochemical defense responses enhanced by soluble silicon in bitter gourd-powdery mildew pathosystem. *Australasian Plant Pathology* 45 (2016): 425–433.

Raza, Ali. Eco-physiological and biochemical responses of rapeseed (Brassica napus L.) to abiotic stresses: consequences and mitigation strategies. *Journal of Plant Growth Regulation* 40, no. 4 (2021): 1368–1388.

Raza, Ali, Javaria Tabassum, Zainab Zahid, Sidra Charagh, Shanza Bashir, Rutwik Barmukh, Rao Sohail Ahmad Khan, Fernando Barbosa Jr., Chong Zhang, Hua Chen, Weijian Zhuang, and Rajeev K. Varshney. Advances in "omics" approaches for improving toxic metals/metalloids tolerance in plants. *Frontiers in Plant Science* 12 (2022): 794373.

Reynolds, Olivia L., Matthew P. Padula, Rensen Zeng, and Geoff M. Gurr. Silicon: potential to promote direct and indirect effects on plant defense against arthropod pests in agriculture. *Frontiers in Plant Science* 7 (2016): 744.

Sadeghi, Elahe, Reza Vafaei Shoushtari, and Hamid Madani. The influence of Tetranychus urticae Koch (Acari: Tetranychidae) life table and reproductive parameters by applying Si on bean at library condition. *Advances in Entomology* 4, no. 5 (2016): 260–267.

Sakurai, Gen, Akiko Satake, Naoki Yamaji, Namiki Mitani-Ueno, Masayuki Yokozawa, François Gabriel Feugier, and Jian Feng Ma. In silico simulation modeling reveals the importance of the Casparian strip for efficient silicon uptake in rice roots. *Plant and Cell Physiology* 56, no. 4 (2015): 631–639.

Savvas, Dimitrios; Giotis, Dimitrios; Chatzieustratiou, Eleni; Bakea, Maria; and Patakioutas, Georgios. Silicon supply in soilless cultivations of zucchini alleviates stress induced by salinity and powdery mildew infections. *Environmental and Experimental Botany* 65, no. 1 (2009): 11–17.

Savvas, Dimitrios, and Georgia Ntatsi. Biostimulant activity of silicon in horticulture. *Scientia Horticulturae* 196 (2015): 66–81.

Sebastian, Dennis, Hugh Rodrigues, Charles Kinsey, Gaspar Korndorfer, Hamilton Pereira, Guilherme Buck, Lawrence Datnoff, Stephen Miranda, and Mary Provance-Bowley. A 5-day method for determination of soluble silicon concentrations in nonliquid fertilizer materials using a sodium carbonate-ammonium nitrate extractant followed by visible spectroscopy with heteropoly blue analysis. Single-laboratory validation. *Journal of AOAC International* 96 (2013): 251–58.

Sharma, Shivani, Muntazir Mushtaq, Sreeja Sudhakaran, Vandana Thakral, Gaurav Raturi, Ruchi Bansal, Virender Kumar, Sanskriti Vats, S. M. Shivaraj, and Rupesh Deshmukh. Silicon uptake, transport, and accumulation in plants. In: Pandey, Sangeeta, Tripathi, Durgesh Kumar, Singh, Vijay Pratap, Sharma, Shivesh, Chauhan, Devendra Kumar (Eds.), *Beneficial Chemical Elements of Plants: Recent Developments and Future Prospects*, pp. 205–226. John Wiley & Sons, 2023.

Silva Júnior, José da, Carlo R. da Cunha, Flávio L. S. de Carvalho, Ubirajara P. Rodrigues Filho, Paulo R. Oliveira, and Marcos A. Segatto Silva. Obtaining high purity silica from rice hulls. *Quimica* Nov*a* 33 (2010): 794–797.

Singhal, Rajesh Kumar, Mahesh Kumar, Bandana Bose, Sananda Mondal, Sudhakar Srivastava, Om Parkash Dhankher, and Rudra Deo Tripathi. Heavy metal (loid) s phytotoxicity in crops and its mitigation through seed priming technology. *International Journal of Phytoremediation* 25, no. 2 (2023): 187–206.

Somapala, Koshala, Darshani Weerahewa, and Sunil Thrikawala. Silicon rich rice hull amended soil enhances anthracnose resistance in tomato. *Procedia Food Science* 6 (2016): 190–193.

Song, Xiu-Peng, Krishan K. Verma, Dan-Dan Tian, Xiao-Qiu Zhang, Yong-Jian Liang, Xing Huang, Chang-Ning Li, and Yang-Rui Li. Exploration of silicon functions to integrate with biotic stress tolerance and crop improvement. *Biological Research* 54, no. 1 (2021): 1–12.

Souri, Zahra, Kanika Khanna, Naser Karimi, and Parvaiz Ahmad. Silicon and plants: current knowledge and future prospects. *Journal of Plant Growth Regulation* 40 (2021): 906–925.

Sun, Hao, Jia Guo, Yaoke Duan, Tiantian Zhang, Heqiang Huo, and Haijun Gong. Isolation and functional characterization of CsLsi1, a silicon transporter gene in Cucumis sativus. *Physiologia Plantarum* 159, no. 2 (2017): 201–214.

Sun, Hao, Yaoke Duan, Xiaocui Qi, Liyang Zhang, Heqiang Huo, and Haijun Gong. Isolation and functional characterization of CsLsi2, a cucumber silicon efflux transporter gene. *Annals of Botany* 122, no. 4 (2018): 641–648.

Swain, Rinny, and Gyan Ranjan Rout. Silicon in agriculture. In: Lichtfouse, Eric (Ed.), *Sustainable Agriculture Reviews*, pp. 233–260. Springer, 2017.

Tang, Zhong, Han-Qing Wang, Jie Chen, Jia-Dong Chang, and Fang-Jie Zhao. Molecular mechanisms underlying the toxicity and detoxification of trace metals and metalloids in plants. *Journal of Integrative Plant Biology* 65, no. 2 (2023): 570–593.

Tayade, Rupesh, Amit Ghimire, Waleed Khan, Liny Lay, John Quarshie Attipoe, and Yoonha Kim. Silicon as a smart fertilizer for sustainability and crop improvement. *Biomolecules* 12, no. 8 (2022): 1027.

Toresano-Sánchez, Fernando, Antonio Valverde-García, and Francisco Camacho-Ferre. Effect of the application of silicon hydroxide on yield and quality of cherry tomato. *Journal of Plant Nutrition* 35, no. 4 (2012): 567–590.

Tripathi, Durgesh Kumar, Swati Singh, Vijay Pratap Singh, Sheo Mohan Prasad, Nawal Kishore Dubey, and Devendra Kumar Chauhan. Silicon nanoparticles more effectively alleviated UV-B stress than silicon in wheat (Triticum aestivum) seedlings. *Plant Physiology and Biochemistry* 110 (2017): 70–81.

Tubaña, Brenda Servaz, and Joseph Raymond Heckman. Silicon in soils and plants. In: Rodrigues, Fabricio A., Datnoff, Lawrence E. (Eds.), *Silicon and Plant Diseases*, pp. 7–51. Springer International Publishing, 2015.

ur Rehman, Muhammad Zia, Muhammad Rizwan, Arslan Rauf, Muhammad Ashar Ayub, Shafaqat Ali, Muhammad Farooq Qayyum, Aisha A. Waris, Asif Naeem, and Muhammad Sanaullah. Split application of silicon in cadmium (Cd) spiked alkaline soil plays a vital role in decreasing Cd accumulation in rice (Oryza sativa L.) grains. *Chemosphere* 226 (2019): 454–462.

Vasanthi, N., L. M. Saleena, and Raj, S. A. Silicon in day today life. *World Applied Sciences Journal* 17, no. 11 (2012): 1425–1440.

Vasilachi, Ionela Cătălina, Vasile Stoleru, and Maria Gavrilescu. Analysis of heavy metal impacts on cereal crop growth and development in contaminated soils. *Agriculture* 13, no. 10 (2023): 1983.

Verma, Krishan K., Kai-Chao Wu, Pratiksha Singh, Mukesh Kumar Malviya, Rajesh Kumar Singh, Xiu-Peng Song, and Yang-Rui Li. The protective role of silicon in sugarcane under water stress: photosynthesis and antioxidant enzymes. *Biomedical Journal of Scientific and Technical Research* 15, no. 2 (2019): 1–7.

Verma, Krishan K., Xiu-Peng Song, Dong-Mei Li, Munna Singh, Vishnu D. Rajput, Mukesh Kumar Malviya, Tatiana Minkina, Rajesh Kumar Singh, Pratiksha Singh, and Yang-Rui Li. Interactive role of silicon and plant–rhizobacteria mitigating abiotic stresses: a new approach for sustainable agriculture and climate change. *Plants* 9, no. 9 (2020): 1055.

Verma, Krishan K., Xiu-Peng Song, Yuan Zeng, Dao-Jun Guo, Munna Singh, Vishnu D. Rajput, Mukesh Kumar Malviya, Kai-Jun Wei, Anjney Sharma, Dong-Ping Li, Gan-Lin Chen, and Yang-Rui Li. Foliar application of silicon boosts growth, photosynthetic leaf gas exchange, antioxidative response and resistance to limited water irrigation in sugarcane (Saccharum officinarum L.). *Plant Physiology and Biochemistry* 166 (2021): 582–592.

Vilela, Michelle, Jair C. Moraes, Eduardo Alves, Terezinha M. Santos-Cividanes, and Fabíola A. Santos. Induced resistance to Diatraea saccharalis (Lepidoptera: Crambidae) via silicon application in sugarcane. *Revista Colombiana de Entomología* 40, no. 1 (2014): 44–48.

Wang, Min, Limin Gao, Suyue Dong, Yuming Sun, Qirong Shen, and Shiwei Guo. Role of silicon on plant–pathogen interactions. *Frontiers in Plant Science* 8 (2017): 701.

Wang, Shihua, Fayuan Wang, and Shuangcheng Gao. Foliar application with nano-silicon alleviates Cd toxicity in rice seedlings. *Environmental Science and Pollution Research* 22 (2015): 2837–2845.

Wang, Yaofeng, Xin Xiao, Yiliang Xu, and Baoliang Chen. Environmental effects of silicon within biochar (sichar) and carbon–silicon coupling mechanisms: a critical review. *Environmental Science & Technology* 53, no. 23 (2019a): 13570–13582.

Wang, Zhigang, Naoki Yamaji, Sheng Huang, Xiang Zhang, Mingxing Shi, Shan Fu, Guangzhe Yang, Jian Feng Ma, and Jixing Xia. OsCASP1 is required for Casparian strip formation at endodermal cells of rice roots for selective uptake of mineral elements. *The Plant Cell* 31, no. 11 (2019b): 2636–2648.

Wanyika, Harrison, Erastus Gatebe, Paul Kioni, Zhiyong Tang, and Yan Gao. Mesoporous silica nanoparticles carrier for urea: potential applications in agrochemical delivery systems. *Journal of Nanoscience and Nanotechnology* 12, no. 3 (2012): 2221–2228.

Wei, Lijuan, Jing Zhang, Chunlei Wang, and Weibiao Liao. Recent progress in the knowledge on the alleviating effect of nitric oxide on heavy metal stress in plants. *Plant Physiology and Biochemistry* 147 (2020): 161–171.

Wiese, Joachim, Heike Wiese, Jana Schwartz, and Sven Schubert. Osmotic stress and silicon act additively in enhancing pathogen resistance in barley against barley powdery mildew. *Journal of Plant Nutrition and Soil Science* 168, no. 2 (2005): 269–274.

Xu, Jiaqi, Lifeng Guo, and Liwei Liu. Exogenous silicon alleviates drought stress in maize by improving growth, photosynthetic and antioxidant metabolism. *Environmental and Experimental Botany* 201 (2022): 104974.

Yan, Guo-Chao, Miroslav Nikolic, Mu-Jun YE, Zhuo-Xi Xiao, and Yong-Chao Liang. Silicon acquisition and accumulation in plant and its significance for agriculture. *Journal of Integrative Agriculture* 17, no. 10 (2018): 2138–2150.

Younas, Hafiza Samra, Muhammad Abid, Muhammad Shaaban, and Muhammad Ashraf. Influence of silicon and chitosan on growth and physiological attributes of maize in a saline field. *Physiology and Molecular Biology of Plants* 27 (2021): 387–397.

Zahedi, Seyed Morteza, Hosseini, Marjan Sadat, Hoveizeh, Narjes Fahadi, Kadkhodaei, Saeid, and Vaculík, Marek. Comparative morphological, physiological and molecular analyses of drought-stressed strawberry plants affected by SiO_2 and SiO_2-NPs foliar spray. *Scientia Horticulturae* 309 (2023): 111686.

Zaid, Abbu, Fahima Gul, Mohammad A. Ahanger, and Parvaiz Ahmad. Silicon-mediated alleviation of stresses in plants. In: Ahmad, Parvaiz, Ahanger, Mohammad Abas, Singh, Vijay Pratap, Tripathi, Durgesh Kumar, Alam, Pravej, Alyemeni, Mohammad Nasser (Eds.), *Plant Metabolites and Regulation under Environmental Stress*, pp. 377–387. Academic Press, 2018.

Zhang, Anyu, Jing Zhang, Sumei Liu, Jiliang Xuan, and Zhuoyi Zhu. Evaluation of silicic acid sources for spring diatom blooms on the continental shelf: insights fromstable silicon isotopes in the East China Sea. *Journal of Geophysical Research: Oceans* 125, no. 2 (2020): e2019JC015478.

Zhang, Tao, Xiaosha Wu, Sabry M. Shaheen, Hamada Abdelrahman, Esmat F. Ali, Nanthi S. Bolan, Yong Sik Ok, Guoxue Li, Daniel C. W. Tsang, and Jörg Rinklebe. Improving the humification and phosphorus flow during swine manure composting: a trial for enhancing the beneficial applications of hazardous biowastes. *Journal of Hazardous Materials* 425 (2022): 127906.

7 Biochemistry of Boron in Plants

Unraveling its Vital Role and Transportation

Arpit Jindal, Cengiz Kaya, Mohammad Wasim Siddiqui, Kamel A. Abd-Elsalam, Saurabh Chandra Saxena, Prabhakar Sharma, and Chandra Shekhar Seth

7.1 INTRODUCTION

Boron (B) is considered a minor element in relation to other nutrients because of its optimal need, i.e., at nanolevels. B has long been regarded to be a toxic element, but it has also been proposed that it has the ability to reduce phenolic chemical toxicity through the development of non-toxic composites, suggesting that B insufficiency was misconstrued as a sign of phenol toxicity (Kohli et al., 2023). B's importance as an important element in agriculture cannot be stressed, since both its lack and toxicity in soils can negatively affect plant growth and development (Thakur et al., 2023). B (a metalloid) can enter the soil naturally or through anthropogenic activities such as B-related mining and processing, fertilization, and irrigation (Kumar et al., 2023; Hua et al., 2021); leaching such as of heavy metals (Misra et al., 2009; Seth et al., 2008; Chaturvedi et al., 2006; Mishra et al., 2006). In many places of the world, a lack of B is viewed as a limiting factor impacting agricultural productivity (Dong et al., 2022; Bargali et al., 2024). Low soil B is most common in high-rainfall areas, such as portions of China, India, Nepal, Bangladesh, and Brazil. Excess soil B is also found in dry and semi-arid locations, including sections of India, the United States, Chile, Serbia, Peru, Pakistan, Malaysia, Hungary, Italy, and the Middle East (Jothi and Takano, 2023; Mushtaq et al., 2024). Excess B can impair physiological processes in plants (for example, CO_2 absorption, photosystem II photochemistry, carbohydrate metabolism, and the antioxidant system), resulting in visual signs such as chlorosis and shoot necrosis (Hua et al., 2021; Mariyam et al., 2023a); like the salinity (Prajapati et al., 2023; Gupta and Seth, 2020; Agnihotri and Seth, 2016) and heavy metal stress (Kumar et al., 2022; Yadav et al., 2022; Gupta and Seth, 2021; Singh et al., 2017). Excess B can also limit root development and, as a result, diminish water intake, exacerbating water stress in dry places (Hua et al., 2021; Kim et al., 2024b).

The optimal quantity of B in tissues varies between 30 and 75 mg kg^{-1}; however, the amounts vary from plant to plant. The deficit of B begins in plants when its levels in the soil fall to 5–25 mg ha^{-1}, which is mostly due to the fact that a few chemicals, such as $Al(OH)_3$, absorb them from the soil in soluble forms, causing the soil to become very acidic. Leaching of B occurs when there is a lot of rain, which increases the solubility of B in water under acidic circumstances. In contrast, in alkaline soil conditions, B is less soluble, resulting in lower absorption (Kohli et al., 2023), and reduced levels under salinity (Gupta and Seth 2023; Mariyam et al., 2023b). B's immobile nature enhances its unavailability in plants, resulting in a wide spectrum of insufficiency symptoms in younger regions of fruit harvests (Thakur et al., 2023; Solanki et al., 2024; Verma et al., 2024).

DOI: 10.1201/9781032673592-7

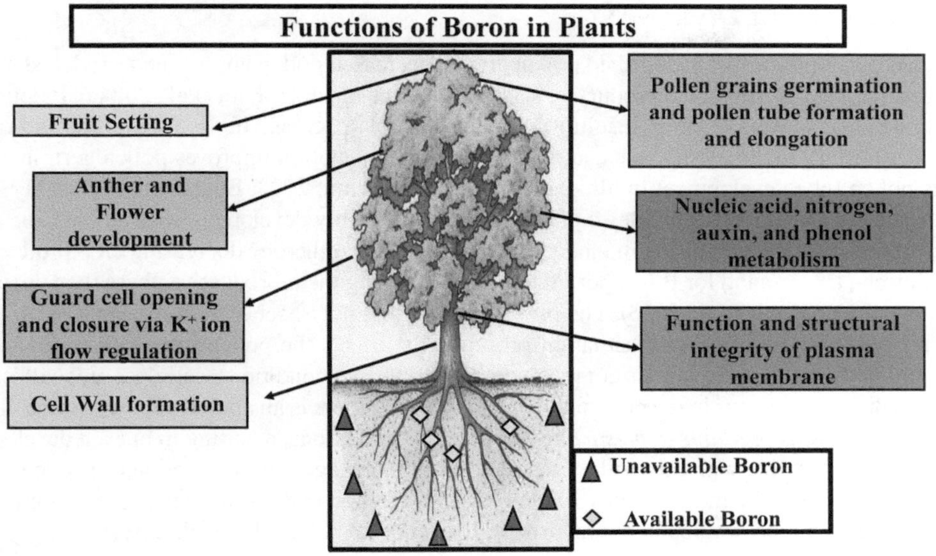

FIGURE 7.1 Functions of B in plants, accompanied by the representation of limited B availability in comparison to the surplus amount of unavailable B.

7.2 IMPORTANCE OF B IN PLANTS

Plants are also classified into three groups based on their B requirements: (1) graminaceous species that require very little B, (2) non-grass monocots and dicots that require moderate levels of B and are severely affected by increases in B concentration, and (3) high-B-demanding plants that contain latex. Many physiological aspects of vascular plants, such as root growth, hormone synthesis, sugar transport, amino acid synthesis, protein metabolism, carbohydrate metabolism, pollen tube formation, nucleotide synthesis, phenol metabolism, and so on, are adversely affected by B deficiency. Furthermore, they impair membrane integrity, plasmodesmata bridges, enzymes, fruit and flower development, ion fluxes, and membrane potential across different membranes (Kohli et al., 2023; Pathak et al., 2024; Hazaimeh et al., 2024). The major functions of B in the plant system are summarized below and represented in Figure 7.1.

7.2.1 Cell Wall Formation

B is important in cell wall production, functioning, and strength since the bulk of B (>90%) is present in plant cell walls (Thakur et al., 2023). Normal cell division occurs in B-deficient plants, but it does not progress to cell differentiation, which would normally result in the creation of cells suitable for the development of leaves or stems. Its absence causes cell wall defects, cytoskeleton polymerization disruption, and differences in plasma membrane permeability. B interacts with glycoproteins, glycolipids, and phosphoinositides in plant plasma membranes, altering the stability of its domains, particularly $B(OH)_3$, and establishing crosslinks in the rhamnogalacturonan II (RG-II) domain of the pectin molecule (Kohli et al., 2023). Under normal B concentrations, the proportion of RG-II dimers in the cell wall was greater than 80%, but in a B-deficient environment, the proportion was only 10%–20%. In addition, the amount of RG-II components in the cell wall increased with an increase in B supply (Li et al., 2023). When compared to dicots, monocots require less B for vegetative development since their cell walls contain less pectin (Sharma et al., 2023; Kumari et al., 2024).

7.2.2 Reproductive Development

B deficiency symptoms include unfilled pollen grains, decreased pollen vitality, and a reduced number of flowers per plant during the reproductive stage (Thakur et al., 2023; Kim et al., 2024a). B deficiency causes buds and flowers to drop, resulting in a considerable decrease in seed, fruit set, and quality characteristics of growing fruits, nuts, and seeds. B supplementation improves pollen germination as well as pollen tube development in almonds (Bons and Sharma, 2023; Bhupenchandra et al., 2024). The B deficiency symptoms in barley and rice are heads not fully developed, while in oats, the pollen tube created is ruptured, resulting in inadequate pollination and thereby decreasing crop productivity. In sunflowers, the demand for B is reported to be larger during the reproductive phase than during the vegetative phase (Kohli et al., 2023). The primary B-binding polysaccharide component of the pollen tube cell wall is RG-II. Excess B situation substantially affects the pollen tube wall component by deteriorating the pollen tube and its composition. Under normal conditions, cellulose and callose synthesis and deposition occur in the pollen tube wall. However, there is an altered deposition of callose in the pollen tube walls of *Malus domestica* under excess B conditions, resulting in uneven development and inability to fertilize with the ovule. Ca and B are both engaged in the creation of pectic polysaccharides in the cell wall, and B toxicity limits Ca's participation in the synthesis of these compounds, resulting in pollen tube development impairment (Behera et al., 2023; Jin et al., 2024).

7.2.3 Plant Metabolism

B is recognized for promoting pollinator activity during blooming because it is involved in sugar translocation to flowers, which increases the sugar content of nectar and progressively improves fruit setting (Thakur et al., 2023). B's limited supply hampers photosynthesis and production of photosynthetic products in leaves, as well as the transfer of carbohydrates to other areas, resulting in sugar and starch buildup in leaves (Li et al., 2023), similar to the decline in photosynthetic rate under heavy metal (Agnihotri and Seth, 2020; Gupta and Seth, 2019; Agnihotri et al., 2018; Seth et al., 2012; Seth et al., 2011) and salinity stress (Gupta et al., 2017). B shortage increases the proteins involved in glucose and energy metabolism, resulting in an increase in glycolysis, which provides more energy to compensate for B deficit. Furthermore, B is involved in the process of N metabolism in plants. B deficiency reduces the expression of plasma membrane H^+-ATPase, which reduces the nitrate absorption capacity of tobacco roots (Li et al., 2023). B is also vital in the transfer of potassium into guard cells and consequently in the control of stomatal opening and closure (Thakur et al., 2023; Kumar et al., 2022). The suppression of DNA synthesis caused by pyrimidine base deficiencies leads to the termination of root development observed under situations of low B availability. In legumes and mycorrhiza, the element is involved in cell division, tissue vegetative development and differentiation, translation and transcription, and nitrogen fixation. B shortage impairs photosynthetic capacity and photosynthesis product transport. In B-deficient *Citrus sinensis* leaves, stomatal conductance and CO_2 absorption decreased, but intercellular CO_2 concentrations increased (Brdar-Jokanović, 2020). The effect of different B treatment dosages on the productivity of different plant species is provided in Table 7.1.

7.2.4 Enzyme Activation

B plays an indirect involvement in the activation and control of plant hormones and dehydrogenase enzymes (Thakur et al., 2023). In B-deficient *Citrus* leaves, inhibition of photosynthetic enzymes such as stromal fructose-1,6-bisphosphatase

(RuBisCO), NADP-glyceraldehyde-3-phosphate dehydrogenase (NADP-GAPDH), and stromal fructose-1,6-bisphosphatase (FBPase) is seen. While, in 'Newhall' navel orange, B deficiency dramatically reduced the concentrations of chlorophyll and carotenoids while increasing the concentrations of total soluble sugar and lignin. Furthermore, B deficiency significantly reduced the

TABLE 7.1
Different B Treatment and Source Effect Across Different Plant Species on Productivity

Plant Name	B Treatment Dosage	Source	Parameters	Effect	References
Cajanus cajan L.	Recommended dose of fertilizer	—	10 q ha^{-1} (mean grain yield)	20% increase in mean gain yield	Dhaliwal et al. (2023)
	10.0 kg ha^{-1}	Borax	12 q ha^{-1} (mean grain yield)		
Vigna radiata L.	Control	—	543 kg ha^{-1} (mean grain yield)	30.9% increase in mean grain yield	Zafar et al. (2023)
	1.1 kg ha^{-1}	NA	711 kg ha^{-1} (mean grain yield)		
Oryza sativa	Control	—	78.3% (filled grain)	5.3% increase in filled grain	Songsriin et al. (2023)
	19.9 kg ha^{-1}	Borax pentahydrate power	82.5% (filled grain)		
Cauliflower	Recommended dose of fertilizer	—	19.09 t ha^{-1} (crop yield)	21.7% increase in crop yield	Bhupenchandra et al. (2022)
	2 kg ha^{-1}	NA	23.25 t ha^{-1} (crop yield)		
Merlot grapes	Control	—	0.9 t ha^{-1} (vineyard productivity)	4.4, and 100% increase at 2 and 8 g L^{-1} compared to control for vineyard productivity	Bredun et al. (2021)
	2 g L^{-1}	Boric acid (foliar application)	1.3 t ha^{-1} (vineyard productivity)		
	8 g L^{-1}	Boric acid (foliar application)	1.8 t ha^{-1} (vineyard productivity)		
Rice cultivars	Control	—	19.9 g pot^{-1} (mean rice cultivar grain yield)	20.6% increase in mean rice cultivar grain yield	Shahid et al. (2018)
	2 kg ha^{-1}	Boric acid	24 g pot^{-1} (mean rice cultivar grain yield)		
Helianthus annuus L.	Recommended dose of fertilizer	—	1.46 t ha^{-1} (pooled seed yield)	Initial 43.1% increase as compared to control at 2 kg ha^{-1}; followed by a decrease in pooled seed yield at 4 kg ha^{-1} compared to 2 kg ha^{-1} by 16%	Bhattacharyya et al. (2015)
	2 kg ha^{-1}	Granubor	2.09 t ha^{-1} (pooled seed yield)		
	4 kg ha^{-1}	Granubor	1.77 t ha^{-1} (pooled seed yield)		

*NA, not available.

expression of photosynthesis-related genes while increasing the expression of genes involved in Chl decomposition, glucose synthesis, and lignin synthesis, but significantly inhibited the expression of carotenoid synthesis-related genes in 'Newhall' navel orange (Sharma et al., 2021).

7.3 TRANSPORTATION OF B IN PLANTS

In nature, B occurs with O_2 as borates as borax $Na_2B_4O_7 10{\bullet}H_2O$, often as H_3BO_3, and infrequently as fluorine as the anion BF_4^-, which is pH dependent (Hamada et al., 2023; Poschenrieder et al., 2019). B appears in soil solution as boric acid (H_3BO_3), borate ion (H_2BO_3), and $B(OH)_4$ (or $H_4BO_4^-$). H_3BO_3 predominates at pH 7.0, while $B(OH)_4$ predominates at pH > 8.5 (Thakur et al., 2023). Plants absorb B in the form of boric acids and borates, which seep into the plant cells as uncharged components in exchange for other necessary ions. Only around 2%–4% of B is absorbed as borate anions. The chemical enters the cell quickly via the lipid bilayer and mercury (Hg)-sensitive channels such as aquaporins and protein channels (Bolan et al., 2023). Because of the variability in B demand by function, site, and growth stage, as well as the limited remobilization opportunity from older leaves for a phloem immobile nutrient, it is critical that supply be determined precisely at the needed site (sink) during a quantitative relationship between function and supply (Rerkasem et al., 2020). As compared to the surrounding external solutions, the B-tolerant barley cultivar 'Sahara' may sustain a lower internal concentration of B (Parveen et al., 2022).

All mineral nutrients are normally taken up by the roots in ionic form with the exception of B, which is taken up in uncharged form (Hamada et al., 2023). In general, Casparian strips are formed in the root endodermis, and exodermis acts as an apoplastic pathway barrier; therefore, mineral nutrients must be taken up and then exported to the xylem via a symplastic pathway. However, B transport is primarily governed by two types of transport molecules located at the plasma membrane: BORs that export borate from cells, and nodulin-26-like intrinsic proteins (NIPs) that facilitate boric acid permeation (Onuh and Miwa, 2021). B can pass through plants via symplastic and apoplastic channels via diffusion and bulk flow pathways (Reid, 2014). There are three ways for B absorption by higher plants: (1) passively across the plasma membrane, (2) facilitatively via particular proteins such as NIPs, and (3) borate transporters (BOR) facilitated by high-affinity active transport. When exposed to a significant quantity of B, the roots absorb it by passive diffusion across the membrane. Boric acid [$B(OH)_3$], an uncharged tiny molecule, is absorbed by plants' subterranean sections via passive diffusion across the plasma membrane. Under adequate B availability, facilitated entry of B occurs via intrinsic proteins such as NIPs and major intrinsic proteins (MIPs) and via high-affinity transporters, such as BOR. The MIP family includes $B(OH)_3$ channels that are aquaporin homologs; these transport water and a limited quantity of uncharged molecules. These intrinsic proteins are abundant in membranes and are found in the root epidermis, cortical cells, and endodermis cells (Kohli et al., 2023). NIP5;1 only plays a role in B absorption when B is rare, and post-transcriptional regulation modulates NIP5;1 mRNA levels in response to cellular B concentration (Hamada et al., 2023).

B is largely delivered to plant roots via mass flow and the transpiration stream via the xylem, which also determines how it is dispersed inside plants (Thakur et al., 2023). Transpiration causes B translocation across xylem cells after B absorption by the roots. B may also be transferred via the phloem, allowing it to move between vegetative and reproductive tissues; however, this varies greatly between species. B complexation with polyols contributes to mobility. As a result, B is generally immobile in phloem in restricted mobility plants; hence, the most common deficiency symptom can be noticed in young leaves and meristematic tissue, which is associated with the death of the apical meristem (Pereira et al., 2021). The administration of ^{10}B foliar to citrus seedlings, together with girdling and shade treatments, demonstrated that foliar-provided B may be transferred from leaves to roots via phloem in *Citrus sinensis.* Plant B status and the synthesis of photosynthetic assimilates influence the translocation of foliar-supplied B to the roots (Du et al., 2020). NIP5;1 transports boric acid to root epidermal, cortical, and endodermal cells, but the Casparian band inhibits B from backflowing. Under low B circumstances, B is collected from soil solution via root cells, exported from xylem parenchyma cells into the stellar apoplasm, and then dispersed in shoots and other plant components (Shireen et al., 2018). Under B deficiency situations, B is delivered to aerial regions of the plant via xylem parenchyma cells into the stellar apoplasm and distributed to stems and leaves. BOR1 is mostly found in the plasma membrane of root, cotyledon, and hypocotyl cells, pointing toward the vascular area. In agreement with NIP5, polar localization of BOR1 is required for efficient B translocation from roots to xylem (Kohli et al., 2023). The bulk movement of water in the apoplast caused by transpiration will transport any dissolved solutes with it. B may therefore swiftly move long distances in the xylem. Concentration variations between the xylem and surrounding tissues (e.g., stem, leaves) provide diffusional gradients for passive B unloading, with or without the participation of boron-permeable channels along the way (Reid, 2014).

Plants subjected to B deprivation retain B concentrations in old mature leaves, but immature plant parts do not acquire enough B to support growth (Tanaka and Fujiwara, 2008). Similarly, under high-B soil conditions, B tends to accumulate in older leaves, where its re-translocation to young tissues is limited due to passive movement along the transpiration stream (Shireen et al., 2018; Vajjiravel et al., 2024).

7.4 CONCLUSION

In conclusion, B is essential for several physiological activities, including cell wall construction, membrane integrity, and nucleic acid metabolism. Its engagement in these key tasks emphasizes its significance in ensuring plant health and production. Boron deficiencies or excesses in the soil can have a negative impact on crop yields, resulting in suboptimal growth and lower agricultural production. The most abundant form of B in soil is $B(OH)_3$, which enters plants by passive diffusion through intrinsic protein and is transported by high-affinity protein channels that need energy and are activated in response to stumpy B signals. Moreover, with the polar expression of membrane transporters, plants are able to change the direction of B flow at the cellular level. At the total plant level, however, the integration of xylem and phloem transfer allows B to be delivered to certain tissues based on developmental stage. Researchers can create techniques to limit the environmental effect of excessive B application in agriculture by identifying the extensive network of transporters involved in B absorption, minimizing the danger of soil and water pollution.

REFERENCES

Agnihotri, Ashish, and Chandra Shekhar Seth. Exogenously applied nitrate improves the photosynthetic performance and nitrogen metabolism in tomato (Solanum lycopersicum L. cv Pusa Rohini) under arsenic (V) toxicity. *Physiology and Molecular Biology of Plants* 22 (2016): 341–349.

Agnihotri, Ashish, and Chandra Shekhar Seth. Does jasmonic acid regulate photosynthesis, clastogenecity, and phytochelatins in Brassica juncea L. in response to Pb-subcellular distribution? *Chemosphere* 243 (2020): 125361.

Agnihotri, Ashish, Praveen Gupta, Anuj Dwivedi, and Chandra Shekhar Seth. Counteractive mechanism (s) of salicylic acid in response to lead toxicity in Brassica juncea (L.) Czern. cv. Varuna. *Planta* 248 (2018): 49–68.

Bargali, Pooja, Ravendra Kumar, Ayush Devrani, Sonu Kumar Mahawer, Himani Karakoti, Om Prakash, Shilpi Rawat et al. Essential oils work synergistically to mitigate pathogenic impact of Meloidogyne incognita, Rhizoctonia solani and Sclerotinia rolfsii. *Biocatalysis and Agricultural Biotechnology* 58 (2024): 103160.

Behera, Biswaranjan, Mrunalini Kancheti, Md Basit Raza, Aalok Shiv, Vikas Mangal, Gajendra Rathod, Muhammad Ahsan Altaf et al. Mechanistic insight on boron-mediated toxicity in plant vis-a-vis its mitigation strategies: a review. *International Journal of Phytoremediation* 25, no. 1 (2023): 9–26.

Bhattacharyya, Kallol, Jajati Mandal, Hirak Banerjee, Ayon Alipatra, Krishnendu Ray, and Amit Phonglosa. Boron fertilization in sunflower (Helianthus annuus L.) in an Inceptisol of West Bengal, India. *Communications in Soil Science and Plant Analysis* 46, no. 4 (2015): 528–544.

Bhupenchandra, Ingudam, Anjali Basumatary, Samiron Dutta, L. Nabachandra Singh, Anup Das, Laishram Kanta Singh, Soibam Helena Devi, Soibam Sinyorita, and Chingakham Premabati Devi. Direct and residual impact of boron fertilization improves the crop yield, nutrient contents, nutrient uptake, and nutrient use efficiencies in cauliflower–cowpea–okra sequence in an acidic Inceptisol of North East India. *Journal of Plant Nutrition* 45, no. 7 (2022): 963–983.

Bhupenchandra, Ingudam, Sunil Kumar Chongtham, Ayam Gangarani Devi, Pranab Dutta, Elangbam Lamalakshmi, Sansuta Mohanty, Anil K. Choudhary et al. Harnessing weedy rice as functional food and source of novel traits for crop improvement. *Plant, Cell & Environment* (2024). https://doi.org/10.1111/pce.14868

Bolan, Shiv, Hasintha Wijesekara, Dhulmy Amarasiri, Tao Zhang, Péter Ragályi, Milka Brdar-Jokanović, Márk Rékási et al. Boron contamination and its risk management in terrestrial and aquatic environmental settings. *Science of the Total Environment* (2023): 164744.

Bons, Harsimrat K., and Anil Sharma. Impact of foliar sprays of potassium, calcium, and boron on fruit setting behavior, yield, and quality attributes in fruit crops: a review. *Journal of Plant Nutrition* 46, no. 13 (2023): 3232–3246.

Brdar-Jokanović, Milka. Boron toxicity and deficiency in agricultural plants. *International Journal of Molecular Sciences* 21, no. 4 (2020): 1424.

Bredun, Maiara Arbigaus, Trilicia Margarida Gomes, Thalita Isabel Assumpção, Alberto Fontanella Brighenti, Eduardo S. Chaves, Carolina Pretto Panceri, and Vívian Maria Burin. Statement of boron application impact on yield, composition and structural properties in Merlot grapes. *Scientia Horticulturae* 288 (2021): 110364.

Chaturvedi, Pranav Kumar, Chandra Shekhar Seth, and Virendra Misra. Sorption kinetics and leachability of heavy metal from the contaminated soil amended with immobilizing agent (humus soil and hydroxyapatite). *Chemosphere* 64, no. 7 (2006): 1109–1114.

Dhaliwal, Salwinder Singh, Vivek Sharma, Arvind Kumar Shukla, Janpriya Kaur, Manmeet Kaur, Vibha Verma, and Prabhjot Singh. Boron application to pigeon pea [Cajanus cajan L.] improves crop productivity and boron nutrition in Typic ustochrept soils of North-Western India. *Journal of Plant Nutrition* 46, no. 11 (2023): 2676–2690.

Dong, Xiaochang, Cuncang Jiang, Shuwei Wei, Huijun Jiao, Kun Ran, Ran Dong, and Shaomin Wang. The regulation of plant lignin biosynthesis under boron deficiency conditions. *Physiologia Plantarum* 174, no. 6 (2022): e13815.

Du, Wei, Zhi-Yong Pan, Syed Bilal Hussain, Zhong-Xing Han, and Yong-Zhong Liu. Foliar supplied boron can be transported to roots as a boron-sucrose complex via phloem in citrus trees. *Frontiers in Plant Science* 11 (2020): 515249.

Gupta, Praveen, and Chandra Shekhar Seth. Nitrate supplementation attenuates As (V) toxicity in Solanum lycopersicum L. cv Pusa Rohini: insights into As (V) sub-cellular distribution, photosynthesis, nitrogen assimilation, and DNA damage. *Plant Physiology and Biochemistry* 139 (2019): 44–55.

Gupta, Praveen, and Chandra Shekhar Seth. Interactive role of exogenous 24 epibrassinolide and endogenous NO in Brassica juncea L. under salinity stress: evidence for NR-dependent NO biosynthesis. *Nitric Oxide* 97 (2020): 33–47.

Gupta, Praveen, and Chandra Shekhar Seth. 24-Epibrassinolide regulates functional components of nitric oxide signalling and antioxidant defense pathways to alleviate salinity stress in Brassica juncea L. cv. Varuna. *Journal of Plant Growth Regulation* 42, no. 7 (2023): 4207–4222.

Gupta, Praveen, Sudhakar Srivastava, and Chandra Shekhar Seth. 24-Epibrassinolide and sodium nitroprusside alleviate the salinity stress in Brassica juncea L. cv. Varuna through cross talk among proline, nitrogen metabolism and abscisic acid. *Plant and Soil* 411 (2017): 483–498.

Gupta, Samta, and Chandra Shekhar Seth. Salicylic acid alleviates chromium (VI) toxicity by restricting its uptake, improving photosynthesis and augmenting antioxidant defense in Solanum lycopersicum L. *Physiology and Molecular Biology of Plants* 27 (2021): 2651–2664.

Hamada, Afaf M., Fatma A. Farghaly, Hussein Kh Salam, and Abeer A. Radi. Excess boron stress and alleviation of its toxicity in plants: mechanisms and strategies. *Journal of Plant Nutrition* 46, no. 11 (2023): 2724–2746.

Hazaimeh, Mohammad, Belal Muneeb Kanaan, Faleh A. AlFaleh, Mahmoud M. Elhaig, Emran Issa Khamaiseh, Qamar Zia, Mohammed Alaidarous, Chandra Shekhar Seth, Noorah Alsowayeh, and Faheem Ahmad. Biodegradation of petroleum hydrocarbons using a novel bacterial strain isolated from hydrocarbons contaminated soil of Saudi Arabia. *Biocatalysis and Agricultural Biotechnology* 57 (2024): 103074.

Hua, Tianwei, Rui Zhang, Hongwen Sun, and Chunguang Liu. Alleviation of boron toxicity in plants: mechanisms and approaches. *Critical Reviews in Environmental Science and Technology* 51, no. 24 (2021): 2975–3015.

Jin, Weihuan, Lan Cheng, Chunyan Liu, Haitao Liu, Qiujuan Jiao, Haoyang Wang, Zhaolong Deng, Chandra Shekhar Seth, Hongxiang Guo, and Yong Shi. Cadmium negatively affects the growth and physiological status and the alleviation effects by exogenous selenium in silage maize (Zea mays L.). *Environmental Science and Pollution Research* 31 (2024): 1–13.

Jothi, Muthuvel, and Junpei Takano. Understanding the regulatory mechanisms of B transport to develop crop plants with B efficiency and excess B tolerance. *Plant and Soil* 487, no. 1 (2023): 1–20.

Kim, Dae-Young, Min Kim, Jung-Suk Sung, Janardhan Reddy Koduru, Shivraj Hariram Nile, Asad Syed, Ali H. Bahkali, Chandra Shekhar Seth, and Gajanan Sampatrao Ghodake. Extracellular synthesis of silver nanoparticle using yeast extracts: antibacterial and seed priming applicationss. *Applied Microbiology and Biotechnology* 108, no. 1 (2024a): 150.

Kim, Dae-Young, Tianxi Yang, Priyanka Srivastava, Shivraj Hariram Nile, Chandra Shekhar Seth, Umesh Jadhav, Asad Syed, Ali H. Bahkali, and Gajanan Sampatrao Ghodake. Alginic acid-functionalized silver nanoparticles: a rapid monitoring tool for detecting the technology-critical element tellurium. *Journal of Hazardous Materials* 465 (2024b): 133161.

Kohli, Sukhmeen Kaur, Harsimran Kaur, Kanika Khanna, Neha Handa, Renu Bhardwaj, Jörg Rinklebe, and Parvaiz Ahmad. Boron in plants: uptake, deficiency and biological potential. *Plant Growth Regulation* 100, no. 2 (2023): 267–282.

Kumar, Dharmendra, Om Parkash Dhankher, Rudra Deo Tripathi, and Chandra Shekhar Seth. Titanium dioxide nanoparticles potentially regulate the mechanism (s) for photosynthetic attributes, genotoxicity, antioxidants defense machinery, and phytochelatins synthesis in relation to hexavalent chromium toxicity in Helianthus annuus L. *Journal of Hazardous Materials* 454 (2023): 131418.

Kumar, Vinod, Shevita Pandita, Ravdeep Kaur, Ashok Kumar, and Renu Bhardwaj. Biogeochemical cycling, tolerance mechanism and phytoremediation strategies of boron in plants: a critical review. *Chemosphere* 300 (2022): 134505.

Kumari, Nisha, Sarita Arya, Monalisha Behera, Chandra Shekhar Seth, and Ritu Singh. Chitosan anchored nZVI bionanocomposites for treatment of textile wastewater: optimization, mechanism, and phytotoxic assessment. *Environmental Research* 245 (2024): 118041.

Li, Shuang, Lei Yan, Munyaneza Venuste, Fangsen Xu, Lei Shi, Philip J. White, Xu Wang, and Guangda Ding. A critical review of plant adaptation to environmental boron stress: uptake, utilization, and interplay with other abiotic and biotic factors. *Chemosphere* 338 (2023): 139474.

Mariyam, Safoora, Renu Bhardwaj, Nafees A. Khan, Shivendra V. Sahi, and Chandra Shekhar Seth. Review on nitric oxide at the forefront of rapid systemic signaling in mitigation of salinity stress in plants: crosstalk with calcium and hydrogen peroxide. *Plant Science* 336 (2023a): 111835.

Mariyam, Safoora, Sudhir K. Upadhyay, Koushik Chakraborty, Krishan K. Verma, Joginder Singh Duhan, Sowbiya Muneer, Mukesh Meena, Rajesh Kumar Sharma, Gajanan Ghodake, and Chandra Shekhar Seth. Nanotechnology, a frontier in agricultural science, a novel approach in abiotic stress management and convergence with new age medicine-a review. *Science of the Total Environment* 912 (2023b): 169097.

Mishra, Seema; Srivastava, Sudhakar; Tripathi, Rudra Deo; Kumar, Rajesh; Seth, Chandra Shekhar; Gupta, Dharmendra K. Lead detoxification by coontail (Ceratophyllum demersum L.) involves induction of phytochelatins and antioxidant system in response to its accumulation. *Chemosphere* 65, no. 6 (2006): 1027–1039.

Misra, Virendra, Anjana Tiwari, Bhaskar Shukla, and Chandra Shekhar Seth. Effects of soil amendments on the bioavailability of heavy metals from zinc mine tailings. *Environmental Monitoring and Assessment* 155 (2009): 467–475.

Mushtaq, Naveed Ul, Seerat Saleem, Inayatullah Tahir, Chandra Shekhar Seth, and Reiaz Ul Rehman. Crosstalk in proline biosynthesis regulates proline augmentation and resilience to salt stress in Panicum miliaceum L. *Environmental and Experimental Botany* 224 (2024): 105810.

Onuh, Amarachukwu Faith, and Kyoko Miwa. Regulation, diversity and evolution of boron transporters in plants. *Plant and Cell Physiology* 62, no. 4 (2021): 590–599.

Parveen, Nishat, Roberto Berni, Sreeja Sudhakaran, Javaid A. Bhat, Suhas Shinde, Naleeni Ramawat, Vijay P. Singh et al. Metalloids in plants: a systematic discussion beyond description. *Annals of Applied Biology* 180, no. 1 (2022): 7–25.

Pathak, Himanshu K., Prabhat K. Chauhan, Chandra Shekhar Seth, Gopal Dubey, and Sudhir K. Upadhyay. Mechanistic and future prospects in rhizospheric engineering for agricultural contaminants removal, soil health restoration, and management of climate change stress. *Science of the Total Environment* 927 (2024): 172116.

Pereira, Greice Leal, João Antonio Siqueira, Willian Batista-Silva, Flávio Barcellos Cardoso, Adriano Nunes-Nesi, and Wagner L. Araújo. Boron: more than an essential element for land plants? *Frontiers in Plant Science* 11 (2021): 610307.

Poschenrieder, Charlotte, Silvia Busoms, and Juan Barceló. How plants handle trivalent (+ 3) elements. *International Journal of Molecular Sciences* 20, no. 16 (2019): 3984.

Prajapati, Priyanka, Praveen Gupta, Ravindra Nath Kharwar, and Chandra Shekhar Seth. Nitric oxide mediated regulation of ascorbate-glutathione pathway alleviates mitotic aberrations and DNA damage in Allium cepa L. under salinity stress. *International Journal of Phytoremediation* 25, no. 4 (2023): 403–414.

Reid, Rob. Understanding the boron transport network in plants. *Plant and Soil* 385 (2014): 1–13.

Rerkasem, Benjavan; Jamjod, Sansanee; Pusadee, Tonapha. Productivity limiting impacts of boron deficiency: A review. *Plant and Soil* 455, no. 1-2 (2020): 23–40.

Seth, Chandra Shekhar, Tom Remans, Els Keunen, Magdalena Jozefczak, Hilde Gielen, Koen Opdenakker, Nele Weyens, Jaco Vangronsveld, and Ann Cuypers. Phytoextraction of toxic metals: a central role for glutathione. *Plant, Cell & Environment* 35, no. 2 (2012): 334–346.

Seth, Chandra Shekhar, Vinod Misra, L. K. S. Chauhan, and R. R. Singh. Genotoxicity of cadmium on root meristem cells of Allium cepa: cytogenetic and Comet assay approach. *Ecotoxicology and Environmental Safety* 71, no. 3 (2008): 711–716.

Seth, Chandra Shekhar, Vinod Misra, R. R. Singh, and Lello Zolla. EDTA-enhanced lead phytoremediation in sunflower (Helianthus annuus L.) hydroponic culture. *Plant and Soil* 347 (2011): 231–242.

Shahid, Mohammad, Amaresh Kumar Nayak, Rahul Tripathi, Jawahar Lal Katara, Priyanka Bihari, Banwari Lal, and Priyanka Gautam. Boron application improves yield of rice cultivars under high temperature stress during vegetative and reproductive stages. *International Journal of Biometeorology* 62 (2018): 1375–1387.

Sharma, Shanti S., Vijay Kumar, and Karl-Josef Dietz. Emerging trends in metalloid-dependent signaling in plants. *Trends in Plant Science* 26, no. 5 (2021): 452–471.

Sharma, Yogesh, Praveen Soni, Gaurav Raturi, Rushil Mandlik, Vinay Kumar Rachappanavar, Manish Kumar, Prafull Salvi, Durgesh Kumar Tripathi, Hasthi Ram, and Rupesh Deshmukh. Regulation of metalloid uptake in plants by transporters and their solute specificity. *Environmental and Experimental Botany* 206 (2023): 105180.

Shireen, Fareeha, Muhammad Azher Nawaz, Chen, Qikai Zhang, Zuhua Zheng, Hamza Sohail, Jingyu Sun, Haishun Cao, Yuan Huang, and Zhilong Bie. Boron: functions and approaches to enhance its availability in plants for sustainable agriculture. *International Journal of Molecular Sciences* 19, no. 7 (2018): 1856.

Singh, Durga, Ashish Agnihotri, and Chandra Shekhar Seth. Interactive effects of EDTA and oxalic acid on chromium uptake, translocation and photosynthetic attributes in Indian mustard (Brassica juncea L. var. Varuna). *Current Science* 112 (2017): 2034–2042.

Solanki, Swati, Surbhi Sinha, Chandra Shekhar Seth, Shivanshi Tyagi, Aarushi Goyal, and Rachana Singh. Enhanced adsorption of Bismark Brown R dye by chitosan conjugated magnetic pectin loaded filter mud: a comprehensive study on modeling and mechanisms. *International Journal of Biological Macromolecules* 270 (2024): 131987.

Songsriin, Jintana, Supapohn Yamuangmorn, Sithisavet Lordkaew, Suchada Jumrus, Jeeraporn Veeradittakit, Sansanee Jamjod, and Chanakan Prom-U-Thai. Efficacy of soil and foliar boron fertilizer on boron uptake and productivity in rice. *Agronomy* 13, no. 3 (2023): 692.

Tanaka, Mayuki, and Toru Fujiwara. Physiological roles and transport mechanisms of boron: perspectives from plants. *Pflügers Archiv-European Journal of Physiology* 456 (2008): 671–677.

Thakur, Shivani, Agnibha Sinha, and Animesh Ghosh Bag. Boron-a critical element for fruit nutrition. *Communications in Soil Science and Plant Analysis* 54, no. 21 (2023): 2899–2914.

Vajjiravel, Prakash, Divya Nagarajan, Varsha Pugazhenthi, Ajay Suresh, Madhan Kumar Sivalingam, Ajila Venkat, Pritam Paramguru Mahapatra et al. Circadian-based approach for improving physiological, phytochemical and chloroplast proteome in Spinacia oleracea under salinity stress and light emitting diodes. *Plant Physiology and Biochemistry* 207 (2024): 108350.

Verma, Krishan K., Abhishek Joshi, Xiu-Peng Song, Shraddha Singh, Aradhna Kumari, Jaya Arora, Santosh Kumar Singh, Manoj Kumar Solanki, Chandra Shekhar Seth, and Yang-Rui Li. Synergistic interactions of nanoparticles and plant growth promoting rhizobacteria enhancing soil-plant systems: a multigenerational perspective. *Frontiers in Plant Science* 15 (2024): 1376214.

Yadav, Manisha, Praveen Gupta, and Chandra Shekhar Seth. Foliar application of α-lipoic acid attenuates cadmium toxicity on photosynthetic pigments and nitrogen metabolism in Solanum lycopersicum L. *Acta Physiologiae Plantarum* 44, no. 11 (2022): 112.

Zafar, Muhammad, Siraj Ahmed, Muhammad Kashif Munir, Nawal Zafar, Muhammad Saqib, Muhammad Aleem Sarwar, Saba Iqbal et al. Application of zinc, iron and boron enhances productivity and grain biofortification of Mungbean. *Phyton* 92, no. 4 (2023): 983–999.

8 Metalloid Toxicity in Plants
A Systemic Assessment on Recent Insights Concerning their Impact on Plants

Safoora Mariyam, Kamel A. Abd-Elsalam, Hrishikesh Upadhyaya, Muthu Thiruvengadam, Asish Kumar Parida, Cengiz Kaya, and Chandra Shekhar Seth

8.1 INTRODUCTION

All strata of our environment are naturally populated with metals and metalloid ions. Numerous living forms may become toxic when their levels are found in excessive quantities due to their accumulation in food chain (Sharma et al., 2021a,b; Tavanappanavar et al., 2024). Aside from being naturally occurring, human activity has played a significant role in their release from these natural sources, which has contaminated the atmosphere (Mariyam et al., 2023a,b). This, in turn, has had an impact on bioaccumulation at different food chain levels and presents serious risks to the well-being of people as a result of the destruction of other ecosystems and agricultural land (Kumar et al., 2023; Mariyam and Seth, 2023). The productivity of agriculture is significantly impacted by harmful metal pollution of the environment, which is compounded by intensive modern farming methods, and human activity has put human health and nutritional security in jeopardy. Increased toxicity of metals/metalloids not only affects agricultural productivity but also alters plant demand and supply balance (Deng et al., 2021; Kumar et al., 2024b). A deeper comprehension of the intricate physiological, metabolic, molecular, cellular plant pathways that could result in higher agricultural productivity is required to counteract the toxicity of toxic metals (TMs) and metalloids (Zhao et al., 2022).

Because many metals and metalloids play vital roles in the structural and functional well-being of all species, their presence in the environment, whether trace or abundant, has an impact on living life (Mishra et al., 2006; Brar et al., 2024). They change the system of cells that are living as part of nutrition via absorption and transport and can be ingested as essential nutrients, micronutrients, or due to over-digestion or the inclusion of an undesirable element unintentionally as hazardous contaminants (Nabi et al., 2021; Upadhyay et al., 2024). Although carbon, hydrogen, and oxygen constitute the fundamental building blocks of life, the efficient operation of metabolism is also dependent on a small number of elements in the form of metals (Seth et al., 2012; Kim et al., 2024). The bioavailability of metalloids in soil determines their impact on crop productivity and metalloid transmission to the food chain. The main factors influencing metalloid bioavailability, speciation, and partition include pH, redox potential (Eh), soil organic matter, soil structure, and soil chemical composition and mix (Batham et al., 2022). However, much focus of research on trace elements, especially transition metals, has been and remains on the geochemical and biogeochemical processes occurring in natural streams. But far less focus of research on trace elements, especially transition metals, has been and remains on the geochemical and biogeochemical processes occurring in natural streams. But far less focus has been placed on the impending "metalloid" pollution, which has what prompted the writing of this chapter.

DOI: 10.1201/9781032673592-8

8.2 METALLOIDS: RELEVANCE IN THE FOOD CHAIN AND INDUCED CONTAMINATION

The problem of trace metal bio-magnification in food chains has only been partially overcome. Not every insignificant heavy metal bio-magnifies in the food web. Methyl mercury (MeHg) has been shown to bio-magnify in food chains (Bali et al., 2021). Within the food system made up of fishes, macro-invertebrate consumers, and producers in general, Cd and lead (Pb) do not bio-magnify (Seth et al., 2008; Jin et al., 2024). These metals, however, are biomagnified in specific food chains. A significant field of environmental study is the trophic transmission of hazardous metals and metalloids in food networks (Parveen et al., 2022; Kumar et al., 2024c). They contaminate food chains by spreading from water or soil to biota, then to animals and humans. Exposure to carcinogenic substances puts millions of people worldwide at risk. Humans can consume these substances through a range of foods and beverages, such as agricultural products, refined foods, fruits, mushrooms, tap water, and livestock products, among others. As hotspots for pollution, Bangladesh and India provide the greatest threat (Xalxo et al., 2021). According to studies, even in generally uncontaminated regions like Australia, rice is farmed with dangerously high concentrations of poisonous chemicals like As. This is a global issue because many countries buy and export rice. The major source of Se for humans is food (Pandey et al., 2019). Seleniferous food consumption has been linked to lethal Se amounts in humans. Se contents in soil are a reliable indicator of Se concentrations in diet (Agnihotri and Seth, 2020). The overall Se content of a meal is influenced by several factors, including weather, protein supply, preparation method, and geographic location (Seth et al., 2007). Se levels throughout the environment and food sources need to be monitored. The diet absorbs the majority of Se, with a 75% rate of absorption as an average. In general, Se in organic form is more absorbent compared to Se in its inorganic state. The key metals are Pb, Cd, and mercury (Hg), while As is the key metalloid found in nutritional supplements. Routine screening detects lethal metal contamination in food and medicines only occasionally (Kumar et al., 2022; Pathak et al., 2024) (Figure 8.1).

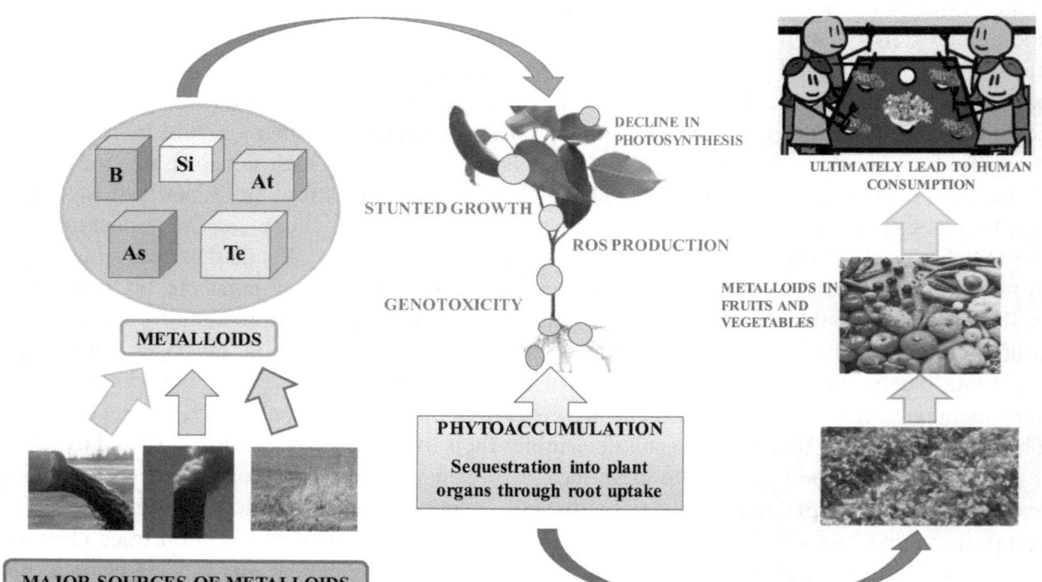

FIGURE 8.1 Major sources of metalloids in the environment, their accumulation in the plants, and transfer to different levels of the food chain.

8.3 METALLOID HAZARDS: IMPACTS ON VARIOUS ASPECTS OF PLANT PHYSIOLOGY AND BIOCHEMISTRY

Metalloids that are non-essential, such as Ge, Sb, and As, are detrimental to all living organisms. Plants' principal reactions to overindulgence of As and Sb cause structural and physiological impairment in addition to the quick emergence of s (ROS), which considerably reduces productivity (Gupta and Seth, 2019). Ge and As share some properties with B and Si at medically relevant pH levels and mostly take the shape of tiny, non-charged molecules. Metalloids impair plant productivity and enhance ROS generation through a variety of mechanisms (Singh et al., 2017). ROS are signalling molecules that participate in numerous physiological progressions such as cell division cycle, stress sensing, hypersensitive response, regulation of gene expression, and senescence, according to several studies. They are also capable of initiating and/or directing plant responses to a variety of environmental stimuli (Silva-Gigante et al., 2023). As generates oxidative stress; it also changes the redox state, causing membranes and proteins to be damaged and eventually leading to cell death. High As levels reduce crop productivity, seed germination, and biomass output (Zhao et al., 2022). It has been demonstrated that As prevents the sprouting of seeds of several crop plants, such as wheat, rice. Other side effects of As include decreased plant height, root and shoot growth, and photosynthetic rate. SbIII was discovered to be more poisonous than SbV (Ren et al., 2014). However, absorption and cytotoxicity of Sb in plants have been the subject of extensive research in the past few decades. More than 10 mg kg^{-1} Sb is considered harmful to plants; nonetheless, 3–4 mg Sb kg^{-1} in plants is assumed to be tolerated. Encountering Sb in both its pentavalent and trivalent forms has been shown to inhibit rice (Yuhong No.1) seed growth, yield, and sprout rates, as well as reduce dry matter transformation ratios (Parveen et al., 2022).

B toxicity symptoms vary between species with limited and extensive phloem mobility. In phloem-immobile species, B travels via the xylem and accumulates at the terminus of the transpiration stream (Chaturvedi et al., 2006). In barley and wheat, stress indications in leaves comprise chlorosis and necrosis that extend from their tips, as well as brownish spots that begin on leaf ends and progress to cover the majority of the leaf surface (Seth et al., 2008). The disease begins with the oldest leaves and progresses to the plant's crown. The three metabolic reactions that cause adverse consequences of B excess are as follows: (1) attachment to ribose as a free sugar and as a constituent of RNA impairs cell division and development; (2) attaching to ribose in ATP or NAD(P)H hinders primary metabolism; and (3) lowering of cytosolic pH affects the shape of proteins and metabolism. Metalloid Ge, which occurs in small amounts in the earth, induces toxicity to plants at high concentrations and to physiological functions at low concentrations (Hua et al., 2021). Its chemical properties are comparable to those of Si. The complimentary ratio of Ge to Si aids in the study of weathering processes under subtropical as well as tropical ecosystems as well as the origins of Si in marine sediments. While Si's extensively reported significance in plants includes, among other functions, improving rigidity, tolerance to pests and diseases, and defence against abiotic stresses, it is uncertain what part Ge plays in metabolic processes (Seth, 2012; Vajjiravel et al., 2024). Another metalloid, Te, is frequently employed in industry as an electronic component in thin films, battery packs, and transference of charge equipment. Te has a strong impact on human health. Te (IV) residues have been traced in water, silt, earth crust, crops, and milk. Te absorption, transport, and metabolism remain unidentified; however, they are most likely related to sulphur or selenium (Se) (Sharma et al., 2021b). The impacts of different metalloids with different dosages on various plant species are provided in Table 8.1.

8.4 METALLOID TOXICITY IN PLANTS: AN OMICS PERSPECTIVE

TMs and metalloids' greater persistence in the natural environment pose a severe threat to plant nutrition and thus crop productivity (Seth et al., 2008). Data sequencing, structural organisation, genetic alignments, linkages, and functions of an entire plant are all part of genomics. Because they

TABLE 8.1
Toxic Impacts of Different Metalloids in Various Plant Species When Applied in Different Concentrations

S. No.	Metalloids	Plant Studied	Concentration Used	Effect on Plant	References
1.	B	*Capsicum annuum* L. (pepper plant)	4 mM	Decrease in total flavonoid, protein, and chlorophyll content	Çatav et al. (2022)
		Solanum lycopersicum L. (tomato plant)	2 mM	• Increase in hydrogen peroxide and malondialdehyde content indicating oxidative damage • Decline in photosystem II efficiency (Fv/Fm)	Kaya et al. (2020)
		Mentha arvensis L. (mint plant)	5 mM	Drop in growth as well as essential oil content	Choudhary et al. (2020)
2.	Silicon nanoparticles (SiNPs)	*Arabidopsis thalliana*	100 mg L^{-1}	Decrease in the adsorption of macro and micronutrients	Slomberg and Schoenfisch (2012)
3.	Antimony (Sb)	*Oryza sativa* (rice plant)	20 mg L^{-1}	• Decrease in shoot and root biomass • Disturbance of O^{2-} dismutation	Zhu et al. (2020)
	Sb (III) Sb (V)	*Oryza sativa* (rice plant)	20 mg L^{-1} 20 mg L^{-1}	• Decline in hemicellulose I • Increase in abscisic acid (ABA) content • Decline in formic acid and acetic acid concentration in root exudates	Feng et al. (2020)
4.	As	*Capsicum annuum L.* (pepper plant)	50 µM	• Decrease in relative water content • Rise in PCs • and activity of glutathione-*S*-transferase	Kaya et al. (2022)
		Vicia faba L. (faba bean)	5 µM	• 1.3-fold increase in endogenous cysteine content • 1.5-times escalation in O-acetylserine lyase activity and 1.7-times upsurge in serine acetyltransferaseactivity	Siddiqui et al. (2021)
		Solanum tuberosum L. (potato)	40 µM	• Limited the starch content • Inhibition of nitrite reductase and nitrite reductase enzymes	Shahid et al. (2019)
		Salvadora persica L.	600 µM	• Bioaccumulation factor increased • Decrease in total GSH content	Patel and Parida (2021)
5.	Ge	*Cucumis sativus L.* (cucumber)	20 mg L^{-1}	• The GeO_2 treatment dramatically reduced shoot and root length and total fresh weight.	Choi et al. (2013)

can quickly discover candidate genes inside large quantitative trait locus (QTL), several trait mapping technologies, including QTL mapping, quantitative trait locus sequencing (QTL-seq), RNA sequencing (RNA-seq), have been developed to change the mapping route. A marker–trait association study for As-associated features was conducted using around 700 SNPs (Raza et al., 2022). There were approximately 9 QTLs with phenotypic variance ranging from 8.5% to 12.5%, with 6 QTLs for As quantity in stem mapped on chromosomes 2, 5, 6, and 9; two QTLs in roots (qAsR8.1 and qAsR8.2) mapped on chromosome 8; and one QTL for relative chlorophyll (Chl) content (qRChlo1) mapped on chromosome 1 (Murugaiyan et al., 2019). These QTL intervals were used to detect 25 genes with As stress-related features, indicating transcription regulation. Transcriptomic research has recently concentrated on the combined toxicity of numerous metals and metalloids in plants. The main objective for such evaluations has been to determine the processes by which plants endure a variety of hazardous and non-toxic substances present in the soil, in addition to the form in which harmful compounds use functioning metal transporters to release detoxifying composites, facilitating them to pass through plant cell (Agnihotri and Seth, 2020). Numerous differentially expressed genes (DEGs) are expected to be convoluted in antioxidant mechanism, transport, and cellular wall formation; thus, regulation of these genes (whether constitutive or triggered by metalloid stress) is crucial for the plant, particularly in its root system through the initial phases of metal stress (Angulo-Bejarano et al., 2021). Locating the number of transporter genes inside cell structure informs regarding the molecular pathways plants prefer (Kim et al., 2023). The preliminary study of transcripts in plants exposed to metal toxicity or tolerance concentrated on recognising key transporters (ATP-binding cassette transporters (ABC transporters), natural resistance-associated macrophage proteins (NRAMPs), and stress-associated proteins (STARs), among others) which assist plant cells in integrating and splitting toxic ions or excreting organic compounds (malate, citrate, and so on) (Mondal et al., 2022).

Metallomics is a modern scientific topic that has seen active development in the previous decade; as a result, this notion established prodigious responsiveness and is thoroughly explored by a number of authors in many review papers and monographs. The polyfunctionality of PCs allows them to play an important part in plant reactions to metal(loid) toxicity and its mitigation. Different phenolic subclass compounds are responsible for their chelating ability towards metalloids, their oxides, and metal(loid)-nanoparticles (NPs) due to binding sites (Misra et al., 2009). The chelating impacts are demonstrated by phenol-containing composites found in a variety of plant tissues and organs along with rhizosphere. Plant–pollinator interactions, antioxidant and pro-oxidant systems, metal decontamination in plant tissues and the rhizosphere, vacuolar sequestration, element mobilisation, phytoavailability—each are reliant on phenolic–metalloid complexation process (Seth et al., 2008). Proteomics is the study of encoded proteins in live animals at a certain time, and it is critical for understanding all biological activities at the molecular level. In many aspects, Se is unusual, and its investigation of biological systems is very intriguing (Feng et al., 2021). Despite the fact that it is non-essential, higher plants absorb and metabolise it through sulphur pathways; nonetheless, higher levels of Se pose lethal indications in plants. Molecular aspects of Se phytotoxicity are merely understood; evidence collected thus far indicates Se toxicity impacts plant proteome (Kolbert et al., 2019). Se absorption disrupts element homeostasis in the organs of Se-supplemented plants by interfering with the uptake of other elements. Applying selenium (Se) at a concentration of 2 mg kg^{-1} to *Triticum aestivum L.* (wheat) has been shown to enhance nitrate reductase (NR) activity and increase the accumulation of nitrogen (N) and sulfur (S) in the plant. When it comes to Se–S interactions, White and Broadley (2005) observed that while exogenous sulphate inhibits selenate absorption in Arabidopsis, rhizosphere selenate stimulates sulphate uptake due to Se's inhibitory action on S transporters in the presence of excess S. Se supplementation has a variety of impacts on other micro- and macro-elements in various plant species, causing disturbance of element homeostasis in plant organs and potentially leading to Se phytotoxicity.

8.5 METALLOID-INDUCED PHYSIOLOGICAL AND BIOCHEMICAL RESPONSES IN PLANTS

Owing to their extensively reported potential hazards to biological systems, toxicology researchers are particularly concerned about metalloids. The majority of metalloids are major environmental pollutants that have a negative influence on crop yield under high concentrations. They are associated with plasma membrane carrier proteins then translocated to different plant parts; triggering alterations in essential biochemical pathways, the destruction of cell biomolecules; and, lastly, the suppression of cell development (Yadav et al., 2022). Due to the adverse consequences on agricultural output, metalloid-contaminated soil is a subject of concern. Metalloids acquired significant prominence in research for clean-up owing to the properties of being high perseverance, lack of degradability, and harmfulness. When metalloid bioavailability in the ground is elevated, it functions as an environmental contaminant, endangering the ecosystem as well as plants and animals. ROS including $_1O^2$, $O^{\bullet 2}$, $OH^{\bullet}$, H_2O_2, and $O_2^{\bullet 2}$ are produced when metal and metalloid concentrations are elevated. Plants are stressed due to high ROS production (Gupta and Seth, 2023). Tripeptide glutathione (GSH, -Glu-Cys-Gly), phytochelatins (PCs), and metallothioneins (MTs) are thiol-containing molecules that can bind to metals and metalloids. GSH is present in plants. GSH can be present in the cell organelles of plant cells. GSH is essential for the formation of PCs, assisting in metal and metalloid decontamination and leaching in plant cells (Agnihotri et al., 2018).

In plants, the difference between optimal and hazardous B concentrations is quite small. Plant-mediated mitigation of B-induced toxic effects in plants varies broadly between species and even between species variations. In plants susceptible to B stress e.g. *Phaseolus vulgaris*, safe B concentrations in irrigation water range from 0.4 to 1.5 mg L–1 (García-Sánchez et al., 2020) for semi-tolerant plants (e.g. *Zea mays* and *Solanum tuberosum*) (Nawaz et al., 2020), 3 mg L–1 for less susceptible plants (e.g. *Daucus carota* and Cuminos melo) (Landi et al., 2012), and 5 mg L–1 for least susceptible plants (e.g. *Solanum lycopersicon*) (Pacci et al., 2023; Mushtaq et al., 2024). B contributes to oxidative damage by creating ROS and causing membrane malfunction and cell death as a result of hyperosmotic and ionic stresses. Despite the fact that a higher amount of B impedes photosynthetic process, evidence of these is still scarce (Mariyam et al., 2023a,b). Lower CO_2 assimilation has been widely connected to stomatal variables in B-deficient leaves. In contrast, the decrease in CO_2 assimilation in B surplus conditions appears to be caused by a mix of variables, including oxidative stress, a drop in photosynthetic enzyme activity, and an impeded electron transport rate. However, the processes behind the change in photosynthesis produced by B stress have yet to be discovered. Researchers discovered that a decline in photosynthetic activity due to exposure of high B was related to an increase in intercellular CO_2, while stomatal conductance remained unchanged. Other writers, on the other hand, observed a decrease in stomatal conductance. Plants exposed to high levels of As lowered their photosynthesis due to the direct influence on the photosynthetic system. The catalytic activity of ribulose-1,5-bisphosphate carboxylase/oxygenase (RuBisCO) is responsible for CO_2 fixation (Siddiqui et al., 2021; Solanki et al., 2024). As interacts exclusively with the larger component of RuBisCO, resulting in the development of a ternary stable complex. This combination may further alter the enzyme structure surrounding the catalytic area under As stress, resulting in overall enzyme inactivation and a decrease in photosynthesis. As, in addition to inducing physiological changes in mitochondria and chloroplasts, destroys the protoplast of plant mesophyll cells. Plant cellular membranes are very sensitive to As attack. As exposure induced increased electrolyte leakage in plants, indicating membrane damage as a result of As toxicity (Mondal et al., 2022; Bargali et al., 2024). As affects the functioning of genes in plant cells along with numerous transporter genes found in the plasma membrane, causing water uptake from soil as well as the transportation of nutrients and water to be disrupted.

Symptoms of metalloid poisoning include growth inhibition, photosynthetic apparatus distortion, biomass accumulation inhibition, and so on. Despite major efforts to identify underlying causes of As-induced harmful consequences in various stages during plant growth, our knowledge of the matter is far from complete (Khanna et al., 2022). Metalloids generate ROS which oxidise cellular macromolecules such as lipids, proteins, nucleic acids, and carbohydrates. This oxidation reaction can disrupt a variety of metabolic pathways, either directly by lowering the activity of essential enzymes or indirectly by producing oxidative stress. Metalloid poisoning has a negative impact on plant quality and productivity as a result.

8.6 CONCLUSIONS AND FUTURE PROSPECTS

Because of human activity and natural erosion, metalloid quantities in the environment have increased. Metalloids have a substantial effect on agricultural land as well as ecosystems. Due to this, agricultural yields are declining, and harmful metalloids are increasingly entering food systems. Passive, or simple diffusion; assisted diffusion; or active, or metabolically active routes are used for metalloid absorption. Metalloids induce gene expression as well as a complex signal transduction system, both of which are important in mediating cellular response and toxicity. Plants modify gene, protein, and metabolite expression levels in reaction for unfavourable conditions including metal or metalloid hazards. To deal with noxiousness of hazardous metalloids under increasingly contemporary society, a thorough understanding of biological processes is essential. Significant advancement has been made in latest years in developing climate-smart plants using leading-edge "omics" technologies. Interestingly, genomewide omics research can identify genes, metabolites, proteins, stress regulators, miRNAs, and metabolic processes associated with stress, which are specifically linked to plant phenotype under stressful conditions. Because the process of metalloid uptake is poorly understood, understanding the mechanism of metalloid uptake is crucial in order to limit its harmful effects on agricultural plants.

REFERENCES

Agnihotri, Ashish, and Chandra Shekhar Seth. Does jasmonic acid regulate photosynthesis, clastogenecity, and phytochelatins in *Brassica juncea* L. in response to Pb-subcellular distribution? *Chemosphere* 243 (2020): 125361.

Agnihotri, Ashish, Gupta, Praveen, Dwivedi, Anuj, and Seth, Chandra Shekhar. Counteractive mechanism(s) of salicylic acid in response to lead toxicity in *Brassica juncea* (L.) Czern. cv. Varuna. *Planta*, 248, no. 1 (2018): 49–68.

Angulo-Bejarano, P. I., Puente-Rivera, J., and Cruz-Ortega, R. Metal and metalloid toxicity in plants: an overview on molecular aspects. *Plants*, *10*(4) (2021): 635.

Bali, A. S., & Sidhu, G. P. S. Arsenic acquisition, toxicity and tolerance in plants—from physiology to remediation: A review. *Chemosphere*, 283 (2021), 131050.

Bargali, Pooja, Ravendra Kumar, Ayush Devrani, Sonu Kumar Mahawer, Himani Karakoti, Om Prakash, Shilpi Rawat et al. Essential oils work synergistically to mitigate pathogenic impact of Meloidogyne incognita, Rhizoctonia solani and Sclerotinia rolfsii. *Biocatalysis and Agricultural Biotechnology* 58 (2024): 103160.

Batham, Manish, Jot Sharma, and Pragya Singh. Toxicity of metalloid-arsenic and its remediation mechanism(s). *International Journal of Life Sciences & Technology* 14, no. 1 (2022).

Brar, Basanti, Baljeet Singh Saharan, Chandra Shekhar Seth, Ajay Kamboj, Kiran Bala, Vishnu Dayal Rajput, Tatiana Minkina et al. Nanobiochar: soil and plant interactions and their implications for sustainable agriculture. *Biocatalysis and Agricultural Biotechnology* 57 (2024): 103077.

Çatav, Serter; Gülçin Özmen Ünal; and Serap Uçarlı. Selenium Supplementation Mitigates Boron Toxicity-Induced Growth Inhibition and Oxidative Damage in Pepper Plants. *South African Journal of Botany* 146 (2022): 375–382.

Chaturvedi, Pranav Kumar, Chandra Shekhar Seth, and Virendra Misra. Sorption kinetics and leachability of heavy metal from the contaminated soil amended with immobilizing agent (humus soil and hydroxyapatite). *Chemosphere* 64, no. 7 (2006): 1109–1114.

Choi, Woo-Jin; Lee, Sang-Mook; and Ro, Hee-Myong. Evaluation of Soil Nitrogen Availability Indices in Upland Soils Amended with Organic Wastes. *Communications in Soil Science and Plant Analysis*, 44, no. 20 (2013): 3006–3019.

Choudhary, Sadaf, Andleeb Zehra, M. Naeem, M. Masroor A. Khan, and Tariq Aftab. Effects of boron toxicity on growth, oxidative damage, antioxidant enzymes and essential oil fingerprinting in Mentha arvensis and Cymbopogon flexuosus. *Chemical and Biological Technologies in Agriculture* 7 (2020): 1–11.

Deng, Fenglin, Fanrong Zeng, Guang Chen, Xue Feng, Adeel Riaz, Xiaojian Wu, Wei Gao, Feibo Wu, Paul Holford, and Zhong-Hua Chen. Metalloid hazards: from plant molecular evolution to mitigation strategies. *Journal of Hazardous Materials* 409 (2021): 124495.

Feng, RenWei, Lei, JunMing Su, RuiRui Zhang, YanMing Zhu, WenXiang Chen, LiZhen Wang et al. Toxicity of different forms of antimony to rice plant: effects on root exudates, cell wall components, endogenous hormones and antioxidant system. *Science of the Total Environment* 711 (2020): 134589.

Feng, RenWei, PingPing Zhao, YanMing Zhu, JiGang Yang, XinQi Wei, Li Yang, Hong Liu, Christopher Rensing, and Yong Zhen Ding. Application of inorganic selenium to reduce accumulation and toxicity of heavy metals (metalloids) in plants: the main mechanisms, concerns, and risks. *Science of the Total Environment* 771 (2021): 144776.

García-Sánchez, Francisco, Silvia Simón-Grao, Juan J. Martínez-Nicolás, Marina Alfosea-Simón, Chunguang Liu, C. Chatzissavvidis, Juan G. Pérez-Pérez, and Jose M. Cámara-Zapata. Multiple stresses occurring with boron toxicity and deficiency in plants. *Journal of Hazardous Materials* 397 (2020): 122713.

Gupta, Praveen, and Chandra Shekhar Seth. Nitrate supplementation attenuates As (V) toxicity in Solanum lycopersicum L. cv Pusa Rohini: insights into As (V) sub-cellular distribution, photosynthesis, nitrogen assimilation, and DNA damage. *Plant Physiology and Biochemistry* 139 (2019): 44–55.

Gupta, Praveen, and Chandra Shekhar Seth. 24-Epibrassinolide regulates functional components of nitric oxide signalling and antioxidant defense pathways to alleviate salinity stress in Brassica juncea L. cv. Varuna. *Journal of Plant Growth Regulation* 42, no. 7 (2023): 4207–4222.

Hua, Tianwei, Rui Zhang, Hongwen Sun, and Chunguang Liu. Alleviation of boron toxicity in plants: mechanisms and approaches. *Critical Reviews in Environmental Science and Technology* 51, no. 24 (2021): 2975–3015.

Jin, Weihuan, Lan Cheng, Chunyan Liu, Haitao Liu, Qiujuan Jiao, Haoyang Wang, Zhaolong Deng, Chandra Shekhar Seth, Hongxiang Guo, and Yong Shi. Cadmium negatively affects the growth and physiological status and the alleviation effects by exogenous selenium in silage maize (*Zea mays* L.). *Environmental Science and Pollution Research* 31 (2024): 1–13.

Kaya, Cengiz, Ali Sarıoglu, Muhammad Ashraf, Mohammed Nasser Alyemeni, and Parvaiz Ahmad. Gibberellic acid-induced generation of hydrogen sulfide alleviates boron toxicity in tomato (Solanum lycopersicum L.) plants. *Plant Physiology and Biochemistry* 153 (2020): 53–63.

Kaya, Cengiz, Ali Sarıoglu, Muhammad Ashraf, Mohammed Nasser Alyemeni, and Parvaiz Ahmad. The combined supplementation of melatonin and salicylic acid effectively detoxifies arsenic toxicity by modulating phytochelatins and nitrogen metabolism in pepper plants. *Environmental Pollution* 297 (2022): 118727.

Khanna, Kanika, Sukhmeen Kaur Kohli, Pankaj Kumar, Puja Ohri, Renu Bhardwaj, Pravej Alam, and Parvaiz Ahmad. Arsenic as hazardous pollutant: perspectives on engineering remediation tools. *Science of the Total Environment* 838 (2022): 155870.

Kim, Dae-Young, Tianxi Yang, Priyanka Srivastava, Shivraj Hariram Nile, Chandra Shekhar Seth, Umesh Jadhav, Asad Syed, Ali H. Bahkali, and Gajanan Sampatrao Ghodake. Alginic acid-functionalized silver nanoparticles: a rapid monitoring tool for detecting the technology-critical element tellurium. *Journal of Hazardous Materials* 465 (2024): 133161.

Kim, T.-W., Park, C. H., Hsu, C.-C., Kim, Y.-W., Ko, Y.-W., Zhang, Z., Zhu, J.-Y., Hsiao, Y.-C., Branon, T., Kaasik, K., Saldivar, E., Li, K., Pasha, A., Provart, N. J., Burlingame, A. L., Xu, S.-L., Ting, A. Y., & Wang, Z.-Y. (2023). Mapping the signaling network of BIN2 kinase using TurboID-mediated biotin labeling and phosphoproteomics. *The Plant Cell*, 35(3), 975–993. https://doi.org/10.1093/plcell/koad013

Kolbert, Zsuzsanna, Árpád Molnár, Gábor Feigl, and Doug Van Hoewyk. Plant selenium toxicity: proteome in the crosshairs. *Journal of Plant Physiology* 232 (2019): 291–300.

Kumar, Amit, Vinod Kumar, Andrés Rodríguez-Seijo, Raj Setia, Sumit Singh, Ashok Kumar, Chandra Shekhar Seth, and Renato Somma. Appraisal of heavy metal(loid)s contamination in rice grain and associated health risks. *Journal of Food Composition and Analysis* 131 (2024a): 106215.

Kumar, Dharmendra, Om Parkash Dhankher, Rudra Deo Tripathi, and Chandra Shekhar Seth. Titanium dioxide nanoparticles potentially regulate the mechanism(s) for photosynthetic attributes, genotoxicity, antioxidants defense machinery, and phytochelatins synthesis in relation to hexavalent chromium toxicity in Helianthus annuus L. *Journal of Hazardous Materials* 454 (2023): 131418.

Kumar, Dharmendra, Ritu Singh, Sudhir K. Upadhyay, Krishan K. Verma, Ravi Mani Tripathi, Haitao Liu, Om Parkash Dhankher, Rudra Deo Tripathi, Shivendra V. Sahi, and Chandra Shekhar Seth. Review on interactions between nanomaterials and phytohormones: novel perspectives and opportunities for mitigating environmental challenges. *Plant Science* 340 (2024b): 111964.

Kumar, Dharmendra, Safoora Mariyam, Kapuganti Jagadis Gupta, Muthu Thiruvengadam, Gajanan Ghodake, Baoshan Xing, and Chandra Shekhar Seth. Comparative investigation on chemical and green synthesized titanium dioxide nanoparticles against chromium (VI) stress eliciting differential physiological, biochemical, and cellular attributes in Helianthus annuus L. *Science of the Total Environment* 930 (2024c): 172413.

Kumar, Saurabh, Arbind Kumar Choudhary, Deep Chandra Suyal, Govind Makarana, and Reeta Goel. Leveraging arsenic resistant plant growth-promoting rhizobacteria for arsenic abatement in crops. *Journal of Hazardous Materials* 425 (2022): 127965.

Kumar, Vinod, Monika Thakur, and Chandra Shekhar Seth. In situ remediation techniques for removal of arsenic in the environment. *Current Opinion in Environmental Science & Health* 38 (2024d): 100538.

Landi, Marco, Elena Degl'Innocenti, Alberto Pardossi, and Lucia Guidi. Antioxidant and photosynthetic responses in plants under boron toxicity: a review. *American Journal of Agricultural and Biological Sciences* 7, no. 3 (2012): 255–270.

Mariyam, Safoora, and Chandra Shekhar Seth. Nitric oxide: a key player in reinforcement of photosynthetic efficiency under abiotic stress. In Khan, M. I. R., Iqbal, N., Poor, P., Ferrante, A. (Eds.), *Nitric Oxide in Developing Plant Stress Resilience* (pp. 157–171). Academic Press, Elsevier, UK, 2023. ISBN: 978-0-323-91209-9

Mariyam, Safoora, Renu Bhardwaj, Nafees A. Khan, Shivendra V. Sahi, and Chandra Shekhar Seth. Review on nitric oxide at the forefront of rapid systemic signaling in mitigation of salinity stress in plants: crosstalk with calcium and hydrogen peroxide. *Plant Science* 336 (2023a): 111835.

Mariyam, Safoora, Sudhir K. Upadhyay, Koushik Chakraborty, Krishan K. Verma, Joginder Singh Duhan, Sowbiya Muneer, Mukesh Meena, Rajesh Kumar Sharma, Gajanan Ghodake, and Chandra Shekhar Seth. Nanotechnology, a frontier in agricultural science, a novel approach in abiotic stress management and convergence with new age medicine-a review. *Science of the Total Environment* 912 (2023b): 169097.

Mishra, Seema, S. Srivastava, R. D. Tripathi, R. Kumar, C. S. Seth, and D. K. Gupta. Lead detoxification by coontail (Ceratophyllum demersum L.) involves induction of phytochelatins and antioxidant system in response to its accumulation. *Chemosphere* 65, no. 6 (2006): 1027–1039.

Misra, Virendra, Anjana Tiwari, Bhaskar Shukla, and Chandra Shekhar Seth. Effects of soil amendments on the bioavailability of heavy metals from zinc mine tailings. *Environmental Monitoring and Assessment* 155 (2009): 467–475.

Mondal, Sayanta, Krishnendu Pramanik, Sudip Kumar Ghosh, Priyanka Pal, Pallab Kumar Ghosh, Antara Ghosh, and Tushar Kanti Maiti. Molecular insight into arsenic uptake, transport, phytotoxicity, and defense responses in plants: a critical review. *Planta* 255, no. 4 (2022): 87.

Murugaiyan, Varunseelan, Jauhar Ali, Anumalla Mahender, Umair M. Aslam, Zilhas Ahmed Jewel, Yunlong Pang, Corinne M. Marfori-Nazarea, Lin-Bo Wu, Michael Frei, and Zhikang Li. Mapping of genomic regions associated with arsenic toxicity stress in a backcross breeding populations of rice (Oryza sativa L.). *Rice* 12 (2019): 1–14.

Mushtaq, Naveed Ul, Seerat Saleem, Inayatullah Tahir, Chandra Shekhar Seth, and Reiaz Ul Rehman. Crosstalk in proline biosynthesis regulates proline augmentation and resilience to salt stress in Panicum miliaceum L. *Environmental and Experimental Botany* 224 (2024): 105810.

Nabi, Aarifa, M. Naeem, Tariq Aftab, M. Masroor A. Khan, and Parvaiz Ahmad. A comprehensive review of adaptations in plants under arsenic toxicity: physiological, metabolic and molecular interventions. *Environmental Pollution* 290 (2021): 118029.

Nawaz, Muhammad, Sabtain Ishaq, Hasnain Ishaq, Naeem Khan, Naeem Iqbal, Shafaqat Ali, Muhammad Rizwan, Abdulaziz Abdullah Alsahli, and Mohammed Nasser Alyemeni. Salicylic acid improves boron toxicity tolerance by modulating the physio-biochemical characteristics of maize (Zea mays L.) at an early growth stage. *Agronomy* 10, no. 12 (2020): 2013.

Pacci, Sena, Muhammet Emin Safli, Mehmet Serhat Odabas, and Orhan Dengiz. Variation of USLE-K soil erodibility factor and its estimation with artificial neural network approach in semi-humid environmental condition. *Brazilian Archives of Biology and Technology* 66 (2023): e23220481.

Pandey, Akhilesh Kumar, Arti Gautam, and Rama Shanker Dubey. Transport and detoxification of metalloids in plants in relation to plant-metalloid tolerance. *Plant Gene* 17 (2019): 100171.

Parveen, Nishat, Roberto Berni, Sreeja Sudhakaran, Javaid A. Bhat, Suhas Shinde, Naleeni Ramawat, Vijay P. Singh et al. Metalloids in plants: a systematic discussion beyond description. *Annals of Applied Biology* 180, no. 1 (2022): 7–25.

Patel, Monika, and Asish Kumar Parida. Salinity alleviates the arsenic toxicity in the facultative halophyte *Salvadora persica* L. by the modulations of physiological, biochemical, and ROS scavenging attributes. *Journal of Hazardous Materials* 401 (2021): 123368.

Pathak, Himanshu K., Prabhat K. Chauhan, Chandra Shekhar Seth, Gopal Dubey, and Sudhir K. Upadhyay. Mechanistic and future prospects in rhizospheric engineering for agricultural contaminants removal, soil health restoration, and management of climate change stress. *Science of the Total Environment* 927 (2024): 172116.

Raza, Ali, Javaria Tabassum, Zainab Zahid, Sidra Charagh, Shanza Bashir, Rutwik Barmukh, Rao Sohail Ahmad Khan et al. Advances in "omics" approaches for improving toxic metals/metalloids tolerance in plants. *Frontiers in Plant Science* 12 (2022): 794373.

Ren, Jing–Hua, Lena Q. Ma, Hong–Jie Sun, Fei Cai, and Jun Luo. Antimony uptake, translocation and speciation in rice plants exposed to antimonite and antimonate. *Science of the Total Environment* 475 (2014): 83–89.

Seth, Chandra Shekhar. A review on mechanisms of plant tolerance and role of transgenic plants in environmental clean-up. *The Botanical Review* 78, no. 1 (2012): 32–62.

Seth, Chandra Shekhar, T. Remans, E. Keunen, M. Jozefczak, H. Gielen, K. Opdenakker, N. Weyens, J. Vangronsveld, and A. Cuypers. Phytoextraction of toxic metals: a central role for glutathione. *Plant, Cell & Environment* 35, no. 2 (2012): 334–346.

Seth, Chandra Shekhar, Virendra Misra, Lalit Kumar Singh Chauhan, and R. R. Singh. Genotoxicity of cadmium on root meristem cells of Allium cepa: cytogenetic and Comet assay approach. *Ecotoxicology and Environmental Safety* 71, no. 3 (2008): 711–716.

Shahid, Muhammad Adnan, Rashad Mukhtar Balal, Naeem Khan, Lincoln Zotarelli, Guodong David Liu, Ali Sarkhosh, Juan C. Fernandez-Zapata, Juan José Martínez Nicolás, and Francisco Garcia-Sanchez. Selenium impedes cadmium and arsenic toxicity in potato by modulating carbohydrate and nitrogen metabolism. *Ecotoxicology and Environmental Safety* 180 (2019): 588–599.

Sharma, Pooja, Ranjna Sirohi, Yen Wah Tong, Sang Hyoun Kim, and Ashok Pandey. Metal and metal(loids) removal efficiency using genetically engineered microbes: applications and challenges. *Journal of Hazardous Materials* 416 (2021a): 125855.

Sharma, Shanti S., Vijay Kumar, and Karl-Josef Dietz. Emerging trends in metalloid-dependent signaling in plants. *Trends in Plant Science* 26, no. 5 (2021b): 452–471.

Siddiqui, Manzer H., Saud Alamri, Soumya Mukherjee, Abdullah A. Al-Amri, Qasi D. Alsubaie, Bander M. A. Al-Munqedhi, Hayssam M. Ali et al. Molybdenum and hydrogen sulfide synergistically mitigate arsenic toxicity by modulating defense system, nitrogen and cysteine assimilation in faba bean (Vicia faba L.) seedlings. *Environmental Pollution* 290 (2021): 117953.

Silva-Gigante, M., L. Hinojosa-Reyes, J. M. Rosas-Castor, P. C. Quero-Jiménez, D. A. Pino-Sandoval, and J. L. Guzmán-Mar. Heavy metals and metalloids accumulation in common beans (Phaseolus vulgaris L.): a review. *Chemosphere* 335 (2023): 139010.

Singh, Durga, Ashish Agnihotri, and Chandra Shekhar Seth. Interactive effects of EDTA and oxalic acid on chromium uptake, translocation and photosynthetic attributes in Indian mustard (Brassica juncea L. var. Varuna). *Current Science* 112 (2017): 2034–2042.

Slomberg, Danielle L., and Mark H. Schoenfisch. Silica nanoparticle phytotoxicity to Arabidopsis thaliana. *Environmental Science & Technology* 46, no. 18 (2012): 10247–10254.

Solanki, Swati, Surbhi Sinha, Chandra Shekhar Seth, Shivanshi Tyagi, Aarushi Goyal, and Rachana Singh. Enhanced adsorption of Bismark Brown R dye by chitosan conjugated magnetic pectin loaded filter mud: a comprehensive study on modeling and mechanisms. *International Journal of Biological Macromolecules* (2024): 131987.

Tavanappanavar, Adinath N., Sikandar I. Mulla, Chandra Shekhar Seth, Zabin K. Bagewadi, Mohamed Rahamathulla, Mohammed Muqtader Ahmed, and Syeda Ayesha Farhana. Phytochemical analysis, GC–MS profile and determination of antibacterial, antifungal, anti-inflammatory, antioxidant activities of peel and seeds extracts (chloroform and ethyl acetate) of Tamarindus indica L. *Saudi Journal of Biological Sciences* 31, no. 1 (2024): 103878.

Upadhyay, Sudhir K., Garima Singh, Nitu Rani, Vishnu D. Rajput, Chandra Shekhar Seth, Padmanabh Dwivedi, Tatiana Minkina, Ming Hung Wong, Pau Loke Show, and Kuan Shiong Khoo. Transforming bio-waste into value-added products mediated microbes for enhancing soil health and crop production: perspective views on circular economy. *Environmental Technology & Innovation* 34 (2024): 103573.

Vajjiravel, Prakash, Divya Nagarajan, Varsha Pugazhenthi, Ajay Suresh, Madhan Kumar Sivalingam, Ajila Venkat, Pritam Paramguru Mahapatra et al. Circadian-based approach for improving physiological, phytochemical and chloroplast proteome in Spinacia oleracea under salinity stress and light emitting diodes. *Plant Physiology and Biochemistry* 207 (2024): 108350.

White, Philip J., and Martin R. Broadley. Biofortifying crops with essential mineral elements. *Trends in Plant Science* 10, no. 12 (2005): 586–593.

Xalxo, Roseline, Jipsi Chandra, Neha Pandey, and Keshavkant Sahu. Physiological and molecular mechanism of metalloid tolerance in plants. In Mirza, H. (Ed.), *Approaches to the Remediation of Inorganic Pollutants* (pp. 197–221). Springer, 2021.

Yadav, Manisha, Praveen Gupta, and Chandra Shekhar Seth. Foliar application of α-lipoic acid attenuates cadmium toxicity on photosynthetic pigments and nitrogen metabolism in *Solanum lycopersicum* L. *Acta Physiologiae Plantarum* 44, no. 11 (2022): 112.

Zhao, Fang-Jie, Zhong Tang, Jia-Jun Song, Xin-Yuan Huang, and Peng Wang. Toxic metals and metalloids: uptake, transport, detoxification, phytoremediation, and crop improvement for safer food. *Molecular Plant* 15, no. 1 (2022): 27–44.

Zhu, Y., Wu, Q., Lv, H., Chen, W., Wang, L., Shi, S., Yang, J., Zhao, P., Li, Y., Rensing, C., Liu, H., & Feng, R. (2020). Toxicity of different forms of antimony to rice plants: Effects on reactive oxidative species production, antioxidative systems, and uptake of essential elements. *Environmental Pollution*, 263(Pt B), 114544. https://doi.org/10.1016/j.envpol.2020.114544

9 Integrating Multi-Omics Approaches to Unravel the Mysteries Underlying Metalloid Stress in Plants

Nahida Rehman Mir, Roohi Sharma, Amandeep Singh, Manu Sharma, Bilal Ahmad Mir, Nitika Kapoor, Puja Ohri, Renu Bhardwaj, and Yogesh K. Ahlawat

9.1 INTRODUCTION

Omics approaches are utilized in an integrated manner to draw comparisons and enhance our understanding of various aspects related to plant health (Misra et al., 2019). As integrated multi-omics has evolved, our understanding of the processes underpinning plant defense signals has deepened, along with our capacity to apply this knowledge for beneficial interventions in controlled natural systems (Crandall et al., 2020). The introduction of omics methods has allowed us to finally comprehend the ecology underlying the suppressive impact (Schlatter et al., 2017). The latest advancements in omics technology enable the pinpointing of molecular and metabolic network alterations specific to different cells and tissues under abiotic stress conditions. This breakthrough deepens our comprehension of the molecular pathways underlying abiotic stress resilience, paving the way for the development of crops which are tolerant to stress (Shelden and Roessner, 2013). Functional genomic approaches have significantly advanced our understanding of the biology and physiology underlying abiotic stress responses. Omics techniques such as genomics, transcriptomics, metabolomics, proteomics, and ionomics have been extensively employed to evaluate the mechanisms of non-living tension tolerance in plants (Cramer et al., 2011).

9.1.1 Multi-Omics Approach for Investigation of Metalloid Stress in Plants

Genomics is the comprehensive examination of an organism's entire genetic makeup, encompassing its structure, function, evolution, mapping, and manipulation of genomes. Functional genomics in tandem with transcript- and prote-omics investigates the expression and roles of genes and proteins across the entire genome or within systems, utilizing genomic information (Sundin et al., 2016). Various techniques, including quantitative trait locus mapping methods (QTL mapping), quantitative trait locus sequencing (QTL-seq), RNA sequencing (RNA-seq) alongside genomewide association studies (GWAS) and the clustered regularly interspaced short palindromic repeats and CRISPR-associated protein 9 (CRISPR/Cas9) system have been instrumental in unraveling the complexities of plant genomes under stressful conditions. For instance, through genomic analyses, genes implicated in salt stress response have been identified, and signaling pathways elucidated, offering promising avenues for enhancing plant salinity resistance (Singhal et al., 2021). Employing molecular breeding techniques like GWAS, Kumar et al. (2015) pinpointed three novel QTLs associated with salt tolerance in *Oryza sativa* on chromosomes 4, 6, and 7. Similarly, Huang et al. (2018) linked salinity tolerance to six genomic loci in soybean, while Dilnur et al. (2019) identified nine single

DOI: 10.1201/9781032673592-9

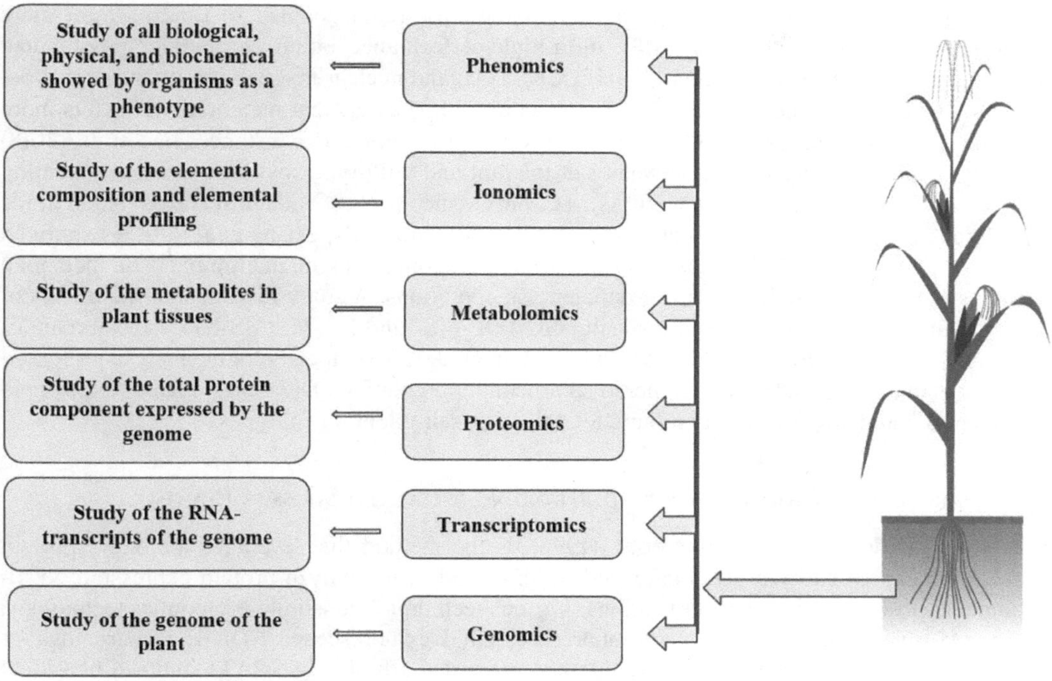

FIGURE 9.1 Multi-omics approaches in plants for stress investigation.

nucleotide polymorphism (SNP)-rich regions in Asian cotton accessions related to plant traits under salt stress conditions, utilizing GWAS. A thorough grasp of how plants tolerate salinity has been attained through high-throughput methods like forward genetics, candidate gene investigations, and next-generation sequencing. This forms the foundation for creating plant varieties resistant to salt (Singhal et al., 2021). Emerging genomic technologies like zinc-finger nuclease (ZFN), transcription activator-like effector nuclease (TALENs), CRISPR-Cas9, and speed breeding offer precise avenues for engineering crops tailored to withstand salt stress (Pennisi, 2013; Li et al., 2018) (Figure 9.1).

Transcriptomics is a branch of "omics" technology focused on analyzing the RNA expression patterns of organisms over time and space. RNA-Seq is a primary technique in transcriptomics, especially for studying gene expression patterns under stress, like salt stress (Qian et al., 2014). By comparing tolerant and sensitive genotypes, RNA-Seq identifies stress-related genes, making it pivotal for gene discovery and understanding stress response. This approach provides significant understanding regarding the molecular processes that contribute to the ability of plants to tolerate high levels of salt aiding in the development of salt-tolerant plant varieties. For instance, a recent study on chickpeas examined the transcriptome response to heavy metal (HM) toxicity, revealing the activation of key metabolic pathways and the identification of nine genes involved in regulating these metabolic responses (Yadav et al., 2019). Similarly, a GWAS of common beans was conducted to recognize the genetic construct implicated in the plant's reaction to salt intensity (Luo et al., 2020). These studies highlight the importance of transcriptomic analysis in determining the genetic basis of various pressure tolerance in plants.

9.2 FUNCTIONAL ASPECTS OF OMICS

9.2.1 Metabolomics in Stress

Metabolomics delves into the metabolome of a specific plant tissue or organ, examining the assortment of metabolites present and providing valuable metabolic fingerprints or profiles based on each metabolite's physiochemical properties. This field utilizes a range of analytical techniques such as mass spectrometry (MS), liquid and gas chromatography (GC) coupled with MS [liquid chromatography (LC)-MS, GC-MS], gas chromatography (GC) coupled to time-of-flight mass spectrometry (TOF-MS) (GC-TOF-MS), ultra-high performance liquid chromatography-tandem mass spectrometry (UHPLC-MS/MS), UHPLC/Q-TOF, and nuclear magnetic resonance spectroscopy (NMR) to profile metabolites. Dos Santos et al. (2017) assert that metabolomics offers more dependable insights compared to transcriptomics and proteomics. For instance, Jiao et al. (2018) explored metabolomic variations between salt-tolerant and wild-type soybean varieties, revealing higher levels of certain organic acids (OAs), TCA cycle metabolites, and amino acids (AAs) in the salt-tolerant variety, conferring greater salinity tolerance. Similarly, Zhang et al. (2017) examined the metabolic reactions of *Camellia sinensis* to Zn stress, observing distinct impacts on metabolic pathways in tea leaves under excess or deficient Zn conditions. A study investigated the effects of copper (Cu) stress on *Cucumis sativus* using GC-TOF-MS- and LC-MS/MS-based metabolomics profiling (Zhao et al., 2016). In another study by Pan et al. (2020), metabolomic analysis of foxtail millet roots under high salinity deciphered significant upregulation of 17 genes related to flavonoid biosynthesis, indicating potential markers for assessing salt-tolerant genotypes.

9.2.2 Proteomics Profiling for Understanding Metalloid Stress in Plants

Proteomics, as defined by Wilkins et al. (1996), is the method that leverages the entire protein complement expressed by a genome or cell of an organism to analyze protein expression, localization, and post-translational modifications. Various techniques constitute proteomics technology, such as MS, two-dimensional gel electrophoresis (2-DIGE), fluorescence 2-DIGE, isobaric tags for relative and absolute quantification (iTRAQ), isotope-coded affinity tag (ICAT), multi-dimensional protein identification technology (MudPIT), among others. Several plant species (spp) that include *O. sativa*, *T. aestivum*, *H. vulgare*, and *Zea mays* have been subjected to proteome studies to understand their response to salt stress (Zhang et al., 2012). Investigations focusing on single-cell-type proteomics have been conducted for *Arabidopsis Glycine max* and *Nicotiana tabacum* (Wan et al., 2005; Amme et al., 2005). Additionally, comprehensive tissue-specific proteomics inquiries have explored *O. sativa* anthers, *T. aestivum* root seedlings, and *O. sativa* plasma membranes (Guo et al., 2012; Sarhadi et al., 2012; Cheng et al., 2009). Kamal et al. (2012) also assessed chloroplast protein content in wheat with regard to various physiological factors. Moreover, a proteomics investigation revealed proteins in xylem sap responsible for detoxifying cadmium (Cd), thereby enhancing plant stress tolerance (Luo and Zhang, 2019) (Table 9.1).

9.2.3 Ionomics Approaches for Stress Resilience in Plants

Ionomics involves the systematic examination of elemental compositions, offering a valuable tool for assessing how plants respond to various stresses such as salinity, nutrient deficiencies, toxicities, or excessive accumulation. Techniques like synchrotron X-ray fluorescence microscopy (SXRF), Fourier transform infrared spectrometry (FTIR), inductively coupled plasma mass spectrometry (ICP-MS), and inductively coupled plasma optical emission spectrometry (ICP-OES) are commonly employed in ionomic studies (Shelden and Roessner, 2013; Punshon et al., 2013; Su et al., 2017; Feng et al., 2017; Stich et al., 2020).

In the realm of stress research, ionomics has been instrumental in unraveling key insights into ion balance and detoxification mechanisms in crops under intense salinity conditions (Sanchez et al., 2011; Wu et al., 2013). A large number of investigations have demonstrated that salt stress alters the levels of essential elements like calcium, magnesium, manganese (Mn), iron (Fe), and zinc in plants. For instance, Arshad et al. (2017) observed that increased expression of the salt-responsive microRNA 156 (miR156) in alfalfa plants led to reduced accumulation of sodium ions. In another study by Stich et al. (2020), the ionome of *Z. mays* shoots was analyzed using ICP-OES under different Fe regimes, revealing substantial genotypic variation across twelve minerals. Similarly, Wang et al. (2020) investigated the ionome of shoots and roots in Brassica species under optimal and low phosphorus conditions, identifying a total of 133 and 123 QTLs, respectively.

TABLE 9.1
Omics Technologies Used in Different Plant Species to Study Abiotic Stress Characteristics

Omics Technique	Technology	Plant Species	Characteristics Studied	References
Genomics	QTL mapping	*Hordeum vulgare*	To elucidate the genetic makeup and regulatory mechanisms involved in barley's response to Cd stress	Derakhshani et al. (2020)
	GWAS	*Oryza sativa*	To identify the novel marker-trait associations with phenotypic variations to improve zinc and Fe tolerance	Bollinedi et al. (2020)
	CRISPR/Cas9 system	*Lycopersicon esculentum*	To develop mutants against Cd toxicity	Kakeshpour (2020)
Transcriptomics	RNA-Seq	*Puccinellia distans*	Transcriptomic analysis for Se tolerance	Kök et al. (2020)
	RNA-Seq	*Fagopyrum tataricum*	Transcriptional profiling of leaves under Pb stress	Wang et al. (2020)
	Translating ribosome affinity purification	*Arabidopsis*	To investigate the transcriptional response in Arabidopsis under Fe stress conditions	Khan et al. (2018)
Proteomics	LC-MS/MS, shotgun	*Brassica napus*	Proteomic analysis of xylem sap in Cd-treated Brassica	Luo and Zhang (2019)
	iTRAQ	*Zea mays*	Root proteome study of maize seedlings in Cd stress	Wen et al. (2019)
Metabolomics	LC-MS/MS and HPLC	*Amaranthus hypochondriacus*	Examination of metabolites and thiol compounds in response to Cd-induced stress	Xie et al. (2019)
	UHPLC-MS/MS	*Apium graveolens* L.	Examination of the metabolome in celery seedlings subjected to Se treatment compared to untreated ones under Se stress	Zhang et al. (2019)
	GC-TOF-MS and LC-MS/MS	*Cucumis sativus*	Analyzing the metabolome to investigate how Cu stress impacts cucumber plants	Zhao et al. (2016)
Ionomics	Inductively coupled plasma optical emission spectrometry	*Z. mays*	Study of root ionome under Fe regime	Stich et al. (2020)
	Fourier transform infrared spectrometry	*Spirodela polyrhiza*	To investigate the subcellular dispersion, accumulation, and changes in metabolism and physiology in response to Cd stress	Su et al. (2017)
	Synchrotron X-ray fluorescence microscopy (SXRF)	*Arabidopsis thaliana*	Phenotype, genotype, and gene identification	Punshon et al. (2013)
Phenomics	PHENOPSIS	*Arabidopsis thaliana*	Phenotyping of plant response against soil water deficit stress	Granier et al. (2006)

9.2.4 Phenomics Approach

A phenome encompasses all the processes pertaining to biology, physics, and biochemistry exhibited by an organism as phenotypes *viz.*, both qualitative and quantitative characteristics—within a specific environmental context. Plant phenomics focuses on understanding the expression of traits and genes in plants under particular environmental conditions (Pratap et al., 2019). Tools such as PHENOPSIS (Granier et al., 2006), *EMPHASIS* (Papoutsoglou et al., 2020), weighing, imaging and watering machines (WIWAM) (Asaari et al., 2019), and the North American plant phenotyping network (NAPPN) (Carroll et al., 2019) offer advanced capabilities for studying plants under stress. These techniques provide benefits as compared to conventional methods (Campbell et al., 2015; Al-Tamimi et al., 2016; Negrão et al., 2017). High-throughput phenotyping is crucial for understanding the impact of salt stress on various plant processes such as photosynthesis, transpiration, ionic interactions, senescence, and yield. It employs automated digital photography, active vision cell technology, deep learning, and other innovative approaches to accurately characterize, measure, and predict plant responses to salinity. Next-generation phenotyping tests like "*PHENOPSIS*" and "*WIWAM*" in *O. sativa* (Meng et al., 2017; Humplík et al., 2015) are instrumental in assessing salinity tolerance.

Falk et al. (2020) utilized a phenomics approach to investigate root system architecture (RSA) traits, aiming to identify genetic diversity in soybeans.

9.2.5 Multi-omics Approach

Combining multiple omics methodologies has emerged as a highly effective and efficient approach, enabling a comprehensive understanding of the intricate connections between an organism's genome and its observable traits across various environmental conditions (Yang et al., 2021). The integration of genomics, transcriptomics, proteomics, and metabolomics has significantly enhanced our comprehension of the molecular mechanisms governing plant responses to abiotic stresses (Cramer et al., 2011; Jogaiah et al., 2013). Researchers have utilized combined omics techniques to assess transcriptomic and metabolic profiles in both potato tubers and Arabidopsis, providing valuable insights (Urbanczyk-Wochniak et al., 2003; Hirai et al., 2004). Furthermore, the amalgamation of GWAS with metabolic profiling has developed as a prevailing strategy for unraveling the intricate interplay between biochemical pathways and genetic regulation in various important crop species such as *O. sativa, Z. mays*, and *Solanum lycopersicum* (Luo, 2015; Matsuda et al., 2015). The investigation of specific biological processes has demonstrated the complementary nature of combined omics techniques, particularly evident in elucidating the differential directive mechanisms in reception to salinity intensity in saline plants (Kumari et al., 2015). These approaches have not only deepened our understanding but have also led to the development of crop genotypes with enhanced resistance to non-living stresses (Jogaiah et al., 2013). To facilitate the analysis of multi-omics data, a plethora of bioinformatics tools and software packages are employed. Consequently, integrating multi-omics approaches holds immense promise in unraveling the functional dynamics of genes and their intricate networks, thereby facilitating advancements in crop genetics, growth, yield, and resilience to physiological and environmental stresses (Yang et al., 2021).

9.3 MECHANISMS THAT REGULATE THE PLANT TOLERANCE AND BIOACCUMULATION OF POTENTIALLY TOXIC ELEMENTS (PTES)

Contamination of food crops often stems from the build-up of PTEs in soil and water. Introducing organic matter into the growing environment has been shown to enhance the redressal of PTEs from soil (Ghanim et al., 2023). Plants possess the ability to uptake various elements from the soil, some essential for their growth and others non-essential, which can accumulate in different plant parts. Examples of these PTEs include essential elements like Cu and Mn and non-essential ones such as Cd, and chromium (Cr). Numerous studies have indicated that plants cultivated in soils contaminated with toxic metals may exhibit altered metabolism and biochemical processes, leading

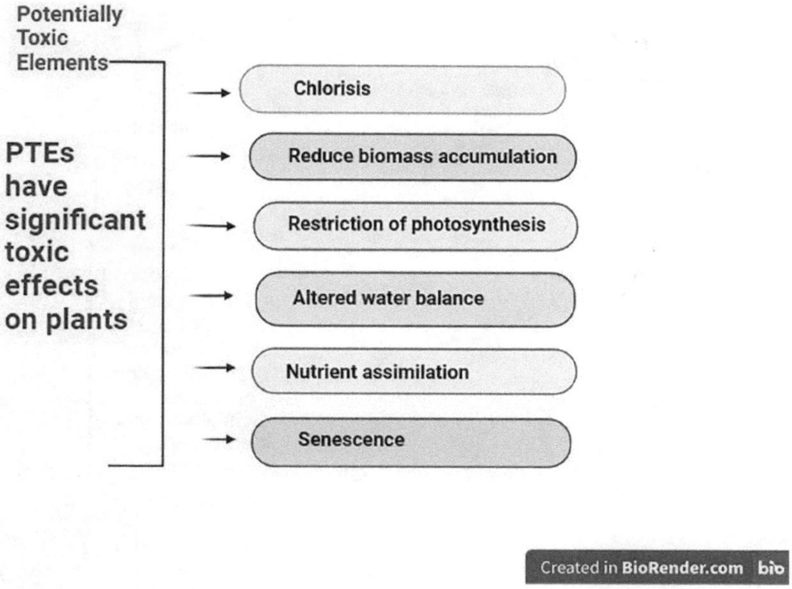

FIGURE 9.2 General effect of PTEs on plants.

to stunted growth, reduced biomass production, and the accumulation of metals (Iftikhar et al., 2021; Alam et al., 2020). Furthermore, the accumulation of toxic metals can trigger the generation of reactive oxygen species (ROS), which can damage crop tissues (Ghanim et al., 2023). PTEs exert harmful effects on soil health, plant development, biomass, and chlorophyll levels (Figure 9.2). The origin and nature of the biochar employed as a raw material could also contribute to reduced accumulation of PTEs and enhanced biomass yield (Ibrahim et al., 2017). Exceeding standard limits of PTEs in the ecology poses a considerable threat to the health of organisms and undermines worldwide environmental conservation efforts. Phytoremediation is recognized to be among the most competent approaches for addressing sites polluted by PTEs. In the process of phytoremediation, plants employ a range of techniques (Alsafran et al., 2022).

Arsenic (As), a metalloid, is a widespread environmental pollutant known to be a Class 1 carcinogen and deemed non-essential for biological processes. Found naturally in various chemical forms such as arsenite [As (III)], arsenate [As(V)], and methylated compounds like methylmalonic acid (MMA) and dimethylamine (DMA), inorganic arsenic (As (III)) is considered the most toxic variant (El-Ghiaty and El-Kadi, 2021). Upon entering the rhizosphere, it competes with essential nutrients for uptake pathways, eventually being transported to grains via different mechanisms. In soil, As(V) closely resembles inorganic phosphate (Pi), thus being absorbed by plant roots through phosphate transporters. Its presence interferes with vital enzyme activities like pyruvate dehydrogenase and disrupts the balance of ROS and antioxidatives such as glutathione (Zulfiqar et al., 2023). As exposure has been linked to adverse effects on rice seedlings, reducing germination rates while simultaneously increasing antioxidant activities. Research by Bailey and Fry (2014) highlights the heightened mortality risks associated with prenatal As exposure, impacting genomic and epigenetic signaling. However, its precise molecular effects on adult diseases remain largely unclear, necessitating further investigation, especially concerning vulnerable populations.

9.3.1 Bioaccumulation of PTEs in Plants

The accumulation of PTEs in the environment beyond certain levels poses a noteworthy threat to the well-being of organisms and challenges global efforts toward ecological sustainability (Tarish et al., 2024). Certain plants known as metal hyperaccumulators have a remarkable ability to absorb

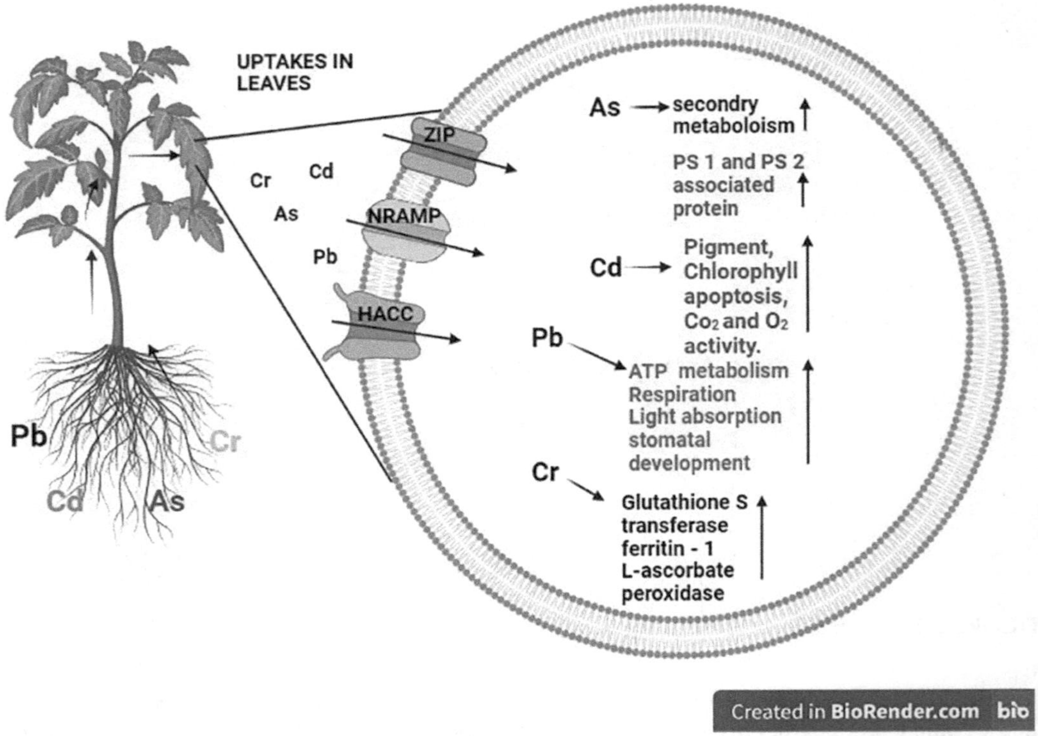

FIGURE 9.3 Schematic diagram of potentially toxic element uptake and plant response in leaves. (Modified the diagram from Alsafran et al, 2022) Abbreviations used: NRAMP, natural resistance-associated macrophage protein; ZIPs, zinc-regulated transporters; HACC, hyperpolarization-activated calcium channels.

PTEs and transport them to their above-ground parts, whereas non-hyperaccumulators have limited ability in this regard. PTEs are absorbed by plant cells primarily through their leaves, triggering various signaling pathways within the cell (Alsafran et al., 2022). These metals enter the cells via different transporters such as HM-transporting ATPases, cation diffusion facilitators (CDF), natural resistance-associated macrophage proteins (NRAMP), multidrug and toxin efflux (MATE), and zinc-iron permease proteins (ZIP). Once inside, they can displace essential ions or hinder various functional groups, leading to biochemical responses (Tripathi and Poluri, 2021).

The miombo plant (*Brachystegia logifolia*), which commonly grows in natural habitats near mining areas, exhibits heightened levels of Cu), zinc, Fe, and Mn. Biochar is employed to mitigate the adverse effects of HM toxicity in plants by stabilizing, retaining, and deactivating toxic HMs in the soil (Mulenga et al., 2023). Additionally, substances like ethylenediaminedisuccinic acid (EDDS), ethylenediamine tetraacetic acid (EDTA), as well as organic and synthetic chelators, are applied to the ground to mitigate the hazardous impact of HMs on plants (Figure 9.3).

9.3.2 Phytoremediation

Improving the effectiveness of phytoremediation can be achieved through two key strategies: selective breeding and the creation of transgenic plants that express genes associated with metal hyperaccumulation (Rascio and Navari-Izzo, 2011; Mohanty et al., 2022). Transgenic plants are deliberately engineered to combine multiple desirable traits, such as accelerated growth, increased biomass production, enhanced tolerance, and the ability to accumulate various contaminants, with the goal of enhancing phytoremediation efficiency (Lu et al., 2017). Academics have established diverse

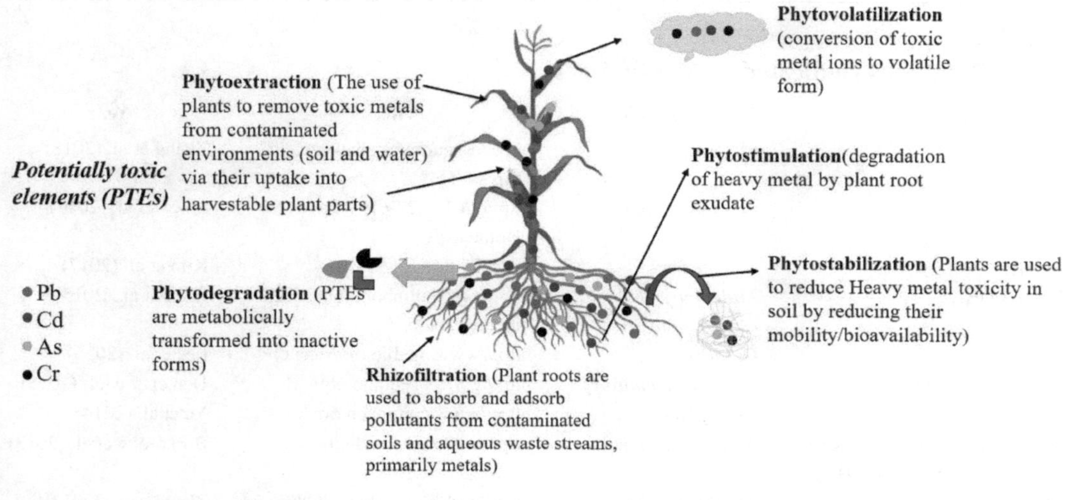

FIGURE 9.4 Phytoremediation techniques involved in PTEs remediation.

transgenic vegetations by introducing target genes into the desired plant species or by creating somatic hybrids, aiming to improve the remediation of both inorganic and organic contaminants (Naz et al., 2022; Rascio and Navari-Izzo, 2011). For example, the creation of somatic hybrids using combinations such as *T. caerulescens* and *B. napus*, as well as *T. caerulescens* and *B. juncea*, has led to increased biomass production, enhanced zinc accumulation, and improved lead phytoextraction, respectively, (Agarwal and Rani, 2022) (Figure 9.4).

9.3.3 Databases for Integrating Omics Platforms in Various Crops to Better Understand Stress Tolerance

A specialized category of databases has been made to encompass large data on pressure responses. These records compile a range of genetic information including expressed sequence tags (ESTs), genes, cDNA sequences, small non-coding RNAs (snoRNAs), small regulatory RNAs (smRNAs), microRNA-targeted transcription factors (miRNTFs), metabolites, hormones, and other proteins that either exhibit upregulation or downregulation in reaction to high levels of salt, heat, HM, water deficit, and freezing stress (Anamika et al., 2019) (Table 9.2). *DRASTIC* (http://www.drastic.org.uk/) stands as a prominent information hub furnishing data on plant ESTs and genes that undergo up- or downregulation in response to various living and non-living stresses and chemical treatments. Its up-to-date iteration encompasses data on cold, salt and drought stress, boasting over 33,400 records sourced from 630+ referenced journals (Anamika et al., 2019; Prabha et al., 2011; Newton et al., 2002). The "*Plant Stress Gene Database*" is another significant resource, housing stress-related genes and ESTs from diverse crop spp. including *Arabidopsis*, *O. sativa*, *T. aestivum*, *Saccharum officinarum*, *H. vulgare*, and *Phaseolus vulgaris*. Additionally, this platform facilitates searches for orthologs and paralogs of stress-related proteins.

9.4 EVOLUTION OF MULTI-OMICS APPROACH IN DEVELOPING STRESS-RESISTANT VARIETY

The evolution of multi-omics approaches in developing stress-resistant varieties has revolutionized agricultural research. By integrating various omics techniques, researchers gain an inclusive understanding of plant responses to stress. This holistic view allows for the identification of key genes, proteins, and metabolic pathways intricated with stress tolerance (Nam et al., 2021; Muthamilarasan

TABLE 9.2
List of Databases Dedicated to Abiotic Stress

Database	Plant Species	Link	Year
Stress2TF	41 species	http://csgenomics.ahau.edu.cn/Stress2TF	Zhang et al. (2018)
PpTFDB	*Cajanus cajan L*	http://14.139.229.199/PpTFDB/Home.aspx	Singh et al. (2017)
SoyNet	*Glycine max*	www.inetbio.org/soynet	Kim et al. (2017)
DroughtDB	*Arabidopsis thaliana* and *O. sativa*	http://pgsb.helmholtzmuenchen.de/droughtdb/	Alter et al. (2015)
RiceNet v2	*O. sativa*	http://www.inetbio.org/ricenet	Lee et al. (2015)
Coffee genome hub	*Coffea canephora*	http://coffee-genome.org/	Dereeper et al. (2015)
CottonGen	*Gossypium*	http://www.cottongen.org	Yu et al. (2014)
Arabidopsis stress-responsive gene database (ASRGDB)	*Arabidopsis thaliana*	http://srgdb.bicpu.edu.in/	Borkotoky et al. (2013)
PASmiR	33 plant species	http://pcsb.ahau.edu.cn:8080/PASmiR	Zhang et al. (2013)
RiceFREND	*O. sativa*	http://ricefrend.dna.affrc.go.jp/	Sato et al. (2013)
Plant stress gene database	11 species	http://ccbb.jnu.ac.in/stressgenes/frontpage.html	Prabha et al. (2011)
DRASTIC	73 species	http://www.drastic.org.uk/	Newton et al. (2002)

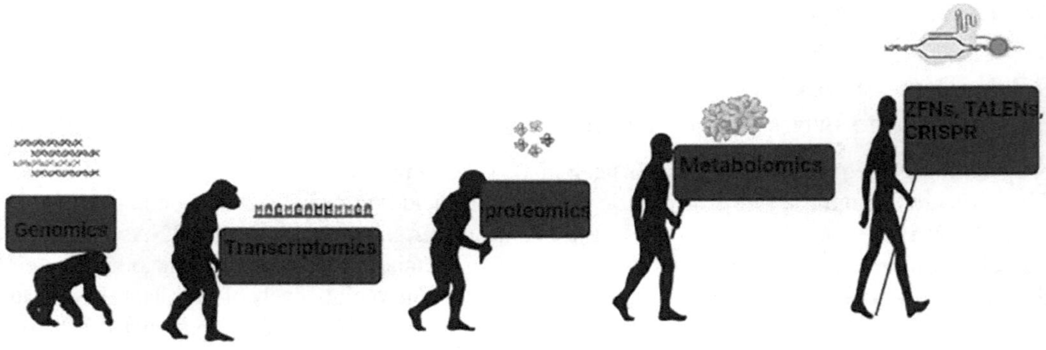

FIGURE 9.5 Evolution of various omics approaches.

et al., 2019). Through sophisticated data analysis techniques, such as machine learning and network modeling, multi-omics data can be leveraged to predict phenotypic outcomes and guide targeted breeding efforts. Overall, the multi-omics approach accelerates the development of resilient crop varieties essential for ensuring food security in the face of environmental challenges (Unel et al., 2023). A plant contends with various biotic and abiotic factors to ensure its survival and quality. Multifaceted omics methodologies are also employed to comprehend variations within plants in order to examine their genetic, proteomic, and metabolic levels (Gupta et al., 2023) (Figure 9.5).

9.4.1 Omic-Based Strategies

In omics, lipidome refers to the comprehensive lipid profile found at the cell, tissue, and organ level within an organism. Lipidomics is an emergent scientific field dedicated to studying both the structure and function of the lipidome, including its interactions with other lipids, proteins,

TABLE 9.3
Functional Aspects of Metabolites on Various Plant Species

Stress	Metabolites	Effect	References
Drought	Gamma-aminobutyric acid (GABA)	Neutralize ROS	Carillo (2018)
Drought	Sugar alcohols	Osmoprotectant	de Miguel et al. (2016)
Drought	Phenols	Antioxidant activity	Griesser et al. (2015)
Salt	TCA	Increase salt tolerance	El-Beltagi et al. (2017)
Salt	Organic acids	Nitrogen cycle	Ni et al. (2015)
Salt	Cyclic acids	Phosphate storage	Sung et al. (2015), Zhang et al. (2011)
Heavy metals	Amino acids	Osmoprotectant, phytochelatins synthesis, polyamine synthesis	Okem et al. (2015)
Heavy metals	Phenolics, flavonoids, phytochelatins	Antioxidant, ROS scavenging, structural integrity	Okem et al. (2015), Pal and Rai (2010)
Heavy metals	Peptides	Antioxidant activity, metal chelators, photoprotection	Sytar et al. (2013), Manivasagaperumal et al. (2011)
Cold stress	Carotenoids and flavonoids	Energy dissipation, antioxidant activity, ultraviolet (UV) absorbent	Neugart et al. (2016), Latowski et al. (2011)
Cold stress	Carbohydrate	Cryoprotectant	Maruyama et al. (2014), Caldana et al. (2011)
Cold stress	Lipids	Membrane stabilization	Degenkolbe et al. (2012)
Heat stress	Amino acids	Antioxidant activity, protein stabilization, signaling	Chebrolu et al. (2016), Shin et al. (2016), Luengwilai et al. (2012)
Heat stress	Sugars	ROS scavenging, osmoprotectant	Chebrolu et al. (2016), Rivero et al. (2014)
Heat stress	Flavonoids	Signaling, ROS scavenging, structural integrity	Chebrolu et al. (2016), Gill and Tuteja (2010)
Heat stress	Fatty acid	Cell ultrastructure reconstruction, isoprenoid synthesis	Mueller et al. (2015), Luengwilai et al. (2012)
Heat stress	Organic acids	Nitrogen cycle	Luengwilai et al. (2012)
Heat stress	Polyamines	Antioxidant activity	Cvikrová et al. (2012)

and metabolites (Pranneshraj et al., 2022). In the context of plant responses to abiotic stress, lipids serve a dual role. They not only act as crucial signaling mediators but also play significant roles in reducing the impact of stress (Buffagni et al., 2022). Signaling lipids (such as fatty acids, sphingolipids, diacylglycerols, lysophospholipids, phosphatidic acid (PA), inositol phosphate, oxylipins, and N-acylethanolamine) are rapidly produced under stressful circumstances. At the same time, lipids are involved in modifying cell membranes during abiotic stress to mitigate cell damage (Zhang et al., 2022) (Table 9.3).

9.4.2 Zinc-Finger Nucleases

ZFNs are synthetic endonucleases that precisely cleave double-stranded DNA at specific sequences, enabling various genetic alterations like insertions, deletions, translocations, inversions, and point mutations (Carroll, 2018; Iqbal et al., 2020; Joung and Sander, 2013). They have been instrumental in genome editing across different organisms (Joung and Sander, 2013; Urnov et al., 2010). Townsend et al. (2009) demonstrated the editing potential of ZFNs in Sulfonylurea Resistance Locus A (*SuRA*) and Sulfonylurea Resistance Locus B (*SuRB*) genes for herbicide resistance. In maize, ZFNs were utilized to stack herbicide resistance genes, Phosphinothricin Acetyltransferase (PAT) and aryloxyalkanoate dioxygenase (AAD1), at a single locus (Ainley et al., 2013).

9.4.2.1 Transcription-Activator-Like Effector Nucleases

TALENs are customized restriction enzymes engineered to cleave precise DNA sequences (Mushtaq et al., 2019). TALENs consist of repetitive sequences made up of approximately 33–34 conserved AAs, with key variations occurring at the 12th and 13th positions, known as repeat variable residues (RVDs). These RVDs are solely responsible for the specific recognition of DNA sequences, serving as the substrates for TALEN molecules (Bogdanove et al., 2010; Ma et al., 2016). Gene editing facilitated by TALENs resulted in the development of blast resistance in rice. TALEN-based disruption of *OsSWEET14* (*Oryza sativa* Suga rs Will Eventually be Exported Transporter 14) led to the incapacity of the TAL effector pathogens to bind to *OsSWEET14*, consequently resulting in disease resistance (Li et al., 2012). Likewise, in the study by Zhou et al. (2015), editing of *OsSWEET13* (*Oryza sativa* Sugars Will Eventually be Exported Transporter 13) resulted in mutants that exhibited resistance to bacterial blight.

9.4.3 CRISPR-Based Editing for Stress

CRISPR represents a promising genetic tool that can be harnessed to enhance specific characters in vegetations of curiosity (Kaur et al., 2019). CRISPR technology facilitates the insertion of specific desired instructions into the genome of a target plant (Mushtaq and Molla, 2021). The CRISPR system has the potential to enhance the phytoremediation capabilities of plants by advancing their abilities in bioaccumulation, complexation, volatilization, and degradation processes (Agarwal and Rani, 2022). This technology holds promise for mitigating the harmful effects of HMs, organic pollutants, radionuclides, and other contaminants in polluted environments. Furthermore, it enables the development of transgenic plants with specific gene expressions. The CRISPR-mediated approach for enhanced phytoremediation primarily focuses on the meticulous examination and activation of target genes, promoting the production of metal-binding ligands such as phytochelatins and metallothioneins (MTs), enhancing metal transporters like *HMA, CDF, MATE, Yellow Stripe-Like,* and *ZIP* protein families, and increasing phytohormones and root exudates (Agarwal and Rani, 2022; Basharat et al., 2018).

9.5 METABOLIC ACTIVITIES AND MECHANISMS UNDERLYING METALLOID STRESS

Numerous molecular elements are required for the control of plants' metalloid toxic effects reactions. In order to comprehend the biological relationships and molecular processes of tolerance to harmful HMs and metalloids, integrated omics techniques have been used extensively. Despite recent advances in genomic techniques for enhancing yields under metallic stress, a considerable emphasis on the analysis of multimodal regulatory systems is necessary to further molecular understanding (Al Mahmud et al., 2019). Omics innovations are useful methods for identifying potential genes, their functions, and key molecular regulators. These methods make it conceivable to determine molecular targets and signaling networks, examine variations in gene activity, and boost agricultural plants' metal tolerance. Enhancing plant resilience has become more and more crucial, thanks to cutting-edge technologies inched to omics (Raza et al., 2022) (Table 9.4). There are certain mechanisms which are attributable to contribute toward metalloids as HMs of plants. There is an array of detoxification techniques employed by plants to combat the menace of HMs (Figure 9.6). When plants are exposed to metal stress, they employ various strategies to mitigate the toxic effects of HMs. Metal binding, vacuole differentiation, and volatilization are bioremediation methods that are involved in the process. Initially, the metals are supposed to be discharged from the plant cells (especially roots). Expulsion transporters in plant roots play a vital role in reducing HM accumulation by expelling these toxic ions from cells (Sarma and Prasad, 2019). Notably, aquaporin low silicon rice transporter 1 (Lsi1) (*OsNIP2:1*) in rice facilitates the efflux of As(III) (Deng et al., 2020). Other aquaporins in various plants, like *Arabidopsis thaliana* Nodulin 26-like Intrinsic

TABLE 9.4
Several Omics' Techniques in Plants' Responses to Metalloid Stress

Metalloid	Plant Species	Omics Platform	Observation	References
As and Cd	*Camellia sinensis*	Transcriptomics	Potential genes increased multi-metal tolerance	Liu et al. (2023)
Methylated Se	*Cardamine violifolia*	Metallomics	Spatial metallomics visualized Se distribution	Xie et al. (2023)
As	*A. thaliana*	Transcriptomics	Transcription factor -SL1M1 controlled As sensitivity	Jobe et al. (2021)
As and Si	*Zea mays*	Ionomics	Si increased antioxidant activity and lowered ROS levels to reduce the harmful effects of As	Mišúthová et al. (2021)
As	*O. sativa*	Ionomics	Variations in transcriptome data and ICP-MS were used to investigate As mitigation.	Xiao et al. (2021)
As	*Pteris vittata*	Metallomics	Applied synchrotron micro XRF imaging to detect As in root sections and studied directional changes in As absorption by roots.	Kashiwabara et al. (2021)
As	*B. napus*	Proteomics	iTRAQ proteomics revealed the synchronization of several mechanisms and crop energy metabolism	Farooq et al. (2021)
Se	*O. sativa*	Proteomics	iTRAQ proteomics expression enabled regulatory mechanism for genes upregulated for growth	Zeng et al. (2020)
As	*B. napus*	Genomics	Diacylglycerol (DGK) genes function in PA formation, hormone response, and plant development.	Tang et al. (2020)
As	*O. sativa*	Ionomics	Ionic profiles in rice plants were enough to identify reactions to changes in their surroundings.	Du et al. (2020)
As	*O. sativa*	miRNAomics	Micro RNA (miR156j) and miR156 families were shown to confer As resistance	Pandey et al. (2020)
As	*B. napus*	miRNAomics	miR159, miR167, and miR268, NRAMP, and ABC transporter interaction provided insights for metal resistance	Ding et al. (2020)
As	*B. juncea*	Transcriptomics	Detoxified the effect of As	Thakur et al. (2019)
As	*Glycine max*	Metabolomics	Identified diphenyleneiodonium chloride (DPI), epibrassinolide (EBL), and proline (Pro) acted as protective agents, promoting plant growth and germination rates.	Chandrakar et al. (2018)
As	*Nicotiana tabacum*	Metabolomics	Proline improved as stimulated BY-2 tobacco cell proliferation	Nahar et al. (2017)

Protein 3;1 (*AtNIP3;1*), *AtNIP5;1*, *AtNIP6;1*, *AtNIP7* in Arabidopsis (Bienert et al., 2017)and *Pteris vittata* Tonoplast Intrinsic Protein 4;1 (PvTIP4;1) in Pteris vittate (He et al., 2016) also mediate bidirectional transport of As(III). Yeast employs As(III) efflux permeases (*ACR3*) to expel As(III) outside the cell (Garbinski et al., 2019). Additionally, *O. sativa* Heavy Metal ATPases 9 *(OsHMA9)*

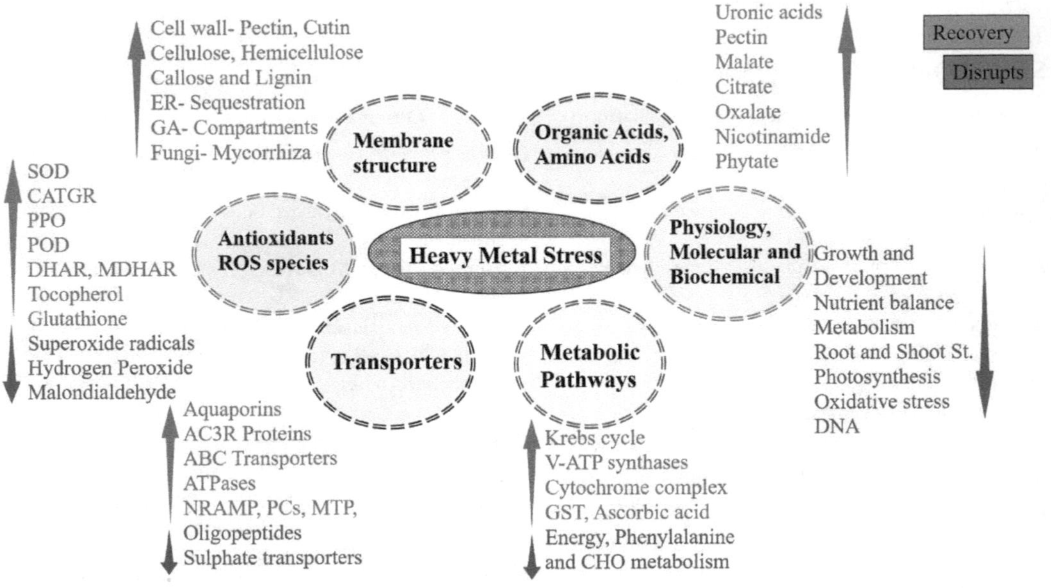

FIGURE 9.6 Various responses to metalloid toxicity in plants.

participate in Zn, lead (Pb), and Cu efflux, while cucumber metal tolerance gene (*CsMTP9)* aids in Mn^{2+} and Cd^{2+} efflux from endodermis into vascular tissue (Migocka et al., 2015).

9.5.1 Role of Phytochelatins (PCs), Metallotheins (MTs), Organic Acids, and Amino Acids in Mitigation of Metalloid Stress

Chelators are the compounds that form complexes with the metal ions. Chelators such as metallotheins (MTs) and phytochelatins (PCs) are plant-based peptides, derived from glutathione and metal-binding AAs and OAs. They are involved in metal binding which provides tolerance toward HMs and creates homeostasis. These chelators aid in the movement of metal ions as well as microbial uptake of HMs by binding to them (Ayangbenro et al., 2017). Metal ions are converted into a volatile state as part of the volatilization process. Only selenium (Se) and mercury (Hg), which have volatile states, are capable of doing this. The MerA (Mercuric Reductase) enzyme is used by bacteria resistant to mercury to convert Hg(II) into the volatile form Hg(0) (Wu et al., 2010). Water and soil contamination have been remedied by reducing Se(V) to elemental Se. In plants exposed to arsenic (As) stress, phytochelatins (PCs) are produced locally using the enzyme phytochelatin synthase, with glutathione as a substrate. For instance, after being taken up by plant cells, As(V) can undergo a swift reduction to As(III) facilitated by arsenate reductases like Arsenate Reductase 2 (*AtACR2*) and histone acetyltransferase (*AtHAC1)* in *Arabidopsis thaliana*, as well as *Oryza sativa* Histone Acetyltransferase 1;1 (*OsHAC1;1*) and *Oryza sativa* Histone Acetyltransferase 1;2 (*OsHAC1;2*) in *O. sativa* afterward which As(III) combines with glutathione or PCs to produce a complex where it is followed by sequestration in vacuoles by ATP-binding cassette (ABC) transporters (Chen et al., 2017). Similarly, in Indian mustard, PCs formed complexes with Cd and Zn (Gasic and Korban, 2007). Also, Cd sequestration in plants involves the formation of low molecular weight (LMW) and high molecular weight (HMW) Cd-PC complexes. The LMW complex carries Cd from the cytosol to the vacuole, whereas the HMW complex serves as the main storage arrangement. Transport is facilitated by multidrug-resistance-associated proteins (MRPs) and tonoplast-localized cation/proton exchangers (CAX), enabling safe Cd storage and detoxification in the case of *B. rapa*, *O. sativa*, and *H. vulgare* (Ahmad et al., 2019).

9.5.2 Alteration in Cell Membrane Structures

The plant cell wall acts as an initial defense against HM intrusion, crucial for safeguarding the plant. Changes in cell wall structure and its modifications have adverse effects on HM transport. Variations in cell wall composition were observed in different soybean varieties with varying Cd tolerance (Jamla et al., 2021). Substantial Cd accumulation, ranging from 48.5% to 75.5%, occurs in the cell wall, primarily composed of polysaccharides like pectin, hemicellulose, and cellulose (Wang et al., 2018). The degree of esterification of pectin affects its metal-binding capacity, with higher demethylation enhancing binding to metals like Al^{3+}, Cd^{2+}, and Cu2+ (Haider et al., 2021). Lignin, part of the secondary cell wall, hardens under high Cd levels, offering protection and influencing metal tolerance (Hodson and Evans, 2020). Additionally, plant stress induces the production of barrier components like callose, cutin, and lignin to counter metal intrusion (Fernández and Eichert, 2009).

Various modifications in cell membrane structures also play a role in HM mitigation. Endoplasmic reticulum (ER) plays a crucial role in HM tolerance in various organisms. For instance, in *Saccharomyces cerevisiae* (yeast), ER-based protein Cadmium Detoxification 1 (*Pca1*) detoxifies the cytosol by sequestering Cd ions due to an endoplasmic-reticulum-associated degradation (Ozturk et al., 2021). In plants such as *Allium cepa,* the ER is important for Pb and Cd detoxification, with Pb accumulating mainly in ER-derived vesicles in certain cells (Wierzbicka et al., 2007). Additionally, aquaglyceroporins like *AtNIP1;1,* located on the ER membrane, control the compartmentalization of metalloids such as arsenite (Ji et al., 2017). The Golgi apparatus and multi-vesicular bodies also play roles in HM tolerance by sequestering and compartmentalizing HMs. Besides, vacuoles are crucial for sequestering HMs and reducing their cytosolic concentration, aiding in HM detoxification (Chandrakar et al., 2020).

9.5.3 Physiological, Biochemical, and Molecular Responses to Metalloid Toxicity in Plants

The physico-chemical and molecular responses of plants to metalloid toxicity vary depending on the specific metalloid and its concentration. The physiological responses of plants to metalloid toxicity involve growth inhibition, oxidative stress, nutrient imbalances, altered photosynthesis, metalloid accumulation, and root morphology changes. For instance, high levels of As inhibited rice plant growth by disrupting various metabolic processes (Khan et al., 2021). As also interfered with nutrient-uptake-replaced essential elements, affecting plant development in *Pteris vittata* (Liu et al., 2018). Also, boron toxicity led to changes in root structure, including reduced root length and increased root branching, as plants attempted to adapt to metalloid stress (Mei et al., 2011). Arsenate competed with essential nutrients like phosphorus and sulfur and hindered ionic balance, sulfur, and nitric oxide metabolism in *Pisum sativum* plants leading to nutrient imbalances that hindered plant growth and impaired nutrient utilization (Rodríguez-Ruiz et al., 2019). Metalloids are known to disrupt photosynthesis, reducing the plant's ability to convert light energy into chemical energy. This results in decreased biomass production (Zaid et al., 2020). Some hyperaccumulators can tolerate high metalloid concentrations but may not thrive in non-contaminated soils (Spiers et al., 2022). In some plants, metalloids can induce oxidative stress by generating ROS that may damage cell membranes, causing DNA/protein oxidation and leading to cellular dysfunction. As and Se induced oxidative stress in *Nicotiana tabacum* that altered antioxidant enzyme activity (Han et al., 2015). Similarly, arsenite disrupted glutathione metabolism, affecting cell wall structure and triggering PCs production as a defense mechanism in *Pistia stratiotes* (Campos et al., 2020). Thus, the toxicity of metalloids in plants triggers a cascade of morphological, physico-chemical, and molecular changes that profoundly influence various aspects of plant biology, ultimately impacting plant growth, nutrient uptake, and stress response.

9.6 CONCLUSION

Understanding the molecular mechanisms underlying plant responses to metalloid stress is essential for developing strategies to enhance plant tolerance and mitigate environmental contamination. Omics approaches are essential for comprehending metalloid stress dynamics, providing insights into the signaling pathways and regulatory networks in plant adaptive responses. Key findings include the upregulation of transporters, the activation of the tricarboxylic acid cycle for OA production, modulation of secondary metabolites, and the involvement of essential phytohormones and thiol-containing molecules. These mechanisms involve complex cross-talk among various metabolic pathways in plant biology. Ultimately, leveraging omics approaches can guide the development of sustainable agricultural practices aimed at enhancing crop resilience in contaminated environments and ensuring global food security.

REFERENCES

Agarwal, Priyanka, and Radha Rani. Strategic management of contaminated water bodies: omics, genome-editing and other recent advances in phytoremediation. *Environmental Technology & Innovation* 27 (2022): 102463.

Ahmad, Javed, Arlene Asthana Ali, Mohd Affan Baig, Muhammad Iqbal, Inamul Haq, and M. Irfan Qureshi. Role of phytochelatins in cadmium stress tolerance in plants. *Cadmium Toxicity and Tolerance in Plants* (2019): 185–212.

Ainley, William M., Lakshmi Sastry-Dent, Mary E. Welter, Michael G. Murray, Bryan Zeitler, Rainier Amora, David R. Corbin et al. Trait stacking via targeted genome editing. *Plant Biotechnology Journal* 11, no. 9 (2013): 1126–1134.

Al Mahmud, Jubayer, M. H. M. Borhannuddin Bhuyan, Taufika Islam Anee, Kamrun Nahar, Masayuki Fujita, and Mirza Hasanuzzaman. Reactive oxygen species metabolism and antioxidant defense in plants under metal/metalloid stress. In: Mirza, Hasanuzzaman, Khalid Rehman Hakeem, Kamrun Nahar, and Hesham F. Alharby (Eds.), *Plant Abiotic Stress Tolerance: Agronomic, Molecular and Biotechnological Approaches*, pp. 221–257. Springer, 2019.

Alam, Mehboob, Zawar Hussain, Anwarzeb Khan, Muhammad Amjad Khan, Abdur Rab, Muhammad Asif, Muhammad Azhar Shah, and Asim Muhammad. The effects of organic amendments on heavy metals bioavailability in mine impacted soil and associated human health risk. *Scientia Horticulturae* 262 (2020): 109067.

Alsafran, Mohammed, Kamal Usman, Bilal Ahmed, Muhammad Rizwan, Muhammad Hamzah Saleem, and Hareb Al Jabri. Understanding the phytoremediation mechanisms of potentially toxic elements: a proteomic overview of recent advances. *Frontiers in Plant Science* 13 (2022): 881242.

Alsafran, M., Usman, K., Ahmed, B., Rizwan, M., Saleem, M. H., and Al Jabri, H. Understanding the phytoremediation mechanisms of potentially toxic elements: a proteomic overview of recent advances. *Frontiers in Plant Science*, 13 (2022): 88124.

Al-Tamimi, Nadia, Chris Brien, Helena Oakey, Bettina Berger, Stephanie Saade, Yung Shwen Ho, Sandra M. Schmöckel, Mark Tester, and Sónia Negrão. Salinity tolerance loci revealed in rice using high-throughput non-invasive phenotyping. *Nature Communications* 7, no. 1 (2016): 13342.

Alter, Svenja, Kai C. Bader, Manuel Spannagl, Yu Wang, Eva Bauer, Chris-Carolin Schön, and Klaus FX Mayer. DroughtDB: an expert-curated compilation of plant drought stress genes and their homologs in nine species. *Database* 2015 (2015): bav046.

Amme, Steffen, Twan Rutten, Michael Melzer, Guido Sonsmann, Johannes P. C. Vissers, Bernhard Schlesier, and Hans-Peter Mock. A proteome approach defines protective functions of tobacco leaf trichomes. *Proteomics* 5, no. 10 (2005): 2508–2518.

Anamika, Sahil Mehta, Baljinder Singh, Anupam Patra, and Md Aminul Islam. Databases: a weapon from the arsenal of bioinformatics for plant abiotic stress research. In: Wani, Shabir Hussain (Eds.), *Recent Approaches in Omics for Plant Resilience to Climate Change*, pp. 135–169. Springer, 2019.

Arshad, Muhammad, Margaret Y. Gruber, Ken Wall, and Abdelali Hannoufa. An insight into microRNA156 role in salinity stress responses of alfalfa. *Frontiers in Plant Science* 8 (2017): 356.

Asaari, Mohd Shahrimie Mohd, Stien Mertens, Stijn Dhondt, Dirk Inzé, Nathalie Wuyts, and Paul Scheunders. Analysis of hyperspectral images for detection of drought stress and recovery in maize plants in a high-throughput phenotyping platform. *Computers and Electronics in Agriculture* 162 (2019): 749–758.

Ayangbenro, Ayansina Segun, and Olubukola Oluranti Babalola. A new strategy for heavy metal polluted environments: a review of microbial biosorbents. *International Journal of Environmental Research and Public Health* 14, no. 1 (2017): 94.

Bailey, Kathryn, and Rebecca C. Fry. Long-term health consequences of prenatal arsenic exposure: links to the genome and the epigenome. *Reviews on Environmental Health* 29, no. 1–2 (2014): 9–12.

Basharat, Zarrin, Luís A. B. Novo, and Azra Yasmin. Genome editing weds CRISPR: what is in it for phytoremediation? *Plants* 7, no. 3 (2018): 51.

Bienert, Manuela Désirée, and Gerd Patrick Bienert. Plant aquaporins and metalloids. In: Chaumont, François and Stephen D. Tyerman (Eds.), *Plant Aquaporins: From Transport to Signaling*, pp. 297–332. Springer, 2017.

Bogdanove, Adam J., Sebastian Schornack, and Thomas Lahaye. TAL effectors: finding plant genes for disease and defense. *Current Opinion in Plant Biology* 13, no. 4 (2010): 394–401.

Bollinedi, Haritha, Ashutosh Kumar Yadav, K. K. Vinod, S. Gopala Krishnan, Prolay Kumar Bhowmick, M. Nagarajan, C. N. Neeraja, Ranjith Kumar Ellur, and Ashok Kumar Singh. Genome-wide association study reveals novel marker-trait associations (MTAs) governing the localization of Fe and Zn in the rice grain. *Frontiers in Genetics* 11 (2020): 213.

Borkotoky, Subhomoi, Vijayakumar Saravanan, Amit Jaiswal, Bipul Das, Suresh Selvaraj, Ayaluru Murali, and P. T. V. Lakshmi. The Arabidopsis stress responsive gene database. *International Journal of Plant Genomics* 2013 (2013): 949564.

Buffagni, Valentina, Leilei Zhang, Biancamaria Senizza, Gabriele Rocchetti, Andrea Ferrarini, Begoña Miras-Moreno, and Luigi Lucini. Metabolomics and lipidomics insight into the effect of different polyamines on tomato plants under non-stress and salinity conditions. *Plant Science* 322 (2022): 111346.

Caldana, Camila, Thomas Degenkolbe, Alvaro Cuadros-Inostroza, Sebastian Klie, Ronan Sulpice, Andrea Leisse, Dirk Steinhauser, Alisdair R. Fernie, Lothar Willmitzer, and Matthew A. Hannah. High-density kinetic analysis of the metabolomic and transcriptomic response of Arabidopsis to eight environmental conditions. *The Plant Journal* 67, no. 5 (2011): 869–884.

Campbell, Malachy T., Avi C. Knecht, Bettina Berger, Chris J. Brien, Dong Wang, and Harkamal Walia. Integrating image-based phenomics and association analysis to dissect the genetic architecture of temporal salinity responses in rice. *Plant Physiology* 168, no. 4 (2015): 1476–1489.

Campos, Fernanda Vidal de, Juraci Alves de Oliveira, Adinan Alves da Silva, Cleberson Ribeiro, Sebastián Giraldo Montoya, and Fernanda dos Santos Farnese. Involvement of glutathione and glutathione metabolizing enzymes in *Pistia stratiotes* tolerance to arsenite. *International Journal of Phytoremediation* 22, no. 4 (2020): 404–411.

Carillo, Petronia. GABA shunt in durum wheat. *Frontiers in Plant Science* 9 (2018): 100.

Carroll, April Agee, Jennifer Clarke, Noah Fahlgren, Malia A. Gehan, Carolyn J. Lawrence-Dill, and Argelia Lorence. NAPPN: who we are, where we are going, and why you should join us! *The Plant Phenome Journal* 2, no. 1 (2019): 1–4.

Chandrakar, Vibhuti, Amit Dubey, and Sahu Keshavkant. Modulation of arsenic-induced oxidative stress and protein metabolism by diphenyleneiodonium, 24-epibrassinolide and proline in *Glycine max* L. *Acta Botanica Croatica* 77, no. 1 (2018): 51–61.

Chandrakar, Vibhuti, Bhumika Yadu, Roseline Xalxo, Meetul Kumar, and S. Keshavkant. Mechanisms of plant adaptation and tolerance to metal/metalloid toxicity. In: Mirza, Hasanuzzaman (Ed.), *Plant Ecophysiology and Adaptation under Climate Change: Mechanisms and Perspectives II: Mechanisms of Adaptation and Stress Amelioration*, pp. 107–135. Springer, 2020.

Chebrolu, Kranthi K., Felix B. Fritschi, Songqing Ye, Hari B. Krishnan, James R. Smith, and Jason D. Gillman. Impact of heat stress during seed development on soybean seed metabolome. *Metabolomics* 12 (2016): 1–14.

Chen, Yanshan, Yong-He Han, Yue Cao, Yong-Guan Zhu, Bala Rathinasabapathi, and Lena Q. Ma. Arsenic transport in rice and biological solutions to reduce arsenic risk from rice. *Frontiers in Plant Science* 8 (2017): 268.

Cheng, Yanwei, Yaocheng Qi, Qian Zhu, Xi Chen, Ning Wang, Xin Zhao, Haiyan Chen, Xiangju Cui, Langlai Xu, and Wei Zhang. New changes in the plasma-membrane-associated proteome of rice roots under salt stress. *Proteomics* 9, no. 11 (2009): 3100–3114.

Cramer, Grant R., Kaoru Urano, Serge Delrot, Mario Pezzotti, and Kazuo Shinozaki. Effects of abiotic stress on plants: a systems biology perspective. *BMC Plant Biology* 11, no. 1 (2011): 1–14.

Crandall, Sharifa G., Kaitlin M. Gold, María del Mar Jiménez-Gasco, Camila C. Filgueiras, and Denis S. Willett. A multi-omics approach to solving problems in plant disease ecology. *PLoS One* 15, no. 9 (2020): e0237975.

Cvikrová, Milena, Lenka Gemperlová, Jana Dobrá, Olga Martincová, Ilja T. Prásil, Jozef Gubis, and Radomira Vanková. Effect of heat stress on polyamine metabolism in proline-over-producing tobacco plants. *Plant Science* 182 (2012): 49–58.

de Miguel, Marina, M. Ángeles Guevara, David Sánchez-Gómez, Nuria De María, Luis Manuel Díaz, Jose A. Mancha, Brígida Fernández De Simón et al. Organ-specific metabolic responses to drought in Pinus pinaster Ait. *Plant Physiology and Biochemistry* 102 (2016): 17–26.

Degenkolbe, Thomas, Patrick Giavalisco, Ellen Zuther, Bettina Seiwert, Dirk K. Hincha, and Lothar Willmitzer. Differential remodeling of the lipidome during cold acclimation in natural accessions of Arabidopsis thaliana. *The Plant Journal* 72, no. 6 (2012): 972–982.

Deng, Fenglin, Xue Liu, Yanshan Chen, Bala Rathinasabapathi, Christopher Rensing, Jian Chen, Jue Bi, Ping Xiang, and Lena Q. Ma. Aquaporins mediated arsenite transport in plants: molecular mechanisms and applications in crop improvement. *Critical Reviews in Environmental Science and Technology* 50, no. 16 (2020): 1613–1639.

Derakhshani, Behnam, Hossein Jafary, Bahram Maleki Zanjani, Karim Hasanpur, Kohei Mishina, Tsuyoshi Tanaka, Yoshihiro Kawahara, and Youko Oono. Combined QTL mapping and RNA-Seq profiling reveals candidate genes associated with cadmium tolerance in barley. *PLoS One* 15, no. 4 (2020): e0230820.

Dereeper, Alexis, Stéphanie Bocs, Mathieu Rouard, Valentin Guignon, Sébastien Ravel, Christine Tranchant-Dubreuil, Valérie Poncet, Olivier Garsmeur, Philippe Lashermes, and Gaëtan Droc. The coffee genome hub: a resource for coffee genomes. *Nucleic Acids Research* 43, no. D1 (2015): D1028–D1035.

Dilnur, Tussipkan, Zhen Peng, Zhaoe Pan, Koffi Kibalou Palanga, Yinhua Jia, Wenfang Gong, and Xiongming Du. Association analysis of salt tolerance in Asiatic cotton (Gossypium arboretum) with SNP markers. *International Journal of Molecular Sciences* 20, no. 9 (2019): 2168.

Ding, Yanfei, Lihong Ding, Yiji Xia, Feijuan Wang, and Cheng Zhu. Emerging roles of microRNAs in plant heavy metal tolerance and homeostasis. *Journal of Agricultural and Food Chemistry* 68, no. 7 (2020): 1958–1965.

Dos Santos, Vanessa Samúdio, Flávio Alves Macedo, Jean Silva Do Vale, Denise Brentan Silva, and Carlos Alexandre Carollo. Metabolomics as a tool for understanding the evolution of Tabebuia sensu lato. *Metabolomics* 13 (2017): 1–11.

Du, Fan, Peng Liu, Kai Wang, Zhaoguang Yang, and Lin Wang. Ionomic responses of rice plants to the stresses of different arsenic species in hydroponics. *Chemosphere* 243 (2020): 125398.

El-Beltagi, Hossam Saad, Sherif Helmy Ahmed, Alia Awad Mahmoud Namich, and Reman Raafat Abdel-Sattar. Effect of salicylic acid and potassium citrate on cotton plant under salt stress. *Fresenius Environmental Bulletin* 26, no. 1A (2017): 1091–1100.

El-Ghiaty, Mahmoud A., and Ayman O. S. El-Kadi. Arsenic: various species with different effects on cytochrome P450 regulation in humans. *EXCLI Journal* 20 (2021): 1184.

Falk, Kevin G., Talukder Zaki Jubery, Jamie A. O'Rourke, Arti Singh, Soumik Sarkar, Baskar Ganapathysubramanian, and Asheesh K. Singh. Soybean root system architecture trait study through genotypic, phenotypic, and shape-based clusters. *Plant Phenomics* 2020 (2020): 1925495.

Farooq, Muhammad Ahsan, Zheyuan Hong, Faisal Islam, Yamna Noor, Fakhir Hannan, Yan Zhang, Ahsan Ayyaz, Theodore M. Mwamba, Weijun Zhou, and Wenjian Song. Comprehensive proteomic analysis of arsenic induced toxicity reveals the mechanism of multilevel coordination of efficient defense and energy metabolism in two Brassica napus cultivars. *Ecotoxicology and Environmental Safety* 208 (2021): 111744.

Feng, Xuemin, Lei Han, Daiyin Chao, Yan Liu, Yajing Zhang, Ruigang Wang, Junkang Guo et al. Ionomic and transcriptomic analysis provides new insight into the distribution and transport of cadmium and arsenic in rice. *Journal of Hazardous Materials* 331 (2017): 246–256.

Fernández, Victoria, and Thomas Eichert. Uptake of hydrophilic solutes through plant leaves: current state of knowledge and perspectives of foliar fertilization. *Critical Reviews in Plant Sciences* 28, no. 1–2 (2009): 36–68.

Garbinski, Luis D., Barry P. Rosen, and Jian Chen. Pathways of arsenic uptake and efflux. *Environment International* 126 (2019): 585–597.

Gasic, Ksenija, and Schuyler S. Korban. Expression of Arabidopsis phytochelatin synthase in Indian mustard (Brassica juncea) plants enhances tolerance for Cd and Zn. *Planta* 225 (2007): 1277–1285.

Ghanim, Abdulnour Ali Jazem, Muhammad Azhar Shah, Mehboob Alam, Anwarzeb Khan, Muhammad Amjad Khan, Saifur Rahman, Mabkhoot A. Alsaiari et al. The influence of compost amendments on bioaccumulation of potentially toxic elements by pea plant cultivated in mine degraded soils. *Arabian Journal of Geosciences* 16, no. 1 (2023): 46.

Gill, Sarvajeet Singh, and Narendra Tuteja. Polyamines and abiotic stress tolerance in plants. *Plant Signaling & Behavior* 5, no. 1 (2010): 26–33.

Granier, Christine, Luis Aguirrezabal, Karine Chenu, Sarah Jane Cookson, Myriam Dauzat, Philippe Hamard, Jean-Jacques Thioux et al. PHENOPSIS, an automated platform for reproducible phenotyping of plant responses to soil water deficit in Arabidopsis thaliana permitted the identification of an accession with low sensitivity to soil water deficit. *New Phytologist* 169, no. 3 (2006): 623–635.

Griesser, Michaela, Georg Weingart, Katharina Schoedl-Hummel, Nora Neumann, Manuel Becker, Kurt Varmuza, Falk Liebner, Rainer Schuhmacher, and Astrid Forneck. Severe drought stress is affecting selected primary metabolites, polyphenols, and volatile metabolites in grapevine leaves (Vitis vinifera cv. Pinot noir). *Plant Physiology and Biochemistry* 88 (2015): 17–26.

Guo, Guangfang, Pei Ge, Chaoying Ma, Xiaohui Li, Dongwen Lv, Shunli Wang, Wujun Ma, and Yueming Yan. Comparative proteomic analysis of salt response proteins in seedling roots of two wheat varieties. *Journal of Proteomics* 75, no. 6 (2012): 1867–1885.

Gupta, B., & Shrestha, J. (2023). Abiotic stress adaptation and tolerance mechanisms in crop plants. Frontiers in Plant Science, 14, 1278895.

Haider, Fasih Ullah, Cai Liqun, Jeffrey A. Coulter, Sardar Alam Cheema, Jun Wu, Renzhi Zhang, Ma Wenjun, and Muhammad Farooq. Cadmium toxicity in plants: impacts and remediation strategies. *Ecotoxicology and Environmental Safety* 211 (2021): 111887.

Han, Dan, Shuanglian Xiong, Shuxin Tu, Jinchang Liu, and Cheng Chen. Interactive effects of selenium and arsenic on growth, antioxidant system, arsenic and selenium species of Nicotiana tabacum L. *Environmental and Experimental Botany* 117 (2015): 12–19.

He, Zhenyan, Huili Yan, Yanshan Chen, Hongling Shen, Wenxiu Xu, Haiyan Zhang, Lei Shi, Yong-Guan Zhu, and Mi Ma. An aquaporin Pv TIP 4; 1 from Pteris vittata may mediate arsenite uptake. *New Phytologist* 209, no. 2 (2016): 746–761.

Hirai, Masami Yokota, Mitsuru Yano, Dayan B. Goodenowe, Shigehiko Kanaya, Tomoko Kimura, Motoko Awazuhara, Masanori Arita, Toru Fujiwara, and Kazuki Saito. Integration of transcriptomics and metabolomics for understanding of global responses to nutritional stresses in Arabidopsis thaliana. *Proceedings of the National Academy of Sciences* 101, no. 27 (2004): 10205–10210.

Hodson, Martin J., and David E. Evans. Aluminium–silicon interactions in higher plants: an update. *Journal of Experimental Botany* 71, no. 21 (2020): 6719–6729.

Huang, Lei, Ailan Zeng, Pengyin Chen, Chengjun Wu, Dechun Wang, and Zixiang Wen. Genomewide association analysis of salt tolerance in soybean [Glycine max (L.) Merr.]. *Plant Breeding* 137, no. 5 (2018): 714–720.

Humplík, Jan F., Dušan Lazár, Alexandra Husičková, and Lukáš Spíchal. Automated phenotyping of plant shoots using imaging methods for analysis of plant stress responses–a review. *Plant Methods* 11, no. 1 (2015): 1–10.

Ibrahim, Muhammad, Gang Li, Sardar Khan, Qiaoqiao Chi, and Yaoyang Xu. Biochars mitigate greenhouse gas emissions and bioaccumulation of potentially toxic elements and arsenic speciation in Phaseolus vulgaris L. *Environmental Science and Pollution Research* 24 (2017): 19524–19534.

Iftikhar, Sumbal, Veysel Turan, Hafiz Muhammad Tauqeer, Bilal Rasool, Muhammad Zubair, Muhammad Asaf Khan, Shamim Akhtar et al. Phytomanagement of As-contaminated matrix: physiological and molecular basis. In: Mirza, Hasanuzzaman and Majeti Narasimha Vara Prasad (Eds.), *Handbook of Bioremediation*, pp. 61–79. Academic Press, 2021.

Iqbal, Zahra, Mohammed Shariq Iqbal, Ausaf Ahmad, Anjuman Gul Memon, and Mohammad Israil Ansari. New prospects on the horizon: genome editing to engineer plants for desirable traits. *Current Plant Biology* 24 (2020): 100171.

Jamla, Monica, Tushar Khare, Shrushti Joshi, Suraj Patil, Suprasanna Penna, and Vinay Kumar. Omics approaches for understanding heavy metal responses and tolerance in plants. *Current Plant Biology* 27 (2021): 100213.

Ji, Chang Yoon, Won-Hyong Chung, Ho Soo Kim, Won Yong Jung, Le Kang, Jae Cheol Jeong, and Sang-Soo Kwak. Transcriptome profiling of sweetpotato tuberous roots during low temperature storage. *Plant Physiology and Biochemistry* 112 (2017): 97–108.

Jiao, Yang, Zhenzi Bai, Jingyu Xu, Mingli Zhao, Yaseen Khan, Yongjun Hu, and Lianxuan Shi. Metabolomics and its physiological regulation process reveal the salt-tolerant mechanism in Glycine soja seedling roots. *Plant Physiology and Biochemistry* 126 (2018): 187–196.

Jobe, Timothy O., Qi Yu, Felix Hauser, Qingqing Xie, Yuan Meng, Tim Maassen, Stanislav Kopriva, and Julian I. Schroeder. The SLIM1 transcription factor regulates arsenic sensitivity in Arabidopsis thaliana. *FEBS Letters* 595 (2021): 1696–1707.

Jogaiah, Sudisha, Sharathchandra Ramsandra Govind, and Lam-Son Phan Tran. Systems biology-based approaches toward understanding drought tolerance in food crops. *Critical Reviews in Biotechnology* 33, no. 1 (2013): 23–39.

Joung, J. Keith, and Jeffry D. Sander. TALENs: a widely applicable technology for targeted genome editing. *Nature Reviews Molecular Cell Biology* 14, no. 1 (2013): 49–55.

Kakeshpour, Tayebeh. *Tomato class II glutaredoxin mutants generated via multiplex CRISPR/Cas9 genome editing technology are susceptible to multiple abiotic stresses.* (Thesis) Kansas State University (2020).

Kamal, Abu Hena Mostafa, Kun Cho, Da-Eun Kim, Nobuyuki Uozumi, Keun-Yook Chung, Sang Young Lee, Jong-Soon Choi, Seong-Woo Cho, Chang-Seob Shin, and Sun Hee Woo. Changes in physiology and protein abundance in salt-stressed wheat chloroplasts. *Molecular Biology Reports* 39 (2012): 9059–9074.

Kashiwabara, Teruhiko, Nobuyuki Kitajima, Ryoko Onuma, Naoki Fukuda, Satoshi Endo, Yasuko Terada, Tomoko Abe, Akiko Hokura, and Izumi Nakai. Synchrotron micro-X-ray fluorescence imaging of arsenic in frozen-hydrated sections of a root of Pteris vittata. *Metallomics* 13, no. 4 (2021): mfab009.

Kaur, Ravdeep, Poonam Yadav, Sukhmeen Kaur Kohli, Vinod Kumar, Palak Bakshi, Bilal Ahmad Mir, Ashwani Kumar Thukral, and Renu Bhardwaj. Emerging trends and tools in transgenic plant technology for phytoremediation of toxic metals and metalloids. In: Prasad, M. N. V. (Ed.), *Transgenic Plant Technology for Remediation of Toxic Metals and Metalloids*, pp. 63–88. Academic Press, 2019.

Khan, M. Iqbal R., Badar Jahan, Mohamed F. AlAjmi, Md Tabish Rehman, Noushina Iqbal, Mohammad Irfan, Zebus Sehar, and Nafees A. Khan. Crosstalk of plant growth regulators protects photosynthetic performance from arsenic damage by modulating defense systems in rice. *Ecotoxicology and Environmental Safety* 222 (2021): 112535.

Khan, Mather A., Norma A. Castro-Guerrero, Samuel A. McInturf, Nga T. Nguyen, Ashley N. Dame, Jiaojiao Wang, Rebecca K. Bindbeutel et al. Changes in iron availability in Arabidopsis are rapidly sensed in the leaf vasculature and impaired sensing leads to opposite transcriptional programs in leaves and roots. *Plant, Cell & Environment* 41, no. 10 (2018): 2263–2276.

Kim, Eiru, Sohyun Hwang, and Insuk Lee. SoyNet: a database of co-functional networks for soybean Glycine max. *Nucleic Acids Research* 45, no. D1 (2017): D1082–D1089.

Kök, Aysu Başak, Mehmet Direnç Mungan, Sami Doğanlar, and Anne Frary. Transcriptomic analysis of selenium accumulation in Puccinellia distans (Jacq.) Parl., a boron hyperaccumulator. *Chemosphere* 245 (2020): 125665.

Kumar, Vinod, Anshuman Singh, SV Amitha Mithra, S. L. Krishnamurthy, Swarup K. Parida, Sourabh Jain, Kapil K. Tiwari et al. Genome-wide association mapping of salinity tolerance in rice (Oryza sativa). *DNA Research* 22, no. 2 (2015): 133–145.

Kumari, Asha, Paromita Das, Asish Kumar Parida, and Pradeep K. Agarwal. Proteomics, metabolomics, and ionomics perspectives of salinity tolerance in halophytes. *Frontiers in Plant Science* 6 (2015): 537.

Latowski, Dariusz, Paulina Kuczyńska, and Kazimierz Strzałka. Xanthophyll cycle–a mechanism protecting plants against oxidative stress. *Redox Report* 16, no. 2 (2011): 78–90.

Lee, Tak, Taeyun Oh, Sunmo Yang, Junha Shin, Sohyun Hwang, Chan Yeong Kim, Hyojin Kim et al. RiceNet v2: an improved network prioritization server for rice genes. *Nucleic Acids Research* 43, no. W1 (2015): W122–W127.

Li, Huihui, Awais Rasheed, Lee T. Hickey, and Zhonghu He. Fast-forwarding genetic gain. *Trends in Plant Science* 23, no. 3 (2018): 184–186.

Li, Ting, Bo Liu, Martin H. Spalding, Donald P. Weeks, and Bing Yang. High-efficiency TALEN-based gene editing produces disease-resistant rice. *Nature Biotechnology* 30, no. 5 (2012): 390–392.

Liu, Shiqi, Xuqian Peng, Xiaojing Wang, and Weibing Zhuang. Transcriptome analysis reveals differentially expressed genes involved in cadmium and arsenic accumulation in tea plant (Camellia sinensis). *Plants* 12, no. 5 (2023): 1182.

Liu, Xue, Hua–Yuan Feng, Jing–Wei Fu, Yanshan Chen, Yungen Liu, and Lena Q. Ma. Arsenic-induced nutrient uptake in As-hyperaccumulator Pteris vittata and their potential role to enhance plant growth. *Chemosphere* 198 (2018): 425–431.

Lu, Hai-Ping, Song-Mei Liu, Shou-Ling Xu, Wu-Yang Chen, Xin Zhou, Yuan-Yuan Tan, Jian-Zhong Huang, and Qing-Yao Shu. CRISPR-S: an active interference element for a rapid and inexpensive selection of genome-edited, transgene-free rice plants. *Plant Biotechnology Journal* 15, no. 11 (2017): 1371.

Luengwilai, Kietsuda, Mikal Saltveit, and Diane M. Beckles. Metabolite content of harvested Micro-Tom tomato (Solanum lycopersicum L.) fruit is altered by chilling and protective heat-shock treatments as shown by GC–MS metabolic profiling. *Postharvest Biology and Technology* 63, no. 1 (2012): 116–122.

Luo, Jie. Metabolite-based genome-wide association studies in plants. *Current Opinion in Plant Biology* 24 (2015): 31–38.

Luo, Jin-Song, and Zhenhua Zhang. Proteomic changes in the xylem sap of Brassica napus under cadmium stress and functional validation. *BMC Plant Biology* 19, no. 1 (2019): 1–14.

Luo, Zinan, Aaron Szczepanek, and Hussein Abdel-Haleem. Genome-wide association study (GWAS) analysis of camelina seedling germination under salt stress condition. *Agronomy* 10, no. 9 (2020): 1444.

Ma, Jing, Chunfang Lv, Minli Xu, Guoxiang Chen, Chuangen Lv, and Zhiping Gao. Photosynthesis performance, antioxidant enzymes, and ultrastructural analyses of rice seedlings under chromium stress. *Environmental Science and Pollution Research* 23 (2016): 1768–1778.

Manivasagaperumal, R., S. Balamurugan, G. Thiyagarajan, and J. Sekar. Effect of zinc on germination, seedling growth and biochemical content of cluster bean (Cyamopsis tetragonoloba (L.) Taub). *Current Botany* 2, no. 5 (2011).

Maruyama, Kyonoshin, Kaoru Urano, Kyouko Yoshiwara, Yoshihiko Morishita, Nozomu Sakurai, Hideyuki Suzuki, Mikiko Kojima et al. Integrated analysis of the effects of cold and dehydration on rice metabolites, phytohormones, and gene transcripts. *Plant Physiology* 164, no. 4 (2014): 1759–1771.

Matsuda, Fumio, Ryo Nakabayashi, Zhigang Yang, Yozo Okazaki, Jun-ichi Yonemaru, Kaworu Ebana, Masahiro Yano, and Kazuki Saito. Metabolome-genome-wide association study dissects genetic architecture for generating natural variation in rice secondary metabolism. *The Plant Journal* 81, no. 1 (2015): 13–23.

Mei, Li, O. U. Sheng, Shu-ang Peng, Gao-feng Zhou, Qing-jiang Wei, and Qiao-hong Li. Growth, root morphology and boron uptake by citrus rootstock seedlings differing in boron-deficiency responses. *Scientia Horticulturae* 129, no. 3 (2011): 426–432.

Meng, Rui, Stephanie Saade, Sebastian Kurtek, Bettina Berger, Chris Brien, Klaus Pillen, Mark Tester, and Ying Sun. Growth curve registration for evaluating salinity tolerance in barley. *Plant Methods* 13 (2017): 1–9.

Migocka, Magdalena, Anna Papierniak, Anna Kosieradzka, Ewelina Posyniak, Ewa Maciaszczyk-Dziubinska, Robert Biskup, Arnold Garbiec, and Tadeusz Marchewka. Retracted: cucumber metal tolerance protein Cs MTP 9 is a plasma membrane H^+-coupled antiporter involved in the Mn^{2+} and Cd^{2+} efflux from root cells. *The Plant Journal* 84, no. 6 (2015): 1045–1058.

Misra, Biswapriya B., Carl Langefeld, Michael Olivier, and Laura A. Cox. Integrated omics: tools, advances and future approaches. *Journal of Molecular Endocrinology* 62, no. 1 (2019): R21–R45.

Mišúthová, Adriana, Ľudmila Slováková, Karin Kollárová, and Marek Vaculík. Effect of silicon on root growth, ionomics and antioxidant performance of maize roots exposed to As toxicity. *Plant Physiology and Biochemistry* 168 (2021): 155–166.

Mohanty, Gargee, Rajashree Das, Ayusman Behera, and Junaid Ahmad Malik. Transgenic approaches for improving phytoremediation potential. In: Malik, Junaid Ahmad (Ed.), *Microbial and Biotechnological Interventions in Bioremediation and Phytoremediation*, pp. 541–567. Springer International Publishing, 2022.

Mueller, Stephanie P., Daniel M. Krause, Martin J. Mueller, and Agnes Fekete. Accumulation of extra-chloroplastic triacylglycerols in Arabidopsis seedlings during heat acclimation. *Journal of Experimental Botany* 66, no. 15 (2015): 4517–4526.

Mulenga, Charles, Darius Phiri, Daigard Ricardo Ortega-Rodriguez, and Martina Meincken. Bioaccumulation of potentially toxic elements by indigenous and exotic trees growing around a copper leaching plant in Mufulira, Zambia. *Environmental Systems Research* 12, no. 1 (2023): 26.

Mushtaq, Muntazir, Aafreen Sakina, Shabir Hussain Wani, Asif B. Shikari, Prateek Tripathi, Abbu Zaid, Aravind Galla et al. Harnessing genome editing techniques to engineer disease resistance in plants. *Frontiers in Plant Science* 10 (2019): 550.

Mushtaq, Muntazir, and Kutubuddin A. Molla. CRISPR technologies for plant biotechnology innovation. In: Ricroch, Agnès, Surinder Chopra, Marcel Kuntz (Eds.), *Plant Biotechnology: Experience and Future Prospects*, pp. 51–67. Springer International Publishing, 2021.

Muthamilarasan, Mehanathan, Nagendra Kumar Singh, and Manoj Prasad. Multi-omics approaches for strategic improvement of stress tolerance in underutilized crop species: a climate change perspective. *Advances in Genetics* 103 (2019): 1–38.

Nahar, Mst Nur-E-Nazmun, Mohammad Muzahidul Islam, Md Anamul Hoque, Anna Yonezawa, Md Yeasin Prodhan, Toshiyuki Nakamura, Yoshimasa Nakamura, Shintaro Munemasa, and Yoshiyuki Murata. Exogenous proline enhances the sensitivity of Tobacco BY-2 cells to arsenate. *Bioscience, Biotechnology, and Biochemistry* 81, no. 9 (2017): 1726–1731.

Nam, Anna S., Ronan Chaligne, and Dan A. Landau. Integrating genetic and non-genetic determinants of cancer evolution by single-cell multi-omics. *Nature Reviews Genetics* 22, no. 1 (2021): 3–18.

Naz, Misbah, Adalberto Benavides-Mendoza, Muhammad Tariq, Jianyu Zhou, Jiahao Wang, Shanshan Qi, Zhicong Dai, and Daolin Du. CRISPR/Cas9 technology as an innovative approach to enhancing the phytoremediation: concepts and implications. *Journal of Environmental Management* 323 (2022): 116296.

Negrão, Sónia, Sandra M. Schmöckel, and Mark Tester. Evaluating physiological responses of plants to salinity stress. *Annals of Botany* 119, no. 1 (2017): 1–11.

Neugart, Susanne, Angelika Krumbein, and Rita Zrenner. Influence of light and temperature on gene expression leading to accumulation of specific flavonol glycosides and hydroxycinnamic acid derivatives in kale (Brassica oleracea var. sabellica). *Frontiers in Plant Science* 7 (2016): 326.

Newton, Alan Charles; Lyon, Gordon David; and Marshall, Brian. DRASTIC: a database resource for analysis of signal transduction in cells. *BSPP Newsletter* 42 (2002): 36–37.

Ni, Jianwei, Xiuyan Yang, Jianfeng Zhu, Zhengxiang Liu, Yuanying Ni, Haiwen Wu, Huaxin Zhang, and Tao Liu. Salinity-induced metabolic profile changes in Nitraria tangutorum Bobr. suspension cells. *Plant Cell, Tissue and Organ Culture (PCTOC)* 122 (2015): 239–248.

Okem, Ambrose, Wendy A. Stirk, Renée A. Street, Colin Southway, Jeffrey F. Finnie, and Johannes Van Staden. Effects of Cd and Al stress on secondary metabolites, antioxidant and antibacterial activity of Hypoxis hemerocallidea Fisch. & CA Mey. *Plant Physiology and Biochemistry* 97 (2015): 147–155.

Ozturk, Munir, Mert Metin, Volkan Altay, Luigi De Filippis, Bengu Turkyilmaz Ünal, Anum Khursheed, Alvina Gul et al. Molecular biology of cadmium toxicity in Saccharomyces cerevisiae. *Biological Trace Element Research* 199 (2021): 1–15.

Pal, Rama, and Jai Prakash Narain Rai. Phytochelatins: peptides involved in heavy metal detoxification. *Applied Biochemistry and Biotechnology* 160 (2010): 945–963.

Pan, Jiaowen, Zhen Li, Shaojun Dai, Hanfeng Ding, Qingguo Wang, Xiaobo Li, Guohua Ding, Pengfei Wang, Yanan Guan, and Wei Liu. Integrative analyses of transcriptomics and metabolomics upon seed germination of foxtail millet in response to salinity. *Scientific Reports* 10, no. 1 (2020): 13660.

Pandey, Akhilesh Kumar, Mallikarjuna Rao Gedda, and Ashok K. Verma. Effect of arsenic stress on expression pattern of a rice specific miR156j at various developmental stages and their allied co-expression target networks. *Frontiers in Plant Science* 11 (2020): 752.

Papoutsoglou, Evangelia A., Daniel Faria, Daniel Arend, Elizabeth Arnaud, Ioannis N. Athanasiadis, Inês Chaves, Frederik Coppens et al. Enabling reusability of plant phenomic datasets with MIAPPE 1.1. *New Phytologist* 227, no. 1 (2020): 260–273.

Pennisi, Elizabeth. The CRISPR craze. *Science* 341 (2013): 833–836.

Prabha, Ratna, Indira Ghosh, and Dhananjaya P. Singh. Plant stress gene database: a collection of plant genes responding to stress condition. *ARPN Journal of Science and Technology* 1, no. 1 (2011): 28–31.

Pranneshraj, Velumani, Manjeet Kaur Sangha, Ivica Djalovic, Jegor Miladinovic, and Maduraimuthu Djanaguiraman. Lipidomics-assisted GWAS (LGWAS) approach for improving high-temperature stress tolerance of crops. *International Journal of Molecular Sciences* 23, no. 16 (2022): 9389.

Pratap, Aditya, Sanjeev Gupta, Ramakrishnan Madhavan Nair, S. K. Gupta, Roland Schafleitner, P. S. Basu, Chandra Mohan Singh et al. Using plant phenomics to exploit the gains of genomics. *Agronomy* 9, no. 3 (2019): 126.

Punshon, Tracy, Felipe Klein Ricachenevsky, Maria N. Hindt, Amanda L. Socha, and Hélène Zuber. Methodological approaches for using synchrotron X-ray fluorescence (SXRF) imaging as a tool in ionomics: examples from Arabidopsis thaliana. *Metallomics* 5, no. 9 (2013): 1133–1145.

Qian, Xi, Yi Ba, Qianfeng Zhuang, and Guofang Zhong. RNA-Seq technology and its application in fish transcriptomics. *Omics: A Journal of Integrative Biology* 18, no. 2 (2014): 98–110.

Rascio, Nicoletta, and Flavia Navari-Izzo. Heavy metal hyperaccumulating plants: how and why do they do it? And what makes them so interesting? *Plant Science* 180, no. 2 (2011): 169–181.

Raza, Ali, Javaria Tabassum, Zainab Zahid, Sidra Charagh, Shanza Bashir, Rutwik Barmukh, Rao Sohail Ahmad Khan et al. Advances in "omics" approaches for improving toxic metals/metalloids tolerance in plants. *Frontiers in Plant Science* 12 (2022): 794373.

Rivero, Rosa M., Teresa C. Mestre, Ron Mittler, Francisco Rubio, Francisco Garcia-Sanchez, and Vicente Martinez. The combined effect of salinity and heat reveals a specific physiological, biochemical and molecular response in tomato plants. *Plant, Cell & Environment* 37, no. 5 (2014): 1059–1073.

Rodríguez-Ruiz, Marta, María V. Aparicio-Chacón, José M. Palma, and Francisco J. Corpas. Arsenate disrupts ion balance, sulfur and nitric oxide metabolisms in roots and leaves of pea (Pisum sativum L.) plants. *Environmental and Experimental Botany* 161 (2019): 143–156.

Sanchez, Diego H., Fernando L. Pieckenstain, Francisco Escaray, Alexander Erban, Ute Kraemer, Michael K. Udvardi, and Joachim Kopka. Comparative ionomics and metabolomics in extremophile and glycophytic Lotus species under salt stress challenge the metabolic pre-adaptation hypothesis. *Plant, Cell & Environment* 34, no. 4 (2011): 605–617.

Sarhadi, Elham, Mitra Mohammadi Bazargani, Andres Godwin Sajise, Shapour Abdolahi, Naireen Aiza Vispo, Marydee Arceta, Ghasem Mohammadi Nejad, Rakesh Kumar Singh, and Ghasem Hosseini Salekdeh. Proteomic analysis of rice anthers under salt stress. *Plant Physiology and Biochemistry* 58 (2012): 280–287.

Sarma, Hemen, and Majeti Narasimha Vara Prasad. Metabolic engineering of rhizobacteria associated with plants for remediation of toxic metals and metalloids. In: Majeti Narasimha Vara Prasad (Ed.) *Transgenic Plant Technology for Remediation of Toxic Metals and Metalloids*, pp. 299–318. Academic Press, 2019.

Sato, Yutaka, Nobukazu Namiki, Hinako Takehisa, Kaori Kamatsuki, Hiroshi Minami, Hiroshi Ikawa, Hajime Ohyanagi et al. RiceFREND: a platform for retrieving coexpressed gene networks in rice. *Nucleic Acids Research* 41, no. D1 (2013): D1214–D1221.

Schlatter, Daniel, Linda Kinkel, Linda Thomashow, David Weller, and Timothy Paulitz. Disease suppressive soils: new insights from the soil microbiome. *Phytopathology* 107, no. 11 (2017): 1284–1297.

Shelden, Megan C., and Ute Roessner. Advances in functional genomics for investigating salinity stress tolerance mechanisms in cereals. *Frontiers in Plant Science* 4 (2013): 123.

Shin, Hyunsuk, Sewon Oh, Keumsun Kim, and Daeil Kim. Proline accumulates in response to higher temperatures during dehardening in peach shoot tissues. *The Horticulture Journal* 85, no. 1 (2016): 37–45.

Singh, Akshay, Ajay Kumar Sharma, Nagendra Kumar Singh, and Tilak Raj Sharma. PpTFDB: a pigeonpea transcription factor database for exploring functional genomics in legumes. *PLoS One* 12, no. 6 (2017): e0179736.

Singhal, Rajesh K., Debanjana Saha, Milan Skalicky, Udit N. Mishra, Jyoti Chauhan, Laxmi P. Behera, Devidutta Lenka et al. Crucial cell signaling compounds crosstalk and integrative multi-omics techniques for salinity stress tolerance in plants. *Frontiers in Plant Science* 12 (2021): 670369.

Spiers, Kathryn M., Dennis Brueckner, Jan Garrevoet, Gerald Falkenberg, and Antony Van Der Ent. Synchrotron XFM tomography for elucidating metals and metalloids in hyperaccumulator plants. *Metallomics* 14, no. 11 (2022): mfac069.

Stich, Benjamin, Andreas Benke, Maria Schmidt, Claude Urbany, Rongli Shi, and Nicolaus von Wirén. The maize shoot ionome: its interaction partners, predictive power, and genetic determinants. *Plant, Cell & Environment* 43, no. 9 (2020): 2095–2111.

Su, Chunlei, Yuji Jiang, Feifei Li, Yaru Yang, Qianqian Lu, Tingting Zhang, Dan Hu, and Qinsong Xu. Investigation of subcellular distribution, physiological, and biochemical changes in Spirodela polyrhiza as a function of cadmium exposure. *Environmental and Experimental Botany* 142 (2017): 24–33.

Sundin, George W., Nian Wang, Amy O. Charkowski, Luisa F. Castiblanco, Hongge Jia, and Youfu Zhao. Perspectives on the transition from bacterial phytopathogen genomics studies to applications enhancing disease management: from promise to practice. *Phytopathology* 106, no. 10 (2016): 1071–1082.

Sung, Jwakyung, Suyeon Lee, Yejin Lee, Sangkeun Ha, Beomheon Song, Taewan Kim, Brian M. Waters, and Hari B. Krishnan. Metabolomic profiling from leaves and roots of tomato (Solanum lycopersicum L.) plants grown under nitrogen, phosphorus or potassium-deficient condition. *Plant Science* 241 (2015): 55–64.

Sytar, Oksana, Abhay Kumar, Dariusz Latowski, Paulina Kuczynska, Kazimierz Strzałka, and Majeti Narasimha Vara Prasad. Heavy metal-induced oxidative damage, defense reactions, and detoxification mechanisms in plants. *Acta Physiologiae Plantarum* 35 (2013): 985–999.

Tang, Fang, Zhongchun Xiao, Fujun Sun, Shulin Shen, Si Chen, Rui Chen, Meichen Zhu et al. Genome-wide identification and comparative analysis of diacylglycerol kinase (DGK) gene family and their expression profiling in Brassica napus under abiotic stress. *BMC Plant Biology* 20, no. 1 (2020): 1–17.

Tarish, Mariam, Rania T. Ali, Muhammad Shan, Zarmeena Amjad, Qingchen Rui, Sayed Abdul Akher, and Abdullah Al Mutery. Plant tissues as biomonitoring tools for environmental contaminants. *International Journal of Plant Biology* 15, no. 2 (2024): 375–396.

Thakur, Sapna, Shruti Choudhary, Preeti Dubey, and Pankaj Bhardwaj. Comparative transcriptome profiling reveals the reprogramming of gene networks under arsenic stress in Indian mustard. *Genome* 62, no. 12 (2019): 833–847.

Townsend, Jeffrey A., David A. Wright, Ronnie J. Winfrey, Fengli Fu, Morgan L. Maeder, J. Keith Joung, and Daniel F. Voytas. High-frequency modification of plant genes using engineered zinc-finger nucleases. *Nature* 459, no. 7245 (2009): 442–445.

Tripathi, Shweta, and Krishna Mohan Poluri. Heavy metal detoxification mechanisms by microalgae: insights from transcriptomics analysis. *Environmental Pollution* 285 (2021): 117443.

Unel, Necdet Mehmet, Mehmet Cengiz Baloglu, and Yasemin Çelik Altunoglu. Comprehensive investigation of cucumber heat shock proteins under abiotic stress conditions: a multi-omics survey. *Journal of Biotechnology* 374 (2023): 49–69.

Urbanczyk-Wochniak, Ewa, Alexander Luedemann, Joachim Kopka, Joachim Selbig, Ute Roessner-Tunali, Lothar Willmitzer, and Alisdair R. Fernie. Parallel analysis of transcript and metabolic profiles: a new approach in systems biology. *EMBO Reports* 4, no. 10 (2003): 989–993.

Urnov, Fyodor D., Edward J. Rebar, Michael C. Holmes, H. Steve Zhang, and Philip D. Gregory. Genome editing with engineered zinc finger nucleases. *Nature Reviews Genetics* 11, no. 9 (2010): 636–646.

Wan, Jinrong, Michael Torres, Ashwin Ganapathy, Jay Thelen, Beverly B. DaGue, Brian Mooney, Dong Xu, and Gary Stacey. Proteomic analysis of soybean root hairs after infection by Bradyrhizobium japonicum. *Molecular Plant-Microbe Interactions* 18, no. 5 (2005): 458–467.

Wang, Pengfei, Ling Su, Huanhuan Gao, Xilong Jiang, Xinying Wu, Yi Li, Qianqian Zhang, Yongmei Wang, and Fengshan Ren. Genome-wide characterization of bHLH genes in grape and analysis of their potential relevance to abiotic stress tolerance and secondary metabolite biosynthesis. *Frontiers in Plant Science* 9 (2018): 64.

Wang, Wei, Guangda Ding, Philip J. White, Meng Wang, Jun Zou, Fangsen Xu, John P. Hammond, and Lei Shi. Genetic dissection of the shoot and root ionomes of Brassica napus grown with contrasting phosphate supplies. *Annals of Botany* 126, no. 1 (2020): 119–140.

Wen, Ren, Liu Ya, Miao-Yi Zhou, Shi Zi, Tian-Yu Wang, Jiu-Ran Zhao, and Li Yu. Dynamic changes of root proteome reveal diverse responsive proteins in maize subjected to cadmium stress. *Journal of Integrative Agriculture* 18, no. 10 (2019): 2193–2204.

Wierzbicka, Maria H.; Przedpełska, Elżbieta; Ruzik, Ryszard; Ouerdane, Laurent; Połeć-Pawlak, Kinga; Jarosz, Małgorzata; Szpunar, Józef; and Szakiel, Anna. Comparison of the toxicity and distribution of cadmium and lead in plant cells. *Protoplasma* 231 (2007): 99–111.

Wilkins, Marc R., Jean-Charles Sanchez, Andrew A. Gooley, Ron D. Appel, Ian Humphery-Smith, Denis F. Hochstrasser, and Keith L. Williams. Progress with proteome projects: why all proteins expressed by a genome should be identified and how to do it. *Biotechnology and Genetic Engineering Reviews* 13, no. 1 (1996): 19–50.

Wu, Dezhi, Qiufang Shen, Shengguan Cai, Zhong-Hua Chen, Fei Dai, and Guoping Zhang. Ionomic responses and correlations between elements and metabolites under salt stress in wild and cultivated barley. *Plant and Cell Physiology* 54, no. 12 (2013): 1976–1988.

Wu, Gang, Hubiao Kang, Xiaoyang Zhang, Hongbo Shao, Liye Chu, and Chengjiang Ruan. A critical review on the bio-removal of hazardous heavy metals from contaminated soils: issues, progress, eco-environmental concerns and opportunities. *Journal of Hazardous Materials* 174, no. 1–3 (2010): 1–8.

Xiao, Weiwei, Peng Liu, Kai Wang, Zhaoguang Yang, and Lin Wang. Relationship between ionomics and transcriptomics of rice plant in response to arsenite stress. *Environmental and Experimental Botany* 189 (2021): 104565.

Xie, Hongxin, Xue Tian, Lina He, Jincheng Li, Liwei Cui, Xin Cong, Bochong Tang et al. Spatial metallomics reveals preferable accumulation of methylated selenium in a single seed of the hyperaccumulator *Cardamine violifolia*. *Journal of Agricultural and Food Chemistry* 71, no. 5 (2023): 2658–2665.

Xie, Mengdi, Wenqing Chen, Xicong Lai, Haibo Dai, Hui Sun, Xiaoyong Zhou, and Tongbin Chen. Metabolic responses and their correlations with phytochelatins in Amaranthus hypochondriacus under cadmium stress. *Environmental Pollution* 252 (2019): 1791–1800.

Yadav, Birendra Singh, Swati Singh, Sameer Srivastava, Nand Kumar Singh, and Ashutosh Mani. Whole transcriptome expression profiling and biological network analysis of chickpea during heavy metal stress. *Journal of Plant Biochemistry and Biotechnology* 28 (2019): 345–352.

Yang, Yaodong, Mumtaz Ali Saand, Liyun Huang, Walid Badawy Abdelaal, Jun Zhang, Yi Wu, Jing Li, Muzafar Hussain Sirohi, and Fuyou Wang. Applications of multi-omics technologies for crop improvement. *Frontiers in Plant Science* 12 (2021): 563953.

Yu, Jing, Sook Jung, Chun-Huai Cheng, Stephen P. Ficklin, Taein Lee, Ping Zheng, Don Jones, Richard G. Percy, and Dorrie Main. CottonGen: a genomics, genetics and breeding database for cotton research. *Nucleic Acids Research* 42, no. D1 (2014): D1229–D1236.

Zaid, Abbu, Javaid Akhter Bhat, and Shabir H. Wani. Influence of metalloids and their toxicity impact on photosynthetic parameters of plants. In: Deshmukh, Rupesh, Durgesh K. Tripathi, and Gea Guerriero (Eds.), *Metalloids in Plants: Advances and Future Prospects*, pp. 113–124. John Wiley & Sons, 2020.

Zeng, Yafei, Jianyu Wen, Weibo Zhao, Qiong Wang, and Wenchao Huang. Rational improvement of rice yield and cold tolerance by editing the three genes OsPIN5b, GS3, and OsMYB30 with the CRISPR–Cas9 system. *Frontiers in Plant Science* 10 (2020): 1663.

Zhang, Anning, Yi Liu, Feiming Wang, Tianfei Li, Zhihao Chen, Deyan Kong, Junguo Bi et al. Enhanced rice salinity tolerance via CRISPR/Cas9-targeted mutagenesis of the OsRR22 gene. *Molecular Breeding* 39 (2019): 1–10.

Zhang, Dayong, Jining Li, Miaomiao Li, Zimeng Cheng, Xiaohui Song, Xiaoguang Shang, and Wangzhen Guo. Overexpression of a cotton nonspecific lipid transfer protein gene, GhLTP4, enhances drought tolerance by remodeling lipid profiles, regulating abscisic acid homeostasis and improving tricarboxylic acid cycle in cotton. *Environmental and Experimental Botany* 201 (2022): 104991.

Zhang, Heng, Bing Han, Tai Wang, Sixue Chen, Haiying Li, Yuhong Zhang, and Shaojun Dai. Mechanisms of plant salt response: insights from proteomics. *Journal of Proteome Research* 11, no. 1 (2012): 49–67.

Zhang, Jingtao, Yong Zhang, Yuanyuan Du, Shiyun Chen, and Huiru Tang. Dynamic metabonomic responses of tobacco (Nicotiana tabacum) plants to salt stress. *Journal of Proteome Research* 10, no. 4 (2011): 1904–1914.

Zhang, Shihua, Yi Yue, Liang Sheng, Yunzhi Wu, Guohua Fan, Ao Li, Xiaoyi Hu, Mingzhu ShangGuan, and Chaoling Wei. PASmiR: a literature-curated database for miRNA molecular regulation in plant response to abiotic stress. *BMC Plant Biology* 13, no. 1 (2013): 1–8.

Zhang, Xiaodan, Chensong Yao, Sicong Fu, Hongdong Xuan, Shuxian Wen, Chao Liu, Fangdong Li et al. Stress2TF: a manually curated database of TF regulation in plant response to stress. *Gene* 638 (2018): 36–40.

Zhang, Yinfei, Yu Wang, Zhaotang Ding, Hui Wang, Lubin Song, Sisi Jia, and Dexin Ma. Zinc stress affects ionome and metabolome in tea plants. *Plant Physiology and Biochemistry* 111 (2017): 318–328.

Zhang, Zhen-Hua, Ting Zhou, Tian-Jiao Tang, Hai-Xing Song, Chun-Yun Guan, Jin-Yong Huang, and Ying-Peng Hua. A multiomics approach reveals the pivotal role of subcellular reallocation in determining rapeseed resistance to cadmium toxicity. *Journal of Experimental Botany* 70, no. 19 (2019): 5437–5455.

Zhao, Lijuan, Yuxiong Huang, Hongjun Zhou, Adeyemi S. Adeleye, Hongtao Wang, Cruz Ortiz, Susan J. Mazer, and Arturo A. Keller. GC-TOF-MS based metabolomics and ICP-MS based metallomics of cucumber (Cucumis sativus) fruits reveal alteration of metabolites profile and biological pathway disruption induced by nano copper. *Environmental Science: Nano* 3, no. 5 (2016): 1114–1123.

Zhou, Junhui, Zhao Peng, Juying Long, Davide Sosso, B. O. Liu, Joon-Seob Eom, Sheng Huang et al. Gene targeting by the TAL effector PthXo2 reveals cryptic resistance gene for bacterial blight of rice. *The Plant Journal* 82, no. 4 (2015): 632–643.

Zulfiqar, Usman, Fasih Ullah Haider, Muhammad Ahmad, Saddam Hussain, Muhammad Faisal Maqsood, Muhammad Ishfaq, Babar Shahzad et al. Chromium toxicity, speciation, and remediation strategies in soil-plant interface: a critical review. *Frontiers in Plant Science* 13 (2023): 1081624.

10 Impacts of Metalloids on Photosynthetic Apparatus in Plants

Twinkle, Ayushi Gautam, Dipanshu Ghosh, Rimpy Dhiman, and Vivek Dogra

10.1 INTRODUCTION

Metalloids are elements found in the periodic table that lie between metals and non-metals in a diagonal zigzag line. They have characteristics in common with both metals and non-metals. In general, components in the bottom left of the line exhibit increasing metallic behavior, whereas those toward the upper right do not. They resemble non-metals chemically and metals physically. Among them, silicon (Si), boron (B), germanium (Ge), arsenic (As), selenium (Se), tellurium (Te), and antimony (Sb) constitute metalloids. Metalloids are found in limited amounts of rocks and water resources; however, their concentration is increasing in the soil due to anthropogenic activities, including pesticide waste from agriculture, vehicle emissions, municipal and industrial waste, expanding mining practices, and smelting processes. It has been reported that around 5 million sites worldwide are contaminated due to metalloids or heavy metal accumulation (He et al. 2015). Since metalloids are non-biodegradable, they tend to accumulate in soil and, ultimately, in plants. Their concentration in the environment determines their beneficial or harmful effects on plants. While some of them are harmful even in very small concentrations, others are necessary in trace amounts for healthy plant metabolism, growth, and development. For example, B is necessary for the continued development and growth of plants, but excessive amounts can be hazardous (Landi et al. 2019). In contrast, As is toxic even in low concentrations (Duffus 2001). Accumulated metalloids in the environment are increasing pollution day by day, resulting in alteration in basic physiological processes of plants and reducing the overall plant production (Anjum et al. 2015; Emamverdian et al. 2015; Singh et al. 2015; Singh et al. 2017; Tripathi et al. 2017). Metalloids disturb normal plant metabolic pathways functioning by replacing enzymes and altering their mode of action. Several key biochemical and physiological events in plants are also affected by metalloids.

Among several physiological processes, photosynthesis is primarily affected by metalloid contaminations. Photosynthesis constitutes a physiochemical process in green organisms that converts light energy into chemical energy utilizing carbon dioxide (CO_2) from the atmosphere and water from the soil to make sugars (Batista-Silva et al. 2020). The photoautotrophic cells contain chloroplasts that accommodate photosynthetic machinery comprising light-harvesting pigment chlorophyll, accessory pigments (carotenoids), pigment binding proteins (light-harvesting complex proteins; LHCPs), and components of the electron transport and Calvin cycle (Orr et al. 2017). Photosynthesis has two phases: light-dependent reaction and light-independent reaction. In light reactions, the light-harvesting antennae complexes harvest light energy and two photosystems (PSII and PSI), to be utilized in the photosynthetic electron transport chain (PETC), to generate nicotinamide adenine dinucleotide phosphate hydrogen (NADPH) and adenosine triphosphate (ATP) (Dogra, Rochaix, and Kim 2018; Halliwell 2006; Nielsen et al. 2013). The NADPH and ATPs thus produced are used in the Calvin cycle to assimilate CO_2 into sugars in a light-independent manner (Nielsen et al. 2013). The photosynthetic process

DOI: 10.1201/9781032673592-10

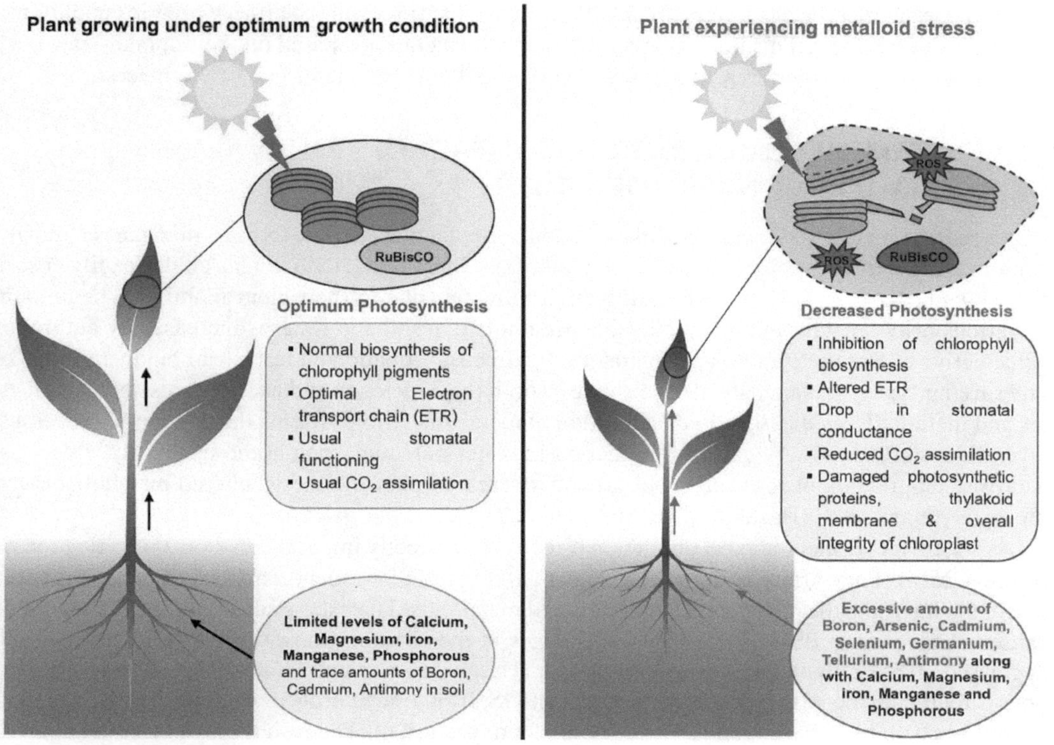

FIGURE 10.1 Effect of metalloid stress on the photosynthetic process of plants. Under normal conditions, plants carry out photosynthesis by converting sunlight into chemical energy in the form of sugars, using CO_2 from the air and water from the soil. The photochemical reaction is catalyzed by chlorophyll pigment housed in the thylakoid membranes of the chloroplast. Water absorbed by roots is transported through the xylem to aerial parts of the plant (leaves). Along with water current, essential minerals are also absorbed and transported into the plant for their growth and metabolism. Metalloids present in soil are also absorbed but are essentially required by plants. If present in excess, the metalloids are absorbed and accumulated in aerial parts. Excess accumulation of metalloids then affects several physiological activities, including photosynthesis. They damage the thylakoid membrane, inhibit chlorophyll biosynthesis, retard electron flow, and alter the structure and function of enzymes. Excess metalloids can even alter the size and number of photosynthetic apparatus. This results in reduced CO_2 assimilation and low photosynthetic output.

is tightly regulated to avoid loss of electrons or energy; however, environmental perturbations caused by fluctuations in temperature and light, drought, salt, metals, and pathogens directly influence the photosynthetic process resulting in photoinhibition and reduced photosynthetic efficiency (Dogra, Duan, et al. 2019; Dogra and Kim 2020; Hitchcock et al. 2022; Kumar et al. 2021; Leister 2019; Kubis and Bar-Even 2019). Similarly, increasing metalloid pollution also has a negative effect on the photosynthetic process, limiting growth and development (Picorel, Alfonso, and Velitchkova 2017). When uptaken and accumulated in the photosynthetic cells, the metalloids interfere with PETC, generating excess ROS and oxidative damage to nearby biomolecules; alter enzyme activities; reduce chlorophyll contents and stomatal conductance; and alter membrane permeability (Nable, Bañuelos, and Paull 1997; Reid et al. 2004; Landi et al. 2013; Simón et al. 2013) (Figure 10.1).

In recent times, several strategies have been developed for the improvement of metalloid tolerance in plants. Genetic engineering is a promising strategy for enhancing plant capability for translocation of metalloids along with limiting their toxic effect. Transgenic plants having overexpression of genes encoding metal transporters, metal chelators, and antioxidant machinery proteins could limit metalloid toxic effects. This book chapter discusses the metalloids based on their uptake, transport, their effects on photosynthesis, and how metalloid-resilient crop plants can be engineered.

10.2 UPTAKE, ACCUMULATION, AND IMPACT OF METALLOIDS ON PHOTOSYNTHESIS

Plants require metals and metalloids like B, copper, Se, Si, nickel, zinc, ferrous, manganese, molybdenum, and cobalt for their growth and development (Ghori et al. 2019). Plants could readily absorb these chemicals from soil or water that is high in metals due to their bioavailability. The amount of various heavy metals and metalloids that accumulate in soil and water is increased by anthropogenic activity (Hasanuzzaman and Fujita 2020), affecting aquatic and terrestrial biodiversity. Like other living systems, plants are also sensitive to both the deficiency and excess accessibility of metals and metalloids, as the higher concentration of metals/metalloids harms the metabolic processes. Overall, metalloid toxicity causes a reduction in water potential, membrane instability, chlorosis, necrosis, suppression of root and shoot growth, decreased photosynthesis, altered metabolism, and ultimately plant death (Hasanuzzaman and Fujita 2020).

As stated above, metalloid accumulation directly or indirectly impacts the photosynthetic process (Meravi 2014). They affect both light-dependent and light-independent reactions of photosynthesis, where the former is the prime target of metals and metalloids. The excess accumulation of metalloids causes a decrease in PSII quantum yield, changes in the electron transport rate (ETR) along with decreased photosynthetic pigment accumulation (Hasanuzzaman and Fujita 2020). These perturbations directly augment the generation of various ROS, including hydrogen peroxide (H_2O_2), superoxide anion (O_2^-), hydroxyl anion (OH^-), and singlet oxygen ($^1O_2$). The resulting overaccumulation of ROS causes oxidative damage to various biomolecules like proteins, nucleic acids, fatty acids, and carbohydrates, decreasing biomass and crop yields (Dogra, Li, et al. 2019; Nahar et al. 2022).

The effect of individual metalloids on the photosynthetic process is discussed in detail in the following sections.

10.2.1 Si

Si is classified as a non-essential metalloid for plants. Yet, it is known for alleviating the harmful effects of diverse abiotic and biotic stresses through various processes, encompassing morphological, physiological, and genetic adaptations. Si exists in the soil solution as monosilicic acid [H_4SiO_4] or alternate silicate constituents, converting into colloidal silicates, polymerized silicic acid, or interchangeable silicates (Etesami and Jeong 2018; Martin 2013). Plants absorb Si from the soil as H_4SiO_4 and transport it from roots to shoots through either passive or active mechanisms (Feng, Yamaji, and Mitani-Ueno 2011). Major intrinsic proteins (MIPs) and nodulin-26-like intrinsic proteins (NIPs) drive the import of Si into plants (Ma and Yamaji 2006). Several factors influence the absorption capacity of Si in plants, including plant-available silicon (PAS) in the soil, plant age, varying Si concentrations in the growing substrate, Si measurement methods (Deshmukh, Sonah, and Belanger 2020), and environmental conditions (Chen et al. 2018).

Si has positively affected the photosynthetic process under different stress conditions. It accomplishes this by safeguarding photosynthetic structures, upregulating the expression of genes involved in photosynthesis, increasing the efficacy of water usage, improving photosynthetic electron transport, and protecting against ROS (Song et al. 2014). Si application protects chlorophyll loss under stress conditions and upregulates crucial photosynthetic genes such as *(Photosystem II reaction center protein B)*, *PsbD (Photosystem II reaction center protein D)*, and *PsbH (Photosystem II reaction center protein H)*, particularly under heat stress, to preserve photosynthesis (Rastogi et al. 2021). The thylakoid membrane, a critical component of the plant photosynthetic apparatus, is responsive to environmental perturbations. It can rapidly adapt to challenges by altering its structure and operations to maximize photosynthetic efficiency (Kaiser et al. 2015). Supporting this notion, Wang et al. (2019) reported that Si application play roles in the absorption, transformation, and transfer of light energy by altering the protein components of thylakoid membranes in rice seedlings experiencing drought stress. This implies that Si might provide protection to plant photosynthesis. So based on existing literature, Si plays a positive and multifaceted role in plants, contributing to resistance against various stresses and positively influencing photosynthesis (Wang et al. 2019).

10.2.2 B

B is a crucial micronutrient necessary for the proper growth and development of plants. Plants absorb B in the form of boric acid through their roots, and this uptake happens via passive diffusion across lipid bilayers from the soil (Dordas and Brown 2000). In conditions of B deficiency, passive diffusion alone cannot meet the demand. Therefore, boric acid channels and borate exporters (BOR) become active in facilitating uptake and translocation to growing cells. In *Arabidopsis thaliana*, a membrane-localized protein called NIP5;1 has been reported as the boric acid channel responsible for enhancing B uptake by roots under B deficiency (Takano et al. 2006). Inside the cytoplasm of root cells, boric acid undergoes conversion into borate anions, which are transported by the borate exporter BOR1 from roots to shoots, and the borate exporter BOR4 facilitates the efflux of B, preventing overaccumulation of B in the xylem and growing cells (Takano et al. 2002; Miwa et al. 2007).

When present in optimum amounts, B stimulates the photosynthetic process by protecting the thylakoid membrane by decreasing the ROS generation. But, when accumulated in excess amounts or is deficient, it has a negative impact on the photosynthetic process. Its accumulation has been reported to cause a reduction in levels of chlorophyll and carotenoids, the rate of photosynthesis, the conductance of stomata, and the concentration of internal CO_2 in several plants such as apple, tomato, citrange, *Vigna radiata*, and sweet wormwood (Paull, Nable, and Rathjen 1992; Kastori et al. 1995; Shah et al. 2017; Aftab et al. 2012; Yusuf, Fariduddin, and Ahmad 2011). Interestingly, in B-deficient sunflower plants, photosynthetic O_2 evolution was reduced, which could be attributed to a reduction in pigment accumulation, photosynthetic ETR, and photophosphorylation (Plesnicar et al. 1997). In addition, B deficiency also leads to starch accumulation in the chloroplast, disrupting chloroplast structure, chlorosis, and reduced CO_2 assimilation (Cave, Tolley, and Strain 1981; Han et al. 2008).

10.2.3 As

As can be found in various chemical forms in the environment, including arsenate [As(V)], arsenite [As(III)], monomethylarsonic acid [MMA], dimethylarsinic acid [DMA], trimethylarsine [TMA], arsenocholine [AsC], arsenobetaine [AsB], and arsenosugars (Quaghebeur and Rengel 2005). Arsenate [As(V)] and arsenite [As(III)] represent inorganic As forms that can be taken up by plant root systems but are very toxic for plants. Arsenate is transported into plants by phosphate transporters. In contrast, arsenite [As(III)] is transported through the involvement of NIPs, aquaporins, and other membrane transporters (Duan et al. 2012; Li et al. 2009). Arsenate, structurally similar to phosphate ions, is absorbed by plants through phosphate transporters, disrupting phosphate homeostasis. This disturbance affects various physiological activities, including reduced photosynthetic efficiency, altered energy metabolism, and compromised overall plant growth (Zhao et al. 2009). Arsenite can interact with enzyme thiol (-SH) groups, hindering diverse metabolic functions and triggering oxidative stress by either directly producing ROS or incapacitating antioxidant molecules through -SH group binding. There is ample evidence suggesting that arsenate substitution for phosphate in ATP molecules disrupts cellular energy flow (Chandrakar, Pandey, and Keshavkant 2018).

If overaccumulated, As exerts a negative effect on the photosynthetic process. Overaccumulation of As hinders the chlorophyll biosynthetic pathway, decreasing chlorophyll concentration in *Ceratophyllum demersum* leaves. The reduction in chlorophyll content is attributed to the degradation of biosynthetic enzymes, δ-aminolevulinic acid dehydratase (ALAD), and protochlorophyllide reductase (POR) (Farooq et al. 2016). In addition to inhibiting pigment biosynthesis and altering pigment composition, accumulation of As results in reduced PSII activity and LHCPs synthesis, inhibition of ribulose-1,5-bisphosphate carboxylase/oxygenase (RuBisCO) activity, and increased damage to thylakoids and envelope membranes (Anjum et al. 2011; Mubarak et al. 2016; Praveen et al. 2020).

10.2.4 Sb

Sb is considered non-essential for plant development. It could take on different forms within the soil, such as antimonite [Sb(III)] and antimonate [Sb(V)] (Sh, Liu, and Wang 2012). Studies are currently being conducted to elucidate how plants absorb, distribute, and detoxify Sb. Based on the reports, glycerol facilitator (GlpF) is implicated in plants' transport of Sb (III). Studies indicated that since $As(OH)_3$, $Sb(OH)_3$, and glycerol exhibit strong similarities, they probably opt for a common transportation mechanism involving aquaglyceroporins (Porquet and Filella 2007).

Sb negatively impacts plant growth, development, and biomass by influencing photosynthetic capacity. Within cells, Sb displaces other metals from binding sites or obstructs functional groups, disrupting the biosynthesis of various compounds (Labancova et al. 2023). In *Acorus calamus*, elevated Sb concentrations in mesophyll cells lead to a reduction in both chlorophyll pigments (a and b), carotenoids, and photosynthetic parameters like net photosynthetic rate, the conductance of stomata, efficiency and quantum yield rate of PSII (Zhou et al. 2018; Pan et al. 2011). To counter the toxicity posed by Sb, plants employ antioxidant and detoxification systems to mitigate the impact of active ROS and reactive nitrogen species (RNS) and reduce Sb absorption or transport (Chu et al. 2023).

10.2.5 Se and Te

Se and Te belong to the chalcogen group of chemical elements. Se is a vital trace element that contributes to the structure of the selenocysteine amino acid, and its chemical behavior resembles sulfur. Se is necessary for both prokaryotes and eukaryotes at low concentrations, but it becomes highly toxic to living systems at higher concentrations (Kessi et al. 1999). Se occurs in the form of selenite (SeO_3^{2-}) and selenate (SeO_4^{2-}), which are mobile, harmful, and soluble in water. The presence of these compounds facilitates their effortless absorption by plants from water sources or Se-enriched soil (Espinosa-Ortiz et al. 2017). SeO_3^{2-} is transported through passive diffusion. Once uptaken in the root cells, it quickly reduces to form organic Se compounds (H2SeO3) (Zhang, Shi, and Wang 2006).

Te, the scarcest element in the Earth's crust, is a by-product of copper electrolytic refining. While the processes of uptake, transportation, and metabolism of Te remain uncertain, they may be linked to sulfur or Se. Se, in the form of selenite (SeO_3^{2-}), and Te, in the form of tellurite (TeO_3^{2-}), are more toxic for living organisms because of their high oxidizing ability. Both the most reactive forms of Te and Se actively enhance ROS accumulation by weakening the cellular antioxidant defense mechanisms (Vavrova et al. 2021). Both forms alter protein function, disrupt optimal enzymatic activity, and can induce oxidative stress, resulting in excessive cellular damage and eventual cell death (McNaughton et al. 2004).

Limited studies have been done to observe the effects of these elements on photosynthesis. It has been reported that the exogenous application of Se positively affected the photosynthetic process. Exogenous application of Se reduced the expression of chlorophyllase (CHLH) enzyme responsible for chlorophyll degradation, thus augmenting the levels of photosynthetic pigments (Germ, Kreft, and Osvald 2005). In *Arabidopsis thaliana*, a TeO_3^{2-} resistance protein, namely, tellurite resistance C (TerC), has been identified as the membrane protein of thylakoids implicated within the PSII complex (Schneider et al. 2014).

10.2.6 Ge

Ge is a scarce element found in the Earth's crust that behaves as a chemical analog to Si. The function of Ge and its metabolic processes remain unclear, and very little data are available on its uptake and transport. Ge is present in both inorganic and organic forms. Germanium dioxide (GeO_2) and elemental Ge represent the inorganic forms of Ge, whereas spiro-germanium, β-carboxyethyl germanium (Ge-132), and proxi-germanium are classified as organic forms (Han et al. 2007; Cheong et al. 2009). Inorganic Ge can be taken up from the soil by plants and stored within their tissues. Subsequently, it undergoes conversion into organic Ge compounds, known for their abilities to scavenge free radicals and detoxify heavy metals (Li et al. 2012).

Ge is found to play both beneficial as well as harmful roles in photosynthesis. It is helpful when present in low concentration and vice versa. In lettuce plants, the high dose of Ge concentration impacts plant growth by inhibiting photosynthetic efficiency (Cheong et al. 2009). On the contrary, low concentrations of Ge increase the PSII efficiency, quantum yield, and the rate of electron flow to improve photosynthetic performance in apple plants (Wang et al. 2018). Furthermore, certain studies have reported that in some plants, Ge can promote the synthesis of sugars, which indicates its role in improving the photosynthetic capacity (Li et al. 2021).

10.3 INHERENT AND INDUCED METALLOID TOLERANCE IN PLANTS ALLEVIATING NEGATIVE IMPACTS ON PHOTOSYNTHESIS

As stated above, the deleterious effects of metalloid toxicity extend to vital biochemical and physiological functions in plants, including photosynthesis, transportation, and overall development. These interferences adversely affect plant growth, leading to compromised development and decreased crop yield. To overcome the negative impacts of metalloids, plants have inherent mechanisms, including chelation, vacuolar sequestration, and antioxidant detoxification, to eradicate or sustain metalloid stress.

Chelation is a process of neutralizing ions by employing chelating agents to neutralize charges and protect the cellular environment. Plant cells use chelating agents like phytochelatins (PCs), metallothioneins (MT), and metal-binding thiols such as glutathione (GSH) (Pandey, Gautam, and Dubey 2019) to chelate metalloids. Following the chelation of metalloids, the next step is their vacuolar sequestration, a process exhibited by plant cells to transport and accumulate certain substances, especially toxic substances, for their degradation in the vacuole. It has been reported that in plant cells, As undergoes vacuolar sequestration after being chelated by GSH or PC (Sinha et al. 2023). The antioxidant defense mechanism is another important type of defense mechanism exhibited by plants. Under metalloid toxicity, various ROS are generated inside plant cells, causing oxidative damage to different biomolecules. To counter this, plants show heightened expression of antioxidant enzymes such as glutathione peroxidase (GPX), glutathione-*S*-transferase (GST), superoxide dismutase (SOD), glutathione reductase (GR), monodehydroascorbate reductase (MDHAR), dehydroascorbate reductase (DHAR), guaiacol peroxidase (GOPX), ascorbate peroxidase (APX), peroxidase (POX), and catalase (CAT) to alleviate oxidative stress and protect the cellular environment. In addition, non-enzymatic antioxidants such as proline, ascorbate, and tocopherols also accumulate to detoxify secondary oxidative derivatives. Such patterns of modulation in antioxidant enzyme activities have been reported in As-tolerant *Holcus lanatus* (Cao, Ma, and Tu 2004; Srivastava et al. 2005) and sunflower under Sb toxicity (Ortega et al. 2017). Rajabpoor et al. (2019) reported Sb to induce an increase in both enzymatic and non-enzymatic systems in sunflower, with APX activity and accumulation of proline showing a stronger induction by Sb[V] compared to Sb[III] (Rajabpoor, Ghaderian, and Schat 2019). Proline, known for its osmoprotective role, accumulates in plants exposed to metalloids, acting as a ROS scavenger and potentially improving plant photosynthetic efficiency by enhancing water status (Signorelli et al. 2014). The interplay of non-enzymatic and enzymatic antioxidants and osmoprotectants like proline constitutes a crucial defense mechanism for plants combating metalloid-induced oxidative stress.

Even though plants have natural defenses against metalloid exposure, rising metalloid concentrations in soil and water are becoming a serious problem. As a result, plans for countering these threats must be created. Three levels of strategies can be developed to address the issue of increasing metalloid resilience: the first is the removal of metalloids from contaminated soil and/or water;

the second is the modification of metalloid uptake and transport within plants; and the third is the detoxification and removal of such metalloids from the cellular system. To eradicate metalloid contaminants from soil, phytoremediation is a valuable technique. Phytoremediation involves utilizing plants in contaminated soil to alleviate toxicity. There are several phytoremediation processes, such as phytoextraction, phytostabilization, phytovolatization, and phytodegradation. Phytoextraction removes metalloids from contaminated soils by extracting and accumulating them in harvestable tissues. Phytostabilization converts transportable forms of metalloids into non-transportable forms, fixing them in the contaminated soil. Phytovolatization involves the uptake of reactive forms of metalloids by plants, converting them into less toxic volatile forms that are diffused into the air. Phytodegradation is a slow process where plants uptake metalloids, immobilize them in roots or shoots, and chemically modify them for their deactivation (Pathak, Agarwal, and Pandey 2020). All these processes are useful for decreasing the toxicity of contaminated soil before cultivating crops.

Metalloid stress can also be dealt with by introducing metalloid tolerance in plants experiencing metalloid toxicity by biotechnological interventions. This involves cis/transgenic overexpression or downregulation of genes involved in chelation, vacuolar sequestration, antioxidant defense, and reduced metalloid transport rate. Enhancing the expression of genes that encode antioxidant enzymes, biosynthetic pathways for non-enzymatic antioxidants, phytochelatin genes, vascular transporters for heavy metals, and introducing heavy metal exporters from other organisms resulted in noteworthy decreases in the toxicity of heavy metals and metalloids in plants (Misra et al. 2009). Genes involved in MT and phytochelatin-glutathione (PC-GSH) cycles, such as *Serine acetyltransferase* (*SAT*), *Glutathione synthetase* (*GSH2*), *Glyoxalases*, *Gamma-glutamylcysteine synthase* (*GHS1*), *ATP sulfurylase* (*APS*), *Phytochelatin synthase* (*PCS*), *Glutathione reductase* (*GR*), and *Cystathionine synthase* (CTS) (Yadav 2010), are potential targets for manipulation. Several MT genes like *MTI*, *MTII*, and *MTIII* identified in soybean, tomato, fescue, and wheat plants (Kim et al. 2013) can also be utilized for metalloid resilience.

Reducing metalloid uptake into plants can be achieved by downregulating metalloid transporters. Studies suggest that As uptake in plants occurs through phosphate transporters. As-tolerant plants typically exhibit a suppressed phosphate uptake system. Significantly, research conducted on Arabidopsis has pinpointed the transporters of phosphate ions Phosphate Transporter 1;1 (Pht1;1) and Phosphate Transporter 1;4 (Pht1;4) as accountable for As absorption. Alteration of the gene expressing Pht1;1 (Phosphate Transporter Helper Protein 1) has been shown to impede As accumulation from the endoplasmic reticulum to the plasma membrane, leading to considerable As tolerance (Kumar et al. 2016). In a comparable report, a rice variant with a knockdown of the As-accumulating *Phosphate transporter* gene (*OsPT4*) demonstrated tolerance to As (Cao et al. 2017). Arabidopsis plants expressing *Arsenate reductase* (*arsC*) and *Glutamylcysteine synthetase* (*GCS*) from *E. coli*, showed better biomass compared to wild-type plants cultivated in medium persisting 125 μM sodium arsenate stress (Dhankher et al. 2002). Similar results were obtained upon overexpression of *AtGCS1* in Arabidopsis itself (Li et al. 2004). Upregulation of Arabidopsis *Plastid-localized APS* that reduces sulfate showed reduced Se accumulation in *Brassica juncea* (Pilon-Smits and LeDuc 2009). Summarily, it is shown convincingly that manipulation of cellular detoxification and metal uptake machinery using biotechnological means provides a precise and promising approach to deal with increasing threats of metalloid stress to plants.

10.4 CONCLUSION AND FUTURE PERSPECTIVE

Photosynthesis is a crucial physiochemical process in green organisms, and its rate and efficiency directly impact crop yield. Unfortunately, this vital process is susceptible to various environmental perturbations, with metalloids being a significant factor resulting in a loss of the photosynthetic rate and crop yields.

Metalloids are necessary for plant development and metabolism in trace amounts, but an excess of them can interfere with several plant functions, including photosynthesis. Metalloids obstruct electron flow, prevent the synthesis of chlorophyll, and inhibit the enzymes that catalyze photosynthesis. This inhibition may occur by inactivating enzymes, blocking their active sites, or inducing changes in their conformation, ultimately decreasing the photosynthetic rate. Collectively, these disruptions contribute to damage to the photosynthetic apparatus. Chelation and vacuolar sequestration are crucial in controlling metalloid toxicity. Modulation of cellular detoxification machinery can be implemented to detoxify or eradicate the metalloids effectively. Although the uptake of the metalloids should be primarily targeted, since many of the metalloids use transporters of essential metal ions, deeper investigations are needed for selective genetic manipulation of the transporters responsible for the uptake and transport of specific metalloids. Though there are plenty of reports of negative impacts of metalloids on photosynthesis, their precise effect on molecular levels are largely unknown. So, thorough investigations are also required to determine how metalloids affect the photosynthetic process. The knowledge gained from such studies could be utilized to induce resilience to metalloids in crop plants.

AUTHOR CONTRIBUTIONS

Twinkle, Ayushi Gautam and Dipanshu Ghosh contributed equally in the literature search, writing and preparation of the first draft of the manuscript. Rimpy Dhiman assisted in the writing. Vivek Dogra conceptualized the idea, designed and finalized the manuscript.

ACKNOWLEDGMENTS

Director, CSIR-Institute of Himalayan Bioresource Technology, Palampur, is acknowledged for providing infrastructural facilities. Twinkle gratefully acknowledges CSIR, New Delhi, for providing the CSIR-JRF fellowship [No. 31/054(0167)/2020-EMR-I]. D.G. gratefully acknowledges UGC, New Delhi, for providing the UGC JRF fellowship [466/(CSIR-UGC NET JUNE 2018)]. This work is supported by financial assistance from Ramalingaswami Re-entry Fellowship, Department of Biotechnology, Government of India (grant number BT/RLF/Re-entry/02/2019) and CSIR-Fundamental & Innovative Research in Science of Tomorrow (CSIR-FIRST) research grant, CSIR, Government of India (grant number 6/1/FIRST/2020-RPPBDD-TMDSeMI) to V.D. The article represents the CSIR-IHBT communication number 5538.

REFERENCES

Aftab, T., M. M. Khan, M. Naeem, M. Idrees, Moinuddin, J. A. Teixeira da Silva, and M. Ram. 2012. Exogenous nitric oxide donor protects Artemisia annua from oxidative stress generated by boron and aluminium toxicity. *Ecotoxicol Environ Saf* 80:60–68. https://doi.org/10.1016/j.ecoenv.2012.02.007

Anjum, N. A., M. Hasanuzzaman, M. A. Hossain, P. Thangavel, A. Roychoudhury, S. S. Gill, M. A. Rodrigo, V. Adam, M. Fujita, R. Kizek, A. C. Duarte, E. Pereira, and I. Ahmad. 2015. Jacks of metal/metalloid chelation trade in plants-an overview. *Front Plant Sci* 6:192. https://doi.org/10.3389/fpls.2015.00192

Anjum, S. A., X. Y. Xie, L. Chang Wang, M. F. Saleem, C. Man, and W. Lei. 2011. Morphological, physiological and biochemical responses of plants to drought stress. *Afr J Agric Res* 6 (9):2026–2032.

Batista-Silva, W., P. da Fonseca-Pereira, A. O. Martins, A. Zsogon, A. Nunes-Nesi, and W. L. Araujo. 2020. Engineering improved photosynthesis in the era of synthetic biology. *Plant Commun* 1 (2):100032. https://doi.org/10.1016/j.xplc.2020.100032

Cao, X., L. Q. Ma, and C. Tu. 2004. Antioxidative responses to arsenic in the arsenic-hyperaccumulator Chinese brake fern (Pteris vittata L.). *Environ Pollut* 128 (3):317–325. https://doi.org/10.1016/j.envpol.2003.09.018

Cao, Y., D. Sun, H. Ai, H. Mei, X. Liu, S. Sun, G. Xu, Y. Liu, Y. Chen, and L. Q. Ma. 2017. Knocking out OsPT4 gene decreases arsenate uptake by rice plants and inorganic arsenic accumulation in rice grains. *Environ Sci Technol* 51 (21):12131–12138. https://doi.org/10.1021/acs.est.7b03028

Cave, G., L. C. Tolley, and B. R. Strain. 1981. Effect of carbon dioxide enrichment on chlorophyll content, starch content and starch grain structure in Trifolium subterraneum leaves. *Physiol Plant* 51 (2):171–174.

Chandrakar, V., N. Pandey, and S. Keshavkant. 2018. Plant responses to arsenic toxicity: morphology and physiology. In Mirza, H., Nahar, K., and Fujita, M. (Eds.), *Mechanisms of Arsenic Toxicity and Tolerance in Plants* (pp. 27–48). Singapore: Springer.

Chen, R., C. Zhang, Y. Zhao, Y. Huang, and Z. Liu. 2018. Foliar application with nano-silicon reduced cadmium accumulation in grains by inhibiting cadmium translocation in rice plants. *Environ Sci Pollut Res Int* 25 (3):2361–2368. https://doi.org/10.1007/s11356-017-0681-z

Cheong, Y. H., S. U. Kim, D. C. Seo, N. I. Chang, J. B. Lee, J. H. Park, K. S. Kim, S. D. Kim, H. T. Kim, and J.-S. Heo. 2009. Effect of inorganic and organic germanium treatments on the growth of lettuce (Lactuca sativa). *J Korean Soc Appl Biol Chem* 52:389–396.

Chu, Y., Q. Bao, Y. Li, H. Sun, Z. Liu, J. Shi, and Y. Huang. 2023. Melatonin alleviates antimony toxicity by regulating the antioxidant response and reducing antimony accumulation in Oryza sativa L. *Antioxidants* 12 (11):1917.

Deshmukh, R., H. Sonah, and R. R. Belanger. 2020. New evidence defining the evolutionary path of aquaporins regulating silicon uptake in land plants. *J Exp Bot* 71 (21):6775–6788. https://doi.org/10.1093/jxb/eraa342

Dhankher, O. P., Y. Li, B. P. Rosen, J. Shi, D. Salt, J. F. Senecoff, N. A. Sashti, and R. B. Meagher. 2002. Engineering tolerance and hyperaccumulation of arsenic in plants by combining arsenate reductase and gamma-glutamylcysteine synthetase expression. *Nat Biotechnol* 20 (11):1140–1145. https://doi.org/10.1038/nbt747

Dogra, V., J. Duan, K. P. Lee, and C. Kim. 2019. Impaired PSII proteostasis triggers an UPR-like response in the var2 mutant of Arabidopsis thaliana. *J Exp Bot* 70 (12):3075–3088. https://doi.org/10.1093/jxb/erz151

Dogra, V., and C. Kim. 2020. Singlet oxygen metabolism: from genesis to signaling. *Front Plant Sci* 10:1640. https://doi.org/10.3389/fpls.2019.01640

Dogra, V., M. Li, S. Singh, M. Li, and C. Kim. 2019. Oxidative post-translational modification of EXECUTER1 is required for singlet oxygen sensing in plastids. *Nat Commun* 10 (1):2834. https://doi.org/10.1038/s41467-019-10760-6

Dogra, V., J. D. Rochaix, and C. Kim. 2018. Singlet oxygen-triggered chloroplast-to-nucleus retrograde signalling pathways: an emerging perspective. *Plant Cell Environ* 41 (8):1727–1738. https://doi.org/10.1111/pce.13332

Dordas, C., and P. H. Brown. 2000. Permeability of boric acid across lipid bilayers and factors affecting it. *J Membr Biol* 175:95–105.

Duan, G., T. Kamiya, S. Ishikawa, T. Arao, and T. Fujiwara. 2012. Expressing ScACR3 in rice enhanced arsenite efflux and reduced arsenic accumulation in rice grains. *Plant Cell Physiol* 53 (1):154–163. https://doi.org/10.1093/pcp/pcr161

Duffus, J. H. 2001. "Heavy metals"–a meaningless term. *Chemistry International-Newsmagazine for IUPAC* 23 (6):163–167.

Emamverdian, A., Y. Ding, F. Mokhberdoran, and Y. Xie. 2015. Heavy metal stress and some mechanisms of plant defense response. *Sci World J* 2015:756120.

Espinosa-Ortiz, E. J., E. R. Rene, F. Guyot, E. D. van Hullebusch, and P. N. L. Lens. 2017. Biomineralization of tellurium and selenium-tellurium nanoparticles by the white-rot fungus Phanerochaete chrysosporium. *Int Biodeter Biodegr* 124:258–266.

Etesami, H., and B. R. Jeong. 2018. Silicon (Si): review and future prospects on the action mechanisms in alleviating biotic and abiotic stresses in plants. *Ecotoxicol Environ Saf* 147:881–896. https://doi.org/10.1016/j.ecoenv.2017.09.063

Farooq, M. A., F. Islam, B. Ali, U. Najeeb, B. Mao, R. A. Gill, G. Yan, K. H. M. Siddique, and W. Zhou. 2016. Arsenic toxicity in plants: cellular and molecular mechanisms of its transport and metabolism. *Environ Exp Bot* 132:42–52.

Feng, J., N. Yamaji, and N. Mitani-Ueno. 2011. Transport of silicon from roots to panicles in plants. *Proc Jpn Acad B* 87 (7):377–385.

Germ, M., I. Kreft, and J. Osvald. 2005. Influence of UV-B exclusion and selenium treatment on photochemical efficiency of photosystem II, yield and respiratory potential in pumpkins (Cucurbita pepo L.). *Plant Physiol Biochem* 43 (5):445–448.

Ghori, N. H., T. Ghori, M. Q. Hayat, S. R. Imadi, A. Gul, V. Altay, and M. Ozturk. 2019. Heavy metal stress and responses in plants. *Int J Environ Sci Technol* 16 (3):1807–1828. https://doi.org/10.1007/s13762-019-02215-8

Halliwell, B. 2006. Reactive species and antioxidants. Redox biology is a fundamental theme of aerobic life. Plant Physiol 141 (2):312–322.

Han, M.-J., S.-U. Kim, D.-C. Seo, Y.-H. Cheong, D.-J. Lee, M.-S. Park, Y.-S. Rim, B.-K. Sohn, J.-S. Heo, and J.-S. Cho. 2007. Uptake properties of germanium to vegetable plants and its effect on seed germination and on early stage growth. *Korean J Environ Agric* 26 (3):217–222.

Han, S., L.-S. Chen, H.-X. Jiang, B. R. Smith, L.-T. Yang, and C.-Y. Xie. 2008. Boron deficiency decreases growth and photosynthesis, and increases starch and hexoses in leaves of citrus seedlings. *J Plant Physiol* 165 (13):1331–1341.

Hasanuzzaman, M., and M. Fujita. 2020. *Plant Responses and Tolerance to Metal/Metalloid Toxicity.* MDPI.

He, Z., J. Shentu, X. Yang, V. C. Baligar, T. Zhang, and P. J. Stoffella. 2015. Heavy metal contamination of soils: sources, indicators and assessment. *J Environ Indicators* 9:17–18.

Hitchcock, A., C. N. Hunter, R. Sobotka, J. Komenda, M. Dann, and D. Leister. 2022. Redesigning the photosynthetic light reactions to enhance photosynthesis–the PhotoRedesign consortium. *Plant J* 109 (1):23–34.

Kaiser, E., A. Morales, J. Harbinson, J. Kromdijk, E. Heuvelink, and L. F. Marcelis. 2015. Dynamic photosynthesis in different environmental conditions. *J Exp Bot* 66 (9):2415–2426. https://doi.org/10.1093/jxb/eru406

Kastori, R., M. Plesničar, D. Panković, and Z. Sakač. 1995. Photosynthesis, chlorophyll fluorescence and soluble carbohydrates in sunflower leaves as affected by boron deficiency. *J Plant Nutr* 18 (9):1751–1763.

Kessi, J., M. Ramuz, E. Wehrli, M. Spycher, and R. Bachofen. 1999. Reduction of selenite and detoxification of elemental selenium by the phototrophic bacterium Rhodospirillum rubrum. *Appl Environ Microbiol* 65 (11):4734–4740. https://doi.org/10.1128/AEM.65.11.4734-4740.1999

Kim, Y. O., S. Jung, K. Kim, and H. J. Bae. 2013. Role of pCeMT, a putative metallothionein from Colocasia esculenta, in response to metal stress. *Plant Physiol Biochem* 64:25–32. https://doi.org/10.1016/j.plaphy.2012.12.009

Kubis, A., and A. Bar-Even. 2019. Synthetic biology approaches for improving photosynthesis. *J Exp Bot* 70 (5):1425–1433. https://doi.org/10.1093/jxb/erz029

Kumar, A., G. Dixit, A. P. Singh, S. Dwivedi, S. Srivastava, K. Mishra, and R. D. Tripathi. 2016. Selenate mitigates arsenite toxicity in rice (Oryza sativa L.) by reducing arsenic uptake and ameliorates amino acid content and thiol metabolism. *Ecotoxicol Environ Saf* 133:350–359.

Kumar, A., S. Guleria, D. Ghosh, V. Dogra, and S. Kumar. 2021. Managing reactive oxygen species - some learnings from high altitude extremophytes. *Environ Exp Bot* 189:104525. https://doi.org/10.1016/j.envexpbot.2021.104525

Labancova, E., Z. Vivodova, K. Siposova, and K. Kollarova. 2023. Silicon actuates poplar calli tolerance after longer exposure to antimony. *Plants (Basel)* 12 (3):689. https://doi.org/10.3390/plants12030689

Landi, M., T. Margaritopoulou, I. E. Papadakis, and F. Araniti. 2019. Boron toxicity in higher plants: an update. *Planta* 250:1011–1032.

Landi, M., A. Pardossi, D. Remorini, and L. Guidi. 2013. Antioxidant and photosynthetic response of a purple-leaved and a green-leaved cultivar of sweet basil (Ocimum basilicum) to boron excess. *Environ Exp Bot* 85:64–75.

Leister, D. 2019. Genetic engineering, synthetic biology and the light reactions of photosynthesis. *Plant Physiol* 179 (3):778–793.

Li, R. Y., Y. Ago, W. J. Liu, N. Mitani, J. Feldmann, S. P. McGrath, J. F. Ma, and F. J. Zhao. 2009. The rice aquaporin Lsi1 mediates uptake of methylated arsenic species. *Plant Physiol* 150 (4):2071–2080. https://doi.org/10.1104/pp.109.140350.

Li, S.-P., W.-L. Xie, H.-H. Cai, J.-Y. Cai, and P.-H. Yang. 2012. Hydroxyl radical scavenging mechanism of human erythrocytes by quercetin–germanium (IV) complex. *Eur J Pharm Sci* 47 (1):28–34.

Li, X., Y. Pan, X. Qi, S. Zhang, C. Zhi, H. Meng, and Z. Cheng. 2021. Effects of exogenous germanium and effective microorganisms on germanium accumulation and nutritional qualities of garlic (Allium sativum L.). *Sci Hortic* 283:110114.

Li, Y., O. P. Dhankher, L. Carreira, D. Lee, A. Chen, J. I. Schroeder, R. S. Balish, and R. B. Meagher. 2004. Overexpression of phytochelatin synthase in Arabidopsis leads to enhanced arsenic tolerance and cadmium hypersensitivity. *Plant Cell Physiol* 45 (12):1787–1797. https://doi.org/10.1093/pcp/pch202

Ma, J. F., and N. Yamaji. 2006. Silicon uptake and accumulation in higher plants. *Trends Plant Sci* 11 (8):392–397. https://doi.org/10.1016/j.tplants.2006.06.007

Martin, K. R. 2013. Silicon: the health benefits of a metalloid. In Sigel, A., Sigel, H., and Sigel, R. K. O. (Eds.), *Interrelations between Essential Metal Ions and Human Diseases* (pp. 451–473). Dordrecht: Springer.

McNaughton, M., L. Engman, A. Birmingham, G. Powis, and I. A. Cotgreave. 2004. Cyclodextrin-derived diorganyl tellurides as glutathione peroxidase mimics and inhibitors of thioredoxin reductase and cancer cell growth. *J Med Chem* 47 (1):233–239. https://doi.org/10.1021/jm030916r

Meravi, N. 2014. Effects of heavy metals/metalloids present in fly ash from coal fired thermal power plant on photosynthetic parameters of Mangifera indica. *Environ Skeptics Critics* 3 (4):88.

Miwa, K., J. Takano, H. Omori, M. Seki, K. Shinozaki, and T. Fujiwara. 2007. Plants tolerant of high boron levels. *Science* 318 (5855):1417–1417.

Mubarak, H., N. Mirza, L. Y. Chai, Z. H. Yang, W. Yong, C. J. Tang, Q. Mahmood, A. Pervez, U. Farooq, S. Fahad, W. Nasim, and K. H. Siddique. 2016. Biochemical and metabolic changes in arsenic contaminated Boehmeria nivea L. *Biomed Res Int* 2016:1423828. https://doi.org/10.1155/2016/1423828

Nable, R. O., G. S. Bañuelos, and J. G. Paull. 1997. Boron toxicity. *Plant Soil* 193:181–198.

Nahar, K., M. S. Rhaman, K. Parvin, K. Bardhan, D. N. Marques, P. García-Caparrós, and M. Hasanuzzaman. 2022. Arsenic-induced oxidative stress and antioxidant defense in plants. *Stresses* 2 (2):179–209.

Nielsen, A. Z., B. Ziersen, K. Jensen, L. M. Lassen, C. E. Olsen, B. L. Moller, and P. E. Jensen. 2013. Redirecting photosynthetic reducing power toward bioactive natural product synthesis. *ACS Synth Biol* 2 (6):308–315. https://doi.org/10.1021/sb300128r

Orr, D. J., A. M. Pereira, P. da Fonseca Pereira, I. A. Pereira-Lima, A. Zsogon, and W. L. Araujo. 2017. Engineering photosynthesis: progress and perspectives. *F1000Res* 6:1891. https://doi.org/10.12688/f1000research.12181.1

Ortega, A., I. Garrido, I. Casimiro, and F. Espinosa. 2017. Effects of antimony on redox activities and antioxidant defence systems in sunflower (Helianthus annuus L.) plants. *PLoS One* 12 (9):e0183991. https://doi.org/10.1371/journal.pone.0183991

Pan, X., D. Zhang, X. Chen, A. Bao, and L. Li. 2011. Antimony accumulation, growth performance, antioxidant defense system and photosynthesis of Zea mays in response to antimony pollution in soil. *Water Air Soil Pollut* 215:517–523.

Pandey, A. K., A. Gautam, and R. S. Dubey. 2019. Transport and detoxification of metalloids in plants in relation to plant-metalloid tolerance. *Plant Gene* 17:100171.

Pathak, S., A. V. Agarwal, and V. C. Pandey. 2020. Phytoremediation—a holistic approach for remediation of heavy metals and metalloids. In Pandey, V. C., and Singh, V. (Eds.), *Bioremediation of Pollutants* (pp. 3–16). Elsevier.

Paull, J. G., R. O. Nable, and A. J. Rathjen. 1992. Physiological and genetic control of the tolerance of wheat to high concentrations of boron and implications for plant breeding. *Plant Soil* 146:251–260.

Picorel, R., M. Alfonso, and M. Velitchkova. 2017. Molecular basis of the response of photosynthetic apparatus to light and temperature stress. *Front Plant Sci* 8:288.

Pilon-Smits, E. A., and D. L. LeDuc. 2009. Phytoremediation of selenium using transgenic plants. *Curr Opin Biotechnol* 20 (2):207–12. https://doi.org/10.1016/j.copbio.2009.02.001

Plesnicar, M., Rudolf Kastori, Z. Sakac, D. Panković, and N. Petrović. 1997. Boron as limiting factor in photosynthesis and growth of sunflower plants in relation to phosphate supply. *Agrochimica* 41 (3–4):144–154.

Porquet, A., and M. Filella. 2007. Structural evidence of the similarity of $Sb(OH)_3$ and $As(OH)_3$ with glycerol: implications for their uptake. *Chem Res Toxicol* 20 (9):1269–1276.

Praveen, A., C. Pandey, K. Ehasanullah, M. Panthri, and M. Gupta. 2020. Silicon-mediated genotoxic alterations in Brassica juncea under arsenic stress: a comparative study of biochemical and molecular markers. *Pedosphere* 30 (4):517–527.

Quaghebeur, M., and Z. Rengel. 2005. Arsenic speciation governs arsenic uptake and transport in terrestrial plants. *Microchim Acta* 151 (3):141–152. https://doi.org/10.1007/s00604-005-0394-8

Rajabpoor, S., S. M. Ghaderian, and H. Schat. 2019. Effects of antimony on enzymatic and non-enzymatic antioxidants in a metallicolous and a non-metallicolous population of Salvia spinosa L. *Plant Physiol Biochem* 144:386–394. https://doi.org/10.1016/j.plaphy.2019.10.011

Rastogi, A., S. Yadav, S. Hussain, S. Kataria, S. Hajihashemi, P. Kumari, X. Yang, and M. Brestic. 2021. Does silicon really matter for the photosynthetic machinery in plants…? *Plant Physiol Biochem* 169:40–48. https://doi.org/10.1016/j.plaphy.2021.11.004

Reid, R. J., J. E. Hayes, A. Post, J. C. R. Stangoulis, and R. D. Graham. 2004. A critical analysis of the causes of boron toxicity in plants. Plant *Cell Environ* 27 (11):1405–1414.

Schneider, A., I. Steinberger, H. Strissel, H.-H. Kunz, N. Manavski, J. Meurer, G. Burkhard, S. Jarzombski, D. Schünemann, and S. Geimer. 2014. The Arabidopsis Tellurite resistance C protein together with ALB3 is involved in photosystem II protein synthesis. *Plant J* 78 (2):344–356.

Sh, T., C.-Q. Liu, and L. Wang. 2012. Antimony coordination to humic acid: nuclear magnetic resonance and X-ray absorption fine structure spectroscopy study. *Microchem J* 103:68–73.

Shah, A., X. Wu, A. Ullah, S. Fahad, R. Muhammad, L. Yan, and C. Jiang. 2017. Deficiency and toxicity of boron: alterations in growth, oxidative damage and uptake by citrange orange plants. *Ecotoxicol Environ Saf* 145:575–582. https://doi.org/10.1016/j.ecoenv.2017.08.003

Signorelli, S., E. L. Coitino, O. Borsani, and J. Monza. 2014. Molecular mechanisms for the reaction between OH radicals and proline: insights on the role as reactive oxygen species scavenger in plant stress. *J Phys Chem B* 118 (1):37–47. https://doi.org/10.1021/jp407773u

Simón, I., L. Díaz-López, V. Gimeno, M. Nieves, W. E. Pereira, V. Martínez, V. Lidon, and F. García-Sánchez. 2013. Effects of boron excess in nutrient solution on growth, mineral nutrition, and physiological parameters of Jatropha curcas seedlings. *J Plant Nutr Soil Sci* 176 (2):165–174.

Singh, S., P. K. Srivastava, D. Kumar, D. K. Tripathi, D. K. Chauhan, and S. M. Prasad. 2015. Morpho-anatomical and biochemical adapting strategies of maize (Zea mays L.) seedlings against lead and chromium stresses. *Biocatal Agric Biotechnol* 4 (3):286–295.

Singh, S., D. K. Tripathi, S. Singh, S. Sharma, N. K. Dubey, D. K. Chauhan, and M. Vaculík. 2017. Toxicity of aluminium on various levels of plant cells and organism: a review. *Environ Exp Bot* 137:177–193.

Sinha, D., S. Datta, R. Mishra, P. Agarwal, T. Kumari, S. B. Adeyemi, A. K. Maurya, S. Ganguly, U. Atique, and S. Seal. 2023. Negative impacts of arsenic on plants and mitigation strategies. *Plants* 12 (9):1815.

Song, A., P. Li, F. Fan, Z. Li, and Y. Liang. 2014. The effect of silicon on photosynthesis and expression of its relevant genes in rice (Oryza sativa L.) under high-zinc stress. *PLoS One* 9 (11):e113782. https://doi.org/10.1371/journal.pone.0113782

Srivastava, M., L. Q. Ma, N. Singh, and S. Singh. 2005. Antioxidant responses of hyper-accumulator and sensitive fern species to arsenic. *J Exp Bot* 56 (415):1335–1342. https://doi.org/10.1093/jxb/eri134

Takano, J., K. Noguchi, M. Yasumori, M. Kobayashi, Z. Gajdos, K. Miwa, H. Hayashi, T. Yoneyama, and T. Fujiwara. 2002. Arabidopsis boron transporter for xylem loading. *Nature* 420 (6913):337–340.

Takano, J., M. Wada, U. Ludewig, G. Schaaf, N. von Wirén, and T. Fujiwara. 2006. The Arabidopsis major intrinsic protein NIP5;1 is essential for efficient boron uptake and plant development under boron limitation. *Plant Cell* 18 (6):1498–1509.

Tripathi, D. K., S. Singh, V. P. Singh, S. M. Prasad, N. K. Dubey, and D. K. Chauhan. 2017. Silicon nanoparticles more effectively alleviated UV-B stress than silicon in wheat (Triticum aestivum) seedlings. *Plant Physiol Biochem* 110:70–81. https://doi.org/10.1016/j.plaphy.2016.06.026

Vavrova, S., E. Struharnanska, J. Turna, and S. Stuchlik. 2021. Tellurium: a rare element with influence on prokaryotic and eukaryotic biological systems. *Int J Mol Sci* 22 (11):5924. https://doi.org/10.3390/ijms22115924

Wang, C., Z. Wang, T. Zhao, Y. Li, G. Huang, B. D. Sumer, and J. Gao. 2018. Optical molecular imaging for tumor detection and image-guided surgery. *Biomaterials* 157:62–75. https://doi.org/10.1016/j.biomaterials.2017.12.002

Wang, Y., B. Zhang, D. Jiang, and G. Chen. 2019. Silicon improves photosynthetic performance by optimizing thylakoid membrane protein components in rice under drought stress. *Environ Exp Bot* 158:117–124.

Yadav, S. K. 2010. Heavy metals toxicity in plants: an overview on the role of glutathione and phytochelatins in heavy metal stress tolerance of plants. *South Afr J Bot* 76 (2):167–179. https://doi.org/10.1016/j.sajb.2009.10.007

Yusuf, M., Q. Fariduddin, and A. Ahmad. 2011. 28-Homobrassinolide mitigates boron induced toxicity through enhanced antioxidant system in Vigna radiata plants. *Chemosphere* 85 (10):1574–1584. https://doi.org/10.1016/j.chemosphere.2011.08.004

Zhang, L., W. Shi, and X. Wang. 2006. Difference in selenite absorption between high-and low-selenium rice cultivars and its mechanism. *Plant Soil* 282:183–193.

Zhao, F. J., J. F. Ma, A. A. Meharg, and S. P. McGrath. 2009. Arsenic uptake and metabolism in plants. *New Phytol* 181 (4):777–794. https://doi.org/10.1111/j.1469-8137.2008.02716.x

Zhou, X., C. Sun, P. Zhu, and F. Liu. 2018. Effects of antimony stress on photosynthesis and growth of Acorus calamus. *Front Plant Sci* 9:579. https://doi.org/10.3389/fpls.2018.00579

11 Cross-Talk between Metalloids and Secondary Metabolism in Plants

Anmol Sidhu, Shruti Kaushik, Gurvarinder Kaur, Isha Madaan, Rashmi Mishra, and Geetika Sirhindi

11.1 INTRODUCTION

Through the past few decades, a rapid rise of industrialization in a quest for economic development has contributed to an enormous exacerbation in environmental pollution. Soil being one of the key elements of terrestrial ecosystem serves as a major role in plant nourishment and biomass degradation. Lately owing to intensive industrialization and chemical-based agricultural practices, heavy metal toxicity in soil has become a grave concern.

All of the many layers of earths' crust naturally include metal and metalloid ions. However, toxicity for several of its living forms including plants and animals may arise when their levels are discovered above the advised threshold quantities which are largely attributed to various anthropogenic activities (Rahman and Singh, 2019; Narendrula-Kotha et al., 2020). As a result of undifferentiating industrial exploitation of these metals and metalloids, the quality and productivity of soil have degraded owing to the deterioration of ecosystems and negative effects on plant growth and possible human health risks due to their entry into the food chain (Anjum et al., 2014; Parveen et al. 2022). From an anthropocentric perspective, the fact that many of these components are now present in greater quantities, together with other types of abiotic stress, poses a threat to world food security (Gong et al., 2020).

Metalloids are considered a class of chemicals which possess their physiochemical properties between the metals and non-metals and are acknowledged as both favourable and hazardous for plant growth (Parveen et al., 2022). Metalloids such as silicon (Si) and boron (B) are helpful to plant growth and development as the immunity of plants is enhanced by optimal levels of these metalloids through a variety of direct and indirect mechanisms. Conversely, even in lower quantities, metalloids such as germanium (Ge) and arsenic (As) are harmful to plants (Angulo-Bejarano et al., 2021; Parveen et al., 2022; Sharma et al., 2023). These harmful substances alter molecular, biochemical, and physiological processes within cells, disturbing cellular homeostasis (Sharma et al., 2023). Regardless of their detrimental influence on plant growth, certain metalloids also play a pivotal role in the growth and development of plants by participating in major metabolic reactions. Also, certain metalloids are known to efficiently control the activity of several enzymes and metabolic pathways when present at optimal amounts. In addition to controlling structural and functional integrity, these metalloids play a crucial role in a number of biosynthetic processes, including those involving proteins, chloroplasts, and nucleic acids (Oves et al., 2016; Dhakate et al., 2019).

Over the years, plants have developed many intricate mechanisms to cope with their toxic soil environment, where the process of evolution is considered to play a key role in acquiring various adaptive mechanisms against stress conditions (Qiao et al., 2019). Thus, to curb various stress conditions, plants have evolved with a plethora of genomes to build up certain tolerance mechanisms to retaliate stressful environments. Furthermore, as more and more cultivable

DOI: 10.1201/9781032673592-11

and arable lands throughout the world became polluted with metals and metalloids, plants have developed the innate potential to flourish in these conditions (Clemens and Ma, 2016). To better understand plant evolution and to use plants' adaptation skills to employ phytoremediation techniques in mitigating the effects of rising metal and metalloid concentrations in agronomically significant soils globally, a number of research groups have been developed to elucidate the complex mechanisms underlying interactions between plants and metalloids (Angulo-Bejarano et al., 2021).

Existence of metalloids in the soil above their threshold value is reported to modify cellular redox homeostasis in various ways: they trigger the production of reactive oxygen species (ROS), disrupt the antioxidant defence mechanism, induce mitochondrial dysfunctions, and so on. Conversely, certain redox-active metalloids have the ability to directly produce ROS through the Fenton-type or Haber–Weiss reaction (Yadav, 2010). The antioxidant defence system (ADS), which consists of a number of enzymes and non-enzymatic components, strictly controls the formation of ROS. Apart from this, some key secondary metabolites of plants also contribute a major share of amelioration towards metalloid toxicity. It is well versed that a number of secondary metabolites found in plants, such as ascorbic acid, jasmonic acid, salicylic acid, glutathione (GSH), and serotonin, protect the plants against abiotic stresses caused by metals and metalloids where redox homeostasis regulation has been suggested as the protective mechanism (Choudhury et al., 2021). Therefore, an insight understanding of the secondary metabolism of plants as well as the stress-regulating secondary metabolites is requisite so as to get a generalized state of the mitigation strategy for selective reduction of the toxic metalloids.

11.2 PLANT SECONDARY METABOLISM: AN OVERVIEW

Plant metabolism may be split into two categories i.e. primary and secondary metabolism. Secondary routes are found in specialized cells and generate a broad range of distinct chemicals and fundamental pathways, which are present in all cells and deal with the manipulation of a uniform set of basic molecules (Rajashekar, 2023). The primary metabolic pathways are chiefly concerned with the processes of glycolysis; the acid cycle; the pentose phosphate shunt; and lipid, protein, and nucleic acid biosynthesis. Whereas on the contrary, the production of a wide range of secondary metabolites (phenols, flavonoids, alkaloids, tannins, terpenes, lignin, etc.) is attributed to malonic acid, shikimic acid, methylerythritol phosphate, and mevalonic acid pathways (Saltveit, 2017; Rajashekar, 2023).

In life sciences, the most basic description of secondary metabolism is the pathways that operate in living organisms to produce small organic molecules which have no direct role in the growth and developmental processes, but rather play a pivotal role in forging ecological interactions between the organism and its environment. Plants are known to produce a large number of compounds via secondary metabolism, which have no contribution towards their basic growth and development, but are actively involved in forming plant–environmental interactions (Nascimento and Fett-Neto, 2010). These routes, which are mostly limited to certain taxonomic groups and are involved in the interaction between the plant and its environment, are generally recruited from important basic metabolic pathways during early gene duplication. Strict temporal and geographical regulation of gene expression guarantees the proper pattern of accumulation of different secondary products (Nascimento and Fett-Neto, 2010).

The large variety of plant secondary metabolites is among the most striking features of plants. There are currently tens of thousands of secondary metabolites identified in plants. It is estimated that there are about 21,000 alkaloids, 22,000 terpenoids, 5,000 flavonoids, and 5,000 tannins in nature (Wu and Chappell, 2008). However, this is most likely just a small portion of the total amount. Primary molecules and the genes involved in primary metabolism are the biochemical and evolutionary sources of secondary metabolites, respectively. Amino acids; nitrogenous bases; and intermediates of respiratory, photosynthetic, carbohydrate, or fatty acid metabolism are frequently the sources of secondary metabolites (Hartmann, 2007). Therefore, by precisely regulating plant growth and development, serving as a reservoir for vital phytochemicals, and protecting plants from a range of environmental obstacles, these compounds are created to enable the plants to compete and enhance their viability in their natural habitat (Ahmad et al., 2018) (Figure 11.1).

Plant secondary metabolism can therefore be viewed as a branch of primary metabolism, forming metabolites substituting as a reservoir for various phytochemicals, as well as orchestrate the intricate plant–environment interactions to increase their viability under dynamic conditions. Thus, plant secondary metabolites are intriguing and significant participants in mediating plant responses to biotic and abiotic environmental stimuli (Yang et al., 2012). The chemical structures of secondary metabolites found in plants are incredibly varied and divided into three main classes based on their structures and biosynthesis: terpenoids, phenolic compounds, and chemicals containing nitrogen (Fang et al., 2011).

The plant secondary metabolite diversity is a myriad of structurally and functionally varied compounds formed as a result of numerous modifications of the common carbon backbone (Wang et al., 2019). Although the number of unique metabolites in the plant kingdom is thought to range from 200,000 to a million, this is only an estimate. It is possible to categorize secondary metabolites according to their chemical structure (e.g. rings, sugar), content (nitrogen or not), solubility in different solvents, or synthesis route (e.g. phenylpropanoid) (Tiwari and Rana, 2015). Some of the major secondary metabolites in plants can be discussed as

- **Phenolic Compounds:** One of the primary and most varied classes of plant secondary metabolites are phenolic compounds, which have one or more hydroxyl groups and at least one aromatic ring in their structures. The most basic types of phenols include C_6H_5OH. Phenolic compounds are a broad class of molecules with a wide range of forms and activities that are categorized as secondary metabolites in plants. They may be divided into two groups: those that are water insoluble (lignins, condensed tannins, and hydroxycinammic acids bound to the cell wall) and those that are water soluble (phenolic acids, flavonoids, phenylpropanoids, and quinones) (Rispail et al., 2005).

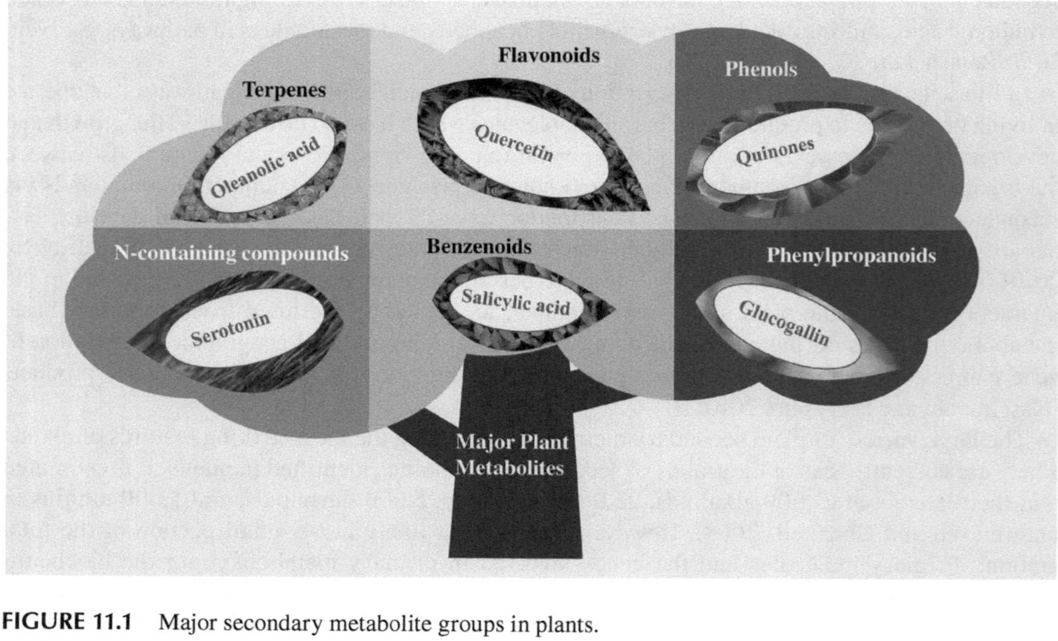

FIGURE 11.1 Major secondary metabolite groups in plants.

When it comes to making up the majority of organic matter, phenolic compounds, mainly lignin, come in second only to cellulose (Saltveit, 2017). Phenolic chemicals are aromatic rings with one or more hydroxyl groups. They can have basic phenolic molecule structures or complicated high-molecular-mass polymer structures which are known to have antioxidant properties. The free radical receives an electron from these antioxidants, which then transforms it into a harmless molecule (Balasundram et al., 2006). This ubiquitous class of plant secondary metabolites is ascribed to be synthesized both during normal conditions as well as under dynamic growth conditions like drought, salt, temperature, or metal and metalloid toxicity where oxidative stress is more pronounced in plant cells (Naczk and Shahidi, 2004; Luo et al., 2011; Haminiuk et al., 2012).

- **Flavonoids:** The majority of phenols that exist naturally as phytochemicals are called flavonoids. They can exist as glycosides, the biggest naturally occurring phenols, or in the free state. These are the polyphenolic chemicals, which are produced via the polypropanoid pathway using phenylalanine as a precursor (Mehta et al., 2022). The functional hydroxy groups of flavonoids can mediate antioxidant protection in plants under environmental stress conditions by stabilizing free radicals by donating electrons through resonance (Šamec et al., 2021; Mehta et al., 2022). Based on their structural characteristics, flavonoids are divided into five main types: flavones, anthocyanins, flavonols, flavanones, and isoflavone (Šamec et al., 2021). Flavonoids are reported to boost the ADS by limiting the metabolic activity of enzymes in ROS-engendering pathways, as the diversity in the structure of flavonoids enables them to interact with variable biomolecules (Mathesius, 2018). Thus, the active role of flavonoids under various stress conditions viz. temperature, metalloids, salt, drought, etc., has been immensely evinced (Pourcel et al., 2007; Di Ferdinando et al., 2012; Shomali et al., 2022; Wu et al., 2023).
- **Terpenes (Terpenoids or Isoprenoids):** The terpenes class of plant secondary metabolites is recognized as the largest and ubiquitous group. Terpenes are actually hydrocarbons synthesized from isoprene (isopentane) C5 units. Terpenoids are produced either by the 2-C-methyl-D-erythritol-4-phosphate (MEP) route inside the plastid or the mevalonate (MVA) pathway within the cytosol (Aharoni et al., 2005). They serve a variety of purposes in plants, including controlling plant growth, functioning as defence chemicals against viruses and herbivores, luring insects for pollination, and influencing (either directly or secondarily) the growth and developmental process of nearby plants (Mazid et al., 2011; Tiwari and Rana, 2015). Over tens of thousands of terpenes are reported in plants, spanning a number of over 30,000 different members (Zwenger and Basu, 2008; Yazaki et al., 2017; Boncan et al., 2020). Many studies conducted on a variety of plants have demonstrated that terpenes are influenced by a range of environmental conditions, including temperature changes, drought, metalloids, and pathogen assault. These circumstances cause the terpenes production and emission to be rearranged (Marichali et al., 2016; Perveen and Al-Taweel, 2018) (Figure 11.2).
- **Alkaloids:** Alkaloids are the predominant class of secondary metabolites found in 20% of plant species, known to include one or more nitrogen atoms that are produced during the decarboxylation of amino acids (Song et al., 2014). Based upon the amino acid, alkaloids are divided into seven types. Approximately 600 of the more than 20,000 identified alkaloids are known to be bioactive (Wangchuk, 2019). Although the type and pace of alkaloid production in plants are determined by genetics; environmental variables like light (UV), temperature; metal toxicity; salt stress; and soil minerals also have an impact (McCulley et al., 2014; Yang et al., 2018) (Figure 11.3).

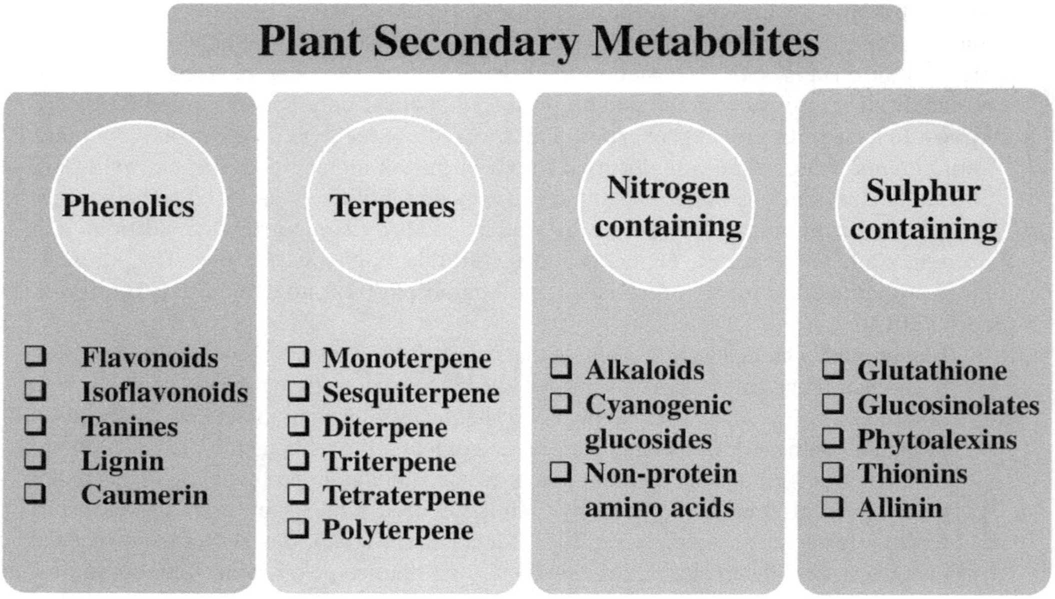

FIGURE 11.2 Major types of plant secondary metabolites.

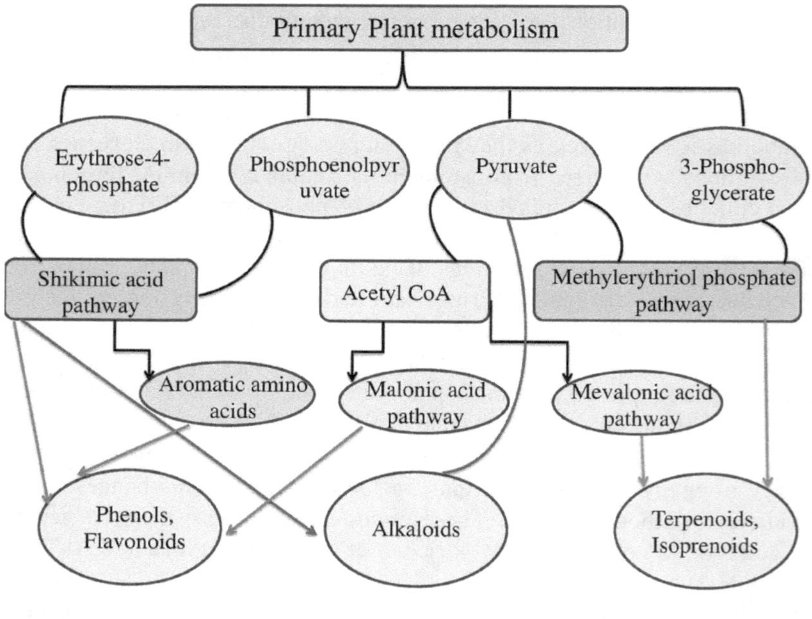

FIGURE 11.3 Major biosynthetic pathways of secondary metabolites.

11.3 ELICITING THE FUNCTIONAL ASPECTS AND INTERPLAY BETWEEN METALLOIDS AND SECONDARY METABOLISM OF PLANTS

Plants have developed special defence mechanisms to withstand extreme climatic conditions, such as high or low temperatures and excess level of metal elements [cadmium (Cd), As, and lead (Pb)] in soil, without compromising their cellular or physiological functions. In order to become accustomed to the changing effects of the environment, plants display a number of metabolic processes that produce secondary metabolites as a retaliation to their dynamic environmental conditions (Movahedi et al., 2021; Asare et al., 2023).

Plants create a vast array of secondary metabolites that arbitrate their interaction with the environment for defence and adaptation, but they have no discernible effect on activities that maintain life, where under unfavourable circumstances, they have long-term consequences on the development and survival of the plants (Anjitha et al., 2021). For instance, plants those have secondary metabolites like flavonoids, isoflavones, terpenes, carotenes, and isoprenoids may be able to withstand metal stress and may be used as instruments to deal with different stressors (Khare et al., 2020). Plants have various mechanisms for detoxifying metalloid toxicity, such as galvanizing various antioxidants, binding to the cell wall, metalloid sequestration and internal compartmentalization, biogenesis of osmoprotectants, trading of metalloid ions as well as intracellular chelation by low-molecular-weight organic acids, metallothioneins, phytochelatins, etc. (Sharma et al., 2016; Anjitha et al., 2021).

Plants may be threatened by high concentrations of As, Sb, and tellurium (Te) in soils that are contributed by both human (such as mining and industrial operations) and natural (geogenic-geochemical composition of source rock) sources (Edreva et al., 2008; Tang et al., 2022). These toxic metalloids have a number of negative consequences on a plant's ability to grow and develop and are therefore considered poisonous to plants (Asare et al., 2023). The majority of risk-enduring metalloids are known to be cumulated in the below-ground biomass of plants, or the root system. Nonetheless, metalloids like As are comparatively toxic to be easily transferred to biomass above ground, such as leaves (Száková et al., 2018). Apart from the obligatory negative effects of most of the metalloids, some like B are regarded to have the imperative role in growth and development of plants where it plays an essential role in maintaining membrane integrity of cell wall and plasma membrane as well as facilitating movement of ions through the membrane (Kohli et al., 2023). Despite the useful outlook of metalloids like B, the accumulation and deposition of metalloid cues is regarded as a critical issue that raises concerns about environmental contamination and nutrition. The relation of some major metalloids with the secondary metabolism of plants can be discussed as

- **As:** As is one of the most common, soluble, mobile, and harmful metalloids found in the environment as arsenate [As(V)] and arsenite [As(III)] forms. Mineralogy, biological processes, and chemical speciation are some of the composite factors that dictate the toxicity, mobility, and destiny of As in the environment (Bowell et al., 2014). Also, recently As has been reported to have certain mutagenic properties with potential carcinogenic properties of its inorganic forms (Nurchi et al., 2020; Ozturk et al., 2022). As's high solubility and translocation qualities allow it to potentially get past soil-plant barriers and into the food chain (Sharma et al., 2022). Plants growing in soil environments with toxic As concentration tend to retaliate the metalloid stress by diminishing As ion content in the cytosol by means of binding with several thiol-rich metalloid-binding chelating compounds like phytochelatins and GSH (Angulo-Bejarano et al., 2021; Wang et al., 2021). Interestingly, some recent studies also demonstrated a similar As ion chelating property of anthocyanins (a kind of flavonoid) in cytosol and vacuole under metalloid toxicity and thus ascribed a major role of secondary metabolites of plants in amelioration of metalloid toxicity (Ahammed and Yang, 2022). Further, under As stress, the MVA metabolic route is reported to be up-regulated in cytosol and plastid, enhancing the synthesis of isoprenoids and terpenes, conferring metalloid tolerance (Kamali et al., 2023). The role of terpenes has also been delineated in metalloid-stressed plants, where the degraded development and photosynthesis reprograms the metabolic pathways to produce secondary metabolites with a high carbon content (Pichersky and Raguso, 2018; Kopaczyk et al., 2020). On the contrary, some studies report that elevated

As concentrations (more than 40 mg kg^{-1} soil) markedly reduced the build-up of secondary metabolites like essential oils and phenols which is attributed to the reduced activities of phenylalanine ammonia lyase (PAL) and deoxy-D-xylulose-5-phosphate reductoisomerase (DXR) (Ahmad et al., 2022). In plants, carbohydrates are produced in chloroplasts and catabolized in the mitochondria and cytoplasm, where they are then directed towards the formation of secondary metabolites viz. phenols, flavonoids, terpenes, etc. Further, the possible conversion of stored carbohydrates to secondary metabolism is observed to be impacted by the disruption of carbon (C) absorption under As toxicity (Singh et al., 2006; Duman et al., 2010; Zhao et al., 2013; Nabi et al., 2019). Thus, As toxicity in soil generally induces the production of numerous secondary metabolites in plants.

- **Si:** In earths' crust, Si is the second most common element which is regarded as an important but non-essential element for plants (Datnoff et al., 2001; Bhat et al., 2019). The International Institute of Plant Nutrition has classified Si as a "quasi-essential" or advantageous nutritional element for plants based on a review of a large body of data (Bhat et al., 2019). Si is taken up by plants primarily in the form of $Si(OH)_4$, where its amount varies among species, ranging from 0.1% to 10% of their dry weight (Yamaji et al., 2008). Although there is a lot of variety among different plant species to absorb Si from the soil and store it in their tissues to benefit from its properties. Most monocots, which are high accumulators, benefit greatly from Si treatment, but many dicots, like *Arabidopsis*, are little or non-accumulators (Montpetit et al., 2012; Deshmukh et al., 2013). Increased activity of enzymes including PAL, peroxidase (POD), and polyphenol oxidase (PPO) involved in the manufacture of secondary metabolites is linked to Si-mediated resistance of certain biotic stresses (Yang et al., 2003; Hayasaka et al., 2008; Ahanger et al., 2020). These are essential enzymes that play a role in the phenylpropanoid pathway, a secondary metabolic process that produces phenolic compounds such as lignin, phytoalexins, and pathogenesis-related proteins (PRPs) (Kurabachew and Wydra, 2014; Ahanger et al., 2020; El-Beltagi et al., 2020). The phenylpropanoid pathway's first committed enzyme, PAL, is known to regulate the diversion of carbon fluxes from primary metabolism to the synthesis of phenolics (Ahanger et al., 2020). Apart from this, the terpenoid synthesis pathway via mevalonic acid is also described to be activated by Si in response to biotic stress (Davis et al., 2000; El-Beltagi et al., 2020; Pech-Kú et al., 2020). Therefore, from various research evidence, it is clear that Si supplementation can actively enhance the abiotic and biotic stress tolerance of plants via influencing secondary metabolism.
- **Ge:** The earth's crust is rich in Ge. Ge, an element cognate of Si, has very comparable chemical properties. In particular, the average worldwide concentration of Ge in soils is recorded to be 2.0 mg kg^{-1} (Wiche et al., 2018). While Ge is present in many rock-forming minerals and soil minerals, its concentrations are often 10,000 times lower than those of Si (Kaiser et al., 2020). Neither the severe toxicity of Ge to plants nor its need for plant growth has been determined up until this point. Unintentionally, plants may collect Ge while acquiring Si. Si has a lot of positive impacts on plant development, unlike Ge. The capacity of plants to withstand salt stress, temperature changes, UV-B damage, drought stress, metal/metalloid toxicity, and pathogen invasion is enhanced by Si (Bakhat et al., 2018; Katz, 2019). Ge has been described to interact with plant secondary metabolites in a course to enhance the accumulation of other target elements from soil essential for plant growth (Wiche et al., 2017; Schwabe et al., 2021) (Figure 11.4).

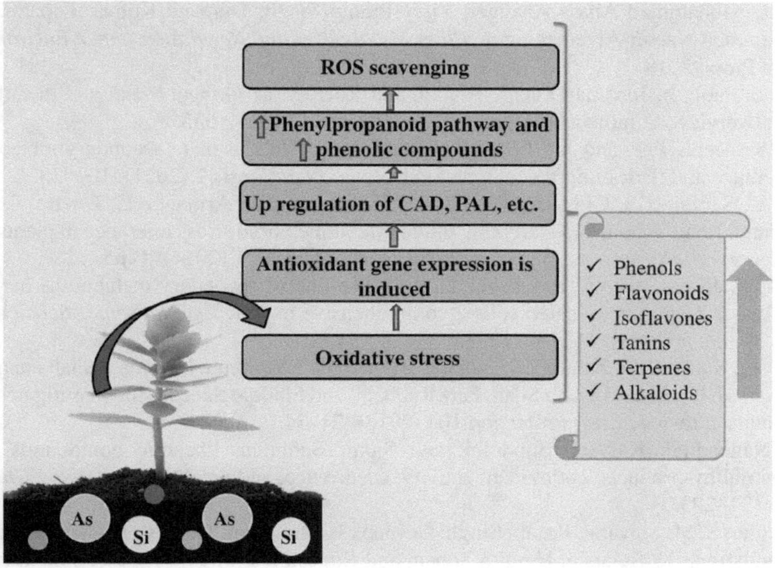

FIGURE 11.4 Plant secondary metabolites in relation to metalloid stress.

11.4 CONCLUSION AND FUTURE PROSPECTS

Plants are exposed to various abiotic and biotic stress conditions in their environment. Along with the heavy metal toxicity, certain metalloids like As are also reported to be lethal for plant growth and development. Over the recent decades, along with the known stress regulatory mechanisms of plants via several morphological, biochemical, and molecular means, a new insight into secondary metabolites had also been explored under metalloid toxicity, and their interrelationship is regarded to have a great impact on stress retaliation and development of plants. Secondary metabolites are now being viewed as metal chelators, antioxidants, and metal precipitators under stress conditions. Although many beneficial roles of metalloid interplay with the secondary metabolism of plants has been largely described over the last decade, the extensive knowledge and mechanism of plant stress regulation by various metalloids via their influence on the secondary metabolism is generally lacking. Thus, a more holistic approach to the subject of the interrelationship between various metalloids and secondary metabolites is requisite for clear mapping of their interplay and interaction during developmental and stress conditions.

REFERENCES

Ahammed, Golam Jalal, and Youxin Yang. Anthocyanin-mediated arsenic tolerance in plants. *Environmental Pollution* 292 (2022): 118475.

Ahanger, Mohammad Abass, Javaid Akhter Bhat, Manzer H. Siddiqui, Jörg Rinklebe, and Parvaiz Ahmad. Integration of silicon and secondary metabolites in plants: a significant association in stress tolerance. *Journal of Experimental Botany* 71, no. 21 (2020): 6758–6774.

Aharoni, Asaph, Maarten A. Jongsma, and Harro J. Bouwmeester. Volatile science? Metabolic engineering of terpenoids in plants. *Trends in Plant Science* 10, no. 12 (2005): 594–602.

Ahmad, Bilal, Tariq Ahmad Dar, M. Khan, Ajaz Ahmad, Jörg Rinklebe, Yinglong Chen, and Parvaiz Ahmad. Oligochitosan fortifies antioxidative and photosynthetic metabolism and enhances secondary metabolite accumulation in arsenic-stressed peppermint. *Frontiers in Plant Science* 13 (2022): 987746.

Ahmad, Parvaiz, Mohammad Abass Ahanger, Vijay Pratap Singh, Durgesh Kumar Tripathi, Pravej Alam, and Mohammed Nasser Alyemeni, eds. *Plant Metabolites and Regulation under Environmental Stress*. Academic Press, 2018.

Angulo-Bejarano, Paola I., Jonathan Puente-Rivera, and Rocío Cruz-Ortega. Metal and metalloid toxicity in plants: an overview on molecular aspects. *Plants* 10, no. 4 (2021): 635.

Anjitha, K. S., Sameena, P. P. and Jos T. Puthur. Functional aspects of plant secondary metabolites in metal stress tolerance and their importance in pharmacology. *Plant Stress* 2 (2021): 100038.

Anjum, Naser A., Sarvajeet S. Gill, Ritu Gill, Mirza Hasanuzzaman, Armando C. Duarte, Eduarda Pereira, Iqbal Ahmad, Renu Tuteja, and Narendra Tuteja. Metal/metalloid stress tolerance in plants: role of ascorbate, its redox couple, and associated enzymes. *Protoplasma* 251 (2014): 1265–1283.

Asare, Michael O., Jiřina Száková, and Pavel Tlustoš. The fate of secondary metabolites in plants growing on Cd-, As-, and Pb-contaminated soils-a comprehensive review. *Environmental Science and Pollution Research* 30, no. 5 (2023): 11378–11398.

Bakhat, Hafiz Faiq, Najma Bibi, Zahida Zia, Sunaina Abbas, Hafiz Mohkum Hammad, Shah Fahad, Muhammad Rizwan Ashraf, Ghulam Mustafa Shah, Faiz Rabbani, and Shafqat Saeed. Silicon mitigates biotic stresses in crop plants: a review. *Crop Protection* 104 (2018): 21–34.

Balasundram, Nagendran, Kalyana Sundram, and Samir Samman. Phenolic compounds in plants and agri-industrial by-products: antioxidant activity, occurrence, and potential uses. *Food Chemistry* 99, no. 1 (2006): 191–203.

Bhat, Javaid Akhter, S. M. Shivaraj, Pritam Singh, Devanna B. Navadagi, Durgesh Kumar Tripathi, Prasanta K. Dash, Amolkumar U. Solanke, Humira Sonah, and Rupesh Deshmukh. Role of silicon in mitigation of heavy metal stresses in crop plants. *Plants* 8, no. 3 (2019): 71.

Boncan, Delbert Almerick T., Stacey S. K. Tsang, Chade Li, Ivy H. T. Lee, Hon-Ming Lam, Ting-Fung Chan, and Jerome H. L. Hui. Terpenes and terpenoids in plants: interactions with environment and insects. *International Journal of Molecular Sciences* 21, no. 19 (2020): 7382.

Bowell, Robert J., Charles N. Alpers, Heather E. Jamieson, D. Kirk Nordstrom, and Juraj Majzlan. The environmental geochemistry of arsenic-an overview. *Reviews in Mineralogy and Geochemistry* 79, no. 1 (2014): 1–16.

Choudhury, Shuvasish, Debojyoti Moulick, and Muhammed Khairujjaman Mazumder. Secondary metabolites protect against metal and metalloid stress in rice: an in silico investigation using dehydroascorbate reductase. *Acta Physiologiae Plantarum* 43 (2021): 1–10.

Clemens, Stephan, and Jian Feng Ma. Toxic heavy metal and metalloid accumulation in crop plants and foods. *Annual Review of Plant Biology* 67 (2016): 489–512.

Datnoff, Lawrence E., George Heft Snyder, and Gaspar H. Korndörfer, eds. *Silicon in Agriculture*. Elsevier, 2001.

Davis, E. M., R. Croteau, F. J. Leeper, and J. C. Vederas. Biosynthesis. *Topics in Current Chemistry*, pp. 53–95. 2000.

Deshmukh, Rupesh K., Julien Vivancos, Valérie Guérin, Humira Sonah, Caroline Labbé, François Belzile, and Richard R. Bélanger. Identification and functional characterization of silicon transporters in soybean using comparative genomics of major intrinsic proteins in Arabidopsis and rice. *Plant Molecular Biology* 83 (2013): 303–315.

Dhakate, Priyanka, Prateek Sharma, Sahil Mehta, Javed Akter, Vacha Bhatt, Sonali Chandanshive, Dhiresh Chakravarty, Mehzabin Rahman, and Md Aminul Islam. Beneficial role of metalloids in plants: molecular understanding and applicability. In Wani, Shabir Hussain ed., *Recent Approaches in Omics for Plant Resilience to Climate Change*, pp. 357–387. Springer, 2019.

Di Ferdinando, Martina, Cecilia Brunetti, Alessio Fini, and Massimiliano Tattini. Flavonoids as antioxidants in plants under abiotic stresses. In Ahmad, Parvaiz, and M. N. V. Prasad eds., *Abiotic Stress Responses in Plants: Metabolism, Productivity and Sustainability*, pp. 159–179. Springer, 2012.

Duman, Fatih, Fatma Ozturk, and Zeki Aydin. Biological responses of duckweed (Lemna minor L.) exposed to the inorganic arsenic species As (III) and As (V): effects of concentration and duration of exposure. *Ecotoxicology* 19 (2010): 983–993.

Edreva, A., Violeta Velikova, T. Tsonev, Soleya Dagnon, A. Gürel, Lale Yildiz Aktaş, and E. Gesheva. Stress-protective role of secondary metabolites: diversity of functions and mechanisms. *General and Applied Plant Physiology*, 34, no. 1–2 (2008): 67–78.

El-Beltagi, Hossam S., Mahmoud R. Sofy, Mohammed I. Aldaej, and Heba I. Mohamed. Silicon alleviates copper toxicity in flax plants by up-regulating antioxidant defense and secondary metabolites and decreasing oxidative damage. *Sustainability* 12, no. 11 (2020): 4732.

Fang, Xin, Chang Qing Yang, YuKun Wei, QiXia Ma, Lei Yang, and Xiao Ya Chen. Genomics grand for diversified plant secondary metabolites. *Plant Diversity and Resources* 33, no. 1 (2011): 53–64.

Gong, Zhizhong, Liming Xiong, Huazhong Shi, Shuhua Yang, Luis R. Herrera-Estrella, Guohua Xu, Dai-Yin Chao et al. Plant abiotic stress response and nutrient use efficiency. *Science China Life Sciences* 63 (2020): 635–674.

Haminiuk, Charles WI, Giselle M. Maciel, Manuel S. V. Plata-Oviedo, and Rosane M. Peralta. Phenolic compounds in fruits–an overview. *International Journal of Food Science & Technology* 47, no. 10 (2012): 2023–2044.

Hartmann, Thomas. From waste products to ecochemicals: fifty years research of plant secondary metabolism. *Phytochemistry* 68, no. 22–24 (2007): 2831–2846.

Hayasaka, T., H. Fujii, and K. Ishiguro. The role of silicon in preventing appressorial penetration by the rice blast fungus. *Phytopathology* 98, no. 9 (2008): 1038–1044.

Kaiser, Sabine, Stephan Wagner, Christin Moschner, Claudia Funke, and Oliver Wiche. Accumulation of germanium (Ge) in plant tissues of grasses is not solely driven by its incorporation in phytoliths. *Biogeochemistry* 148 (2020): 49–68.

Kamali, Soheila, Alireza Iranbakhsh, Mostafa Ebadi, Zahra Oraghi Ardebili, and Setareh Haghighat. Methyl jasmonate conferred arsenic tolerance in Thymus kotschyanus by DNA hypomethylation, stimulating terpenoid metabolism, and upregulating two cytochrome P450 monooxygenases. *Journal of Hazardous Materials* 465 (2023): 133163.

Katz, Ofir. Silicon content is a plant functional trait: implications in a changing world. *Flora* 254 (2019): 88–94.

Khare, Shubhra, N. B. Singh, Ajey Singh, Imtiyaz Hussain, K. M. Niharika, Vijaya Yadav, Chanda Bano, Ravi Kumar Yadav, and Nimisha Amist. Plant secondary metabolites synthesis and their regulations under biotic and abiotic constraints. *Journal of Plant Biology* 63 (2020): 203–216.

Kohli, Sukhmeen Kaur, Harsimran Kaur, Kanika Khanna, Neha Handa, Renu Bhardwaj, Jörg Rinklebe, and Parvaiz Ahmad. Boron in plants: uptake, deficiency and biological potential. *Plant Growth Regulation* 100, no. 2 (2023): 267–282.

Kopaczyk, Joanna Maria, Joanna Warguła, and Tomasz Jelonek. The variability of terpenes in conifers under developmental and environmental stimuli. *Environmental and Experimental Botany* 180 (2020): 104197.

Kurabachew, Henok, and Kerstin Wydra. Induction of systemic resistance and defense-related enzymes after elicitation of resistance by rhizobacteria and silicon application against Ralstonia solanacearum in tomato (Solanum lycopersicum). *Crop Protection* 57 (2014): 1–7.

Luo, Wei, Mouming Zhao, Bao Yang, Jiaoyan Ren, Guanglin Shen, and Guohua Rao. Antioxidant and antiproliferative capacities of phenolics purified from Phyllanthus emblica L. fruit. *Food Chemistry* 126, no. 1 (2011): 277–282.

Marichali, Ahmed, Sana Dallali, Saloua Ouerghemmi, Houcine Sebei, Herve Casabianca, and Karim Hosni. Responses of Nigella sativa L. to zinc excess: focus on germination, growth, yield and yield components, lipid and terpene metabolism, and total phenolics and antioxidant activities. *Journal of Agricultural and Food Chemistry* 64, no. 8 (2016): 1664–1675.

Mathesius, Ulrike. Flavonoid functions in plants and their interactions with other organisms. *Plants* 7, no. 2 (2018): 30.

Mazid, M., T. A. Khan, and F. Mohammad. Role of secondary metabolites in defense mechanisms of plants. *Biology and Medicine* 3, no. 2 (2011): 232–249.

McCulley, Rebecca L., Lowell P. Bush, Anna E. Carlisle, Huihua Ji, and Jim A. Nelson. Warming reduces tall fescue abundance but stimulates toxic alkaloid concentrations in transition zone pastures of the US. *Frontiers in Chemistry* 2 (2014): 88.

Mehta, Shuchi Dave, Sukirti Upadhyay, and Gopal Rai. Importance of flavonoid as secondary metabolites. In Abbas, Hafiz Muhammad Khalid, and Aqeel Ahmad eds., *Flavonoid Metabolism-Recent Advances and Applications in Crop Breeding*. IntechOpen, 2022.

Montpetit, Jonatan, Julien Vivancos, Namiki Mitani-Ueno, Naoki Yamaji, Wilfried Rémus-Borel, François Belzile, Jian Feng Ma, and Richard R. Bélanger. Cloning, functional characterization and heterologous expression of TaLsi1, a wheat silicon transporter gene. *Plant Molecular Biology* 79 (2012): 35–46.

Movahedi, Ali, Amir Almasi Zadeh Yaghuti, Hui Wei, Paul Rutland, Weibo Sun, Mohaddeseh Mousavi, Dawei Li, and Qiang Zhuge. Plant secondary metabolites with an overview of populus. *International Journal of Molecular Sciences* 22, no. 13 (2021): 6890.

Nabi, Aarifa, M. Naeem, Tariq Aftab, M. Masroor, and A. Khan. Arsenic toxicity induced changes in growth, photosynthetic pigments, antioxidant machinery, essential oil, menthol and other active constituents of menthol mint (Mentha arvensis L.). *Journal of Essential Oil Bearing Plants* 22, no. 5 (2019): 1333–1348.

Naczk, Marian, and Fereidoon Shahidi. Extraction and analysis of phenolics in food. *Journal of Chromatography A* 1054, no. 1–2 (2004): 95–111.

Narendrula-Kotha, Ramya, Gabriel Theriault, Melanie Mehes-Smith, Kersey Kalubi, and Kabwe Nkongolo. Metal toxicity and resistance in plants and microorganisms in terrestrial ecosystems. *Reviews of Environmental Contamination and Toxicology* 249 (2020): 1–27.

Nascimento, Naíla Cannes do, and Arthur Germano Fett-Neto. Plant secondary metabolism and challenges in modifying its operation: an overview. In Arthur Germano Fett-Neto ed., *Plant Secondary Metabolism Engineering: Methods and Applications*, pp. 1–13. Humana Press, 2010.

Nurchi, Valeria M., Aleksandra Buha Djordjevic, Guido Crisponi, Jan Alexander, Geir Bjørklund, and Jan Aaseth. Arsenic toxicity: molecular targets and therapeutic agents. *Biomolecules* 10, no. 2 (2020): 235.

Oves, M., M. Saghir Khan, A. Huda Qari, M. Nadeen Felemban, and T. Almeelbi. Heavy metals: biological importance and detoxification strategies. *Journal of Bioremediation and Biodegradation* 7, no. 2 (2016): 1–15.

Ozturk, Munir, Mert Metin, Volkan Altay, Rouf Ahmad Bhat, Mahnoor Ejaz, Alvina Gul, Bengu Turkyilmaz Unal et al. Arsenic and human health: genotoxicity, epigenomic effects, and cancer signaling. *Biological Trace Element Research* 200 (2022): 1–14.

Parveen, Nishat, Roberto Berni, Sreeja Sudhakaran, Javaid A. Bhat, Suhas Shinde, Naleeni Ramawat, Vijay P. Singh et al. Metalloids in plants: a systematic discussion beyond description. *Annals of Applied Biology* 180, no. 1 (2022): 7–25.

Pech-Kú, Roberto, Beatriz Rodas-Junco, José A. Muñoz-Sánchez, and SM Teresa Hernández-Sotomayor. Linking secondary metabolism and signaling response to aluminum and silicon stress. In Deshmukh, Rupesh, Gea Guerriero, and Durgesh Kumar Tripathi eds., *Metalloids in Plants: Advances and Future Prospects*, pp. 367–379. John Wiley & Sons, 2020.

Perveen, Shagufta, and Areej Al-Taweel, eds. *Terpenes and Terpenoids*. BoD–Books on Demand, 2018.

Pichersky, Eran, and Robert A. Raguso. Why do plants produce so many terpenoid compounds? *New Phytologist* 220, no. 3 (2018): 692–702.

Pourcel, Lucille, Jean-Marc Routaboul, Véronique Cheynier, Loïc Lepiniec, and Isabelle Debeaujon. Flavonoid oxidation in plants: from biochemical properties to physiological functions. *Trends in Plant Science* 12, no. 1 (2007): 29–36.

Qiao, Xin, Qionghou Li, Hao Yin, Kaijie Qi, Leiting Li, Runze Wang, Shaoling Zhang, and Andrew H. Paterson. Gene duplication and evolution in recurring polyploidization–diploidization cycles in plants. *Genome Biology* 20, no. 1 (2019): 1–23.

Rahman, Zeeshanur, and Ved Pal Singh. The relative impact of toxic heavy metals (THMs) (arsenic (As), cadmium (Cd), chromium (Cr) (VI), mercury (Hg), and lead (Pb)) on the total environment: an overview. *Environmental Monitoring and Assessment* 191 (2019): 1–21.

Rajashekar, Channa B. Dual role of plant phenolic compounds as antioxidants and prooxidants. *American Journal of Plant Sciences* 14, no. 1 (2023): 15–28.

Rispail, Nicolas, Phillip Morris, and K. Judith Webb. Phenolic compounds: extraction and analysis. In Márquez, A. J., and J. Stougaard eds. *Lotus Japonicus Handbook*, pp. 349–354. Springer, 2005.

Saltveit, Mikal E. Synthesis and metabolism of phenolic compounds. In Yahia, Elhadi M. ed., *Fruit and Vegetable Phytochemicals: Chemistry and Human Health,* 2nd Edition, pp. 115–124. John Wiley & Sons Ltd, 2017.

Šamec, Dunja, Erna Karalija, Ivana Šola, Valerija Vujčić Bok, and Branka Salopek-Sondi. The role of polyphenols in abiotic stress response: the influence of molecular structure. *Plants* 10, no. 1 (2021): 118.

Schwabe, Ringo, Christine Dittrich, Julian Kadner, Christoph Helmut Rudi Senges, Julia Elisabeth Bandow, Dirk Tischler, Michael Schlömann, Gloria Levicán, and Oliver Wiche. Secondary metabolites released by the rhizosphere bacteria Arthrobacter oxydans and Kocuria rosea enhance plant availability and soil–plant transfer of germanium (Ge) and rare earth elements (REEs). *Chemosphere* 285 (2021): 131466.

Sharma, Resham, Renu Bhardwaj, Neha Handa, Vandana Gautam, Sukhmeen Kaur Kohli, Shagun Bali, Parminder Kaur et al. Responses of phytochelatins and metallothioneins in alleviation of heavy metal stress in plants: an overview. In Ahmad, Parvaiz ed., *Plant Metal Interaction*, pp. 263–283. Elsevier, 2016.

Sharma, Renu, Rupali Nanda, and Nisha Bhagat. Background level, occurrence, speciation, bioavailability, uptake, detoxification mechanisms and management of arsenic polluted soil. In Kumar, Vinod, Anket Sharma, and Raj Setia eds., *Appraisal of Metal (Loids) in the Ecosystem*, pp. 221–254. Elsevier, 2022.

Sharma, Yogesh, Praveen Soni, Gaurav Raturi, Rushil Mandlik, Vinay Kumar Rachappanavar, Manish Kumar, Prafull Salvi, Durgesh Kumar Tripathi, Hasthi Ram, and Rupesh Deshmukh. Regulation of metalloid uptake in plants by transporters and their solute specificity. *Environmental and Experimental Botany* 206 (2023): 105180.

Shomali, Aida, Susmita Das, Namira Arif, Mohammad Sarraf, Noreen Zahra, Vaishali Yadav, Sasan Aliniaeifard, Devendra Kumar Chauhan, and Mirza Hasanuzzaman. Diverse physiological roles of flavonoids in plant environmental stress responses and tolerance. *Plants* 11, no. 22 (2022): 3158.

Singh, Nandita, Lena Q. Ma, Mrittunjai Srivastava, and Bala Rathinasabapathi. Metabolic adaptations to arsenic-induced oxidative stress in Pteris vittata L and Pteris ensiformis L. *Plant Science* 170, no. 2 (2006): 274–282.

Song, Myoung Chong, Eun Ji Kim, Eunji Kim, Kris Rathwell, Sang-Jip Nam, and Yeo Joon Yoon. Microbial biosynthesis of medicinally important plant secondary metabolites. *Natural Product Reports* 31, no. 11 (2014): 1497–1509.

Száková, Jiřina, Miroslava Dziaková, Anna Kozáková, and Pavel Tlustoš. The risk element uptake by chamomile (Matricaria recutita (L.) Rauschert) growing in four different soils. *Archives of Environmental Protection* 44, no. 4 (2018): 12–21.

Tang, Haiying, Guiyuan Meng, Junqing Xiang, Athar Mahmood, Guohong Xiang, Ying Liu, and Guoqin Huang. Toxic effects of antimony in plants: reasons and remediation possibilities-a review and future prospects. *Frontiers in Plant Science* 13 (2022): 3705.

Tiwari, Ruby, and C. S. Rana. Plant secondary metabolites: a review. *International Journal of Engineering Research and General Science* 3, no. 5 (2015): 661–670.

Wang, Kang, Yaqi Wang, Yanan Wan, Zidong Mi, Qiqi Wang, Qi Wang, and Huafen Li. The fate of arsenic in rice plants (Oryza sativa L.): influence of different forms of selenium. *Chemosphere* 264 (2021): 128417.

Wang, Shouchuang, Saleh Alseekh, Alisdair R. Fernie, and Jie Luo. The structure and function of major plant metabolite modifications. *Molecular Plant* 12, no. 7 (2019): 899–919.

Wangchuk, Phurpa. Plant alkaloids: classification, isolation and drug development. In Swamy, M. K., J. K. Patra, and G. R. Rudramurthy, eds., *Medicinal Plants: Chemistry, Pharmacology and Therapeutic Applications*, pp. 131–137. CRC Press/Taylor & Francis Group, 2019.

Wiche, Oliver, Balázs Székely, Christin Moschner, and Hermann Heilmeier.Germanium in the soil-plant system-a review. *Environmental Science and Pollution Research* 25 (2018): 31938–31956.

Wiche, Oliver, Viola Zertani, Werner Hentschel, Roland Achtziger, and Pavol Midula. Germanium and rare earth elements in topsoil and soil-grown plants on different land use types in the mining area of Freiberg (Germany). *Journal of Geochemical Exploration* 175 (2017): 120–129.

Wu, Jieting, Sidi Lv, Lei Zhao, Tian Gao, Chang Yu, Jianing Hu, and Fang Ma. Advances in the study of the function and mechanism of the action of flavonoids in plants under environmental stresses. *Planta* 257, no. 6 (2023): 108.

Wu, Shuiqin, and Joe Chappell. Metabolic engineering of natural products in plants; tools of the trade and challenges for the future. *Current Opinion in Biotechnology* 19, no. 2 (2008): 145–152.

Yadav, S. K. Heavy metals toxicity in plants: an overview on the role of glutathione and phytochelatins in heavy metal stress tolerance of plants. *South African Journal of Botany* 76, no. 2 (2010): 167–179.

Yamaji, Naoki, Namiki Mitatni, and Jian Feng Ma. A transporter regulating silicon distribution in rice shoots. *The Plant Cell* 20, no. 5 (2008): 1381–1389.

Yang, Chang-Qing, Xin Fang, Xiu-Ming Wu, Ying-Bo Mao, Ling-Jian Wang, and Xiao-Ya Chen. Transcriptional regulation of plant secondary metabolism F. *Journal of Integrative Plant Biology* 54, no. 10 (2012): 703–712.

Yang, Li, Kui-Shan Wen, Xiao Ruan, Ying-Xian Zhao, Feng Wei, and Qiang Wang. Response of plant secondary metabolites to environmental factors. *Molecules* 23, no. 4 (2018): 762.

Yang, Y. F., Y. C. Liang, and Y. S. Lou. Influences of silicon on peroxide dismutase activity and lignin content of wheat (Tritium Aestivum L.) and its relation to resistance to powdery mildew. *Scientia Agricultura Sinica* 36, no. 7 (2003): 813–817.

Yazaki, Kazufumi, Gen-ichiro Arimura, and Toshiyuki Ohnishi. 'Hidden' terpenoids in plants: their biosynthesis, localization and ecological roles. *Plant and Cell Physiology* 58, no. 10 (2017): 1615–1621.

Zhao, Lishan, Wei-Chen Chang, Youli Xiao, Hung-Wen Liu, and Pinghua Liu. Methylerythritol phosphate pathway of isoprenoid biosynthesis. *Annual Review of Biochemistry* 82 (2013): 497–530.

Zwenger, Sam, and Chhandak Basu. Plant terpenoids: applications and future potentials. *Biotechnology and Molecular Biology Reviews* 3, no. 1 (2008): 1.

12 Biofortification of Crops with Metalloids

Importance for Human Health and Sustainable Plant Growth

Nirupma Singh and Kajal Nagre

12.1 INTRODUCTION

The energy needs of the impoverished masses in developing nations have mostly been satisfied by modern agriculture. Over the last four decades, agricultural research in developing nations has addressed the malnutrition problem by prioritizing greater cereal output. But to produce more food and provide it with higher quality, agriculture needs to adopt a new paradigm. In a world facing the dual challenges of feeding a growing global population and addressing the persistent problem of nutrient deficiencies, biofortification has emerged as a promising and innovative agricultural approach. To combat malnutrition, the World Health Organization (WHO) places a strong emphasis on food fortification and vitamin supplements. Through the Harvest Plus challenge program, CGIAR has increased its focus on biofortification.

As = Arsenic
P = Phosphorus
Si = Silicon
S = Sulfur
Se = Selenium
B = Boron

Furthermore, the micronutrient content of staple crops such as beans, sweet potatoes, millets, cereals, and cassava has been increased by breeding and biotechnology methods (Khush et al., 2012). Biofortification, a process aimed at increasing the concentration of essential nutrients, such as vitamins and minerals, in edible parts of crops, has the potential to address these critical issues. Unlike conventional approaches, which require changes in dietary habits or additional infrastructure for supplementation programs, biofortified crops are breed or engineered to naturally accumulate higher levels of essential nutrients. This implies that the food itself becomes nutrient-dense, which is advantageous to individuals who eat it. A variety of vital elements are targeted by biofortification, such as zinc, iron, selenium (Se), and vitamins C and A. These nutrients are selected based on local dietary habits and the prevalence of specific nutrient deficiencies in the region.

Among the various nutrients targeted for biofortification, metalloids have gained prominence for their role in enhancing human health and promoting sustainable plant growth. The earth's crust contains a significant amount of metalloid particles in addition to other elements. They are naturally occurring elements found in various environmental sources, including rocks, soils, and minerals. Their presence in the environment can influence their uptake into the food chain and their possible effect on human well-being. The effects of metalloid chemical elements on human health are complicated because they have characteristics that bridge metals and non-metals. Their dual nature

DOI: 10.1201/9781032673592-12

gives them special characteristics and makes them significant in the context of nutrition and health. The periodic table lists the metalloids polonium (Po), tellurium (Te), antimony (Sb), germanium (Ge), silicon (Si), arsenic (As), boron (B), and Se. Within the non-metal and metal elements, these components are positioned diagonally. They can control the ideal development and growth of all plants and animals, just like many other minerals. Nonetheless, it is also widely known that elevated metalloid concentrations negatively impact plant health by interfering with several biological processes essential to plant metabolic processes (Adrees et al., 2015; Afshan et al., 2015; Nagajyoti et al., 2010). Certain metalloids are important micronutrients for human health, but when taken in excess, some of them can be quite hazardous. This book chapter delves into the multifaceted world of metalloid biofortification, exploring its importance in addressing global health challenges, supporting sustainable agriculture, and safeguarding our environment. In addition to increasing agricultural yields and lowering agriculture's environmental impact, it may also improve plant health.

12.2 BASIC CONCEPTS

In an increasingly interconnected world grappling with issues of global health, food security, and sustainable agriculture, the practice of crop biofortification has emerged as a beacon of hope. Crop biofortification is an innovative approach that aims to increase the nutritional content of staple crops, providing essential vitamins and minerals to those who rely on them as their primary source of sustenance. This process involves breeding or genetically modifying crops to increase the levels of essential nutrients, such as vitamins and minerals, in their edible parts. The goal is to provide a sustainable and cost-effective solution to address malnutrition and improve overall well-being. Here are some key aspects of crop biofortification:

a. **Nutrient Targets**
   It is possible to biofortify different crops to meet different nutrient deficits. Increasing the concentrations of vitamins, minerals, and other beneficial substances in crops is a common goal.
b. **Breeding *vs.* Genetic Modification**
   Crop biofortification can be achieved through traditional breeding methods or genetic modification (GMO). In traditional breeding, plants with desired features are chosen, and over several generations, they are crossbreed. GMO allows for more precise control over the insertion of specific genes to enhance nutrient content.
c. **Selection of Target Crops**
   The choice of crops for biofortification depends on the dietary habits and nutritional needs of the target population. Common biofortified crops include rice, wheat, maize, beans, sweet potatoes, and cassava.
d. **Bioavailability and Absorption**
   Besides increasing nutrient content, attention must be given to the bioavailability of these nutrients. Several variables, including plant genetics, food processing techniques, and soil conditions, can impact how well the body absorbs nutrients.
e. **Adaptation to Local Conditions**
   Biofortified crops need to be well-adapted to local environmental conditions and farming practices. This includes considerations for drought resistance, pest resistance, and suitability for various soil types.
f. **Consumer Acceptance and Palatability**
   Biofortified crops should be culturally acceptable, palatable, and visually appealing to ensure that they are adopted by the target population. Efforts to educate and create awareness among consumers about the nutritional benefits are crucial.
g. **Monitoring and Evaluation**
   Programs for biofortification must be regularly observed and evaluated to determine their effect on public health. This includes measuring changes in nutrient intake, addressing

any unintended consequences, and making necessary adjustments to the biofortification process.

h. **Collaboration and Partnerships**

Successful crop biofortification programs often involve collaboration between agricultural scientists, nutritionists, policymakers, and local communities. Partnerships with non-governmental organizations (NGOs) and international agencies can also contribute to the success of these initiatives.

i. **Regulatory Considerations**

Compliance with local and international regulations is critical, especially in the case of genetically modified crops. Adherence to safety standards and transparent communication about the biofortification process are essential.

j. **Long-Term Sustainability**

Sustainable biofortification programs should be designed to ensure long-term benefits. This includes considerations for the economic viability of farming practices, environmental impact, and community empowerment.

12.3 APPROACHES IN CROP BIOFORTIFICATION

The methodology for metalloid biofortification with Se, B, and Si involves several steps, including selection of appropriate plant species, optimization of growth conditions, and application of metalloid supplements. Here's a general outline of the methodology:

a. **Selection of Plant Species:**
   i. Choose plant species that are suitable for biofortification and are commonly consumed by humans or animals.
   ii. Consider the ability of the selected species to accumulate metalloids efficiently without adverse effects on growth and development.
b. **Optimization of Growth Conditions:**
   i. Conduct soil analysis to assess the baseline levels of Se, B, and Si, as well as other essential nutrients.
   ii. Adjust soil pH, organic matter content, and other soil properties to optimize plant uptake of metalloids.
   iii. To encourage healthy plant development and metalloid buildup, regulate environmental parameters like temperature, humidity, and light intensity.
c. **Application of Metalloid Supplements:**
   i. Select appropriate sources of Se, B, and Si for supplementation.
   ii. Apply metalloid supplements to the soil or foliage at the appropriate growth stages, considering factors such as plant uptake efficiency and potential environmental contamination.
   iii. Monitor metalloid concentrations in plant tissues and soil to ensure optimal levels for biofortification without exceeding toxic thresholds.
   iv. Consider the use of foliar sprays, seed treatments, or soil amendments for efficient delivery of metalloid supplements.
d. **Assessment of Metalloid Content and Bioavailability:**
   i. Use methods like colorimetric tests, atomic absorption spectroscopy, *etc.*, to analyze plant tissues for the presence of Si, B, and Se.
   ii. Evaluate the bioavailability of metalloids in biofortified crops by assessing their absorption, distribution, metabolism, and excretion in humans or animals.
   iii. Conduct feeding trials or nutritional studies to determine the impact of metalloid biofortification on human or animal health outcomes, such as nutrient status, disease risk, and physiological function.
e. **Field Trials and Scaling Up:**

i. Conduct field trials to evaluate the performance of biofortified crops under diverse environmental conditions and agricultural practices.
ii. Assess the agronomic traits, yield potential, and nutritional quality of biofortified crops compared to conventional varieties.
iii. Collaborate with farmers, researchers, and policymakers to scale up metalloid biofortification programs and promote the adoption of biofortified crops for improved nutrition and food security.

Throughout the methodology, it's essential to adhere to good agricultural practices, ensure environmental sustainability, and prioritize safety and regulatory compliance in the use of metalloid supplements for biofortification. Additionally, ongoing research and monitoring are necessary to optimize biofortification strategies and address potential challenges and limitations.

12.4 GENETIC/BREEDING BIOFORTIFICATION

Among the various nutrients targeted for biofortification, metalloids have gained recognition for their multifaceted role in enhancing both plant health and human nutrition. Elements that share chemical characteristics with both metals and non-metals are known as metalloids. Only a few numbers of metalloids, including Si, Se, and B, have been thoroughly investigated and are now known to have advantageous effects on plant growth and metabolism. Conventional breeding techniques generate various genetic variations in crops, aiming to achieve specific traits like elevated micronutrient levels through multiple generations of crossing different varieties; the goal is to develop biofortified crops with enhanced nutrient content alongside other desirable characteristics. Breeding biofortification *via* fertilization is a secure enrichment technique for raising Se levels in food for humans and animals because there is very little danger of overdosage from plant-based sources.

A viable option for addressing the issue of Se insufficiency in food has been made possible by genetic variances in the intake of Se by different species, including varieties of the same species. Genetic biofortification is a handy way to build structured breeding programs, as an alternative to agronomic techniques might be to have enough genetic variations between crop species for Se enrichment (Lyons et al., 2005).

When Lyons et al. (2005) examined the Se concentrations of wild wheat, Australian and Mexican landraces, and the progenitors of different commercial varieties, they discovered no discernible variations in the Se of commercial wheat varieties. Researchers have discovered that diploid wheat grain contains more Se than contemporary wheat. This finding suggests the alluring possibility of specific breeding initiatives for the biofortification of Se. Variations in genotype might not be much in low-Se soils; however, they might be significant in high-Se soils. According to Garvin et al. (2006), research studies carried out in the United States on different types of red winter wheat revealed notable differences in grain Se as a result of increased influences from environmental variables. Moreover, variations in Se improvement have been noted in soybean (Yang et al., 2003), onion (Kopsell and Randle, 1997), and tomato (Pezzarossa et al., 1999).

Si breeding biofortification can be achieved by identifying and selecting genotypes with high Si uptake and transport capacity and incorporating these traits into elite varieties through conventional breeding. The development of Si-biofortified rice and wheat varieties has been achieved through conventional breeding by crossing cultivars with high Si uptake and transport capacity with elite varieties. Si breeding biofortification can increase crops' resilience to different stresses and enhance their food value.

However, further research is needed to optimize the breeding strategies for Si biofortification and to evaluate their effects on the productivity and quality of crops (Garg et al., 2018).

One example of Si genetic biofortification is the development of transgenic rice plants with enhanced Si uptake and transport. This has been achieved by overexpressing the Si transporter gene OsLsi1, which is responsible for Si uptake in rice roots. The transgenic rice plants showed increased Si concentration in the shoots and grains, which resulted in improved grain yield and quality. Another example of Si genetic biofortification is the development of transgenic wheat by overexpressing the TaLsi1 gene, which is responsible for Si uptake in wheat roots. The transgenic wheat plants showed increased Si concentration in the shoots and grains, which resulted in improved grain yield and quality (Ofori et al., 2022).

Boron genetic biofortification is a promising approach to address B deficiencies in crops. GMO techniques can be used to identify and manipulate genes that control B uptake and accumulation in plants, providing a sustainable and cost-effective solution to Boron deficiency. However, there is limited research on B genetic biofortification, and more research is required to completely grasp this strategy's potential.

12.5 MICROBIOLOGICAL BIOFORTIFICATION

Microbial biopropagation aims to enhance the soil microbiome to guarantee that the soil contains the micronutrients and nutrients required by plants. With the application of microbial biofertilizers, edible crops with higher micronutrient concentrations will yield more crops in soils with lower fertility. Many studies have shown that in the rhizosphere plant-promoting rhizobacteria have an impact on the growth and development of plants (Barret et al., 2011; Jog et al., 2016; Martínez-Viveros et al., 2010; Yang et al., 2009). It is well known that plants use a variety of signaling pathways to influence yield and biomass production. It has been reported that plants inoculated with plant growth-promoting rhizobacteria (PGPR) can produce IAA and activate the 1-Aminocyclopropane-1-Carboxylate (ACC) deaminase enzyme, which increases plant biomass and yield in response to wind heat, drought, salinity, and heavy metals (Arshad et al., 2007; Barret et al., 2011; Yang et al., 2009; Zhuang et al., 2007).

Lack of Se, a trace element that is vital to human health, is a major global health issue. The chemical form and bioavailability of Se in soil can be altered by microorganisms like fungi and bacteria, which encourage the uptake and accumulation of Se in crops (Schiavon et al., 2020). Its tolerance and the ability to accumulate Se were demonstrated by various bacteria, which were isolated from soils low in Se. According to the investigation, tissue Se buildup was increased when wheat was infused with Se-tolerant bacteria obtained from soils poor in Se. According to the studies, compared to conventional methods, induced microbial aid and agronomic uses of Se can increase Se concentration in eatable parts of plants (Yang et al., 2021).

According to Lyons et al. (2003), cereals, particularly wheat, are an excellent source of bioavailable Se, seleno methionine, which is the predominant organic form of Se in grains. According to Broadley et al. (2006), there is a greater correlation between the ideal Se status in a person's diet and the Se concentration of wheat grains. Similarly, a study by Durán et al. (2015) evaluated the relationship between Se-containing bacteria and arbuscular mycorrhizal fungi (AMF) in Se uptake and transport in wheat; these plants were inoculated with both *G. claroideum* and Se-containing bacteria, which greatly improved grain Se concentration.

Boron microbial biofortification is a process that involves the use of B-tolerant bacteria to enhance plant growth and nutrient uptake, particularly in B-deficient soils. One study found that seed inoculation with B-tolerant bacteria as well as B seed coating improved grain biofortification and yield of chickpeas and rice (Hayat et al., 2023). The interaction between B seed coating and B-resistant bacteria with a treatment of biochar and inoculation of seed increased yield contributing traits as compared to without coating of seed. B application, combined with B-tolerant bacteria, enhanced plant growth, nodulation, and crop production, such as in chickpeas and wheat (Mehboob et al., 2021). These studies highlight the potential of microbial interventions to enhance the biofortification of essential nutrients like B in crops.

Microbial biofortification has been achieved in forage crops such as cowpea and sorghum, with B playing a crucial role in mechanical resilience and strengthening the cell membrane (Dhaliwal et al., 2022). For plants to grow and develop healthily, Si acts as a necessary element. It also helps them become more resilient to abiotic and biotic stresses (Verma et al., 2022). Si is absorbed by roots, transported by plants, and accumulated in the cell walls and intercellular spaces in the form of Si dioxide phytoliths. However, the efficiency of Si accumulation varies greatly among plant species, with some species being less affected by Si application than others. The interaction of Si and PGPRs can stimulate crop enhancement and improve plant adaptation to adverse environmental conditions (Verma et al., 2022). Si can control the presence of many things in soil by competing with soil materials, improving the use of nutrients, and promoting plant growth and development. Si can regulate the availability of various elements in soils by competing for binding on soil particles, and it can enhance nutrient use efficiency and promote plant growth and development. The use of microorganisms to help crops in the translocation of essential nutrients, such as Si, is a promising option for biofortification. Endophytic microorganisms, in particular, are considered more promising agents for enhancing nutrient uptake and translocation (Singh et al., 2020).

The advantageous metalloids guarantee optimal plant growth, development, and productivity by acting as vital micronutrients. Metalloids efficiently control the activity of numerous enzymes and metabolic pathways when present at optimal amounts. In addition to controlling the structural and functional integrity of cellular membranes, these metalloids play a crucial role in several biosynthetic pathways, including those involving proteins, chloroplasts, and nucleic acids (Adrees et al., 2015). Researchers Wagner (1993) found that high levels of metalloids negatively affect plant growth, photosynthesis, accumulation, and changes in essential elements. Therefore, in the last 20 years, extensive research on metalloid accumulation and prevention has been carried out worldwide to evaluate the effects of metalloids on the environment, humans, animals, and plants (Adrees et al., 2015).

12.6 AGRONOMIC BIOFORTIFICATION

Increasing the nutrient, vitamin, and mineral density of crops is called agricultural biofortification. This method is used in India to address nutrient deficiencies in crops and increase micronutrients in plants grown in soil deficient in certain micronutrients or minerals. According to Kauffman et al. (2007), biostimulants include chemicals that are not fertilizers or granules and that affect the regulation of various metabolic processes such as cell division, photosynthesis, and ion uptake. Boron has an important role in regulating the composition and function of membranes of many plant species. This biostimulant interacts with contracting cells, gradually reducing the effects of different stress factors and promoting healthy growth (Brown and Saa, 2015).

Wollastonite is a calcium (Ca) silicate obtained from slag residues from P production in industries and is widely used as a sealant for crops grown in soils containing less silica content. It is one of the most widely used silicas in agricultural soils, but its use as fertilizer may be limited due to its high cost (Haynes, 2014; Wang et al., 2004). Two other minerals that can be used as Silicon (Si) sources in agriculture, with Si contents of approximately 20% and 38%–42% by weight, respectively, are olivine ($MgSiO_3$) and diatomaceous earth (Haynes, 2014). The slag used as a Si source contains different amounts of silicates of Ca and Mg and acts like a lime material by releasing hydroxyl ions into the water.

In addition to inorganic Si products, biological residues from crops can be used as botanical Si sources for cultivated crops. However, these Si products require a lot of materials because Si release is slow in crops. Therefore, the soil solution will not be able to provide sufficient Si to the plant at that moment (Cristina et al., 2019). However, foliar application did not increase systemic resistance to *Podosphera xanthii* infections in cucumbers on leaf surfaces (Liang et al., 2007). Nanosilica materials are more effective than traditional Si materials in reducing stress when using a foliar spray, due to the smaller size of nanosilica particles and thus greater absorption of water by the leaf (Wang et al., 2017).

When biostimulants were used simultaneously with low concentrations of Se, fresh/dry biomass increased, but without biostimulants, fresh/dry biomass remained unchanged. Similar findings regarding the positive effects of low Se on plant products are frequently encountered in the literature (Cartes et al., 2010; Ishtiaq et al., 2023). The explanation for these results lies in the various mechanisms by which Se plays an important role, including reducing metal toxicity, increasing photosynthetic capacity, and maintaining cell structure and function (Khan et al., 2023). This was seen in the study, as treatment with 10 mg Se kg^{-1} soil resulted in a decrease in fresh/dry biomass and therefore in NDVI/NDRE and SPAD values. In addition to the application of biostimulants, the Se level in the tissues of plants receiving high amounts continues to increase, causing their weight and physical strength to decrease further, which shows the association of biostimulants with this result.

Boron (B) biofortification can be achieved through agronomic practices such as foliar sprays, seed priming, and soil treatments. Foliar application is considered the best way to increase crop micronutrient availability as the uptake of nutrients into leaves at an appropriate growth stage promotes changes in nutrients that benefit and prevent poor soil. Research shows that the combination of $ZnSO_4{\cdot}7H_2O$ (0.5%)+$FeSO_4{\cdot}7H_2O$ @ 0.5% and borax @ 0.1% can increase the yield of bean seeds, B, Zn, and Fe content, absorption, and increase industrial benefits. B biofortification can solve mineral deficiency problems while also providing sustainable reproductive services (Dhaliwal et al., 2023).

Si agricultural biofortification is a process that involves applying Si to crops to increase the Si content in plants. This can be done through various methods such as the application of solution on foliage and root or a combination of both (Bhardwaj et al., 2022). Si application has been shown to increase nitrogen and phosphorus (P) use efficiency and improve grain biofortification. Agronomic biofortification of crops with Si is effective in enriching plants and their consumable parts with the element (Bhardwaj et al., 2022).

This approach has been applied to various crops such as wheat, quinoa, and rice with good results (Lata-Tenesaca et al., 2022). For example, the use of Si in quinoa cultivation has been shown to promote biofortification, increase ascorbic acid content and dry matter, and reduce post-harvest losses (Lata-Tenesaca et al., 2022). The use of Si in agriculture can also help reduce the negative effects of other elements, such as phosphorus toxicity in red eggplant (Dhaliwal et al., 2022). Studies have also shown that silica application methods affect nutrient uptake by maize plants.

12.7 ROLE OF METALLOIDS IN BOTH PLANT GROWTH AND HUMAN/ANIMAL HEALTH

12.7.1 Selenium

12.7.1.1 Impact on Plants' Growth and Development

Se is a necessary mineral element that both humans and animals get from their food. One conceivable method to increase the amount of Se in food is biofortification. Increasing the variety of foods that are naturally high in Se through plant breeding, genetic and agronomic biofortification, or giving humans or livestock direct supplements of organic Se are some ways to enhance the amount of Se dose. (Huang et al., 2020; Lyons et al., 2005). Due to food crops' and animal feeds' inability to extract enough Se from the soils in which they are grown, people who live in areas with low endemic levels of Se typically do not ingest enough of the mineral. Regarding plants, the soil concentration of Se determines the amount of Se in the plants (Figure 12.1).

Nearly everything on earth is home to the naturally occurring Se element. This metalloid has been deemed hazardous and poisonous to both human and animal health for several years. But recently, the importance of Se as a facet of vitamin essential for humans and animals has been determined (Fernández-Martínez, 2009). When it is identified as a component of thioredoxin reductases, GSH-Px, the Se-containing enzymes and proteins whose roles are unclear but which contribute to preserving the redox potential of cells, it is considered important (Rayman, 2019). As of right now, 25 selenoproteins related to Se as an enzyme cofactor have been identified in human metabolism (Rayman, 2019). Accordingly, a lack of Se raises the risk of oxidative damage and major co-morbidities like HIV and cancer (Facompre and El-Bayoumy, 2009; Méplan and Hesketh, 2012; Rayman, 2019).

Although selenoproteins have been found in bacteria and animals, with many of them having their physiological purpose established, attempts to identify these proteins in plants have not been successful in experiments, therefore the question is still open. However, several studies have demonstrated that Se enhances several physiological parameters, promotes tolerance to biotic and abiotic

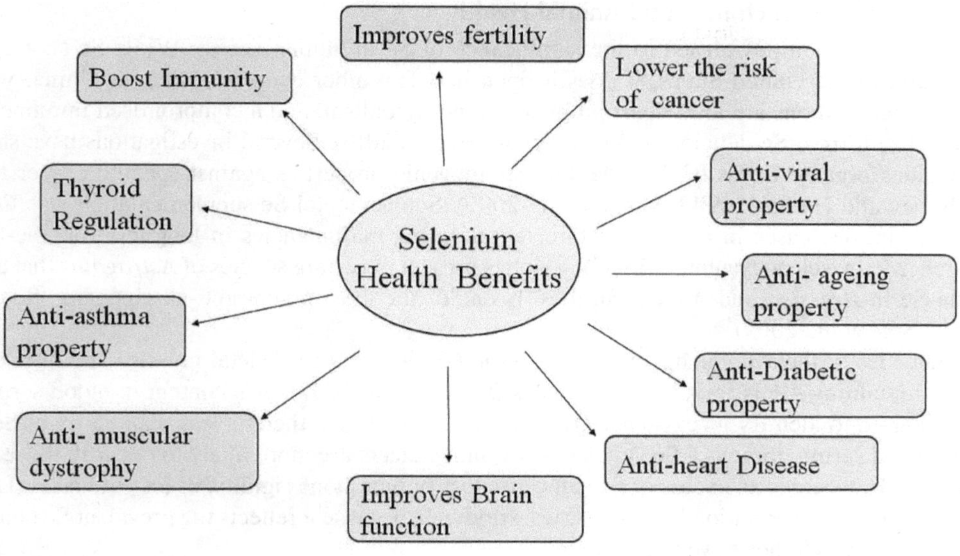

FIGURE 12.1 Benefits of Selenium.

stress, and enhances plant growth (Pilon-Smits and Quinn, 2010). This is intriguing since the soil-plant system is main source of Se nutrition for us, and a mineral imbalance can result in food that is Se deficient, which can have negative effects on our health (Govasmark and Salbu, 2011).

In most nations worldwide, plant foods represent the main dietary source of Se (Rayman, 2008). The reason for the enormous attention given to the biofortification of this element could be explained by the understanding that the first phase of the food chain's entry for Se is the consumption of plants with desired, non-toxic Se levels. Applying Se to the field could be thought of as a relatively safe, secure method of achieving this goal in areas lacking in Se (Dinh et al., 2018; Wang et al., 2019).

According to Bañuelos et al. (2012), agronomic biofortification with Se fertilizers is extensively used in nations including the UK, US, Australia, New Zealand, and Finland. The agronomic Se biofortification is part of a National Farm Input Subsidy Programme in Africa (Chilimba et al., 2011) and successfully examined in maize boosted with Se. Similar tests of Se biofortification have been conducted on many other plant crop species, including pea (Poblaciones et al., 2014), wheat (Carucci et al., 2022; Hart et al., 2011; Hawkesford and Zhao, 2007; Lyons et al., 2005; Shinmachi et al., 2010; Wang et al., 2019), mushrooms (Bhatia et al., 2013), etc.

According to Mora et al. (2008), an effective way to strengthen against Al toxicity and to make it more available to plants at the same time in pastures is to enrich fertilizer applications with Se through coated seeds or foliar applications in crops.

As, P, and Al are examples of both toxic and necessary elements that must be present in the growth media for Se to have any positive effects on higher plants. Cartes et al. (2006) assessed the impact of elements S and Se present at the same time on the antioxidant system of perennial ryegrass. Elevating S levels decreased GSH-Px activity and increased lipid membrane degradation. Greater sulfate concentrations caused more S to be absorbed than Se, as previously discussed because Se is carried into the cell by the S transporter, which has a higher affinity for S. When P and lime were applied, the enzymes APX and POD were activated, while the level of Se concentrations in a shoot for peroxidation of lipid decreased. Mora et al. (2008) studied effects on white clover when treated with selenite on acidity of soil and fertilization of P. These findings unambiguously show that various mixes of P fertilization and soil acidity may affect how much Se plants absorb and may be in charge of the favorable or unfavorable effects on peroxidation of lipids, POD, and APX activity.

12.7.1.2 Impact on Human and Animal Health

With time, attention has turned to the significance of Se to human health. While Se shortage is uncommon in the United States, it does happen in a few other countries, such as China, where soil Se concentrations are low. Heart problems, hypothyroidism, and a compromised immune system can result from Se deficiency (Weekley and Harris, 2013). Several investigations have shown that various organic forms of Se have anti-carcinogenic properties against specific cancer types (Gladyshev and Hatfield, 1999; Micke et al., 2009). Supplemental Se supplementation was linked to significant decreases in prostate, colorectal, and lung malignancies in long-term double-blind research. Methylselenocysteine (MSeC), which is present in certain species of *Astragalus* that accumulate Se in *Brassica* and *Allium* families, is one of the most potent anti-carcinogenic Se compounds (Sors et al., 2009).

About 47% of the average human's 3–5 mg total Se is found in skeletal muscles and fiber cells, and the remaining 4% is found in the kidneys (Lyons et al., 2007). Se's content in blood serum is typically used to identify its existence (Kipp et al., 2015). When there is less than 85 μg of Se per liter of blood serum, there is a Se deficiency. Certain cancers are more likely to occur in those with low Se levels, whereas an excess of Se can cause loss of hair, bone rigidity, or loss of vision (Lyons et al., 2007). Se concentration in foods varies worldwide because it reflects the prevalent Se quantity of the soil on which they have been grown.

The health effects of Se are harmful, whether it is present in excess or insufficiently. Higher dosages of Se can have hazardous effects. The safe and appropriate intake of Se and overconsumption that results in toxicity are separated by a small margin. The fact that commonly consumed foods don't accumulate Se may be the reason for the paucity of data on Se toxicity in humans. Signs of Se toxicity include loss of hair, lesions on the skin and nails, low blood pressure, fatigue, gastrointestinal distress, and discomfort in the muscles (Navarro-Alarcon, 2008). Acute Se poisoning, in severe circumstances, result in fatalities, heart attacks, kidney failure, and major digestive and neurological issues (Vinceti et al., 2018). Prolonged Se insufficiency impairs cardiovascular health and may be a contributing factor to heart attack (Shahid, 2018). Keshan and Kashin-Beck disorders are also brought on by a shortage in Se and mostly affect women who are pregnant as well as children who reside in low-Se areas (Oropeza-Moe, 2015). Moderate Se deficit in the daily diet lowers immunity, can damage the neurological system, and can result in congenital hypothyroidism in developing fetuses (Steinbrenner and Sies, 2009). Se might aid in the suppression of HIV and delay the onset of AIDS (Lipinski, 2005).

12.7.2 Silicon

Approximately 25% of the crust and second highest abundance on the earth is Si. Among the elements, it is most beneficial to humans. Given its presence in the earth's crust, it can build up in a wide variety of plants in quantities comparable to macronutrients and at high concentrations. Since Si is a quasi-essential or non-essential element for plant growth and development, it has been discovered to provide several advantages, especially in stressful environments in plants.

In some crops, Si has been shown to boost resistance to fungal pathogen assault, prolong wilting, strengthen stems, and improve drought tolerance. Moreover, it has been discovered to reduce transpiration and boost drought resilience, which may enhance the yield of floriculture crops. Under stress, certain plants appear to gain from Si. It has been demonstrated to strengthen resistance to drought and delay withering in some crops when irrigation is stopped and may strengthen the plant's resistance to micronutrients and other metal toxicities (such as those containing Al, Cu, Fe, Mn, Zn, *etc.*). Furthermore, Si has been shown to aid in boosting stem strength. For instance, studies have shown that Si-deficient rice and wheat cause their stems to weaken and collapse readily in wind or rain- a characteristic known as lodging. It has been discovered that Si strengthens a few plant species' defenses against fungal infections. In Si-treated zinnia, rose, sunflower, and phytophthora cases, the disease attack was postponed.

12.7.2.1 Impact on Plant Growth and Development

Supplementation of Si may help with growth improvement in greenhouse production, where most plants are cultivated on soilless surfaces with low Si concentration. According to a study, Si supplementation raised sunflower stem diameter and enhanced gerbera flower quality.

There are benefits in utilizing Si with some agricultural products (rice, wheat, sugarcane, *etc.*), especially if they are produced in low-quality soils, however, studies on greenhouse crops are scarce. For the crops that were examined, Si treatment for accumulator and intermediary plants could be beneficial. There is contradictory evidence regarding the benefits of non-accumulators. For instance, when fed more Si, the non-accumulator tomato showed enhanced flowering and fruit set. Studies on nutrient toxicity and powdery mildew, however, revealed that Si treatment of non-accumulators did not significantly decrease either.

According to one study, adding Si might shield leaves' photosynthetic pigments, lessen the harm done to the chloroplast, and increase the photosynthesis-related gene expression in rice subjected to severe zinc stress. The study also discovered that stomatal restriction was responsible for the decrease in photosynthesis in zinc-stressed rice and that in response to severe zinc stress, Si activated and controlled certain genes associated with photosynthesis, hence enhancing the photosynthesis process (Song et al., 2014).

In another study, it was shown that under drought and ultraviolet-B (UV-B) stress, Si supplementation enhanced the concentration and fluorescence of chlorophyll, rate of transpiration, and conductance in stomata and intercellular CO_2 concentration in soybean seedlings. Additionally, the study discovered that Si may be able to lessen the stress on seedlings (Shen et al., 2010).

In a review paper that examined the interactions between Si and photosynthesis during stress and signaling, it was discovered that Si efficiently shields the photosynthetic apparatus by preserving photosynthetic pigments, reducing damage to the ultrastructure of chloroplasts, and boosting the expression of genes related to photosynthesis in stressed plants (Rastogi et al., 2021). Several key macro- and micronutrients are crucial for plant development and metabolism. Si is not regarded by many plant scientists as a necessary nutrient for plants. Nonetheless, a wealth of data has emerged in recent years that points to Si's critical function in regulating plant development and quality. Si plays a key role in controlling vital plant biological activities like transpiration and photosynthesis in addition to helping plants adapt to harsh environmental factors like temperature, drought, and nutrient shortage. In many plant species, silica supplementation has been shown to have favorable impacts on plant growth and development.

The same study discovered that the amount of Si that accumulated in various plant tissues varied among species, indicating that there are species-specific differences in the tissues' and plant species' preferences for Si uptake and localization, respectively. According to Hasanuzzaman et al. (2017), plants grown on soils enriched with Si that had higher concentrations of cadmium showed fewer ROS species than control plants, indicating a role for Si in antioxidant defense mechanisms.

According to Kanto et al. (2009), the incidence of powdery mildew disease significantly lowered with an increase in the Si content of leaves. In a similar vein, a deficit in Si was linked to poor development and vulnerability to powdery mildew in wheat and barley (Zeyen et al., 2002). According to research by Menzies et al. (1992), applying Si foliar sprays to cucumber, muskmelon, and grape plants can prevent powdery mildew disease. According to Datnoff (2005), using Si foliar sprays to turf grass can prevent several diseases. There have also been reports of Si supplementation reducing the effects of several chemical stressors, including salinity, nutritional imbalance, and metal toxicity. Si supplementation was proven to help plants with P deficiencies in rice and barley.

12.7.2.2 Impact on Animal and Human Health

It has been shown that Si, an essential trace element in animal diet, is crucial for connective tissue, especially for the formation of bones and cartilage. A shortage in Si can impact the synthesis of collagen and glycosaminoglycan, as Si is involved in the creation of the organic matrix in bone and cartilage. These cells' mitochondria also contain relatively large concentrations of Si, suggesting that

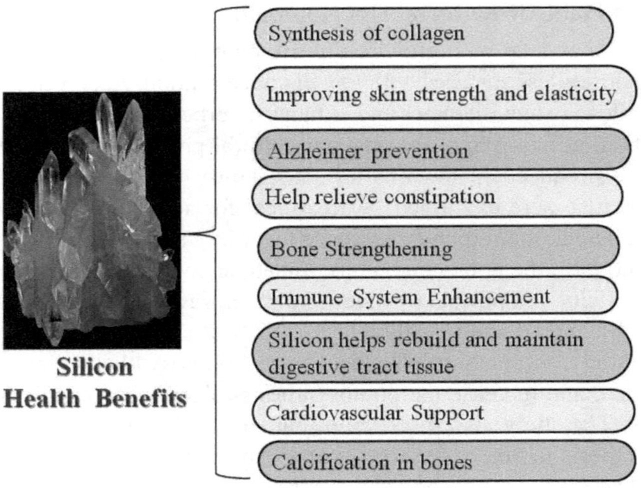

FIGURE 12.2 Health benefits of Silicon.

it may play a role in subcellular structures that house enzymes (Carlisle, 2007). The recommended daily intake of Si for humans has not been established as a daily minimum requirement. However, it has been estimated to be 10–25 mg day^{-1} based on 24-hour urinary excretion (Figure 12.2).

The mean daily consumption of Si has been determined by the European Food Safety Authority (EFSA) to be 20–50 mg or 0.3–0.8 mg kg^{-1} of body weight (Boqué et al., 2021). The German Federal Institute for Risk Assessment (BfR) estimated the advised threshold for everyday consumption of Si is 760 mg at a body weight of 60 kg. It is important to note that the exact mechanism of gastrointestinal Si uptake is still unknown, and the Si source matters for absorption and retention.

12.7.2.3 Silicon and Bone

Due to its capacity to affect bone development indicators and bone mineral density (BMD), Si is essential for maintaining healthy human bones. Research has connected higher BMD in people with Si intake, indicating a favorable correlation between dietary Si intake and ideal bone health. It has been demonstrated that Si has a role in the manufacture and/or stabilization of collagen, especially in the extracellular matrix, which is crucial for the health of bones and connective tissues (Jugdaohsingh et al., 2015). Greater Si concentrations are linked to strong connective tissues, including bone, and the proteins that make them up, such as collagen and elastin. Furthermore, it has been discovered that Si supplementation influences mineral metabolism, which may enhance animal bone strength and indicators of bone formation. Nevertheless, there is still much to learn about the precise process of gastrointestinal Si absorption and how it interacts with other minerals. Although some studies found that Si supplementation did not significantly increase bone density, shape, or breaking strength, other parameters that were shown to be possibly more relevant in the development of bone quality included growth rate and mechanical stress (Scholey et al., 2018).

12.7.2.4 Skin, Hair, and Nail Health

Si is an essential mineral that has a major impact on nail, hair, and skin wellness. Collagen is essential to increase the elasticity and strength of skin, and Si is needed to synthesize it. To improve the elasticity and strength of skin, collagen is required, and for its synthesis, Si is needed. In terms of hair, greater Si concentration leads to brighter hair and a slower rate of hair loss (Araujo et al., 2016). Si is the most common mineral in nails; its presence also has an impact on them. The bones of humans contain silica, a substance that occurs in nature. Rapidly growing cells like hair, nails, and skin rely particularly on Si for durability, adaptability, and generation of collagen. It's crucial for stimulating hydroxylation enzymes, promoting collagen synthesis, enhancing the texture of the skin, stopping hair thinning, fortifying nails, and enhancing the color of the nails, hair, and skin in both men and women. Glycosaminoglycans (GAGs) are a specific protein in our bodies that is capable of holding as much as 1,000 times their weight in water. Silica stimulates and binds to GAGs to help promote skin hydration. It also supports the growth and health of hair and nails (Barel et al., 2006).

12.7.2.5 Cardiovascular Health

Si may support cardiovascular health by maintaining the integrity of blood vessel walls and reducing arterial stiffness, thus potentially lowering the risk of cardiovascular diseases. Atherosclerosis is a chronic inflammatory disease of the arterial wall and is the most common cause of death in developed countries due to its complications, such as heart attack and cerebral stroke. According to research, Si may have a role in maintaining the healthy function of the central nervous system, skeletal system, and cardiovascular system. This possibility of Si mitigating atherosclerosis dates back to the 1970s. It appears that soluble Si compounds found in food and water have significant biological implications. Research also indicates that Si may protect against atherosclerosis by lowering the amounts of Mono-unsaturated fatty acid (MUFA) and Polyunsaturated fatty acid (PUFA) in the arterial wall (Dudek et al., 2024).

12.7.2.6 Reproductive Health

Si has been studied for its potential impact on reproductive health, with some studies suggesting that exposure to Si nanoparticles (SiNPs) may have negative effects on male reproductive health. A study published in the journal Food Chemical Toxicology found that inflammation and oxidative stress were brought on by SiNP exposure during puberty, leading to DNA damage and cell apoptosis in the testis and epididymis of mice (Sun et al., 2022). This resulted in decreased sperm concentration and motility and increased sperm abnormalities. Another study published in the journal Biological Trace Element Research found that when SiNPs were toxic to male brown rats, the effects were seen in biochemical analyses, hepato-renal function, ethological behavior changes, hematological disorders, and reproductive toxicity. Si has been explored for its potential role in male fertility, as it may influence sperm motility and viability (Robbins, 2012). However, there is also evidence that Si may have a role in promoting reproductive health. According to a published study, biomaterials containing Si may prove beneficial for treating bone-related disorders in female patients (Swingle et al., 2023).

In summary, the function of Si in human bone health is complicated and not entirely understood. Although more study is required to fully understand its methods and effects on human bone health, it may influence BMD, indicators of bone formation, and perhaps collagen production and stabilization, all of which might contribute to optimal bone health.

12.7.3 Boron

B is probably necessary for both human and animal health, as it has a wide variety of physiological impacts on biological systems. Both people and animals depend on B, a vital trace element. Nevertheless, compared to plant growth, the precise processes and functions of B in human and animal health are less well understood. Many foods, such as leafy green vegetables, cereals, raisins, prunes, non-citrus fruits, and nuts, contain B (Krans, www.healthline.com). It is estimated that the average daily consumption of B in the population is 1.5–3 milligrams (mg). The recommended daily allowances (ULs) for B in kids between 1 and 3 years is 3 mg, for 4–8 years old is 6 mg, 11 mg for kids between 9 and 13 years old, 17 mg for teens ages 14–18, and 20 mg for adults ages 19 and older (Krans, www.healthline.com). It is significant to remember that when B levels rise over threshold amounts, B toxicity can affect both plants and animals, and that a key component in regulating B bioavailability is the pH of the soil.

12.7.3.1 Impact on Plants' Growth and Development

All plant development processes require B, and sufficient B nutrition is essential for both large crop yields and high-quality harvests. It is the most utilized metalloid in agriculture. Even in trace levels of 1 μg^{-1} to 1 mg g^{-1}, it is essential for healthy growth and development (Brdar-Jokanović, 2020).

B promotes fruit and seed growth, pollen tube elongation and germination, and flower production and retention. For example, a B deficit might result in partial pollination of maize or limit the number of pods that can form in soybeans (500/US Borax, https://agriculture.borax.com). The main role of B is in the synthesis of cell walls. Through the formation of ester linkages between the apiose residues of rhamnogalacturonan II (RGII) monomers, B helps in the construction of the cell wall (Brdar-Jokanović, 2020). B is necessary for the structural integrity of plant cell walls throughout growth, especially when it comes to stabilizing and reinforcing plants in harsh environments. Additionally, by improving adsorption on cell walls, reducing oxidative stress, and changing the subcellular distribution of hazardous metals, it lessens the toxicity of aluminum, cadmium, and vanadium (Kesawat et al., 2020; Zhang et al., 2024).

Additionally, in cell division, B is necessary for the formation of new leaves and buds as well as the actively developing parts of plants, such as the tips of their roots. The most common indicator of a B deficiency is a shift in the structure of the plants in these actively developing areas, which can cause problems like rosetting (stunting) of plants in the apical (higher) growth parts of legumes like lucerne and clovers (500/US Borax, https://agriculture.borax.com).

However, prior studies have demonstrated B's capacity to mitigate biotic and abiotic stressors, particularly in harsh environments. B may reduce the toxicities of several metals, most notably Al. Its buildup by lessening the demethylation inside the cell wall decreases the number of Al^{3+} ion-binding sites while encouraging the production of callose and thicker cell walls of pectin, which oversees Al-binding; this modification influences the uptake and elimination of Al^{3+} (Yan et al., 2018). Furthermore, the Nrat1 gene that codes for an Al transporter, is downregulated in the presence of B, which lowers the accumulation of Al in plant roots. By inducing the plant's antioxidant defense systems, B increases the chelation of vanadium (V) within the cell wall. B efficiently reduces V translocation to aerial tissues by molecularly upregulating the expression of genes involved in V transport and sequestration within the root cell vacuole (Shireen et al., 2021). Supplementing with B can mitigate the oxidative damage resulting from abiotic stress. For example, grafting watermelon onto pumpkin rootstock, which is renowned for its capacity to absorb B, enhanced the fruit's resistance to oxidative stress. Plant photosynthesis and respiration typically deteriorate when exposed to salt, drought, and high temperature stressors. B accumulation, on the other hand, can serve as a buffer, reducing plant damage by keeping cell wall elasticity and cation concentrations in check (García-Sánchez et al., 2020).

Increased B concentrations cause poor development of roots and shoots, disruptions to metabolic processes, marginal and tip necrosis in leaves, and decreased levels of lignin and chlorophyll in plants. B deficiency can impair Ca transport, proper cell wall formation, and cell division, resulting in problems like a "hollow-heart" in peanuts.

12.7.3.2 Impact on Human/Animal Health

In both humans and animals, B and its compounds are involved in the formation of bones, the preservation of cell membranes, the control of various enzyme activities, embryonic development, a steroid hormone, and the metabolism of carbohydrates and minerals. B has a wide range of physiological impacts on biological systems, making it an element that is probably necessary for both human and animal health. It contributes to bone health by prolonging the half lives of two important nutrients for bone health: vitamin D and estrogen. Additionally, B influences testosterone and estrogen levels and aids in the body's metabolism of important vitamins and minerals (Krans, www.healthline.com).

Results of various research papers suggest the positive influences of B in the bone metabolism of animals. In vitamin D-scarce chicks, dietary deficiencies of vitamin B inhibit bone formation and may result in abnormalities of the bones (Bai et al., 1996). The mineralization, growth, and development of bones are all significantly influenced by B (Hakki et al., 2010; Uysal et al., 2009). Together with Ca, magnesium (Mg), and vitamin D, B helps in the metabolism of bones (Devirian and Volpe, 2003). Deficit in vitamin D and addition of vitamin B to rats' diet enhance P and Ca uptake and retention while boosting the amount of Mg in the femur (Hegsted et al., 1991). Furthermore, B's biological action may influence the metabolism of steroid hormones like testosterone and 17-b-estradiol hormone. According to Armstrong et al. (2000), there is no responsive impact of B inclusion in any of the dependent characteristics in pigs. One of the important nutrients in an animal's diet is soybeans and the byproducts of soybean protein. Therefore, plant legumes with high B content, including soybeans and their protein byproducts, may be responsible for the observed results.

In addition to protecting bone integrity by halting the depletion of Ca and Mg, B is involved in the control of bodily minerals, including Ca and vitamin D. There have been sporadic findings since 1981 indicating that B may help prevent or treat arthritis. According to research on animals, B can prevent inflammatory reactions to substances injected to create the symptoms of arthritis (Pietrzkowski et al., 2014).

Consuming excessive amounts of B can harm the stomach, intestines, liver, kidney, and brain. It can also irritate the eyes, nose, and throat (B, https://wwwn.cdc.gov). In addition, dermatitis, baldness, anorexia, hypothermia, restlessness, fatigue, headaches, and indigestion can all be symptoms of boric acid poisoning. For healthy adults, the Food and Nutrition Board (FNB) sets B upper limit values based on levels linked to reproductive and developmental consequences.

12.8 FURTHER PROSPECTS FOR METTALOID BIOFORTIFICATION

Future research in metalloid biofortification can explore various avenues to address existing challenges and further enhance the efficacy and safety of biofortified crops. Some key areas for future research include:

Bioavailability Enhancement: Look into cutting-edge methods to increase metalloids' bioavailability in crops that have been biofortified. This might entail creating cutting-edge agronomic techniques that encourage the transformation of metalloids into more bioavailable forms for plant uptake, like soil amendments or biofertilizers.

Genetic Engineering: Investigate cutting-edge genetic engineering methods to maximize the distribution and accumulation of metalloid accumulation in crop plants. This could entail genetically altering important transporters, enzymes, or elements in the metabolism and translocation of metalloids within plants.

Crop Selection and Trait Improvement: Determine and describe crop species or varieties that exhibit improved accumulation, tolerance, and uptake of metalloids. Use cutting-edge breeding methods to create high-yielding biofortified crop varieties with better nutritional quality and metalloid content, such as genomic selection or marker-assisted selection.

Effects on Nutrition and Health: Carry out in-depth research to evaluate the effects on nutrition and health resulting from eating biofortified crops high in metalloids. Analyze the bioavailability, absorption, and metabolic fate of metalloids in humans to find out how well they work to fill in specific micronutrient gaps and enhance general health.

Environmental Sustainability: Examine how soil health, biodiversity, and ecosystem functioning are affected by metalloid biofortification. To ensure sustainable agricultural practices, evaluate the long-term impacts of biofortification techniques on soil chemistry, microbial communities, and nutrient cycling.

Regulatory Aspects: Take care of any obstacles in the form of laws and regulations that could prevent the use and sale of biofortified crops. Collaborate with regulatory organizations to develop precise protocols and safety requirements for metalloid biofortification, including the upper limit at which metalloid accumulation in food crops is permitted.

Consumer Acceptance and Behavior: Examine how consumers feel about foods that have been biofortified with metalloids, as well as their attitudes and preferences. To encourage the adoption of biofortified crops, conduct consumer research to identify the variables influencing acceptance and willingness to buy and consume these crops. Then, create focused communication strategies.

Global Programme Scaling and Impact Evaluation: Expand metalloid biofortification initiatives worldwide and evaluate the results in terms of nutrition, food security, and public health. Work together with governments, NGOs, and international organizations to carry out extensive biofortification programs in areas impacted by micronutrient deficiencies.

By focusing on these research priorities, scientists and decision-makers can make progress in the field of metalloid biofortification and help create long-term solutions that will enhance global food security and fight malnutrition.

12.9 CONCLUSION

Metalloids offer a promising approach to improving the nutritional content of crops, particularly in regions with health concerns. In contrast to other living organisms, plants possess more robust defense mechanisms against metalloid invasion and can control their metabolism to maintain their viability and ability to reproduce. Different substrate specifics, expression trends, and membrane localizations are exhibited by such transporters, which regulate the transport of the corresponding metalloids throughout the plant. The metalloid biofortification approach supports sustainable agriculture and safeguards the environment. However, careful consideration of health, environmental, and regulatory factors is necessary to ensure benefits outweigh the risks. It bridges the gap between human health and agricultural advancement. While metalloid biofortification holds promise for addressing nutritional deficiencies, it must be approached with careful consideration of safety factors. Rigorous testing, monitoring, education, and collaboration are essential components of successful and safe biofortification programs.

REFERENCES

500|U.S. Borax. Boost Crop Quality and Your Bottom Line with U.S. Borax|U.S. Borax. Accessed April 13, 2024. https://agriculture.borax.com/resources/agronomy-notes/boron-in-plant-nutrition/functions-of-boron-in-plant-nutrition

Adrees, Muhammad, Shafaqat Ali, Muhammad Rizwan, Muhammad Zia-ur-Rehman, Muhammad Ibrahim, Farhat Abbas, Mujahid Farid, Muhammad Farooq Qayyum, and Muhammad Kashif Irshad. Mechanisms of silicon-mediated alleviation of heavy metal toxicity in plants: a review. *Ecotoxicology and Environmental Safety* 119 (2015): 186–197. https://doi.org/10.1016/j.ecoenv.2015.05.011

Afshan, Sehar, Shafaqat Ali, Saima Aslam Bharwana, Muhammad Rizwan, Mujahid Farid, Farhat Abbas, Muhammad Ibrahim, Muhammad Aamer Mehmood, and Ghulam Hasan Abbasi. Citric acid enhances the phytoextraction of chromium, plant growth, and photosynthesis by alleviating the oxidative damages in Brassica napus L. *Environmental Science and Pollution Research* 22, no. 15 (2015): 11679–11689. https://doi.org/10.1007/s11356-015-4396-8

Araujo, Lidiane Advincula de, Flavia Addor, and Patrícia Maria Berardo Gonçalves Maia Campos. Use of silicon for skin and hair care: an approach of chemical forms available and efficacy. *Anais brasileiros de dermatologia* 91 (2016): 0331–0335.

Armstrong, Todd A., Jerry W. Spears, Thomas D. Crenshaw, and Forrest H. Nielsen. Boron supplementation of a semipurified diet for weanling pigs improves feed efficiency and bone strength characteristics and alters plasma lipid metabolites. *The Journal of Nutrition* 130, no. 10 (2000), 2575–2581. https://doi.org/10.1093/jn/130.10.2575

Arshad, Muhammad, Muhammad Saleem, and Sarfraz Hussain. Perspectives of bacterial ACC deaminase in phytoremediation. *Trends Biotechnology* 25, no. 8 (2007). https://doi.org/10.1016/J.TIBTECH.2007.05.005

Bai, Yisheng, and Curtiss D. Hunt. Dietary boron enhances efficacy of cholecalciferol in broiler chicks. *The Journal of Trace Elements in Experimental Medicine* 9, no. 3 (1996), 117–132. https://doi.org/10.1002/(SICI)1520-670X(1996)9:3<117::AID-JTRA4>3.0.CO;2-P

Bañuelos, G., Dhillon, K., & Walse, S. Agricultural practices to biofortify selenium in the food chain. *Journal of Agricultural and Food Chemistry* 60, no. 36 (2012): 8947–8962.

Barel, Andre, M. Calomme, A. Timchenko, K. De Paepe, N. Demeester, Vera Rogiers, Peter Clarys, and D. Vanden Berghe. Effect of oral intake of choline-stabilized orthosilicic acid on skin, nails and hair in women with photodamaged skin. *Archives in Dermatological Research* 297, no. 4 (2006): 147–153. https://doi.org/10.1007/S00403-005-0584-6

Barret, Matthieu, John P. Morrissey, and Fergal O'Gara. Functional genomics analysis of plant growth-promoting rhizobacterial traits involved in rhizosphere competence. *Biology and Fertility of Soils* 47, no. 7 (2011): 729–743. https://doi.org/10.1007/S00374-011-0605-X

Bhardwaj, Ajay Kumar, Sukirtee Chejara, Kapil Malik, Raj Kumar, Ashwani Kumar, and Rajender Kumar Yadav. Agronomic biofortification of food crops: an emerging opportunity for global food and nutritional security. *Frontiers in Plant Science* 13 (2022). https://doi.org/10.3389/fpls.2022.1055278

Bhatia, Poonam, Federica Aureli, and Marilena D'Amato. Selenium in mushrooms. *Mushroom Science* 19 (2013): 531–536.

Boqué, Noemi, Rosa-Maria Valls, Anna Pedret, F. Puiggrós, Lluis Arola, and Rosa Solà. Relative absorption of silicon from different formulations of dietary supplements: a pilot randomized, double-blind, crossover post-prandial study. *Scientific Reports* 11, no. 1 (2021): 16479. https://doi.org/10.1038/s41598-021-95220-2

Brdar-Jokanović, Milka. 2020. Boron toxicity and deficiency in agricultural plants. *International Journal of Molecular Sciences* 21, no. 4 (2020): 1424. https://doi.org/10.3390/ijms21041424

Broadley, Martin R., Philip J. White, Rosie J. Bryson, Mark C. Meacham, Helen C. Bowen, Sarah E. Johnson, Malcolm J. Hawkesford, Steve P. McGrath, Fang-Jie Zhao, Neil Breward, et al. 2006. Biofortification of UK food crops with selenium. *Proceedings of the Nutrition Society* 65 (2006): 169–181. https://doi.org/10.1079/PNS2006490

Brown, Patrick, and Sebastian Saa. Biostimulants in agriculture. *Frontiers in Plant Science* 6 (2015). https://doi.org/10.3389/fpls.2015.00671

Carlisle, Edith M. Silicon as an essential trace element in animal nutrition. *Ciba Foundation Symposium 121 - Silicon Biochemistry*, 121 (2007): 123–139. https://doi.org/10.1002/9780470513323.ch8

Cartes, Paula, Alejandra A. Jara, Ligia Pinilla, Anali Rosas, and Maria Luz Mora. Selenium improves the antioxidant ability against aluminium-induced oxidative stress in ryegrass roots. *Annals of Applied Biology* 156, no. 2 (2010): 297–307. https://doi.org/10.1111/j.1744–7348.2010.00387.x

Cartes, Paula, Carolina Shene, and Maria L. Mora. Selenium distribution in ryegrass and its antioxidant role as affected by sulfur fertilization. *Plant and Soil* 285 (2006): 187–195.

Carucci, Federica, Gustavo Moreno-Martín, Yolanda Madrid-Albarrán, Giuseppe Gatta, Pasquale De Vita, and Marcella M. Giuliani. Selenium agronomic biofortification of durum wheat fertilized with organic products: Se content and speciation in grain. *Agronomy* 12, no. 10 (2022): 2492. https://doi.org/10.3390/agronomy12102492

Chilimba, Allan D., Scott D. Young, Colin R. Black, Katie B. Rogerson, E. L. Ander, Michael J. Watts, Joachim Lammel, and Martin R. Broadley. Maize grain and soil surveys reveal suboptimal dietary selenium intake is widespread in Malawi. *Scientific Reports* 1, no. 1 (2011): 72. https://doi.org/10.1038/srep00072

Datnoff, Lawrence E. Silicon in the life and performance of turfgrass. *Applied Turfgrass Science* 2, no. 1 (2005): 1–6. https://doi.org/10.1094/ats-2005-0914-01-rv

Devirian, Tara A., and Stella L. Volpe. The physiological effects of dietary boron. *Critical Reviews in Food Science and Nutrition* 43, no. 2 (2003): 219–231. https://doi.org/10.1080/10408690390826491

Dhaliwal, Salwinder S., Vivek Sharma, Arvind K. Shukla, Manmeet Kaur, Janpriya Kaur, Vibha Verma, Prabhjot Singh, Viliam Barek, Ahmed Gaber, and Akbar Hossain. Biofortification of mungbean (Vigna radiata L. (Wilczek)) with boron, zinc and iron alters its grain yield and nutrition. *Scientific Reports* 13, no. 1 (2023): 3506. https://doi.org/10.1038/s41598-023-30539-6

Dhaliwal, Salwinder S., Vivek Sharma, Arvind K. Shukla, Vibha Verma, Manmeet Kaur, Yashbir S. Shivay, Shahida Nisar, et al. Biofortification—a frontier novel approach to enrich micronutrients in field crops to encounter the nutritional security. *Molecules* 27, no. 4 (2022): 1340. https://doi.org/10.3390/molecules27041340

Dinh, Quang T., Mengke Wang, Thi A. Tran, Fei Zhou, Dan Wang, Hui Zhai, Qin Peng, et al. Bioavailability of selenium in soil-plant system and a regulatory approach. *Critical Reviews in Environmental Science and Technology* 49, no. 6 (2018): 443–517. https://doi.org/10.1080/10643389.2018.1550987

Dudek, Łukasz, Wacław Kochman, and Ewelina Dziedzic. Silicon in prevention of atherosclerosis and other age-related diseases. *Frontiers in Cardiovascular Medicine* 11 (2024): 1370536. https://doi.org/10.3389/fcvm.2024.1370536

Durán, P., J. J. Acuña, L. Gianfreda, R. Azcón, V. Funes-Collado, and M. L. Mora. Endophytic selenobacteria as new inocula for selenium biofortification. *Applied Soil Ecology* 96 (2015): 319–326. https://doi.org/10.1016/j.apsoil.2015.08.016

Facompre, Nicole, and Karam El-Bayoumy. Potential stages for prostate cancer prevention with selenium: implications for cancer survivors. *Cancer Research* 69, no. 7 (2009): 2699–2703. https://doi.org/10.1158/0008–5472.can-08–4359

Fernández-Martínez, Alejandro, and Laurent Charlet. Selenium environmental cycling and bioavailability: a structural chemist point of view. *Reviews in Environmental Science and Bio/Technology* 8, no. 1 (2009), 81–110. https://doi.org/10.1007/s11157-009-9145-3

García-Sánchez, Francisco, Silvia Simón-Grao, Juan J. Martínez-Nicolás, Marina Alfosea-Simón, Chunguang Liu, C. Chatzissavvidis, Juan G. Pérez-Pérez, and José M. Cámara-Zapata. Multiple stresses occurring with boron toxicity and deficiency in plants. *Journal of Hazardous Materials* 397 (2020): 122713. https://doi.org/10.1016/j.jhazmat.2020.122713

Garg, Monika, Natasha Sharma, Saloni Sharma, Payal Kapoor, Aman Kumar, Venkatesh Chunduri, and Priya Arora. Biofortified crops generated by breeding, agronomy, and transgenic approaches are improving lives of millions of people around the world. *Frontiers in Nutrition* 5 (2018): 12. https://doi.org/10.3389/fnut.2018.00012

Garvin, David F., Ross M. Welch, and John W. Finley. Historical shifts in the seed mineral micronutrient concentration of US hard red winter wheat germplasm. *Journal of the Science of Food and Agriculture* 86, no. 13 (2006): 2213–2220. https://doi.org/10.1002/jsfa.2601

Gladyshev, Vadim N., and Dolph L. Hatfield. Selenocysteine-containing proteins in mammals. *Journal of Biomedical Science* 6, no. 3 (1999): 151–160. https://doi.org/10.1159/000025383

Govasmark, Espen, and Brit Salbu. Translocation and re-translocation of selenium taken up from nutrient solution during vegetative growth in spring wheat. *Journal of the Science of Food and Agriculture* 91, no. 8 (2011): 1367–1372. https://doi.org/10.1002/jsfa.4387

Hakki, Sema S., Buket S. Bozkurt, and Erdogan E. Hakki. Boron regulates mineralized tissue-associated proteins in osteoblasts (MC3T3-E1). *Journal of Trace Elements in Medicine and Biology* 24, no. 4 (2010): 243–250. https://doi.org/10.1016/j.jtemb.2010.03.003

Hart, D. J., S. J. Fairweather-Tait, M. R. Broadley, S. J. Dickinson, I. Foot, P. Knott, S. P. McGrath, et al. Selenium concentration and speciation in biofortified flour and bread: retention of selenium during grain biofortification, processing and production of Se-enriched food. *Food Chemistry* 126, no. 4 (2011): 1771–1778. https://doi.org/10.1016/j.foodchem.2010.12.079

Hasanuzzaman, Mirza, Kamrun Nahar, Taufika I. Anee, and Masayuki Fujita. Exogenous silicon attenuates cadmium-induced oxidative stress in Brassica napus L. by modulating AsA-GSH pathway and glyoxalase system. *Frontiers in Plant Science* 8 (2017): 1061. https://doi.org/10.3389/fpls.2017.01061

Hawkesford, Malcolm and Fang-Jie Zhao. Strategies for increasing the selenium content of wheat. *Journal of Cereal Science* 46, no.3 (2007): 282–292. https://doi.org/10.1016/j.jcs.2007.02.006

Hayat, Hafiz S., Atique-ur Rehman, Shahid Farooq, Muhammad Naveed, Hayssam M. Ali, and Mubshar Hussain. Boron seed coating combined with seed inoculation with boron tolerant bacteria (Bacillus sp. MN-54) and maize stalk biochar improved growth and productivity of maize (Zea mays L.) on saline soil. *Heliyon* 9, no. 11 (2023): e22075. https://doi.org/10.1016/j.heliyon.2023.e22075

Haynes, Richard J. A contemporary overview of silicon availability in agricultural soils. *Journal of Plant Nutrition and Soil Science* 177, no. 6 (2014): 831–844. https://doi.org/10.1002/jpln.201400202

Hegsted, Maren, Michael James Keenan, Frances Catherine Siver, and P. Wozniak. Effect of boron on vitamin D deficient rats. *Biological Trace Element Research* 28, no. 3 (1991), 243–255. https://doi.org/10.1007/bf02990471

Huang, Yanyu, Yanyu Huang, Yuanting Fu, Mengting Li, Dawei Jiang, Christopher J. Kutyreff, Jonathan W. Engle, Xiaoli Lan, Weibo Cai, and Tianfeng Chen. Chirality-driven transportation and oxidation prevention by chiral selenium nanoparticles. *Angewante Chemie (International ed. in English)* 59, no. 11 (2020): 4406–4414. https://doi.org/10.1002/ANIE.201910615

Ishtiaq, Muhammad, Muhammad Waqas Mazhar, Mehwish Maqbool, Tanveer Hussain, Syed Atiq Hussain, Ryan Casini, Ahmed M. Abd-ElGawad, and Hosam O. Elansary. Seed priming with the selenium nanoparticles maintains the redox status in the water stressed tomato plants by modulating the antioxidant defense enzymes. *Plants* 12, no. 7 (2023): 1556. https://doi.org/10.3390/plants12071556

Jog, R., G. Nareshkumar, and S. Rajkumar. Enhancing soil health and plant growth promotion by actinomycetes. In Subramaniam, Gopalakrishnan, Sathya Arumugam, and Vijayabharathi Rajendran eds., *Plant Growth Promoting Actinobacteria*, pp. 33–45. Springer, 2016. https://doi.org/10.1007/978-981-10–0707-1_3

Jugdaohsingh, Ravin, Abigail I. Watson, Liliana D. Pedro, and Jonathan J. Powell. The decrease in silicon concentration of the connective tissues with age in rats is a marker of connective tissue turnover. *Bone* 75 (2015): 40–48. https://doi.org/10.1016/j.bone.2015.02.004

Kanto, T., K. Matsuura, M. Yamada, T. Usami, and Y. Amemiya. UV-B radiation for control of strawberry powdery mildew. *Acta Horticulturae* 842, no. 842 (2009): 359–362. https://doi.org/10.17660/actahortic.2009.842.68

Kauffman, Gordon L., Daniel P. Kneivel, and Thomas L. Watschke. Effects of a biostimulant on the heat tolerance associated with photosynthetic capacity, membrane thermostability, and polyphenol production of perennial ryegrass. *Crop Science* 47, no. 1 (2007): 261–267. https://doi.org/10.2135/cropsci2006.03.0171

Kesawat, Mahipal S., S. M. Shivaraj, Dong K. Kim, Manu Kumar, Bum-Soo Hahn, and Rupesh Deshmukh. Metalloids and their role in the biological system. In Deshmukh, Rupesh, Durgesh K. Tripathi, and Gea Guerriero eds., *Metalloids in Plants,* pp. 1–17. John Wiley & Sons Ltd, 2020. https://doi.org/10.1002/9781119487210.ch1

Khan, Zesmin, Thorny C. Thounaojam, Devasish Chowdhury, and Hrishikesh Upadhyaya. The role of selenium and nano selenium on physiological responses in plant: a review. *Plant Growth Regulation* 100, no. 2 (2023): 409–433. https://doi.org/10.1007/s10725-023-00988-0

Khush, Gurdev S., Sichul Lee, Jung-Il Cho, and Jong-Seong Jeon. Biofortification of crops for reducing malnutrition. *Plant Biotechnology Reports* 6, no. 3 (2012): 195–202. https://doi.org/10.1007/s11816-012-0216-5

Kipp, A. P., D. Strohm, R. Brigelius-Flohé, L. Schomburg, A. Bechthold, E. Leschik-Bonnet, and H. Heseker. Revised reference values for selenium intake. *Journal of Trace Elements in Medicine and Biology* 32 (2015): 195–199. https://doi.org/10.1016/j.jtemb.2015.07.005

Kopsell, Dean A., and William M. Randle. Selenate concentration affects selenium and sulfur uptake and accumulation by 'Granex 33' onions. *Journal of the American Society for Horticultural Science* 122, no. 5 (1997): 721–726. https://doi.org/10.21273/jashs.122.5.721

Krans, Brian. Brains, Bones, and Boron. Healthline. Accessed April 13, 2024. https://www.healthline.com/health/brains-bones-boron

Lata-Tenesaca, Luis F., Renato De Mello Prado, Marisa De Cássia Piccolo, Dalila L. Da Silva, José L. Da Silva, and Gabriela E. Ajila-Celi. Forms of application of silicon in quinoa and benefits involved in the association between productivity with grain biofortification. *Scientific Reports* 12, no. 1 (2022): 12732. https://doi.org/10.1038/s41598-022-17181-4

Liang, Yongchao, Wanchun Sun, Yong-Guan Zhu, and Peter Christie. Mechanisms of silicon-mediated alleviation of abiotic stresses in higher plants: a review. *Environmental Pollution* 147, no. 2 (2007): 422–428. https://doi.org/10.1016/j.envpol.2006.06.008

Lipinski, P. Selenium in the natural environment and essentiality. *Polish Journal of Environmental Studies* 14, no. 1 (2005): 3–9.

Lyons, Graham H., Yusuf Genc, James C. Stangoulis, Lyndon T. Palmer, and Robin D. Graham. Selenium distribution in wheat grain, and the effect of postharvest processing on wheat selenium content. *Biological Trace Element Research* 103, no. 2 (2005): 155–168. https://doi.org/10.1385/bter:103:2:155

Lyons, Graham, James Stangoulis, and Robin Graham. High-selenium wheat: biofortification for better health. *Nutrition Research Reviews* 16, no. 01 (2003): 45.

Lyons, M. P., T. T. Papazyan, and P. F. Surai. Selenium in food chain and animal nutrition: lessons from nature -review-. *Asian-Australasian Journal of Animal Sciences* 20, no. 7 (2007): 1135–1155. https://doi.org/10.5713/ajas.2007.1135

Martínez-Viveros, O., M. A. Jorquera, D. E. Crowley, G. Gajardo, and M. L. Mora. Mechanisms and practical considerations involved in plant growth promotion by rhizobacteria. *Journal of soil science and plant nutrition* 10, no. 3 (2010). https://doi.org/10.4067/s0718-95162010000100006

Mehboob, Noman, Mubshar Hussain, Waqas A. Minhas, Tauqeer A. Yasir, Muhammad Naveed, Shahid Farooq, Saleh Alfarraj, and Ali T. Zuan. Soil-applied boron combined with boron-tolerant bacteria (Bacillus sp. MN54) improve root proliferation and nodulation, yield and agronomic grain biofortification of chickpea (Cicer arietinum L.). *Sustainability* 13, no. 17 (2021): 9811. https://doi.org/10.3390/su13179811

Menzies, Jim, Pat Bowen, David Ehret, and Anthony D. Glass. Foliar applications of potassium silicate reduce severity of powdery mildew on cucumber, muskmelon, and zucchini squash. *Journal of the American Society for Horticultural Science* 117, no. 6 (1992): 902–905. https://doi.org/10.21273/jashs.117.6.902

Méplan, Catherine, and John Hesketh. The influence of selenium and selenoprotein gene variants on colorectal cancer risk. *Mutagenesis* 27, no. 2 (2012): 177–186. https://doi.org/10.1093/mutage/ger058

Micke, Oliver, Lutz Schomburg, Jens Buentzel, Klaus Kisters, and Ralph Muecke. Selenium in oncology: from chemistry to clinics. *Molecules* 14, no. 10 (2009), 3975–3988. https://doi.org/10.3390/molecules14103975

Mora, María D., Ligia Pinilla, Analí Rosas, and Paula Cartes. Selenium uptake and its influence on the antioxidative system of white clover as affected by lime and phosphorus fertilization. *Plant and Soil* 303, no. 1–2 (2008): 139–149. https://doi.org/10.1007/s11104-007-9494-z

Nagajyoti, P. C., Lee, K. D and T. V. Sreekanth. Heavy metals, occurrence and toxicity for plants: a review. *Environmental Chemistry Letters* 8, no. 3 (2010): 199–216. https://doi.org/10.1007/s10311-010-0297-8

Navarro-Alarcon, Miguel, and Carmen Cabrera-Vique. Selenium in food and the human body: a review. *Science of the Total Environment* 400, no. 1–3 (2008): 115–141. https://doi.org/10.1016/j.scitotenv.2008.06.024

Ofori, Kelvin F., Sophia Antoniello, Marcia M. English, and Alberta N. Aryee. Improving nutrition through biofortification–a systematic review. *Frontiers in Nutrition* 9 (2022): 1043655. https://doi.org/10.3389/fnut.2022.1043655

Oropeza-Moe, Marianne, Helene Wisløff, and Aksel Bernhoft. Selenium deficiency associated porcine and human cardiomyopathies. *Journal of Trace Elements in Medicine and Biology* 31 (2015): 148–156. https://doi.org/10.1016/j.jtemb.2014.09.011

Pezzarossa, B., F. Malorgio, and P. Tonutti. Effects of selenium uptake by tomato plants on senescence, fruit ripening and ethylene evolution. In Kanellis, A. K., C. Chang, H. Klee, A. B. Bleecker, J. C. Pech, and D. Grierson eds., *Biology and Biotechnology of the Plant Hormone Ethylene II*, pp. 275–276. Kluwer Academic Publishers, 1999. https://doi.org/10.1007/978-94-011-4453-7_51

Pietrzkowski, Z, M. J. Phelan, R. Keller, C. Shu, R. Argumedo, T. Reyes-Izquierdo. Short-term efficacy of calcium fructoborate on subjects with knee discomfort: a comparative, double-blind, placebo-controlled clinical study. *Clinical Interventions in Aging* 9 (2014): 895–899. https://doi.org/10.2147/CIA.S64590. Erratum in: *Clinical Interventions in Aging* 16 (2021): 203. PMID: 24940052; PMCID: PMC4051624.

Pilon-Smits, Elizabeth A., and Colin F. Quinn. Selenium metabolism in plants. *Plant Cell Monographs* 17 (2010): 225–241. https://doi.org/10.1007/978-3-642-10613-2_10

Poblaciones, M. J., S. Rodrigo, O. Santamaría, Y. Chen, and S. P. McGrath. Agronomic selenium biofortification in Triticum durum under Mediterranean conditions: from grain to cooked pasta. *Food Chemistry* 146 (2014): 378–384. https://doi.org/10.1016/j.foodchem.2013.09.070

Rastogi, Anshu, Saurabh Yadav, Sajad Hussain, Sunita Kataria, Shokoofeh Hajihashemi, Pragati Kumari, Xinghong Yang, and Marian Brestic. Does silicon really matter for the photosynthetic machinery in plants…? *Plant Physiology and Biochemistry* 169 (2021): 40–48. https://doi.org/10.1016/j.plaphy.2021.11.004

Rayman, Margaret P. Food-chain selenium and human health: emphasis on intake. *British Journal of Nutrition* 100, no. 2 (2008): 254–268. https://doi.org/10.1017/s0007114508939830

Rayman, Margaret P. Selenium intake, status, and health: a complex relationship. *Hormones* 19, no. 1 (2019): 9–14. https://doi.org/10.1007/s42000-019-00125-5

Robbins, Wendie A., Lin Xun, Leah Z. FitzGerald, Samantha Esguerra, Susanne M. Henning, and Catherine L. Carpenter. Walnuts improve semen quality in men consuming a western-style diet: randomized control dietary intervention trial. *Biology of Reproduction* 87, no. 4 (2012): 101. https://doi.org/10.1095/biolreprod.112.101634

Schiavon, Michela, Serenella Nardi, Francesca Dalla Vecchia, and Andrea Ertani. Selenium biofortification in the 21st century: status and challenges for healthy human nutrition. *Plant and Soil* 453, no. 1–2 (2020): 245–270. https://doi.org/10.1007/s11104-020-04635-9

Scholey, Dawn V., D. J. Belton, Emily J. Burton, and Carole C. Perry. Bioavailability of a novel form of silicon supplement. *Scientific Reports* 8, no. 1 (2018): 17022. https://doi.org/10.1038/s41598-018-35292-9

Shahid, Siddique A. Nutritional supplements and serum selenium in the progression of myocardial infarction. *The American Journal of Cardiology* 78, no. 8 (2018): 1103–1105.

Shen, Xuefeng, Yuyi Zhou, Liusheng Duan, Zhaohu Li, A. E. Eneji, and Jianmin Li. Silicon effects on photosynthesis and antioxidant parameters of soybean seedlings under drought and ultraviolet-B radiation. *Journal of Plant Physiology* 167, no. 15 (2010): 1248–1252. https://doi.org/10.1016/j.jplph.2010.04.011

Shinmachi, Fumie, Peter Buchner, Jacqueline L. Stroud, Saroj Parmar, Fang-Jie Zhao, Steve P. McGrath, and Malcolm J. Hawkesford. Influence of sulfur deficiency on the expression of specific sulfate transporters and the distribution of sulfur, selenium, and molybdenum in wheat. *Plant Physiology* 153, no. 1 (2010): 327–336. https://doi.org/10.1104/pp.110.153759

Shireen, Fareeha, Muhammad A. Nawaz, Junyang Lu, Mu Xiong, Mohsin Kaleem, Yuan Huang, and Zhilong Bie. Application of boron reduces vanadium toxicity by altering the subcellular distribution of vanadium, enhancing boron uptake and enhancing the antioxidant defense system of watermelon. *Ecotoxicology and Environmental Safety* 226 (2021): 112828. https://doi.org/10.1016/j.ecoenv.2021.112828

Singh, Devendra, and Radha Prasanna. Potential of microbes in the biofortification of Zn and Fe in dietary food grains: a review. *Agronomy for Sustainable Development* 40, no. 2 (2020): 15. https://doi.org/10.1007/s13593-020-00619-2

Song, Alin, Ping Li, Fenliang Fan, Zhaojun Li, and Yongchao Liang. The effect of silicon on photosynthesis and expression of its relevant genes in rice (Oryza sativa L.) under high-zinc stress. *PLoS One* 9, no. 11 (2014): e113782. https://doi.org/10.1371/journal.pone.0113782

Sors, Thomas G., Catherine P. Martin, and David E. Salt. Characterization of selenocysteine methyltransferases from *Astragalus* species with contrasting selenium accumulation capacity. *The Plant Journal* 59, no. 1 (2009): 110–122. https://doi.org/10.1111/j.1365-313x.2009.03855.x

Steinbrenner, Holger, and Helmut Sies. Protection against reactive oxygen species by selenoproteins. *Biochimica et Biophysica Acta (BBA) - General Subjects* 1790, no. 11 (2009): 1478–1485. https://doi.org/10.1016/j.bbagen.2009.02.01

Sun, Fanli, Xuying Wang, Pinzheng Zhang, Ziyun Chen, Zhiyi Guo, and Xuan Shang. Reproductive toxicity investigation of silica nanoparticles in male pubertal mice. *Environmental Science and Pollution Research* 29, no. 24 (2022): 36640–36654. https://doi.org/10.1007/s11356-021-18215-6

Swingle, Kelsey L., Adele S. Ricciardi, William H. Peranteau, and Michael J. Mitchell. Delivery technologies for women's health applications. *Nature Reviews Bioengineering* 1, no. 6 (2023): 408–425. https://doi.org/10.1038/s44222-023-00040-w

Uysal, Tancan, Ayca Ustdal, Mehmet F. Sonmez, and Figen Ozturk. Stimulation of bone formation by dietary boron in an orthopedically expanded suture in rabbits. *The Angle Orthodontist* 79, no. 5 (2009): 984–990. https://doi.org/10.2319/112708-604.1

Verma, Krishan K., Xiu-Peng Song, Dong-Mei Li, Munna Singh, Jian-Ming Wu, Rajesh K. Singh, Anjney Sharma, Bao-Qing Zhang, and Yang-Rui Li. Silicon and soil microorganisms improve rhizospheric soil health with bacterial community, plant growth, performance and yield. *Plant Signaling & Behavior* 17, no. 1 (2022): 2104004. https://doi.org/10.1080/15592324.2022.2104004

Vinceti, Marco, Tommaso Filippini, and Kenneth J. Rothman. Selenium exposure and the risk of type 2 diabetes: a systematic review and meta-analysis. *European Journal of Epidemiology* 33, no. 9 (2018): 789–810. https://doi.org/10.1007/s10654-018-0422-8

Wagner, George J. Accumulation of cadmium in crop plants and its consequences to human health. *Advances in Agronomy* 51 (1993): 173–212. https://doi.org/10.1016/s0065-2113(08)60593-3

Wang, Hailong, Gujja N. Magesan, Mark O. Kimberley, Tim W. Payn, Peter J. Wilks, and Craig R. Fisher. Environmental and nutritional responses of a Pinus radiata plantation to biosolids application. *Plant and Soil* 267, no. 1–2 (2004): 255–262. https://doi.org/10.1007/s11104-005-0107-4

Wang, Min, Fayaz Ali, Mengke Wang, Quang T. Dinh, Fei Zhou, Gary S. Bañuelos, and Dongli Liang. Understanding boosting selenium accumulation in wheat (Triticum aestivum L.) following foliar selenium application at different stages, forms, and doses. *Environmental Science and Pollution Research* 27, no. 1 (2019): 717–728. https://doi.org/10.1007/s11356-019-06914-0

Wang, Min, Limin Gao, Suyue Dong, Yuming Sun, Qirong Shen, and Shiwei Guo. Role of silicon on plant–pathogen interactions. *Frontiers in Plant Science* 8 (2017): 701. https://doi.org/10.3389/fpls.2017.00701

Weekley, Claire M., and Hugh H. Harris. Which form is that? The importance of selenium speciation and metabolism in the prevention and treatment of disease. *Chemical Society Reviews* 42, no. 23 (2013): 8870. https://doi.org/10.1039/c3cs60272a

Yan, Lei, Muhammad Riaz, Xiuwen Wu, Chenqing Du, Yalin Liu, and Cuncang Jiang. Ameliorative effects of boron on aluminum induced variations of cell wall cellulose and pectin components in trifoliate orange (Poncirus trifoliate (L.) Raf.) rootstock. *Environmental Pollution* 240 (2018): 764–774. https://doi.org/10.1016/j.envpol.2018.05.022

Yang, Dandan, Chengxiao Hu, Xu Wang, Guangyu Shi, Yanfeng Li, Yuchen Fei, Yinran Song, and Xiaohu Zhao. Microbes: a potential tool for selenium biofortification. *Metallomics* 13, no. 10 (2021): mfab054. https://doi.org/10.1093/mtomcs/mfab054

Yang, Fangmei, Licheng Chen, Qiuhui Hu, and Genxing Pan. Effect of the application of selenium on selenium content of soybean and its products. *Biological Trace Element Research* 93, no. 1–3 (2003): 249–256. https://doi.org/10.1385/bter:93:1-3:249

Yang, Jungwook, Joseph W. Kloepper, and Choong-Min Ryu. Rhizosphere bacteria help plants tolerate abiotic stress. *Trends in Plant Science* 14, no. 1 (2009): 1–4. https://doi.org/10.1016/j.tplants.2008.10.004

Zeyen, R. J, W. M Kruger, M. F Lyngkjær, and T. L. W Carver. Differential effects of D-mannose and 2-deoxy-D-glucose on attempted powdery mildew fungal infection of inappropriate and appropriate G ramineae. *Physiological and Molecular Plant Pathology* 61, no. 6 (2002), 315–323. https://doi.org/10.1006/pmpp.2003.0444

Zhang, Lingxiao, Zhengyan Liu, Yun Song, Junkang Sui, and Xuewen Hua. Advances in the involvement of metals and metalloids in plant defense response to external stress. *Plants* 13, no. 2 (2024): 313. https://doi.org/10.3390/plants13020313

Zhuang, Xuliang, Jian Chen, Hojae Shim, and Zhihui Bai. New advances in plant growth-promoting rhizobacteria for bioremediation. *Environment International* 33, no. 3 (2007): 406–413. https://doi.org/10.1016/j.envint.2006.12.005

13 Phytohormones and Metalloids

Crosstalk for Plant Growth and Development

Shruti Kaushik, Alok Ranjan, Anmol Sidhu, Gurvarinder Kaur, Isha Madaan, Rashmi Mishra, Anil Kumar Singh, and Geetika Sirhindi

13.1 INTRODUCTION

Environmental stresses, specifically abiotic factors, have been noted to be significant possible barriers to maintaining the normal growth of plants and sustaining agricultural productivity. In the last few decades, metalloids have surpassed pesticides and prevalent pollutants like sulfur dioxide and carbon dioxide. It is predicted that metalloids may elevate to the status of the most hazardous contaminant, potentially exceeding both solid and nuclear waste (Gong et al. 2020). Metalloids are not required by plants for their growth and can even be detrimental to them (Bandara et al. 2020). These chemicals are frequently found in polluted soils, posing dangers to human health. Human activities like industry and agriculture are the most common sources of contamination, but natural processes such as rock weathering and volcanic eruptions can also contribute (Balali-Mood et al. 2021). The detrimental impact of metalloids on plants and other living organisms is determined by their ability to compete with naturally occurring ions crucial for key enzymes in primary and secondary metabolism (Chmielowska-Bak and Deckert 2021). When these metalloid ions interact with sulfhydryl groups, it causes disruptions in protein functions and increases the oxidative state of the plant. Additionally, they can substitute essential elements from cell walls and membranes (like Ca^{2+} and Mg^{2+}); for example, elements such as Cu, Cd, Al, Pb, and Zn have a greater affinity than Ca for binding to cell wall pectins (Rasafi et al. 2022). As a result, these interactions have a significant effect on germination, yield productivity, and growth patterns. This scenario raises concerns, and it has been projected that by 2050, the rising world population could suffer a catastrophic scarcity of staple foods (Yadav et al. 2022).

In recent years, environmental and biological experts have worked together to find potential remedies and alternative techniques to reducing the detrimental impacts of metalloid pollution. The pollution of agricultural soils with metalloids poses substantial environmental issues. One interesting method in this area is the use of phytohormones to regulate molecular and biochemical processes, allowing plants to heal contaminated soil (Rahman and Singh 2019). When plants encounter stress, the cells in roots can rapidly produce phytohormones, which can alter the structure of the roots, nutrient uptake, and the expression of metal transporters (Zhang et al. 2020). Thus, phytohormones such as cytokinins (CKs), jasmonates, abscisic acid (ABA), polyamines, ethylene (ET), brassinosteroids (BRs), and salicylic acid (SA) mitigate metalloid toxicity in plants and support their growth and development. Both exogenous treatments and alteration of the endogenous level of phytohormones by regulating their signaling/biosynthesis-related genes are effective strategies for protecting plants against metal-induced toxic effects. The application of phytohormones' environmentally

DOI: 10.1201/9781032673592-13

friendly strategies for endorsing heavy metal toxicity resistance in plants has attained significant importance in recent years. Phytohormones are essential for controlling the growth and development of plants. For instance, hormone application can lessen the accumulation of metals in plants (Verma et al. 2020). Thus, applying phytohormones may help plants grow in the face of abiotic stress, which will ultimately increase crop yield. Phytohormones are tiny secondary compounds in plants that play critical roles in managing plant growth, development, and defense against metalloid stress (Sytar et al. 2019). Classical plant hormones that control a plant's defensive reactions in the presence of environmental stress include auxin, ABA, gibberellic acid (GA), ET, CK, BRs, SA, and jasmonic acid (JA) (N. Li et al. 2021). These hormones can exert their effects either in proximity to or at a distance from their sites of synthesis, regulating responses to environmental stimuli. In such way, they play a crucial role in managing metalloid stress in plants by influencing cell division, nutrient processes, antioxidant systems, and the absorption of metalloid elements.

The goal of this chapter is to investigate the roles and processes of phytohormones in stimulating plant growth while reducing metalloid stress. The insights presented in this chapter may serve as a theoretical foundation and practical direction for minimizing metalloid-induced stress, allowing for the application of a variety of techniques to increase crop output and ensure food safety.

13.2 PHYTOHORMONES: A KEY REGULATOR OF PLANT GROWTH

Phytohormones, commonly known as plant growth regulators, are natural organic substances that occur in minute concentrations and exert their action either locally, at their site of synthesis, or be transported to other parts of the plant. These chemical messengers possess various chemical properties and specific structures that directly impact the overall growth and development of the plant through various biochemical processes (Figure 13.1). These growth regulators coordinate the plant's biotic and abiotic stress responses. Plant hormones are ubiquitous, extending across all higher and lower plant forms (Sabagh et al. 2021).

As the first plant hormone to be discovered, indole-3-acetic acid (IAA) is essential for controlling several growth-related processes in plants, including cell elongation, vascular tissue development, and apical dominance. IAA is well known for its impact on plant growth and development, but it also shows potential in controlling how plants respond to stress and coordinating growth in stressful environments (Jurado et al. 2010). Conversely, environmental fluctuations leading to alterations in auxin homeostasis can result in distorted growth and development, causing changes in plant morphogenesis. These stress-induced morphogenic responses assist to prevent or mitigate the negative effects of environmental stressors. Many studies have shown that stress causes endogenous changes in IAA levels in roots and shoots, demonstrating both positive and negative associations between stress and IAA levels. It has also been noted that the control of IAA-producing genes may regulate the local buildup of IAA in response to stress (Kapulnik et al. 2011). Exogenous application of IAA to a variety of plants, either directly or through plant growth-promoting rhizobacteria (PGPR) that produce IAA, has shown noteworthy improvements in plant growth (Jurado et al. 2010). Hence, it can be summarized that the optimum amount of IAA seems to be beneficial in regulating heavy metal-induced oxidative burst in several plant species.

CKs serve as crucial for several processes that are necessary for the growth and development of plants. These processes include vascular differentiation, anthocyanin synthesis, apical dominance, nutrient mobilization, cell division, leaf senescence, chloroplast biogenesis, and photo-morphogenic development (Danilova et al. 2016). CKs are adenine derivatives with aromatic or isoprenoid side chains that are found in plants in lower quantities and function as endogenous plant hormones. Isoprenoid CKs, including isopentenyl adenine (iP), trans-zeatin (tZ), cis-zeatin (cZ), and dihydrozeatin-type derivatives, are distinguished by the hydroxylation and reduction of the side chains. The rate-limiting step in isoprenoid CK biosynthesis is catalyzed by isopentenyl transferase (IPTs). CKs are produced in apical meristems and roots and then move to the aerial parts with the aid of absorbent substances like minerals through the xylem (S. M. Li et al. 2021). Based on these findings,

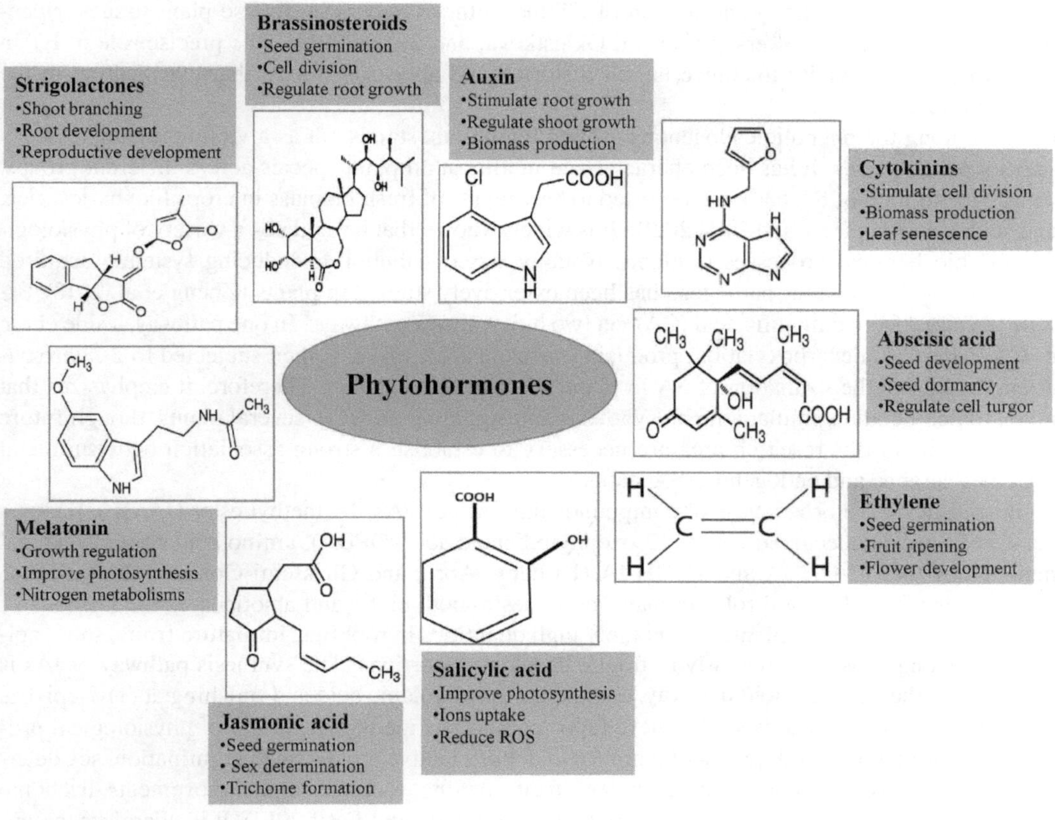

FIGURE 13.1 Multifaceted impact of phytohormones on plant growth and development.

it can be assumed that numerous CK-mediated interaction mechanisms may act simultaneously in a signal transduction pathway that is not well recognized. A better understanding of these mechanisms is important to improve our knowledge of the management of plant metal homeostasis.

ABA plays a crucial role in stress responses and adaptation, serving as an essential component at several stages of the plant life cycle, such as seed development, dormancy, and reactions to environmental stressors (Dar et al. 2017). ABA, referred to as the stress hormone, is the main internal signal that allows plants to survive under harsh environmental conditions. ABA is biosynthesized in the roots and transported through the xylem to the leaves, where it raises the concentration within the leaves (Arkhipova et al. 2020). This elevated ABA acts as a signal, initiating a signaling cascade in guard cells to regulate cell turgor. Notably, ABA interacts with other plant hormones to govern physiological responses as well as regulate them via signaling pathways. It is important to emphasize that ABA activates and mobilizes a number of biochemical defenses, such as proline production, antioxidants, enzymes that detoxify reactive oxygen species (ROS), heat shock proteins, and unsaturated fatty acids, as well as the strengthening of cuticular waxes (Danilova et al. 2016). However, the ABA-mediated heavy metal tolerance in plants is still unclear and needs further mechanical studies to be better understood.

As a gaseous hormone, ET plays a vital role in many morph-physiological processes, including triple responses in seed germination, fruit ripening, flower development, and instigating plant responses to a range of environmental stimuli (Dubois, Van den Broeck, and Inzé 2018). Additionally, ET regulates various stress-related biochemical responses in plants tackled with abiotic stresses like drought, chilling, heat, heavy metals, salinity, waterlogging, and flooding conditions. Moreover, it is essential in altering how plants react to pathogen invasions, mechanical wounds, UV radiation, and nutrient deficits. The quantification of ET biosynthesis occurs in diverse plant tissues, ripening fruits, and wilting flowers (Jiroutova, Oklestkova, and Strnad 2018). The precise role of ET in regulating metal toxicity tolerance is still distorted and appealing for further clarification in the coming future.

SA, among the phenolic endogenous growth regulators, stands out as a vital regulator found in various plant species. It has been characterized in almost all plant species across different groups. The concentration of SA has been reported to be 1 μg g^{-1} of fresh biomass in crops like barley, rice, and soybean (Koo, Heo, and Choi 2020). It is widely known that it regulates a variety of physiological and biochemical processes in plants. Notably, its contribution to inducing systemic acquired resistance against various pathogens has been extensively studied in plants (Zheng et al. 2018). SA is synthesized from cinnamic acid (CA) via two biosynthesis pathways. In one pathway, a side chain of CA undergoes decarboxylation, producing benzoic acid, which is then subjected to 2-hydroxylation, leading to the formation of SA in plants (Sharma et al. 2020). Therefore, it emphasizes that metal toxicity could stimulate the biosynthesis and signaling of SA in several plants, though future investigations in this research area are necessary to establish a strong association between metal toxicity tolerance and endogenous SA status.

Jasmonates are a broad class of compounds that include JAs, JA methyl ester (JAME), the precursor of JAs, octadecanoid cis (+) 12 oxophytodienoic acid (OPDA), amino acid conjugates, and metabolites like 12-OH-JA and 11-OH-JA (Farhangi-Abriz and Ghassemi-Golezani 2019). These compounds often play vital roles in plant responses to both biotic and abiotic stresses. Jasmonates are found throughout the plant, with notably high quantities in root tips, immature fruits, shoot apices, and young leaves, particularly in fragile developing portions. The synthesis pathway of JAs is initiated via the octadecanoid pathway, commencing at linolenic acid and finishing at (+)-7-epi-JAs (Huang et al. 2017). Various studies have reported that JAs mediate a variety of physiological processes, including the development of embryos and reproductive organs, seed germination, sex determination, root growth, seedling development, fruit ripening, senescence, leaf movements, trichome formation, gravitropism, and tuber development (Mohamed and Latif 2017). It is, therefore, essential to conduct an in-depth study in this area to elucidate the biochemical pathways that underlie JA-mediated metal toxicity tolerance in plants.

Melatonin (*N*-acetyl-5-methoxytryptamine) was first detected in plants in 1995 when two research groups simultaneously noticed its presence in vascular plants. Melatonin is a complex molecule that acts in a variety of ways on plants (Elsayed et al. 2020). It is an antioxidant that is essential for regulating harmful oxidative molecules found in plant cells, such as reactive nitrogen species (RNS) and ROS. Beyond its antioxidant properties, plant melatonin is involved in various physiological actions, including seed germination, growth regulation, rooting photosynthesis, and defense against both abiotic and biotic stressors. The latest recognition of the first plant melatonin receptor has unlocked the possibility of considering this regulatory molecule as a new plant hormone (Hamid et al. 2021). Furthermore, because of its diverse functions, melatonin has been proposed as a master regulator in plants. Melatonin can affect a variety of cellular and physiological activities, including membrane permeability via ion transporters, intracellular Ca^{2+}, glucose, lipid, stomatal opening and/or shutting, nitrogen metabolisms, and the synthesis of osmoprotectant metabolites. Its effects extend to processes like growth stimulation, tropism, seed germination promotion, photosynthetic efficiency optimization, root induction, and leaf water/CO_2 exchange. Apart from that, melatonin regulates the internal biological clock, senescence, ripening, and parthenocarpy. At last, melatonin functions as a natural plant biostimulator that protects against stresses (Arnao and Hernández-Ruiz 2015).

Strigolactones (SLs) are plant hormones derived from carotenoids and are produced by approximately 80% of plants that have a symbiotic relationship with arbuscular soil mycorrhiza. These hormones were first identified when strigol, a naturally occurring SL, stimulated the germination of *Striga lutea* (Brewer, Koltai, and Beveridge 2013). They were then identified as an "ecological signal" for parasitic seed germination and a component of the symbiotic relationship between plants and beneficial microbes. The collective term "strigolactone" refers to these compounds because these obligate hemiparasite plants need a living host for germination and initial development (Alvi et al. 2022). Several plants have been used to characterize SL analogs that have been chemically synthesized, such as GR5, GR7, and GR24 (which are known for having high activity). Subsequent research has clarified their functional roles in a variety of developmental processes, including shoot branching, root development, reproductive development, leaf senescence, and the regulation of plant organ architecture. Similar to other phytohormones, SLs are essential for development, and auxin interaction is the primary mechanism through which SL-regulated developmental processes. Several research works confirm the significance of SLs in controlling a range of physiological and molecular functions when plants adjust to diverse abiotic stressors (Rani et al. 2023). Due to their multiple beneficial effects, SLs have attained great importance in the plant scientist's community.

BRs are polyhydroxylated steroid plant hormones that participate in a regulatory role in diverse physiological and biochemical processes, like cell division, cell elongation, photomorphogenesis, growth, xylem differentiation, and reproduction (Divi, Rahman, and Krishna 2016). These hormones are found in both free and conjugated forms; there are about 69 conjugated and five free BRs. BRs exhibit diversity in both nature and biological activity. Among the various BRs, brassinolide (BL) is recognized as the most active and has been isolated and purified from Brassica napus pollen (Fàbregas et al. 2018). The BRs and auxins are closely related because they regulate the latter's transport, help plant organs coordinate their tropic responses, and encourage the initiation of lateral roots' primordial stages during lateral root development (Sabagh et al. 2021). In conclusion, BRs play a vital role in plant growth and development. For metal toxicity tolerance in plants, enduring investigation efforts are still needed to uncover the different hidden biochemical elements involved in the interaction between BRs and metal stress.

13.3 METALLOIDS IN ACTION: DECODING THEIR IMPACT ON PLANTS

Metalloids are naturally occurring substances with characteristic feature representative of intermediates between metals and non-metals. Metalloids, like arsenic, silicon, boron, germanium, selenium, tellurium (Te), antimony, and others, play a vital function in plant development. While silicon, boron, and selenium are important in small quantities, and germanium, Te, antimony, and arsenic are nonessential and can be hazardous to plants even in small concentrations (Martínez-Alcalá and Bernal 2020). Roots take up these metalloids when their concentration in the soil rises above acceptable levels. The extensive contagion of soil with metalloids is a global concern, contributing to a decline in agricultural yield worldwide. Consumption of grains or plant parts that have high levels of metalloids poses serious health risks to both humans and animals. High concentrations of metalloids in the air, water, or soil cause them to enter the food chain. The potential danger of metalloid exposure from the food we eat is increased by the absorption of metalloid particles from the soil by plant roots via transporters, which subsequently transfer the metalloid particles to different plant sections (Dutta et al. 2020). We summarize the impacts of metalloids on plants in Table 13.1.

Besides that, these metalloids are also toxic because they resemble biological components involved in different metabolic processes. For example, arsenic is one of the most dangerous substances that come from human activity. It is a nonessential metal compound. When it comes to elements that negatively affect plants, it takes the top spot. Arsenite (As (III)), arsenate (As (V)), which is the most common form in soil, and methylated forms like monomethylarsinic acid (MMA) and dimethylarsinic acid (DMA) are the primary three kinds of arsenic that can be found in minerals and water. Each kind of ion has a unique uptake mechanism (Mishra, Mattusch, and Wennrich 2017). Arsenate can mimic phosphate, moving it around in metabolic pathways and lowering ATP

TABLE 13.1
Metalloids' Toxic Effects and Accumulation in Various Plant Organs

Metal	Plant Model	Accumulation	Effects	References
Zn	*Zea mays, Arabidopsis thaliana*	Tonoplast, vacuoles during senescence from dividing cells and roots (YS) leaves to seeds (YSL)	Growth inhibition, photosynthesis, increased ROS	Jalil et al. (2023), Chaturvedi, Seth, and Misra (2007)
Sb	*Arabidopsis thaliana*	Roots, stems, leaves	Obstruct the glycolytic pathway's ability to catabolism glucose, and affect the synthesis of ATP	Mishra, Mattusch, and Wennrich (2017)
Se	*Arabidopsis thaliana*	Roots, leaves	Stunted growth and changes in the anatomy and morphology of leaves	Angulo-Bejarano, Puente-Rivera, and Cruz-Ortega (2021)
As	*Arabidopsis thaliana, Oryza sativa*	Shoots, roots, and leaves	Lipid peroxidation and ROS increase	Zhu et al. (2019)
Al	*Arabidopsis thaliana*	Roots, leaf, cell wall, tonoplast, and vacuoles	Lipid peroxidation, growth inhibition, and ROS increase	Schmitt, Watanabe, and Jansen (2016)
B	*Lycopersicon esculentum L.*	Roots, pollen grains, and cell wall	Reduced growth, chlorosis of the leaves, and harm to reproductive organs	Kohli et al. (2023)
Cd	*Arabidopsis thaliana*	Not identified	Photosynthesis suppression, chlorosis, nutrition imbalance, growth inhibition, and ROS increase	Haider et al. (2021)
Pb	*Triticum aestivum, Tetraenaqatarans, Lactuca sativa*	Inactivation in vacuoles	ROS stimulation interferes with electron transport and reduces photosystem II's efficiency in energy conversion	Seraj and Rahman (2018); Mishra et al. (2006)
Cu	*Arabidopsis thaliana, Oryza sativa*	Leaves, roots, and stem	ROS levels rise along with lipid peroxidation, ionic leakage, damage to proteins and nucleic acids, and modifications to chloroplasts	Adrees et al. (2015)

levels (Abbas et al. 2018). Moreover, arsenate inhibits respiration by targeting pyruvate dehydrogenase. Arsenite binds to cysteine residues' -SH group in its trivalent form, impeding the catalytic activities of many different enzymes. This interaction breaks down the citric acid cycle and pyruvate metabolism, depletes carbohydrates, and decreases the synthesis of coenzymes. Arsenate and arsenite both cause tissues to produce ROS, which damage proteins, lipids, and nucleic acids through oxidative stress (Clemens and Ma 2016).

There are more than 100 recognized minerals that contain antimony, a common metalloid. The more common forms of antimony are found in sulfides and sulfur salts, although the element itself is rare (Angulo-Bejarano, Puente-Rivera, and Cruz-Ortega 2021). As a byproduct of burning coal or smelting antimony-containing ores, antimony trioxide is a significant source of antimony emissions. Due to the close resemblance of antimony's chemical behavior to that of arsenic, it is possible that plants and algae that can accumulate bismuth and arsenic can also accumulate antimony. Numerous investigations examine the effects of antimony on microbes, exposing its capacity to create complexes with carbohydrates, target proteins' sulphydryl groups, alter enzyme activity, obstruct the glycolytic pathway's ability to catabolism glucose, and affect the synthesis of ATP (Seraj et al. 2018). Boron has a wide range of beneficial physiological effects. It affects the metabolism of proteins, carbohydrates, and nucleic acids and is essential for maintaining the integrity of cell walls and membranes (Nejad and Etesami 2020). Despite these advantageous features, the molecular mechanisms by which boron toxicity affects critical cellular functions are still not entirely understood. In agronomic practices, managing boron toxicity shows major obstacles (Kohli et al. 2023).

Selenium is a rare element that is usually found with metals such as copper, zinc, and lead, as well as sulfides. Notable forms of selenium are inorganic forms, such as selenides, selenates, and selenites, as well as elemental Se (Hasanuzzaman et al. 2022). Selenium is a naturally occurring element in the environment that is released by both human activity and natural processes like rock weathering. As a byproduct of burning coal and oil, selenium dioxide is the main form of selenium that enters the atmosphere (Feng et al. 2021). This material can be adsorbed onto dust particles and change into other forms of selenium, such as methyl derivatives or selenium acid. Selenium exposure can cause morphological changes in plants, including reduced size of reproductive organs (flowers), stunted growth, and changes in the anatomy and morphology of leaves, especially in terms of venation, shape, and lack of trichomes. Notably, significant alterations were brought about by selenium in plants, such as a decrease in phenolic acids (cafferic, ferulic, and sinapic) and secondary metabolites of sulfur. Elevated selenium concentrations in plants cause oxidative stress and alter the structure and function of proteins (Feng et al. 2021). Metalloid Te is a multipurpose material that is widely used in semiconductors, thin films, rechargeable batteries, photovoltaic devices, and charge transfer systems (Babula et al. 2008). Te has a substantial effect on human health despite its industrial importance. Te (IV) residues have been identified in milk samples, water, sediments, soils, and plants. Te uptake, transport, and metabolism are not well understood, although it is likely related to selenium or sulfur (Ogra 2017). Germanium is a chemical analog of silicon and a significant metalloid in industry. Once absorbed by the plants, it influences numerous metabolic processes and causes senescence (Wiche et al. 2018).

13.4 ROLE OF PHYTOHORMONES IN MITIGATING METALLOID TOXICITY

A key component in the mitigation of metalloid stress in plants are phytohormones, which have an array of functions to play, including boosting plant growth, strengthening antioxidant defense systems, improving nutrient uptake, and helping in the synthesis of cell walls (see Table 13.2 for a detailed summary). Their complex mechanisms provide a promising means of mitigating plant toxicity by considerably reducing metalloid accumulation. This chapter explores the intricate processes by which phytohormones regulate vital aspects of plant physiology, such as growth, defense against free radicals, nutrient absorption, and the synthesis of cell walls. It also highlights the potential uses of phytohormones in mitigating the adverse impacts of metalloid stress on plants.

TABLE 13.2
Different Phytohormones' Impacts on Diminishing Metalloid Stress

Hormone	Plant Species	Adaptive Reaction of Plants	References
Auxin	*Oryza sativa*	Increase plant growth and biomass	He et al. (2022); Betti et al. (2021)
BR	*Glycine max*	Improve plant growth and membrane stability	Guedes et al. (2021); Sytar et al. (2019)
Methyl jasmonate	*Oryza sativa*	Improve plant growth, chlorophyll, and antioxidant enzymes	Verma et al. (2020)
Cytokinin	*Zea mays*	Improve growth, productivity, and ADS system	Mohan et al. (2016)
Salicylic acid	*Triticum aestivum*	Improve photosynthesis and plant biomass	Zulfiqar and Ashraf (2022); Jia et al. (2021)
ABA	*Oryza sativa*	Improve plant growth and reduce oxidative stress	Saha, Hasanuzzaman, and Adak (2021); Fan et al. (2014)
SLs	*Oryza sativa*	Improve plant growth and antioxidants	Marzec (2016); Tai et al. (2017)

13.4.1 Auxin

Plants often show reduced root growth when under stress caused by metalloids, a phenomenon that has been extensively studied by (Tu et al. 2021). Vezza et al. (2018) observed similar results in soybeans, where exposure to metalloid toxicity led to a reduction in root hydraulic conductivity. Additionally, noted that as the concentration of metalloids increased, both root hydraulic conductivity and the expression of water channel protein genes decreased. This inhibition is also closely associated with reduced auxin accumulation in plant meristematic tissue. Since roots are the main pathway through which water and nutrients are absorbed from the soil, roots' flexibility, or root plasticity, becomes an important characteristic of adaptation (Ogura et al. 2019). Auxin is a key regulator of the dynamic root growth alterations that plants make in response to abiotic stress. Particularly, in the proximity of the root tips, optimal auxin levels promote frequent cell division, which in turn promotes elongation and differentiation in *A. thaliana*.

In contrast, higher auxin levels lead cells to quiescence, highlighting the important role of auxin in coordinating the delicate balance of root growth in response to environmental stimuli (Betti et al. 2021). Auxin levels decrease in plants under metalloid-induced stress, which severely inhibits the growth of lateral roots. An effective way to mitigate the negative effects of metalloid stress and promote strong root growth is to strategically apply auxin. Through the utilization of auxin's profound influence, plants can augment the formation of lateral roots, which promotes a more robust and adaptive response to environmental stressors. This emphasizes the role that auxin plays as a critical phytohormone in controlling the dynamics of lateral roots under metalloid stress (He et al. 2022).

13.4.2 Brassinosteroids

BRs also play a crucial role in increasing antioxidant enzyme activities, providing plants with a critical line of defense against oxidative stress caused by metalloid stress. The feature of BRs encompasses a wide range of abiotic stressors and goes beyond metalloid stress (Guedes et al. 2021). Interestingly, 0.01 μM BR application greatly increases rice tolerance in the presence of 100 μM Cr VI, boosting the activities of important enzymes like SOD, POD, CAT, and APX. Moreover, bean plants benefit from a deliberate administration of 5 μM BRs in response to 1 mM Cd stress, which lowers lipid peroxidation. The protective effect is ascribed to enhanced SOD, CAT, and GR activity, emphasizing BRs' role in plant defense against a variety of environmental stresses. BR-induced elevation of antioxidant enzyme activities occurs in the activation of their gene expression (Sytar et al. 2019). When 0.01–1 μM BRs is applied to rice that has been exposed to 0.5 mM Cr stress, it significantly boosts the expression of important enzymes like SOD, CAT, APX, and GR. Kour et al. (2021) noted that the exogenous application of BRs causes temporary apoplastic H_2O_2 formation, which in turn causes the upregulation of defensive enzymes and stress proteins as an adaptive response to counter metalloid-induced stress. Nazir et al. (2021) have also demonstrated that ROS-scavenging enzyme activities are effectively enhanced by the application of 0.01 μM BRs in tomatoes under 0.15–1.5 mM Cu stress.

13.4.3 Salicylic Acid

The cell wall, serving as the first line of defense against metal influx, becomes essential to the plant's metal detoxification process. The interaction of metals with pectin can reduce their entry into plants due to their negative charges; that's the reason the composition of the cell wall is crucial in controlling metal uptake. To adapt, plants thicken their cell walls by accumulating cellulose, hemicelluloses, and lignin, which together form complexes that efficiently bind metals. Additionally, phytohormones modulate the composition of the cell wall, which affects the deposition of metals on the wall (Zulfiqar and Ashraf 2022). Recent studies have demonstrated SA's valuable function in improving plants' resistance to cadmium (Cd). For example, 100 μM SA applied to tomato roots promotes cell wall synthesis, which in turn lowers the amount of Cd accumulated in the plant (Jia et al. 2021). Similarly, Xu et al. (2015) found that under 200 μM Cd stress, 100 μM SA increases Cd accumulation in peanut cell walls, highlighting the complex interaction between phytohormones and cell wall dynamics in response to metal stress. The application of SA causes a notable change in *Brachypodium distachyon's* cell wall's lignin and hemicellulose composition. Ferulic and p-coumaric acids, which can cross-link with hemicellulose and lignin to strengthen cell walls against digestibility, are increased in this process.

SA treatment is an effective modulator that significantly raises the glutathione (GSH) content of peppermint plants under Cd stress. This increased GSH is caused by a complex reaction that includes: (1) increased transcription in the precursor gene serine acetyltransferase, which is essential for catalyzing cysteine formation, (2) increased absorption of sulfur, which is essential for GSH synthesis, and (3) increased activities of GSH synthetase and GSH reductase, which ensure that GSH remains reduced. This complex interaction highlights SA's multipurpose function in coordinating a strong defense against Cd-induced (Ahmad et al. 2022).

13.4.4 Abscisic Acid

ABA is a powerful regulator that reduces metalloid accumulation in plants. ABA is effective in plants because it modifies stomatal closure, which lowers transpiration rate and, in turn, the accumulation of metalloids in shoots. On the other hand, plants with fluridone-inhibited ABA biosynthesis experience higher transpiration rates and a higher metalloid content (Saha et al. 2021). Fan et al. (2014) explore the complex molecular mechanisms and find that ABA treatment is correlated with the down-regulation of iron-regulated transporter 1 (*IRT1*), an essential factor in metalloid uptake, in *A. thaliana* roots, which is inhibited by ABA. Furthermore, a major role for the basic region/leucine zipper transcription factor (TF) ABA-insensitive5 (*ABI5*) in metalloid accumulation is revealed. In *A. thaliana*, *ABI5* overexpression reduces metalloid accumulation, but the ABI5 mutant shows the opposite effect, providing insight into the intricate molecular mechanisms underlying ABA's regulatory role in metalloid dynamics. Using ABA on plants has shown promise in reducing metalloid accumulation and boosting plant tolerance to metalloids (Vishwakarma et al. 2017). When ABA was sprayed on Pingyi sweet tea plants under metalloid stress, it significantly decreased cell death, as well as the buildup of H_2O_2 and malondialdehyde in the root system (Deng, Zhang, and Yang 2022). Additionally, ABA reduced leaf transpiration rate, metalloid influx in the roots, content in plants, and transport from roots to shoots. Moreover, when a key ABA synthesis gene, MhNCED3, was introduced into transgenic *Arabidopsis thaliana* and apple (*Malus pumila Mill*) calli, it led to reduced expression of metalloid-uptake-related genes (*NRAMP* and *IRT*) and decreased metalloid influx and content (Zhang et al. 2019). Inoculating pakchoi plants with ABA-producing bacteria like *Azospirillum brasilense* and *Bacillus subtilis* in metalloid-contaminated soil also lowered metalloid levels (by 40%–79% and 43%–77%, respectively), reduced metalloid-induced photosynthetic inhibition and oxidative damage, and increased pakchoi (*Brassica chinensis* L.) biomass (by 28%–281% and 26%–255% for the two bacteria, respectively) (Zhang et al. 2019). Furthermore, ABA has demonstrated alleviating effects on metalloid stress in various plants. These findings suggest that using ABA to mitigate Cd stress holds promising applications.

13.4.5 Strigolactones

SLs, a newly discovered class of phytohormones produced from carotenoids, have piqued interest due to their potential applications in phytoextraction and metalloid stress. Although there aren't many studies examining how SLs affect plant growth when exposed to metalloid stress, there is growing evidence concerning the mechanisms involved in removing Cd toxicity by applying exogenous phytohormones (Marzec 2016). GR24, a synthetic SL phytohormone, is particularly noteworthy because it has been shown to decrease Cd phytotoxicity and uptake in switchgrass, a warm-season C4 perennial grass. As a positive regulator of chlorophyll synthesis, GR24 effectively counteracts the reduction in chlorophyll content caused by Cd stress (Tai et al. 2017). On the other hand, research by Hye et al. (2004) shows opposite findings, demonstrating that SL-deficient mutants (max2/ore9) showed delayed symptoms of leaf senescence. Furthermore, the detrimental effect of SL on chlorophyll synthesis is emphasized by the dad1 mutant, which is SL-deficient in petunias. These discoveries deepen our knowledge of the complex function that SLs play in regulating plant reactions to metalloid stress, highlighting their potential to be significant players in the search for efficient methods of phytoremediation and stress reduction.

13.5 APPLICATION OF OTHER PHYTOHORMONES

Exogenous ET becomes a critical component during senescence and plays a major role in horticultural crops' ripening process. Especially, exogenous ET application provides a viable way to improve photosynthetic efficiency while under the influence of Cd stress. The observed effect is explained by a significant rise in sulfur-containing amino acid concentration and increased thiol production, highlighting the potential of this strategy to protect plants from the harmful effects of metalloids (Khan et al. 2016).

A survey of the literature revealed that other phytohormones like Me-JA, CKs, and GAs also improve plant growth and stress tolerance while also having a positive impact on the amount of metal accumulation. When these phytohormones were applied exogenously under metalloid stress, the activity of antioxidants was modulated (Chen, Yan, and Li 2014). Exogenous application preserved endogenous hormone levels, regulated stomatal aperture, decreased transpiration rate, inhibited metal uptake, and decreased photosynthetic damage (Per et al. 2018). The type-2 metallothionein gene (*MT2*) was also found to have increased expression in plants treated with metalloids, but exogenously applied phytohormones significantly restored the expression in leaves.

13.6 PHYTOHORMONAL CROSSTALK UNDER METALLOID TOXICITY

For plants to survive, phytohormones are essential in reducing oxidative stress caused by metalloids (Sytar et al. 2019). According to Saini, Kaur, and Pati (2021), there is evidence that metalloid toxicity affects the production and distribution of phytohormones, causing plants to produce higher amounts of SA, ET, JA, and ABA and lower levels of auxin and GA3. This intricate crosstalk between phytohormones and signaling molecules is fundamental for plants facing metalloid stress (Figure 13.2).

13.6.1 Ethylene Crosstalk under Metalloid Stress

Upregulation of ET biosynthesis genes such as *AtACO* (AtACO1 and 2) and *AtACS* (*AtACS2, 6,* and *8*) has been observed in *Arabidopsis* plants under aluminum (Al) stress. When exposed to Al stress, the wild-type (WT) plants express higher levels of PINFORMED2 (*AtPIN2*) and AUXIN RESISTANT1 (*AtAUX1*). It is interesting to note that the overexpression of *AtPIN2* and *AtAUX1* in WT plants caused by Al can be effectively eradicated by using ET biosynthesis inhibitors like AVG and Co (Sun et al. 2010). Interestingly, *Arabidopsis* plants exhibit notably less Al-modulated root inhibition than WT plants do. This is mainly because of the auxin-unresponsive single mutant's *pin2* and *aux1–7*. These results highlight the function of ET biosynthesis in plants under Al stress, which affects auxin redistribution by regulating *PIN2* and *AUXAE*, two essential elements of the auxin polar transport network. This demonstrates how auxin and ET have complex interactions when plants are exposed to metalloid toxicity (Sun et al. 2010).

In *Arabidopsis*, crosstalk between ET and JA was identified at two ET-alleviated TFs, *EIL1* and *EIN3*. These elements act together to coordinate the expression of several stress-related genes. Notably, subsequent research efforts have provided additional insight into the complex regulatory mechanisms of ethylene response factors (ERFs) and JA by establishing that the former's

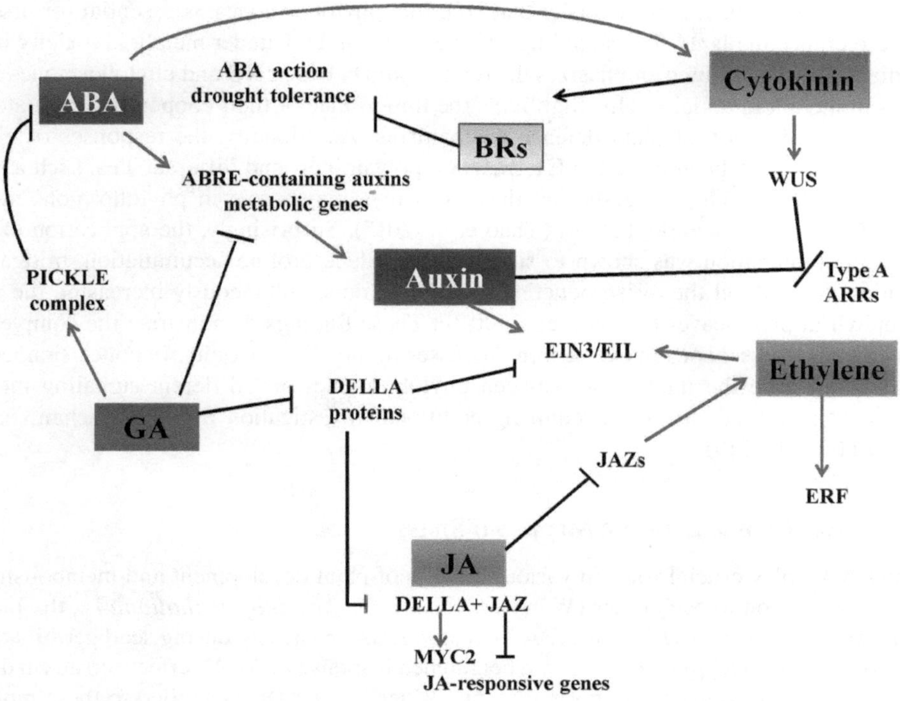

FIGURE 13.2 Interactions between different plant hormones in response to metalloid stress.

posttranslational expression acts independently of *EIN3*/EIL1 (Does et al. 2013). When Cd stress is applied to *Arabidopsis* plants, the ET and JA signaling pathways are activated, which causes the *NRT1.8* (nitrate transporter 1.8) gene to be upregulated and the *NRT1.5* gene to be downregulated simultaneously. This coordinated reaction enhances Arabidopsis's tolerance to Cd by mediating the redistribution of nitrate to the roots caused by stress (Zhang, Yi, and Gong 2014). In *Arabidopsis*, ET biosynthesis is enhanced by the GA-insensitive ET overexpressed 2–1 double mutant, which is particularly linked to the GA3 signaling pathway. This suggests that there may be communication between GA3 and ET (Asgher et al. 2017). Another study showed that endogenous GA3 application, with or without sulfur (S) at optimal concentrations, reduces stress-induced ET overexpression and, as a result, reduces Cd-induced photosynthetic inhibition. This process is closely related to sulfur absorption, which begins the biosynthesis of cysteine, which is necessary for the biosynthesis of GSH and ET in a normal environment (Moravcová et al. 2018). ET-induced GSH enhancement was found to be useful in reducing lead (Pb) toxicity in *Arabidopsis*. This was rendered possible by the ein2-1 mutant's increased expression of GSH1, a gene linked to GSH biosynthesis. This emphasizes how ET signaling regulates metalloid stress reduction through the GSH pathway. In the same way, ET, SA, JA, IAA, and AsA acid all worked together to enhance Pb tolerance in *Arabidopsis* plants treated with $AuCl_4$ by promoting the expression of GSH-independent *PDR12* (pleiotropic drug resistance transporter 12). This demonstrates how different phytohormones work together in a complex signaling network to increase plant resistance to lead stress (Masood, Iqbal, and Khan 2012).

13.6.2 Nitric Oxide Crosstalk under Metalloid Stress

Phytohormones and nitric oxide (NO), a major regulator of various plant responses such as ET synthesis, cell division, growth, and responses to biotic and abiotic stress, interact in a complex way. According to studies by Thao et al. (2015), NO also affects senescence, stomatal conductance, and programmed cell death. It is noteworthy that NO, like phytohormones, is essential for promoting metalloid tolerance in plants. The signaling cascade of JA and ET under metalloid toxicity involves NO-mediated *MAPK6*, which emphasizes the relationship between NO and phytohormones in plant responses to metalloid toxicity. This highlights the importance of their cooperative regulatory network to the coordination of plant defense mechanisms. Additionally, the responses of plants to metalloid toxicity have been linked to ET, *MAPKs*, polyamines, and different TFs, such as *DOF1, MYBZ2*, and *bZIP62*. This suggests that these TFs may play a role in phytohormone-mediated pathways that regulate metalloid stress (Thao et al. 2015). Surprisingly, the application of phytohormones in combination was shown to significantly reduce proline accumulation, mitigate lipid peroxidation, and inhibit the overproduction of ROS while simultaneously increasing the amount of chlorophyll in plant leaves (Kazemi et al. 2010). These findings demonstrate the complex interplay between phytohormones in overseeing processes to metalloid toxicity. In conclusion, complex mechanisms underlie the interaction between phytohormones and different signaling molecules in the face of metalloid toxicities, requiring additional investigation through mechanistic-based research (Bali et al. 2019).

13.6.3 Auxin Crosstalk under Metalloid Stress

Auxin and ABA play crucial roles in various aspects of plant development and metabolism, often working in opposition to each other (Wang et al. 2021). In *Arabidopsis thaliana* L., the induction of auxin response factor 2 (*ARF2*) by ABA increases ABA sensitivity during seed germination and root growth. Mutants lacking *ARF2* show a heightened response to ABA's effects on auxin distribution and cell division compared to normal plants. When extra ABA is applied to these mutants, it results in a reduction in root length. ABA limits primary root growth by inhibiting cell proliferation in the root tips (Wang et al. 2013). Furthermore, ABA negatively affects seedling development by enhancing auxin signaling. The gene expression network responsible for regulating seed dormancy involves the TF *ABI3*. During seed germination, auxin influences the expression of the *ABI3* gene through ARFs 10 and 16. The interplay between ABA and auxin in seed dormancy suggests that these two hormones interact to influence the process, where ABA decreases seed germination while auxin is essential for seed dormancy (Kozaki and Aoyanagi 2022). GA and ABA have a negative interaction that regulates various aspects of plant development. Unlike GA, which promotes seed germination and seedling growth, ABA inhibits seed germination by promoting dormancy (Nguyen, Tuan, and Ayele 2022). Maintaining a balance in the production and breakdown of GA and ABA is crucial for proper germination and dormancy. In *Arabidopsis thaliana* L., providing extra ABA hampers primary root development, while GA enhances root growth. This is similar to how ABA and BR work against each other in influencing plant processes—ABA hinders seed germination, while BR promotes it (Yang et al. 2016). ABA also regulates BR signaling elements, interacting with components like *BIN2*, *BRI1* receptors, *ABI1*, and *ABA12*. Interestingly, the presence of metalloids inhibits the increase in phenolic compounds, but the combination of GA3 and ABA hormones with Zn and Pb shows potential to boost overall phenolic content. ABA's role in the interplay between CKs and BR involves inhibiting BR production during metal toxicity (Ahmad, Hadi, and Ali 2015). However, more research is needed to fully understand the specific connections between ABA and GA, as well as ABA and BR, in governing plant physiological processes.

13.6.4 Jasmonic Acid Crosstalk under Metalloid Stress

JA doesn't work alone but collaborates with other plant hormones in a complex signaling network. Jasmonates, including methyl jasmonate (Me-JA), JA, JA-isoleucine (JA-Ile), and the JA precursor 12-oxophytodienoic acid (OPDA), play roles in regulating plant growth and responding to stress. JAs, including Me-JA, also help plants deal with various abiotic stresses, including those caused by metalloid exposure (Per et al. 2018). *In Arabidopsis*, when exposed to metalloid, the expression of genes promoting endogenous JA synthesis quickly increases, leading to higher JA levels in the roots. However, when JAs are either naturally formed or applied externally, they reduce the metalloid concentration in the root cell sap (Fattorini et al. 2018). This happens by decreasing the expression of genes that facilitate metalloid uptake and long-distance transport, ultimately lessening the impact of metalloid stress on the plant. ET and JA signaling are closely linked to the auxin circuit in root development, mainly through TFs that act as key communication nodes (Ellis and Turner 2002). Two important TFs, *EIN3* and its homolog *EIL1*, control most of the ET responses. The JASMONATE ZIM-DOMAIN (JAZ) proteins are targeted by the COI1 protein, and the COI1–JAZ combination serves as a co-receptor for JA-Ile (Ruan et al. 2019). Additionally, EIN3 and EIL1 directly interact with many JAZs, inhibiting their function. When JA-Ile is present, JAZs break down, releasing EIN3/EIL1. However, for these released proteins to remain stable, they still require ET. In line with this, the ein3eil1 mutant in Arabidopsis shows insensitivity to both JA and ET, a fact confirmed during the formation of adventitious roots in Arabidopsis TCLs (Qin et al. 2017).

13.6.5 Brassinosteroids Crosstalk under Metalloid Stress

BRs play various roles in plant growth and development by interacting with phytohormones. They regulate processes such as ripening in bananas and influence ET production (Ohri et al. 2019). The DELLA gene inhibits BZR1, but GA induces DELLA degradation. Both GA and BR impact the expression of the target gene BRZ1, promoting GA production and DELLA degradation. Direct interaction between DELLA and BZR1 inhibits BZR1 binding to DNA, affecting seedling etiolation and cell elongation. Under metalloid stress, the interaction between DELLA and BRZ1 increases ROS-scavenging enzyme levels (Hu et al. 2017). CKs stimulate defense mechanisms against metalloid, enhancing photosynthesis efficiency. Exogenous BR application leads to CK-related effects, increasing leaf and root length and promoting anthocyanin accumulation (Yuan et al. 2015). ABA also plays a role in the BR-CK interaction, suppressing BR synthesis during metal toxicity. Overall, the crosstalk between BR, GA, CKs, and ABA influences various aspects of plant physiology, offering potential applications in crop yield enhancement (Wang et al. 2018).

13.7 CONCLUSION AND FUTURE OUTLOOK

The escalating impact of metalloid stress has emerged as a pressing concern for plant biologists, posing a significant threat to sustainable agricultural growth. This chapter gives a comprehensive overview of various aspects, such as the roles of phytohormones, the impact of metalloids on plant growth, and the intricate interactions between phytohormones and metalloids in plants. It is crucial to comprehend how multiple hormones work simultaneously to assist plants in recovering from metalloid stress. A full investigation of the regulatory network of phytohormone signaling pathways under metalloid stress is currently absent, while some understanding of the complicated interaction between them does exist. As a result, addressing the negative effects of metalloids requires understanding the molecular basis of phytohormone-metalloid interaction. An efficient strategy for utilizing phytohormones to counter metalloid toxicity will be made possible by a greater comprehension of the molecular networks regulating phytohormone-mediated metalloid stress regulation. Exploring the interactions among various phytohormones may help to better understand the complexities of metalloid stress tolerance and ultimately lead to improved crop productivity. In future decades, one fascinating method for producing metalloid stress-tolerant plants in modern agriculture will be by monitoring phytohormone levels as well as coordinating their actions in particular tissues/organs during relevant developmental stages.

REFERENCES

Abbas, Ghulam, Behzad Murtaza, Irshad Bibi, Muhammad Shahid, Nabeel Khan Niazi, Muhammad Imran Khan, Muhammad Amjad, Munawar Hussain, and Natasha. 2018. "Arsenic Uptake, Toxicity, Detoxification, and Speciation in Plants: Physiological, Biochemical, and Molecular Aspects." *International Journal of Environmental Research and Public Health*. doi:10.3390/ijerph15010059

Adrees, Muhammad, Shafaqat Ali, Muhammad Rizwan, Muhammad Ibrahim, Farhat Abbas, Mujahid Farid, Muhammad Zia-ur-Rehman, Muhammad Kashif Irshad, and Saima Aslam Bharwana. 2015. "The Effect of Excess Copper on Growth and Physiology of Important Food Crops: A Review." *Environmental Science and Pollution Research*. doi:10.1007/s11356-015-4496-5

Ahmad, Ayaz, Fazal Hadi, and Nasir Ali. 2015. "Effective Phytoextraction of Cadmium (Cd) with Increasing Concentration of Total Phenolics and Free Proline in Cannabis Sativa (L) Plant under Various Treatments of Fertilizers, Plant Growth Regulators and Sodium Salt." *International Journal of Phytoremediation*. doi :10.1080/15226514.2013.828018

Ahmad, Zishan, Sabaha Tahseen, Adla Wasi, Irfan Bashir Ganie, Anwar Shahzad, Abolghassem Emamverdian, Muthusamy Ramakrishnan, and Yulong Ding. 2022. "Nanotechnological Interventions in Agriculture." *Nanomaterials*. doi:10.3390/nano12152667

Alvi, Ameena Fatima, Zebus Sehar, Mehar Fatma, Asim Masood, and Nafees A. Khan. 2022. "Strigolactone: An Emerging Growth Regulator for Developing Resilience in Plants." *Plants*. doi:10.3390/plants11192604

Angulo-Bejarano, Paola I., Jonathan Puente-Rivera, and Rocío Cruz-Ortega. 2021. "Metal and Metalloid Toxicity in Plants: An Overview on Molecular Aspects." *Plants*. doi:10.3390/plants10040635

Arkhipova, Tatiana, Elena Martynenko, Guzel Sharipova, Ludmila Kuzmina, Igor Ivanov, Margarita Garipova, and Guzel Kudoyarova. 2020. "Effects of Plant Growth Promoting Rhizobacteria on the Content of Abscisic Acid and Salt Resistance of Wheat Plants." *Plants*. doi:10.3390/plants9111429

Arnao, M. B., and J. Hernández-Ruiz. 2015. "Melatonin: Synthesis from Tryptophan and Its Role in Higher Plant." In *Amino Acids in Higher Plants*. doi:10.1079/9781780642635.0390

Asgher, Mohd, Tasir S. Per, Shagufta Anjum, M. Iqbal R. Khan, Asim Masood, Susheel Verma, and Nafees A. Khan. 2017. "Contribution of Glutathione in Heavy Metal Stress Tolerance in Plants." In *Reactive Oxygen Species and Antioxidant Systems in Plants: Role and Regulation under Abiotic Stress*. doi:10.10 07/978-981-10-5254-5_12

Babula, Petr, Vojtech Adam, Radka Opatrilova, Josef Zehnalek, Ladislav Havel, and Rene Kizek. 2008. "Uncommon Heavy Metals, Metalloids and Their Plant Toxicity: A Review." *Environmental Chemistry Letters*. doi:10.1007/s10311-008-0159-9

Balali-Mood, Mahdi, Kobra Naseri, Zoya Tahergorabi, Mohammad Reza Khazdair, and Mahmood Sadeghi. 2021. "Toxic Mechanisms of Five Heavy Metals: Mercury, Lead, Chromium, Cadmium, and Arsenic." *Frontiers in Pharmacology*. doi:10.3389/fphar.2021.643972

Bali, Shagun, Vijay Lakshmi Jamwal, Sukhmeen Kaur Kohli, Parminder Kaur, R. Tejpal, Vandana Bhalla, Puja Ohri, et al. 2019. "Jasmonic Acid Application Triggers Detoxification of Lead (Pb) Toxicity in Tomato through the Modifications of Secondary Metabolites and Gene Expression." *Chemosphere*. doi:10.1016/j.chemosphere.2019.06.188

Bandara, Tharanga, Ashley Franks, Jianming Xu, Nanthi Bolan, Hailong Wang, and Caixian Tang. 2020. "Chemical and Biological Immobilization Mechanisms of Potentially Toxic Elements in Biochar-Amended Soils." *Critical Reviews in Environmental Science and Technology*. doi:10.1080/10 643389.2019.1642832

Betti, Camilla, Federica Della Rovere, Diego Piacentini, Laura Fattorini, Giuseppina Falasca, and Maria Maddalena Altamura. 2021. "Jasmonates, Ethylene and Brassinosteroids Control Adventitious and Lateral Rooting as Stress Avoidance Responses to Heavy Metals and Metalloids." *Biomolecules*. doi:10.3390/biom11010077

Brewer, Philip B., Hinanit Koltai, and Christine A. Beveridge. 2013. "Diverse Roles of Strigolactones in Plant Development." *Molecular Plant*. doi:10.1093/mp/sss130

Chaturvedi, Pranav Kumar, Chandra Shekhar Seth, and Virendra Misra. 2007. "Selectivity Sequences and Sorption Capacities of Phosphatic Clay and Humus Rich Soil towards the Heavy Metals Present in Zinc Mine Tailing." *Journal of Hazardous Materials*. doi:10.1016/j.jhazmat.2007.01.064

Chen, Jun, Zhongzheng Yan, and Xiuzhen Li. 2014. "Effect of Methyl Jasmonate on Cadmium Uptake and Antioxidative Capacity in Kandelia Obovata Seedlings under Cadmium Stress." *Ecotoxicology and Environmental Safety*. doi:10.1016/j.ecoenv.2014.01.022

Chmielowska-Bak̨, Jagna, and Joanna Deckert. 2021. "Plant Recovery after Metal Stress—A Review." *Plants*. doi:10.3390/plants10030450

Clemens, Stephan, and Jian Feng Ma. 2016. "Toxic Heavy Metal and Metalloid Accumulation in Crop Plants and Foods." *Annual Review of Plant Biology*. doi:10.1146/annurev-arplant-043015–112301

Danilova, M. N., N. V. Kudryakova, A. S. Doroshenko, D. A. Zabrodin, N. S. Vinogradov, and V. V. Kuznetsov. 2016. "Molecular and Physiological Responses of Arabidopsis Thaliana Plants Deficient in the Genes Responsible for ABA and Cytokinin Reception and Metabolism to Heat Shock." *Russian Journal of Plant Physiology*. doi:10.1134/S1021443716030043

Dar, Niyaz A., Insha Amin, Wasia Wani, Shafiq A. Wani, Asif B. Shikari, Shabir H. Wani, and Khalid Z. Masoodi. 2017. "Abscisic Acid: A Key Regulator of Abiotic Stress Tolerance in Plants." *Plant Gene*. doi:10.1016/j.plgene.2017.07.003

Deng, Bo, Weiwei Zhang, and Hongqiang Yang. 2022. "Abscisic Acid Decreases Cell Death in Malus Hupehensis Rehd. Under Cd Stress by Reducing Root Cd2+ Influx and Leaf Transpiration." *Journal of Plant Growth Regulation*. doi:10.1007/s00344-021-10327-0

Divi, Uday K., Tawhidur Rahman, and Priti Krishna. 2016. "Gene Expression and Functional Analyses in Brassinosteroid-Mediated Stress Tolerance." *Plant Biotechnology Journal*. doi:10.1111/pbi.12396

Does, Dieuwertje Van der, Antonio Leon-Reyes, Annemart Koornneef, Marcel C. Van Verk, Nicole Rodenburg, Laurens Pauwels, Alain Goossens, et al. 2013. "Salicylic Acid Suppresses Jasmonic Acid Signaling Downstream of SCFCOI1-JAZ by Targeting GCC Promoter Motifs via Transcription Factor ORA59." *Plant Cell*. doi:10.1105/tpc.112.108548

Dubois, Marieke, Lisa Van den Broeck, and Dirk Inzé. 2018. "The Pivotal Role of Ethylene in Plant Growth." *Trends in Plant Science*. doi:10.1016/j.tplants.2018.01.003

Dutta, Sutapa, Arindam Datta, Abbu Zaid, and Javaid Akhter Bhat. 2020. "Metalloids and Their Impact on the Environment." In *Metalloids in Plants: Advances and Future Prospects*. doi:10.1002/9781119487210.ch2

Ellis, Christine, and John G. Turner. 2002. "A Conditionally Fertile Coi1 Allele Indicates Cross-Talk between Plant Hormone Signalling Pathways in Arabidopsis Thaliana Seeds and Young Seedlings." *Planta*. doi:10.1007/s00425-002-0787-4

Elsayed, Abdelaleim I., Moncef Boulila, Mohammed S. Rafudeen, Azza H. Mohamed, Sonali Sengupta, Mostafa Rady, and Ahmad A. Omar. 2020. "Melatonin Regulatory Mechanisms and Phylogenetic Analyses of Melatonin Biosynthesis Related Genes Extracted from Peanut under Salinity Stress." *Plants*. doi:10.3390/plants9070854

Fàbregas, Norma, Fidel Lozano-Elena, David Blasco-Escámez, Takayuki Tohge, Cristina Martínez-Andújar, Alfonso Albacete, Sonia Osorio, et al. 2018. "Overexpression of the Vascular Brassinosteroid Receptor BRL3 Confers Drought Resistance without Penalizing Plant Growth." *Nature Communications*. doi:10.1038/s41467-018-06861-3

Fan, Shi Kai, Xian Zhi Fang, Mei Yan Guan, Yi Quan Ye, Xian Yong Lin, Shao Ting Du, and Chong Wei Jin. 2014. "Exogenous Abscisic Acid Application Decreases Cadmium Accumulation in Arabidopsis Plants, Which Is Associated with the Inhibition of IRT1-Mediated Cadmium Uptake." *Frontiers in Plant Science*. doi:10.3389/fpls.2014.00721

Farhangi-Abriz, Salar, and Kazem Ghassemi-Golezani. 2019. "Jasmonates: Mechanisms and Functions in Abiotic Stress Tolerance of Plants." *Biocatalysis and Agricultural Biotechnology*. doi:10.1016/j.bcab.2019.101210

Fattorini, Laura, Bettina Hause, Laurent Gutierrez, Angela Veloccia, Federica Della Rovere, Diego Piacentini, Giuseppina Falasca, and Maria Maddalena Altamura. 2018. "Jasmonate Promotes Auxin-Induced Adventitious Rooting in Dark-Grown Arabidopsis Thaliana Seedlings and Stem Thin Cell Layers by a Cross-Talk with Ethylene Signalling and a Modulation of Xylogenesis." *BMC Plant Biology*. doi:10.1186/s12870-018-1392–4

Feng, Ren Wei, Ping Ping Zhao, Yan Ming Zhu, Ji Gang Yang, Xin Qi Wei, Li Yang, Hong Liu, Christopher Rensing, and Yong Zhen Ding. 2021. "Application of Inorganic Selenium to Reduce Accumulation and Toxicity of Heavy Metals (Metalloids) in Plants: The Main Mechanisms, Concerns, and Risks." *Science of the Total Environment*. doi:10.1016/j.scitotenv.2020.144776

Gong, Zhizhong, Liming Xiong, Huazhong Shi, Shuhua Yang, Luis R. Herrera-Estrella, Guohua Xu, Dai Yin Chao, et al. 2020. "Plant Abiotic Stress Response and Nutrient Use Efficiency." *Science China Life Sciences* 63 (5): 635–74. doi:10.1007/s11427-020-1683-x

Guedes, Flávia Raphaela Carvalho Miranda, Camille Ferreira Maia, Breno Ricardo Serrão da Silva, Bruno Lemos Batista, Mohammed Nasser Alyemeni, Parvaiz Ahmad, and Allan Klynger da Silva Lobato. 2021. "Exogenous 24-Epibrassinolide Stimulates Root Protection, and Leaf Antioxidant Enzymes in Lead Stressed Rice Plants: Central Roles to Minimize Pb Content and Oxidative Stress." *Environmental Pollution*. doi:10.1016/j.envpol.2021.116992

Haider, Fasih Ullah, Cai Liqun, Jeffrey A. Coulter, Sardar Alam Cheema, Jun Wu, Renzhi Zhang, Ma Wenjun, and Muhammad Farooq. 2021. "Cadmium Toxicity in Plants: Impacts and Remediation Strategies." *Ecotoxicology and Environmental Safety*. doi:10.1016/j.ecoenv.2020.111887

Hamid, Basharat, Muzafar Zaman, Shabeena Farooq, Sabah Fatima, Riyazali Z. Sayyed, Zahoor Ahmad Baba, Tahir Ahmad Sheikh, et al. 2021. "Bacterial Plant Biostimulants: A Sustainable Way towards Improving Growth, Productivity, and Health of Crops." *Sustainability (Switzerland)*. doi:10.3390/su13052856

Hasanuzzaman, Mirza, Kamrun Nahar, Pedro García-Caparrós, Khursheda Parvin, Faisal Zulfiqar, Naznin Ahmed, and Masayuki Fujita. 2022. "Selenium Supplementation and Crop Plant Tolerance to Metal/Metalloid Toxicity." *Frontiers in Plant Science*. doi:10.3389/fpls.2021.792770

He, Yue, Ting Zhang, Yu Sun, Xuejing Wang, Qingqing Cao, Zhigang Fang, Ming Chang, Qingsheng Cai, and Laiqing Lou. 2022. "Exogenous IAA Alleviates Arsenic Toxicity to Rice and Reduces Arsenic Accumulation in Rice Grains." *Journal of Plant Growth Regulation*. doi:10.1007/s00344-021-10336-z

Hu, Songlin, Cuiling Wang, Darlene L. Sanchez, Alexander E. Lipka, Peng Liu, Yanhai Yin, Michael Blanco, and Thomas Lübberstedt. 2017. "Gibberellins Promote Brassinosteroids Action and Both Increase Heterosis for Plant Height in Maize (Zea Mays L.)." *Frontiers in Plant Science*. doi:10.3389/fpls.2017.01039

Huang, Huang, Bei Liu, Liangyu Liu, and Susheng Song. 2017. "Jasmonate Action in Plant Growth and Development." *Journal of Experimental Botany*. doi:10.1093/jxb/erw495

Hye, Ryun Woo, Hee Kim Jin, Gil Nam Hong, and Ok Lim Pyung. 2004. "The Delayed Leaf Senescence Mutants of Arabidopsis, Ore1, Ore3, and Ore9 Are Tolerant to Oxidative Stress." *Plant and Cell Physiology*. doi:10.1093/pcp/pch110

Jalil, Sanaullah, Muhammad Mudassir Nazir, Qurban Ali, Faisal Zulfiqar, Anam Moosa, Muhammad Ahsan Altaf, Abbu Zaid, Muhammad Nafees, Jean Wan Hong Yong, and Xiaoli Jin. 2023. "Zinc and Nano Zinc Mediated Alleviation of Heavy Metals and Metalloids in Plants: An Overview." *Functional Plant Biology*. doi:10.1071/FP23021

Jia, Hong Lei, Xiao Hong Wang, Ting Wei, Min Wang, Xun Liu, Li Hua, Xin Hao Ren, Jun Kang Guo, and Jisheng Li. 2021. "Exogenous Salicylic Acid Regulates Cell Wall Polysaccharides Synthesis and Pectin Methylation to Reduce Cd Accumulation of Tomato." *Ecotoxicology and Environmental Safety*. doi:10.1016/j.ecoenv.2020.111550

Jiroutova, Petra, Jana Oklestkova, and Miroslav Strnad. 2018. "Crosstalk between Brassinosteroids and Ethylene during Plant Growth and under Abiotic Stress Conditions." *International Journal of Molecular Sciences*. doi:10.3390/ijms19103283

Jurado, Silvia, Zamira Abraham, Concepción Manzano, Gema López-Torrejón, Luis F. Pacios, and Juan C. del Pozo. 2010. "The Arabidopsis Cell Cycle F-Box Protein SKP2A Binds to Auxin." *Plant Cell*. doi:10.1105/tpc.110.078972

Kapulnik, Yoram, Natalie Resnick, Einav Mayzlish-Gati, Yulia Kaplan, Smadar Wininger, Joseph Hershenhorn, and Hinanit Koltai. 2011. "Strigolactones Interact with Ethylene and Auxin in Regulating Root-Hair Elongation in Arabidopsis." *Journal of Experimental Botany*. doi:10.1093/jxb/erq464

Kazemi, Nader, Ramazan Ali Khavari-Nejad, Hamid Fahimi, Sara Saadatmand, and Taher Nejad-Sattari. 2010. "Effects of Exogenous Salicylic Acid and Nitric Oxide on Lipid Peroxidation and Antioxidant Enzyme Activities in Leaves of Brassica Napus L. under Nickel Stress." *Scientia Horticulturae*. doi:10.1016/j.scienta.2010.07.037

Khan, Nafees A., Mohd Asgher, Tasir S. Per, Asim Masood, Mehar Fatma, and M. I. R. Khan. 2016. "Ethylene Potentiates Sulfur-Mediated Reversal of Cadmium Inhibited Photosynthetic Responses in Mustard." *Frontiers in Plant Science*. doi:10.3389/fpls.2016.01628

Kohli, Sukhmeen Kaur, Harsimran Kaur, Kanika Khanna, Neha Handa, Renu Bhardwaj, Jörg Rinklebe, and Parvaiz Ahmad. 2023. "Boron in Plants: Uptake, Deficiency and Biological Potential." *Plant Growth Regulation*. doi:10.1007/s10725-022-00844-7

Kour, J., Kohli, S.K., Khanna, K., Bakshi, P., Sharma, P., Singh, A.D., Ibrahim, M., Devi, K., Sharma, N., Ohri, P. and Skalicky, M., 2021. "Brassinosteroid signaling, crosstalk and, physiological functions in plants under heavy metal stress." *Frontiers in Plant Science*, *12*, p.608061.

Koo, Young Mo, A. Yeong Heo, and Hyong Woo Choi. 2020. "Salicylic Acid as a Safe Plant Protector and Growth Regulator." *Plant Pathology Journal*. doi:10.5423/PPJ.RW.12.2019.0295

Kozaki, Akiko, and Takuya Aoyanagi. 2022. "Molecular Aspects of Seed Development Controlled by Gibberellins and Abscisic Acids." *International Journal of Molecular Sciences*. doi:10.3390/ijms23031876

Li, Ning, Dejuan Euring, Joon Yung Cha, Zeng Lin, Mengzhu Lu, Li Jun Huang, and Woe Yeon Kim. 2021. "Plant Hormone-Mediated Regulation of Heat Tolerance in Response to Global Climate Change." *Frontiers in Plant Science*. doi:10.3389/fpls.2020.627969

Li, Si Min, Hong Xiang Zheng, Xian Sheng Zhang, and Na Sui. 2021. "Cytokinins as Central Regulators during Plant Growth and Stress Response." *Plant Cell Reports*. doi:10.1007/s00299-020-02612-1

Martínez-Alcalá, Isabel, and Maria Pilar Bernal. 2020. "Environmental Impact of Metals, Metalloids, and Their Toxicity." In *Metalloids in Plants: Advances and Future Prospects*. doi:10.1002/9781119487210.ch21

Marzec, Marek. 2016. "Strigolactones as Part of the Plant Defence System." *Trends in Plant Science*. doi:10.1016/j.tplants.2016.08.010

Masood, Asim, Noushina Iqbal, and Nafees A. Khan. 2012. "Role of Ethylene in Alleviation of Cadmium-Induced Photosynthetic Capacity Inhibition by Sulphur in Mustard." *Plant, Cell and Environment*. doi:10.1111/j.1365–3040.2011.02432.x

Mishra, Seema, Jürgen Mattusch, and Rainer Wennrich. 2017. "Accumulation and Transformation of Inorganic and Organic Arsenic in Rice and Role of Thiol-Complexation to Restrict Their Translocation to Shoot." *Scientific Reports*. doi:10.1038/srep40522

Mishra, Seema, S. Srivastava, R. D. Tripathi, R. Kumar, C. S. Seth, and D. K. Gupta. 2006. "Lead Detoxification by Coontail (Ceratophyllum Demersum L.) Involves Induction of Phytochelatins and Antioxidant System in Response to Its Accumulation." *Chemosphere*. doi:10.1016/j.chemosphere.2006.03.033

Mohamed, Heba Ibrahim, and Hanan Helmy Latif. 2017. "Improvement of Drought Tolerance of Soybean Plants by Using Methyl Jasmonate." *Physiology and Molecular Biology of Plants*. doi:10.1007/s12298-017-0451-x

Mohan, Thotegowdanapalya C., Gabriel Castrillo, Cristina Navarro, Sonia Zarco-Fernández, Eswarayya Ramireddy, Cristian Mateo, Angel M. Zamarreño, et al. 2016. "Cytokinin Determines Thiol-Mediated Arsenic Tolerance and Accumulation." *Plant Physiology*. doi:10.1104/pp.16.00372

Moravcová, Šárka, Jiří Tůma, Zuzana Kovalíková Dučaiová, Piotr Waligórski, Monika Kula, Diana Saja, Aneta Słomka, Wojciech Bąba, and Marta Libik-Konieczny. 2018. "Influence of Salicylic Acid Pretreatment on Seeds Germination and Some Defence Mechanisms of Zea Mays Plants under Copper Stress." *Plant Physiology and Biochemistry*. doi:10.1016/j.plaphy.2017.11.007

Nazir, Faroza, Qazi Fariduddin, Anjuman Hussain, and Tanveer Alam Khan. 2021. "Brassinosteroid and Hydrogen Peroxide Improve Photosynthetic Machinery, Stomatal Movement, Root Morphology and Cell Viability and Reduce Cu- Triggered Oxidative Burst in Tomato." *Ecotoxicology and Environmental Safety*. doi:10.1016/j.ecoenv.2020.111081

Nejad, Seyed Ali Ghaffari, and Hassan Etesami. 2020. "The Importance of Boron in Plant Nutrition." In *Metalloids in Plants: Advances and Future Prospects*. doi:10.1002/9781119487210.ch20

Nguyen, Tran Nguyen, Pham Anh Tuan, and Belay T. Ayele. 2022. "Jasmonate Regulates Seed Dormancy in Wheat via Modulating the Balance between Gibberellin and Abscisic Acid." *Journal of Experimental Botany*. doi:10.1093/jxb/erac041

Ogra, Yasumitsu. 2017. "Biology and Toxicology of Tellurium Explored by Speciation Analysis." *Metallomics*. doi:10.1039/c7mt00022g

Ogura, Takehiko, Christian Goeschl, Daniele Filiault, Madalina Mirea, R. Slovak, Bonnie Wolhrab, Santosh B. Satbhai, and Wolfgang Busch. 2019. "Root System Depth in Arabidopsis Is Shaped by EXOCYST70A3 via the Dynamic Modulation of Auxin Transport." *Cell*. doi:10.1016/j.cell.2019.06.021

Ohri, Puja, Renu Bhardwaj, Ravinderjit Kaur, Shivam Jasrotia, Ripu Daman Parihar, Anjali Khajuria, and Nandni Sharma. 2019. "Emerging Trends on Crosstalk of BRS with Other Phytohormones." In *Brassinosteroids: Plant Growth and Development*. doi:10.1007/978–981-13–6058–9_16

Per, Tasir S., M. Iqbal R. Khan, Naser A. Anjum, Asim Masood, Sofi J. Hussain, and Nafees A. Khan. 2018. "Jasmonates in Plants under Abiotic Stresses: Crosstalk with Other Phytohormones Matters." *Environmental and Experimental Botany*. doi:10.1016/j.envexpbot.2017.11.004

Qin, Hua, Zhijin Zhang, Juan Wang, Xinbing Chen, Pengcheng Wei, and Rongfeng Huang. 2017. "The Activation of OsEIL1 on YUC8M Transcription and Auxin Biosynthesis Is Required for Ethylene-Inhibited Root Elongation in Rice Early Seedling Development." *PLoS Genetics*. doi:10.1371/journal.pgen.1006955

Rahman, Zeeshanur, and Ved Pal Singh. 2019. "The Relative Impact of Toxic Heavy Metals (THMs) (Arsenic (As), Cadmium (Cd), Chromium (Cr)(VI), Mercury (Hg), and Lead (Pb)) on the Total Environment: An Overview." *Environmental Monitoring and Assessment*. doi:10.1007/s10661-019-7528-7

Rani, Varsha, R. S. Sengar, Sanjay Kumar Garg, Pragati Mishra, and Pradeep Kumar Shukla. 2023. "Physiological and Molecular Role of Strigolactones as Plant Growth Regulators: A Review." *Molecular Biotechnology*. doi:10.1007/s12033-023-00694-2

Rasafi, Taoufik El, Abdallah Oukarroum, Abdelmajid Haddioui, Hocheol Song, Eilhann E. Kwon, Nanthi Bolan, Filip M. G. Tack, Abin Sebastian, M. N. V. Prasad, and Jörg Rinklebe. 2022. "Cadmium Stress in Plants: A Critical Review of the Effects, Mechanisms, and Tolerance Strategies." *Critical Reviews in Environmental Science and Technology*. doi:10.1080/10643389.2020.1835435

Ruan, Jingjun, Yuexia Zhou, Meiliang Zhou, Jun Yan, Muhammad Khurshid, Wenfeng Weng, Jianping Cheng, and Kaixuan Zhang. 2019. "Jasmonic Acid Signaling Pathway in Plants." *International Journal of Molecular Sciences*. doi:10.3390/ijms20102479

Sabagh, Ayman E.L., Akbar Hossain, Mohammad Sohidul Islam, Muhammad Aamir Iqbal, Khizer Amanet, Muhammad Mubeen, Wajid Nasim, et al. 2021. "Prospective Role of Plant Growth Regulators for Tolerance to Abiotic Stresses." In *Plant Growth Regulators: Signalling under Stress Conditions*. doi:10.1007/978-3-030-61153-8_1

Saha, Indraneel, Mirza Hasanuzzaman, and Malay Kumar Adak. 2021. "Abscisic Acid Priming Regulates Arsenite Toxicity in Two Contrasting Rice (Oryza Sativa L.) Genotypes through Differential Functioning of Sub1A Quantitative Trait Loci." *Environmental Pollution*. doi:10.1016/j.envpol.2021.117586

Saini, Shivani, Navdeep Kaur, and Pratap Kumar Pati. 2021. "Phytohormones: Key Players in the Modulation of Heavy Metal Stress Tolerance in Plants." *Ecotoxicology and Environmental Safety*. doi:10.1016/j.ecoenv.2021.112578

Schmitt, Marco, Toshihiro Watanabe, and Steven Jansen. 2016. "The Effects of Aluminium on Plant Growth in a Temperate and Deciduous Aluminium Accumulating Species." *AoB Plants*. doi:10.1093/aobpla/plw065

Seraj, Ferdous, and Tania Rahman. 2018. "Heavy Metals, Metalloids, Their Toxic Effect and Living Systems." *American Journal of Plant Sciences*. doi:10.4236/ajps.2018.913191

Sharma, Anket, Gagan Preet Singh Sidhu, Fabrizio Araniti, Aditi Shreeya Bali, Babar Shahzad, Durgesh Kumar Tripathi, Marian Brestic, Milan Skalicky, and Marco Landi. 2020. "The Role of Salicylic Acid in Plants Exposed to Heavy Metals." *Molecules*. doi:10.3390/molecules25030540

Sun, Pei, Qiu Ying Tian, Jie Chen, and Wen Hao Zhang. 2010. "Aluminium-Induced Inhibition of Root Elongation in Arabidopsis Is Mediated by Ethylene and Auxin." *Journal of Experimental Botany*. doi:10.1093/jxb/erp306

Sytar, Oksana, Pragati Kumari, Saurabh Yadav, Marian Brestic, and Anshu Rastogi. 2019. "Phytohormone Priming: Regulator for Heavy Metal Stress in Plants." *Journal of Plant Growth Regulation*. doi:10.1007/s00344-018-9886–8

Tai, Zhenglan, Xinqiang Yin, Zhigang Fang, Gaoling Shi, Laiqing Lou, and Qingsheng Cai. 2017. "Exogenous GR24 Alleviates Cadmium Toxicity by Reducing Cadmium Uptake in Switchgrass (Panicum Virgatum) Seedlings." *International Journal of Environmental Research and Public Health*. doi:10.3390/ijerph14080852

Thao, Nguyen Phuong, M. Iqbal R. Khan, Nguyen Binh Anh Thu, Xuan Lan Thi Hoang, Mohd Asgher, Nafees A. Khan, and Lam Son Phan Tran. 2015. "Role of Ethylene and Its Cross Talk with Other Signaling Molecules in Plant Responses to Heavy Metal Stress." *Plant Physiology*. doi:10.1104/pp.15.00663

Tu, Tianli, Shuangshuang Zheng, Panrong Ren, Xianwen Meng, Jiuhai Zhao, Qian Chen, and Chuanyou Li. 2021. "Coordinated Cytokinin Signaling and Auxin Biosynthesis Mediates Arsenate-Induced Root Growth Inhibition." *Plant Physiology*. doi:10.1093/PLPHYS/KIAA072

Verma, Giti, Dipali Srivastava, Shiv Narayan, Pramod Arvind Shirke, and Debasis Chakrabarty. 2020. "Exogenous Application of Methyl Jasmonate Alleviates Arsenic Toxicity by Modulating Its Uptake and Translocation in Rice (Oryza Sativa L.)." *Ecotoxicology and Environmental Safety*. doi:10.1016/j.ecoenv.2020.110735

Vezza, Mariana E., Analía Llanes, Claudia Travaglia, Elizabeth Agostini, and Melina A. Talano. 2018. "Arsenic Stress Effects on Root Water Absorption in Soybean Plants: Physiological and Morphological Aspects." *Plant Physiology and Biochemistry*. doi:10.1016/j.plaphy.2017.11.020

Vishwakarma, Kanchan, Neha Upadhyay, Nitin Kumar, Gaurav Yadav, Jaspreet Singh, Rohit K. Mishra, Vivek Kumar, et al. 2017. "Abscisic Acid Signaling and Abiotic Stress Tolerance in Plants: A Review on Current Knowledge and Future Prospects." *Frontiers in Plant Science*. doi:10.3389/fpls.2017.00161

Wang, Haijiao, Jie Tang, Jing Liu, Jin Hu, Jingjing Liu, Yuxiao Chen, Zhenying Cai, and Xuelu Wang. 2018. "Abscisic Acid Signaling Inhibits Brassinosteroid Signaling through Dampening the Dephosphorylation of BIN2 by ABI1 and ABI2." *Molecular Plant*. doi:10.1016/j.molp.2017.12.013

Wang, Yanping, Lin Li, Tiantian Ye, Yuming Lu, Xi Chen, and Yan Wu. 2013. "The Inhibitory Effect of ABA on Floral Transition Is Mediated by ABI5 in Arabidopsis." *Journal of Experimental Botany*. doi:10.1093/jxb/ers361

Wang, Yun, Salma Mostafa, Wen Zeng, and Biao Jin. 2021. "Function and Mechanism of Jasmonic Acid in Plant Responses to Abiotic and Biotic Stresses." *International Journal of Molecular Sciences* 22 (16). doi:10.3390/ijms22168568

Wiche, Oliver, Balázs Székely, Christin Moschner, and Hermann Heilmeier. 2018. "Germanium in the Soil-Plant System—A Review." *Environmental Science and Pollution Research*. doi:10.1007/s11356-018-3172-y

Xu, L. L., Z. Y. Fan, Y. J. Dong, J. Kong, and X. Y. Bai. 2015. "Effects of Exogenous Salicylic Acid and Nitric Oxide on Physiological Characteristics of Two Peanut Cultivars under Cadmium Stress." *Biologia Plantarum*. doi:10.1007/s10535-014-0475–9

Yadav, Bindu, Vikender Kaur, Om Prakash Narayan, Shashank Kumar Yadav, Ashok Kumar, and Dhammaprakash Pandhari Wankhede. 2022. "Integrated Omics Approaches for Flax Improvement under Abiotic and Biotic Stress: Current Status and Future Prospects." *Frontiers in Plant Science*. doi:10.3389/fpls.2022.931275

Yang, Xiaorui, Yang Bai, Jianxiu Shang, Ruijiao Xin, and Wenqiang Tang. 2016. "The Antagonistic Regulation of Abscisic Acid-Inhibited Root Growth by Brassinosteroids Is Partially Mediated via Direct Suppression of ABSCISIC ACID INSENSITIVE 5 Expression by BRASSINAZOLE RESISTANT 1." *Plant Cell and Environment*. doi:10.1111/pce.12763

Yuan, L. B., Z. H. Peng, T. T. Zhi, Z. Zho, Y. Liu, Q. Zhu, X. Y. Xiong, and C. M. Ren. 2015. "Brassinosteroid Enhances Cytokinin-Induced Anthocyanin Biosynthesis in Arabidopsis Seedlings." *Biologia Plantarum*. doi:10.1007/s10535-014-0472-z

Zhang, Guo Bin, Hong Ying Yi, and Ji Ming Gong. 2014. "The Arabidopsis Ethylene/Jasmonic Acid-NRT Signaling Module Coordinates Nitrate Reallocation and the Trade-off between Growth and Environmental Adaptation." *Plant Cell*. doi:10.1105/tpc.114.129296

Zhang, Qiao Ming, Minggui Gong, Kaiyang Liu, Yanlan Chen, Jiangfeng Yuan, and Qingshan Chang. 2020. "Rhizoglomus Intraradices Improves Plant Growth, Root Morphology and Phytohormone Balance of Robinia Pseudoacacia in Arsenic-Contaminated Soils." *Frontiers in Microbiology*. doi:10.3389/fmicb.2020.01428

Zhang, Weiwei, Zepeng Wang, Jianfei Song, Songqing Yue, and Hongqiang Yang. 2019. "Cd^{2+} Uptake Inhibited by MhNCED3 from Malus Hupehensis Alleviates Cd-Induced Cell Death." *Environmental and Experimental Botany*. doi:10.1016/j.envexpbot.2019.103802

Zheng, Jian, Xiaohua Ma, Xule Zhang, Qingdi Hu, and Renjuan Qian. 2018. "Salicylic Acid Promotes Plant Growth and Salt-Related Gene Expression in Dianthus Superbus L. (Caryophyllaceae) Grown under Different Salt Stress Conditions." *Physiology and Molecular Biology of Plants*. doi:10.1007/s12298-017-0496-x

Zhu, Yi, Fang Xu, Qin Liu, Ming Chen, Xianli Liu, Yanyan Wang, Yan Sun, and Lili Zhang. 2019. "Nanomaterials and Plants: Positive Effects, Toxicity and the Remediation of Metal and Metalloid Pollution in Soil." *Science of the Total Environment*. doi:10.1016/j.scitotenv.2019.01.234

Zulfiqar, Faisal, and Muhammad Ashraf. 2022. "Antioxidants as Modulators of Arsenic-Induced Oxidative Stress Tolerance in Plants: An Overview." *Journal of Hazardous Materials*. doi:10.1016/j.jhazmat.2021.127891

14 Phytoremediation of Metalloids

Alok Ranjan, Shruti Kaushik,
Anil Kumar Singh, and Geetika Sirhindi

14.1 INTRODUCTION

Rapid urbanization and industrialization caused serious metalloids and heavy metal (HM) pollution in the environment. Uncontrolled disposal and excessive accumulation of metalloids and HMs such as arsenic (As), polonium (Po), selenium (Se), cadmium (Cd), mercury (Hg), and lead (Pb) in food crops poses significant risks to human health (Sharma, Katnoria, and Nagpal 2016; Cooper et al. 2020; S. Sharma, Kaur, and Nagpal 2021; de Vries, Römkens, and Schütze 2007). These environmental pollutants in the form of solid, liquid, and gaseous wastes enter the food chain through various routes of exposure and have become the major cause of increasing health problems across the world (Cooper et al. 2020). Uncontrolled production of these metalloids/HMs has critically hampered the soil health too through soil degradation processes including erosion, salinization, and HM contamination (Bakshi, Banik, and He 2018). It has been reported that over recent decades, the global discharge of metalloids/HMs into the environment has reached 22,000 metric tons for Cd (Singh et al. 2003), 939,000 metric tons for copper (Poggere et al. 2023), and 783,000 metric tons for Pb (Beardsley et al. 2021). This huge dispose results in the accumulation of metalloids/HMs in the agricultural soils and water resources, which eventually pose a potential risk to human health (S. Sharma, Kaur, and Nagpal 2021). As (metalloids), Pb, Hg, and Cd (HMs) contamination of land and water resources has now turned out to be a worldwide health risk due to its immutable characteristics, persistence, and biomagnifications. In recent years, approximately ten million people around the world suffer from health problems due to metalloids and HM pollution in soil (Rai et al. 2019). Restoration of land/sites/soil contaminated with metalloids and HM pollutants requires cost effective and environmentally friendly methods of remediation. In the last decades, significant efforts have been employed to eliminate these toxic elements, which are based on physical, chemical, and biological methods (Michael-Igolima, Abbey, and Ifelebuegu 2022). The use of conventional methods is more expensive, especially when compared to more environmentally friendly and cheaper biological methods. Besides the extra cost, physical and chemical methods can negatively alter the physicochemical conditions of the soil and alter the microbial soil community. Biological methods use the capacity of organisms such as plants and/or microorganisms to remove HMs and metalloids from the soil. In pursuit for an economically and environmentally feasible approach toward remediation of metalloids/HMs from soil, biological remediation (bioremediation) has gained much attention in the last decade due to its eco-friendly nature and has been considered as one of the most adequate approaches available (Chon, Lee, and Lee 2011). This approach includes varied techniques, including biodegradation, bioventing, bioleaching, bioaugmentation, biofiltration, biostimulation, and phytoremediation (Chon, Lee, and Lee 2011). Phytoremediation is a plant-based technology, which exploits various natural abilities of plants, namely, selective uptake, translocation, accumulation, and degradation of contaminants, for restoring contaminated soil as well as water resources. This relatively recent and advanced approach has exhibited enormous potential by employing naturally existing or genetically engineered plants and is being considered a green alternative solution against the global hazard of metalloid/HM pollution (Bhat et al. 2022).

DOI: 10.1201/9781032673592-14

14.2 EFFECT OF METALS AND METALLOIDS IN SOILS

HMs and metalloids such as As and Se pose risks when they are deposited in soils due to anthropogenic influences (Sharma, Nagpal, and Kaur 2018). HMs like Zn, Cu, Ni, Cr, and Mo are essential for living organisms, and therefore deficiency situations occur either due to very low amounts of these metals in soil or either low bioavailability in soil. In these cases, when exogenously these metals are supplied, there may be positive biological responses in terms of growth and health of organisms. However, these metals, along with nonessential metals such as Pb, As, Hg, and Cd, tend to accumulate in soils, and when their bioavailability becomes high, toxicity can result. These negative effects can occur in soil microbes, soil fauna, higher animals, plants, and humans (Laishram, Singh, and Alam 2023). A further threat is from radionuclides such as those of U, ^{137}Cs, ^{90}Sr, and ^{3}H in soil and water (Chandra et al. 2023). Of course, these elements may occur at elevated concentrations quite naturally in soils and water bodies and effect plants, animals, and human health. In these situations, it may not be possible or desirable to clean up the soils, but there may be an important role for plants in reducing the exposure of biota to these toxic elements, for instance, by reduced uptake and exclusion from tissues or removing elements like Se in geogenically laden water (Werkneh, Gebretsadik, and Gebru 2023). In fact, where these natural hot spots occur, there may be specialized fauna and flora, like metallophyte vegetation, which may need preservation. Metals and metalloids enter soils and waters due to many processes, including atmospheric deposition from industrial activities or power generation; disposal of wastes such as sewage sludge, animal manures, ash, domestic and industrial wastes, or byproducts; irrigation and flood or seepage waters; and the utilization of fertilizers, lime, or agrochemicals. Radionuclides may build up in some areas due to deliberate or accidental releases related to their use for energy production or for military purposes (Chandra et al. 2023). Unlike nitrate or chloride, many of these elements are relatively strongly retained in the surface of soils and do not readily leach, causing the accumulation that may ultimately pose a threat to humans and biota. However, under some conditions, small amounts of these elements do leach and can be an issue in waters, particularly those used for irrigation or drinking. Key examples here would be radionuclides, As, Se, and Cr (Chandra et al. 2023). Under these conditions, phytoremediation is an important developing technology for the removal of these elements from either soil or water. It has the potential to be low cost and to be applicable to large areas where other methods may be too expensive and where the concentrations of contaminants are too small for other methods to be effective or economically viable.

14.3 BIOREMEDIATION

Bioremediation is the environmentally sustainable process by which living beings such as plants and microorganisms are used to remediate, reduce, or remove HM contamination from the soil and water resources. Bioremediation can be performed by either treating the contaminants directly on the site (*in situ*) or by collecting them, followed by treating them somewhere else (*ex situ*) (S. Sharma, Kaur, and Nagpal 2021). Some of the remediation mechanisms involved are extracellular complexation, precipitation, leaching, intracellular accumulation, and biosorption. Moreover, siderophores and biosurfactants produced by specialized microorganisms act as chelating and complex forming agents, respectively, thereby reducing the solubility and bioavailability of HMs from bioaccumulation (Chon, Lee, and Lee 2011). Although bioremediation methods are diverse (Figure 14.1), most studies on bioremediation are focused on hydrocarbons on account of frequent pollution of soil and groundwater with this particular type of pollutant.

14.3.1 *Ex Situ* Bioremediation

Ex situ bioremediation techniques involve excavation of contaminated soils and thereafter transporting them to another site for treatment.

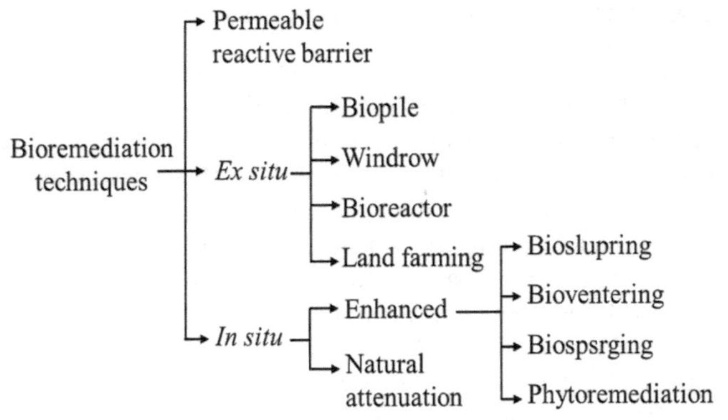

FIGURE 14.1 Bioremediation techniques.

14.3.1.1 Biopile

Biopile is an ex situ bioremediation method. In this method, polluted soil is excavated and followed by nutrient amendment to enhance microbial activities. The application of biopiles to polluted sites limits the volatilization of low-molecular-weight pollutants.

14.3.1.2 Windrows

As an ex situ bioremediation method, windrows rely on periodic turning of polluted soil to enhance bioremediation by increasing degradation activities of indigenous and/or transient hydrocarbon clastic bacteria present in polluted soil. The periodic turning of polluted soil, together with the addition of water brings about an increase in aeration, uniform distribution of pollutants, nutrients, and microbial degradative activities, thus speeding up the rate of bioremediation, which can be accomplished through assimilation, biotransformation, and mineralization (Hawari et al. 2000).

14.3.1.3 Bioreactor

A bioreactor is a vessel in which raw or pollutant materials are converted to specific product(s) following a series of biological reactions. There are several operating modes of bioreactors, which include batch, fed-batch, sequencing batch, continuous, and multistage. Conditions in a bioreactor support the natural process of cells by mimicking and maintaining their natural environment to provide optimum growth conditions (Robles-González, Fava, and Poggi-Varaldo 2008). Polluted samples can be fed into a bioreactor either as dry matter or slurry; in either case, the use of a bioreactor in treating polluted soil has several advantages compared to other *ex situ* bioremediation techniques. The ability to control and manipulate process parameters (temperature, pH, agitation and aeration rates, substrate, and inoculum concentrations) in a bioreactor implies that biological reactions within can be enhanced to effectively reduce bioremediation time. Importantly, it can be used to treat soil or water polluted with volatile organic compounds such as benzene, toluene, ethylbenzene, and xylenes.

14.3.1.4 Land Farming

Land farming is the simplest and most economical bioremediation method. Pollutant depth plays an important role as to whether land farming can be carried out *ex situ* or *in situ*. In land farming, polluted soils are usually excavated and/or tilled, but the site of treatment apparently determines the type of bioremediation. When excavated polluted soil is treated on-site, it can be regarded as *in situ*; otherwise, it is *ex situ*, as it has more in common with other *ex situ* bioremediation techniques. When pollutant lies 1 m below ground surface, bioremediation might proceed without excavation, while pollutant lying more than 2 m needs to be transported to ground surface for bioremediation (Kaur et al. 2020). Land farming is an effective method for remediation of hydrocarbon-polluted sites, including polyaromatic hydrocarbons (Kaur et al. 2020). Land farming systems can be used in any climate and location. Over all, the land farming bioremediation method is very simple to design and implement, requires low capital input, and can be used to treat large volumes of polluted soil with minimal environmental impact and energy requirement (Macci et al. 2021). The major advantages of *ex situ* bioremediation methods are that they do not require a preliminary assessment of the polluted site before remediation. In ex situ bioremediation processes allow chemical modification of pollutants, which is necessary for effective and efficient bioremediation. Importantly, the major influence of soil structure, which regulates transport processes during remediation, can be reduced when polluted soils are excavated. *Ex situ* bioremediation techniques are unlikely to be used in some sites, such as under buildings and working sites.

14.3.2 *In Situ* Bioremediation Method

These methods involve treating polluted materials at the site of pollution. It does not need any excavation; therefore, it is accompanied by little or no disturbance to the soil structure. These methods are to be less expensive compared to *ex situ* bioremediation methods. Some *in situ* bioremediation techniques might be increased biosparging, bioventing, and phytoremediation, while others might proceed without any form of enhancement like intrinsic bioremediation or natural attenuation. The *in situ* bioremediation methods have been successfully used to treat HMs, chlorinated solvents, dyes, and hydrocarbons polluted sites (P. Sharma et al. 2021). Importantly, the status of electron acceptor/donor, moisture content, surrounding pH, nutrient compositions, and temperature are among the important environmental factors that need to be suitable for a successful *in situ* bioremediation to be achieved (P. Sharma et al. 2021). Unlike *ex situ* bioremediation, soil texture strongly influences the application of *in situ* bioremediation to any polluted site.

14.3.3 Enhanced *In Situ* Bioremediation

14.3.3.1 Bioventing

In this method, oxygen is supplied in a controlled way to increase the activities of indigenous microbes. By adding nutrients and moisture, enhance the bioremediation process by enhancing microbial activity to transform pollutants to a harmless state (Sayqal and Ahmed 2021). This is common and most popular among other *in situ* bioremediation methods, especially in restoring sites polluted with light spilled petroleum products (Anekwe and Isa 2022).

14.3.3.2 Bioslurping

This method combines bioventing and vacuum-enhanced pumping soil vapor extraction to achieve soil and groundwater remediation by indirect provision of oxygen and stimulation of contaminant biodegradation (Kim et al. 2014). Bioslurping can also be used to remediate soils contaminated with volatile and semi-volatile organic compounds. The system uses a slurp that extends into the free product layer, which draws liquids from this layer in a manner similar to that of how a straw draws liquid from any vessel. In this method, excessive soil moisture limits air permeability and decreases oxygen transfer rate, in turn reducing microbial activities. Although the method is not suitable for remediating soil with low porocity, it saves cost due to the less amount of groundwater resulting from the operation, thus minimizing storage, treatment, and disposal costs (Akubude et al. 2022).

14.3.3.3 Biosparging

Pollutant removal by using the biosparging method is very similar to bioventing in that air is injected into the soil subsurface to stimulate and enhance microbial activities in order to increase pollutant removal from polluted sites. Although, unlike bioventing, oxygenated air is injected at the saturated zone, which can cause upward movement of volatile organic compounds to the unsaturated zone to promote biodegradation. The effectiveness of biosparging depends on two important factors, namely: soil porocity, which determines pollutant bioavailability to microorganisms, and pollutant biodegradability (Machackova et al. 2012). As with bioventing and soil vapor extraction, biosparing is similar in operation with a closely related method known as *in situ* air sparging, which relies on high airflow rates to achieve pollutant volatilization, whereas biosparging promotes biodegradation (P. Sharma et al. 2021). This method has been largely used in treating aquifers contaminated with hydrocarbons, especially petroleum.

14.3.3.4 Permeable Reactive Barrier

A permeable reactive barrier is a physical method for remediating contaminated groundwater due to its design and mechanism of pollutant removal. Nevertheless, several researchers reported that biological reaction is one of the several mechanisms (degradation, precipitation, and sorption) of pollutant removal by this method. However, alternative terms such as biological permeable reactive barrier, passive bioreactive barrier, bio-enhanced permeable reactive barriers have been proposed to accommodate the bioremediation or biotechnology aspect of the technique; the role of many microorganisms has been reported to be mostly enhancement rather than an independent biotechnology (Budania and Dangayach 2023). Permeable reactive barrier is an *in situ* method used for remediating groundwater polluted with HMs and chlorinated compounds (Miller, Menashe, and Dosoretz 2022). The permanent or semi-permanent reactive barrier, mostly made up of a zero-valent iron, is submerged in the polluted groundwater. As polluted water flows through the barrier under its natural gradient, pollutants become trapped and undergo a series of reactions resulting in clean water in the flow through. Ideally, the barriers are usually reactive enough to trap pollutants, permeable to allow the flow of water but not pollutants, passive with little energy input, inexpensive, readily available, and accessible. The effectiveness of this method depends mostly on the type of media used, which is influenced by pollutant type, biogeochemical and hydrogeological conditions, environmental and health influence, mechanical stability, and cost (Miller, Menashe, and Dosoretz 2022).

14.3.3.5 Intrinsic Bioremediation

Intrinsic bioremediation or natural attenuation is an *in situ* bioremediation method, which involves passive remediation of polluted sites, without any external intervention. The process depends on both microbial aerobic and anaerobic processes to biodegrade pollutants, including those that are recalcitrant. The absence of external intervention implies that the method is less expensive compared to other *in situ* methods. However, the process must be monitored in order to facilitate that bioremediation is ongoing and sustainable; hence, the term monitored natural attenuation. One of the major limitations of intrinsic bioremediation is that it might take a longer time to achieve the target level of pollutant concentration, given that no external intervention is incorporated to expedite the remediation process. It thus follows that prior to application of intrinsic bioremediation, risk assessment needs to be carried out to ensure that remediation time is less than the time stipulated for pollutant to reach exposure point relative to the closest human and animal populations. Moreover, it was reported that intrinsic bioremediation does not result in adequate polyaromatic hydrocarbon removal and corresponding reduction in polluted soil ecotoxicity.

14.4 PHYTOREMEDIATION

Phytoremediation is a plant-based bioremediation technique, which relies on the ability of plants to take up, translocation, accumulation, and degradation of contaminants for restoring contaminated sites. There are several steps and mechanisms involved in phytoremediation, which include accumulation, extraction, degradation, filtration, stabilization, volatilization, and degradation of contaminants. Pollutants like toxic HMs and radionuclides are mostly removed by extraction, transformation, and sequestration. However, organic pollutants like hydrocarbons and chlorinated compounds are mostly removed by degradation, rhizoremediation, stabilization, and volatilization

(Yan et al. 2020). Phytoremediation can be further categorized according to the mechanism involved in HM uptake, which includes phytoextraction, phytofiltration, phytostabilization, phytovolatilization, and phytotransformation (Yan et al. 2020). Various plant species, such as *Azolla caroliniana*, *Ipomea carnea*, *Typha latifolia*, *Ricinus communis*, and *Saccharum munja*, have been identified to remediate HMs from polluted sites (Kafle et al. 2022). When considering a plant as a phytoremediator, which depends on some important factors, include: root system, length of tap root on the depth of pollutant, above ground biomass, which should not be available for animal consumption, toxicity of pollutant to plant, plant survival, and its adaptability to prevailing environmental conditions. In addition, the plant should be resistant to diseases and pests.

14.4.1 Biochemical and Cellular Mechanism of Phytoremediation

Exposure of plants to HM pollutants triggers detoxification mechanisms in the forms of diverse cellular and molecular perturbations (Ranjan, Sinha, Sharma, et al. 2021; Ranjan, Sinha, Lal, et al. 2021). Strategies involve various plant parts that help in uptake, transport, sequestration, and detoxification of pollutants (Kaushik et al. 2022). Moreover, plant microbe interactions in the rhizosphere as well as plant chelator-based sequestrations also form part of the cellular and biological machinery (Aljerf 2018). The uptake of HMs from contaminants soil by plants is the main task for the remediation process. For enhanced uptake of HMs, plants release nutrients, organic acids (citrate, malate, and oxalate), oxygen, and moisture in the soil (Ranjan, Sinha, Sharma, et al. 2021). This alters the pH and redox potential and creates a rich environment for microbial activity, thereby enhancing the tolerance of plants toward pollutants and at the same time making the pollutant bioavailable for uptake by the plant (Han et al. 2021). The HMs enter into the root cells via two routes, apoplastic and symplastic pathways, marking the first step in phytoremediation (Song, Jin, and Wang 2017). Different types of metal transporters, which are involved in HM transport and homeostasis were identified in the plasma membrane, such as the Zn/Fe permeases (ZIP) (Khouja et al. 2013), the natural resistance-associated macrophage proteins (Nramp) (Moore et al. 2020), the cation exchangers (CAXs), cation diffusion facilitators (CDF) (Kotp et al. 2019), and ATP-binding cassettes (ABC) (Guo et al. 2022). One of the major consequences of metalloid/HM stress is oxidative stress that results from excessive production of reactive oxygen species (ROS), viz., superoxide radicals ($O2^{\cdot-}$), hydroxyl radicals (OH^-), hydrogen peroxide (H_2O_2), etc. In that situation, plants possess some anti-oxidative defense systems to scavenge toxic free radicals such as β-carotene, glutathione (GSH), superoxide dismutase (SOD), ascorbate peroxidase (APX), and catalase (CAT) (Ranjan, Sinha, Sharma, et al. 2021). However, some plants possess specialized anatomical structures to eliminate metals from their tissues. In this case, the excess of metals can excrete through salt glands in the form of non-toxic crystals. This method is called phytoexcretion (Roychowdhury et al. 2021). Once transported into the different tissues of the shoot, the toxic HMs are sequestered or compartmentalized in the vacuoles (Andosch et al. 2015).

14.4.2 Different Strategy of Phytoremediation

Some plants accumulate HMs such as Ag, Cd, As, Hg, and Pb, which have no recognized biological functions. However, accumulation of these HMs can be toxic to most plant species. There are five categories of phytoremediation methods applied in remediation of contaminated soils. These include phytoextraction, phytotransformation, rhizofiltration, phytostabilization, and phytovolatilization.

Phytoextraction: Plant species, which have the capacity to absorb and accumulate HMs are planted in polluted area. The criteria of selection of these plants are based on the degree of absorption and translocation of the contaminant from the roots to the shoot (Corzo Remigio et al. 2020). In phytoextraction methods, two types of strategies are used: continuous phytoextraction and induced phytoextraction. The continuous phytoextraction method uses natural hyperaccumulator plant species (endemic species), which have the natural abilities to accumulate high contents of metalloids and HMs. These plant species are called hyperaccumulators. Hyperaccumulator plant species are capable of accumulating metalloids/HMs more than 100-fold than non-accumulators (normal plant species). Induced phytoextraction enhanced plant metal accumulation by exogenous application of chemical substances like chelates (Garbisu and Alkorta 2001). Phytoextraction is enhanced by plants with a high growth rate and deeper root systems.

Phytotransformation: Phytotransformation, also known as phytodegradation, involves the degradation of pollutants assimilated or accumulated by plants through metabolic processes or the degradation of pollutants outside the plant through enzymes produced by plants mainly through roots (Ashraf et al. 2019). Transgenic yellow poplar (*Liriodendron tulipifera*) grows under *in vitro* conditions with high Hg concentrations and transforms the latter from highly toxic Hg^{2+} to the much less toxic form Hg^0 (Rugh, Merkle, and Meagher 2019). Through this method, plants are able to transform toxic contaminants into non-hazardous components.

Rhizofiltration: Rhizofiltration is mainly applied to eliminate pollutants from water waste by the precipitation of toxins onto surface roots or the absorption of soluble pollutants into the roots (Kristanti et al. 2021). The larger fibrous root system or root architecture with large surface areas plays a significant role in this technique. Due to the large surface area of roots, plants are able to absorb a large quantity of water with contaminants from the soil, which is appropriate for this purpose. Plant species like *Typha latifolia* could be efficacious to remove methyl parathion (MeP) from hydromorphic soils (Papadopoulos and Zalidis 2019). *Phaseolus vulgaris* efficiently removes uranium (Ur) and cesium (Cs) from groundwater (Yang, Jawitz, and Lee 2015). *Arundo donax* is an effective poaceae to rhizofiltrate Cu from wetlands (Oustriere et al. 2017). *Eichhornia crassipes*, *Salvinia molesta*, and *Pistia stratiotes* are promising terrestrial candidates for removing HMs, mainly Cu, Cr, Cd, Ni, and Zn, from industrial sewage sludge (Ashraf et al. 2019).

Phytostabilization: In this strategy, plant root exudates precipitate soil pollutants. Thus, this method minimizes the movement of the pollutants and reduces their migration to groundwater. The halophytes plant species *Atriplex nummularia* and *Tamarix smyrnensis* are reported to have potential for phytostabilization of HMs/metalloids. Similarly, some plant species from the Poaceae family, such as *Spartina argentinensis* and *Spartina densiflora*, have the ability for phytostabilization and accumulation of HMs/metalloids such as As, Pb, and Cd (Nedjimi and Daoud 2009). The root system architecture and depth of roots were the principal characteristics of plants suitable for phytostabilization (Mendez and Maier 2008). Maximum contact with the soil matrix was provided by fibrous root system extents throughout the soil. Although root depth varies significantly according to plant species, moisture, soil texture, calcareous encrust, dry conditions, and soil amendments. Through the excretion of root exudates, some plant species can change metal solubility. *Sorghum bicolor* and *Carthamus tinctorius* are able to uptake Ni, Pb, and Zn, and their root exudate changes metal solubility (Moreira et al. 2021). Furthermore, roots accumulated more metals than leaves. *Erica australis* possesses the ability to uptake Cu, Cd, and Pb via its roots without any sign of damage and is suggested as a phytostabilizator species (Abreu, Tavares, and Batista 2008). *Helichrysum microphyllum* is suitable for phytostabilization of Cd and Pb (Mandaresu et al. 2023).

14.4.2.1 Phtovolatization

In the phytovolatilization process, plants uptake pollutants and convert them into less toxic, volatile forms. The phytovolatilization process involves uptake and assimilation of pollutants by roots, translocation of less toxic forms to the shoot system, and ultimate release of less toxic vapors of pollutants into the atmosphere through leaves stomata. Indian mustard (Brassica napus) and Arabidopsis thaliana potentially converted organic Hg salts to the volatile elemental form. Brassica juncea grown hydroponically was reported to remove up to 95% of Hg by both phytovolatilization and plant accumulation. Indian mustard (*Brassica napus*) and *Arabidopsis thaliana* converted organic Hg salts to the volatile elemental form. *Brassica juncea* grown hydroponically was reported to remove up to 95% of Hg by both phytovolatilization and plant accumulation (Bortoloti and Baron 2022).

14.4.3 Next Generation and Innovative Methods for Phytoremediation

14.4.3.1 Microbe-Based Phytoremediation

Bioremediation is the remediation of pollutants from pollutant sites by using microbes. Plant growth-promoting bacteria (PGPB) are able to colonize the plant rhizosphere and stimulate the growth and mineral uptake in plants. These bacteria have a potential to degrade toxic pollutants or to convert them to less toxic forms (Raklami et al. 2022). Various PGPB have been reported to enhance the phytoremediation capacity of plants by allowing the roots to uptake HMs. These bacteria play an important role in HM decontamination by secreting different chemical substances such as siderophores and organic acids, which enhance the bioavailability of HMs by decreasing the soil pH (Kong and Glick 2017). Some other bacteria have been reported to secrete polymeric compounds such as polysaccharides, which contribute to phytostabilization of HMs by reducing their mobility (Kong and Glick 2017). Plant growth-promoting rhizobacteria (PGPR) play a vital role in the phytoremediation processes by improving the detoxification rates of plants, enhancing enzymes in root secretion, leading to accelerated pollutant degradation, and modifying soil pH (Gupta et al. 2023).

Thus, many strains of bacteria were found to increase HM tolerance of plants. Arthrobacter inoculated *Ocimum gratissimum*, inducing the phytoextraction of Cd by roots (Choden et al. 2021). *Bradyrhizobium* sp. promotes Cd uptake by *Lolium multiflorum* and Glycine max (Guo and Chi 2014). *Paxillus involutus* strain enhances growth and promotes Pb tolerance in *Populus canescens* (Szuba et al. 2017). *Pteris vittata* accumulates about 170% of As in its organs when inoculated with *Cupriavidus basilensis* strain (Singh et al. 2006).

14.4.3.2 Mycorrhizal Fungi-Assisted Phytoremediation

The arbuscular mycorrhizal fungi (AMF) are symbiotic fungi that association with plant root and increase mineral availability (Rajtor and Piotrowska-Seget 2016). The mycorrhizal fungi decontaminate HMs by immobilizing and adsorbing them to fungal cell walls. It also promotes phytoextraction of HMs by improving plant growth and increasing HM uptake in the rhizosphere by modifying the chemical composition of root exudate and/or reducing the soil pH (Rajtor and Piotrowska-Seget 2016). The inoculation of *Cassia italica* by AMF prevents Cd translocation to shoot (Raza et al. 2020). The *Trigonella foenumgraecum* plants inoculated with AMF were found to be Cd hyperaccumulative and improve soil quality by neutralizing soil pH and reducing Cd content in soil (Arias et al. 2015). The Glycine *max* plants inoculated with *Bradyrhizobium* and japonicum E109 and *Azospirillum brasilense* Az39 exhibit tolerance to As stress. Similarly, exogenous inoculation of *Artemisia annua* by *Piriformospora indica* enhances tolerance against As stress (Armendariz et al. 2019).

14.4.3.3 Use of Earthworm for Phytoremediation

Earthworms are the main macrofauna biomass in soil. Earthworms promote organic matter decomposition and mineralization, which results in the increased soil organic matter dynamics. Earthworms have an important role in the mineralization of nitrogen and phosphorus and alter microbial functional genes, which modifies soil functions (Wang et al. 2022). Earthworms secrete some organic acids, such as fulvic and humic acids, through their gut microflora and help in the decrease of soil pH, which enhances the nutrient and HM bioavailability in rhizosphere. Integration of earthworms in soil enhances the phytoremediation capacity of Cd in *Solanum nigrum,* Cr in Pontoscolex *corethrurus*, and Ni in *Acacia mangium*. In combination, earthworm *Eisenia andrei* with AMF *Rhizoglomus clarum* increases the phytoextraction capacity of *Canavalia ensiformis* plants contaminated with Cu. *Eisenia fetida* earthworm inoculated with *Brassica juncea* plants enhances detoxification efficiency of Cd (Xiao et al. 2022). The addition of vermicompost using *Ensenia andrei* to HM contaminated soil increases the ability of *Avena strigosa* plants to remediate Pb, Cd, and Cr (Wang et al. 2022).

14.4.3.4 Phytohormone-Assisted Phytoremediation

Phytohormones such as auxins (IAA), cytokinins, gibberellins, jasmonic acid, and abscisic acid (ABA) are known to positively increase the accumulation of HMs and also provide HM tolerance abilities to plants. The exogenous treatment of phytohormones during the earlier growth stage can help plants escape toxicity upon exposure to HMs. Exogenous treatment of auxin offers a promising strategy to improve tolerance against Cd with minor damages. The exogenous application of ABA reduces Cd accumulation in shoots of some plants and enhances the tolerance of plants to Cd. ABA-catabolizing bacteria *R.qingshengii* facilitates Cd and Ni uptake from soil by host plants by stimulation of HM/metalloids uptake-related transporters. *R. qingshengii* produce ABA and enhance the phytoextraction ability of plants from contaminant soil (Cheng et al. 2022).

14.4.3.5 Application of Nanoparticles in Phytoremediation

The addition of nanoparticles in rhizospheres enhances the removal efficiency of HMs by plants (Akubude et al. 2022). The nanoparticles chemically interact with HMs by adsorption reaction and stimulate plant growth and facilitate phytoremediation. Nanoparticles promote plant growth by increasing the nutrient uptake efficiency of plants simultaneously with HMs (Zhou et al. 2021). The addition of silver nanoparticles and nano-TiO_2 in rhizospheric soils enhances the removal potential of Cd in maize and soybean (*Glycine max*) plants (Sardar, Ahmed, and Yasin 2022). Supplementation of salicylic acid nanoparticles improves As remediation potential of *Isatis cappadocica* (Souri et al. 2017). The combination between nanoparticles such as hydroxyapatite, hematite, and maghemite NPs decreased the bioavailability of HMs in soils (Zhou et al. 2021).

14.4.4 Transgenic Approaches

Genetic engineering is one of the vital tools of biotechnology. Genetic engineering is used to modify plants and microorganisms by DNA manipulation and genome transformation to integrate new genes to enhance the uptake and translocation of HMs (Gunarathne et al. 2018). A variety of plants have been engineered for enhanced uptake and translocation of HMs. Hyperaccumulator plants can additionally be improved by using genetic engineering approaches to overcome some limitations of phytoremediation. Genes isolated from microbes, fungi, and plants involved in sequestration and degradation of HMs were used to make transgenic plants. Two bacterial genes, namely, mercuric ion reductase (merA) and organomercurial lyase (merB), are known to be involved in Hg detoxification (Krout et al. 2022). The merB converts toxic Hg into the less toxic ionic form Hg^{2+}, and further merA reduces Hg^{2+} into volatilizing from Hg. The SbMT-2 gene isolated from *Salicornia brachiate* showed Cu, Zn, and Cd tolerance in transgenic *Nicotiana tabacum* (Chaturvedi et al. 2014). Macrophage protein (Nramp) is known to play a key role in HM decontamination. The overexpressed SaNramp6 (*Sedum alfredii*) increases Cd accumulation in transgenic *Arabidopsis thaliana* (Chen et al. 2017). Similarly, the AtACR2 gene (arsenic reductase 2) isolated from *Arabidopsis thaliana* has an important role in As decontamination in transgenic tobacco (Nahar et al. 2017). The protein transporters such as COPT5-transporter; VIT-transporter, H^+-ATPase; Na^+/H^+ antiporter regulate uptake, xylem/phloem loading and unloading, and sequestration of HMs (Kaur et al. 2021). It was observed that expression of VIT, COPT5, IRT1, HMA3, and MATE55 is responsible for HM uptake, sequestration, and exclusion through vacuole. *Sedum plumbizincicola* is a hyperaccumulator crassulaceae, in which P1B-type ATPase (HMA4 and HMA2) transporters were overexpressed under the Cd stress (Zhao et al. 2019). Metallothioneins (MTs), phytochelatins (PCs), and GSH play an important role in HM detoxification through the chelating of toxic metals and their transfer to vacuoles (Cobbett and Goldsbrough 2002). The overexpression of genes such as GSH synthase (GS) and PC synthase (PCS) involved in PC biosynthetic pathway has contributed to enhancing detoxification of HMs (Cobbett and Goldsbrough 2002).

14.4.5 Advantages and Disadvantages of Phytoremediation

Phytoremediation is a biological technique, which has several advantages over physical or chemical remediation methods, but this technique also has some disadvantages. The major advantages of using plants for remediation of contaminant soil are the benefits provided by the plant itself, such as a feeling of wellbeing compared to the landscape alteration. The plants provide a permanent reduction of *in situ* contamination due to wind and water soil erosion. Moreover, the plants biomass can be used as a source of energy that can be obtained through its combustion or anaerobic decomposition (Ali Shah et al. 2014). The disadvantages over physical and chemical methods are the time required, as plants require more time to grow. Another disadvantage is that hyperaccumulator plants can be consumed by herbivores, which may transfer these elements into the food chain (Gavrilescu 2022). In addition, the introduction of exogenous species into an ecosystem can induce its invasion and disturb the existing ecological equilibrium (Prabakaran et al. 2019). Though, considering the endurance of metalloids in the environment, methods based on the use of plant species are still interesting. The application of genetic engineering is a possibility to establish species capable of remediating larger areas of soil in a shorter duration. Simultaneously, combination with agronomic practices such as use of co-culture, crop rotation, and use of biofertilizer can lead to beneficial effects on phytoremediation.

14.5 FUTURE PROSPECTS

Phytoremediation is a promising technique for the remediation and eco-rehabilitation of contaminated sites. For efficient phytoremediation, some promising plant species need to be identified; further investigations and research are also necessary to enhance our knowledge in efficient phytoremediation. It has become an essential need to identify genes, mechanisms, and metabolites using omics techniques. Omics analysis can identify new genes and metabolites implicated in the stabilization or extraction of HMs. This can be achieved by finding a new hyperaccumulator species or by genetic engineering. Hyperaccumulator plant species are able to absorb and accumulate HMs and metalloids and neutralize these toxic elements through specific plant processes. Therefore, identifying new genes and pathways, which may also play a role in the elevation of metals and metalloids phytoremediation. Overexpression of foreign genes in plants for HM remediation of soil is practicable. However, numerous plants and shrubs were established to decontaminate HMs through genetic engineering, no perfect model can be established for HM detoxification until the availability of whole genome data is certified. Intense works for the role of plant hormones, such as auxin, jasmonate, cytokine, and ABA were also needed to understand their role to increasing the potential of HM detoxification plants.

Additional advanced techniques, including the study of the hologenome of the symbiotic microbes of plants, could be applied in the manipulation of microbial niches to further enhance the resistance of microbes against HMs. Nanoremediation is a recent technology assisted to enhance remediation of contaminated soils. Still, the relation between the biochemical and molecular approaches of phytoremediation and nanoparticles needs to be elucidated to expand the prospect of polluted soil remediation.

Finally, the success of phytoremediation can be achieved through the contribution and coordination of researchers, industrial sectors, farmers, and environment authorities through educational programs to ensure the long-term sustainability of this eco-friendly green technology.

14.6 CONCLUSION

HM pollution is a global concern and a major threat for agriculture and health worldwide. The phytoremediation process is a less expensive and environmentally friendly technology compared to other methods of HM remediation. Phytoremediation applied different mechanisms, such as HM uptake, breakdown, and transformation, emission in the atmosphere and their stabilization in the plant root system.

To increase the phytoremediation potential of the plants, different bioremediation strategies like genetic engineering and the application of phytohormones, microbes, and nanoparticles have been widely used. More comprehensive research under field conditions, selection of suitable plants, identification of new genes, and development of genetically modified plants and microbes will be useful to increase the knowledge and understanding of the metabolic activities involved in HM tolerance in hyperaccumulator plants and unlock new directions for phytoremediation.

ACKNOWLEDGMENTS

AR thanks to the Department of Biotechnology, Patna Women's College, Patna, India, for financial help under the College with Potential for Excellence (CPE) and Basic Science Research (BSR) Project. SK thanks to the Department of Botany Punjabi University, Patiala, for providing research facilities. AKS thanks the Indian Council of Agricultural Research (ICAR), India, for funding. GS thanks to DSA-I of UGC New Delhi for financial support.

REFERENCES

Abreu, M. M., M. T. Tavares, and M. J. Batista. 2008. "Potential Use of Erica Andevalensis and Erica Australis in Phytoremediation of Sulphide Mine Environments: São Domingos, Portugal." *Journal of Geochemical Exploration*. doi:10.1016/j.gexplo.2007.04.007

Akubude, V. C., T. F. Oyewusi, V. C. Okafor, P. C. Obumseli, and A. O. Igwe. 2022. "Application of Nanomaterials in the Bioaugmentation of Heavily Polluted Environment." In *Bioaugmentation Techniques and Applications in Remediation*. doi:10.1201/9781003187622-7

Ali Shah, Fayyaz, Qaisar Mahmood, Mohammad Maroof Shah, Arshid Pervez, and Saeed Ahmad Asad. 2014. "Microbial Ecology of Anaerobic Digesters: The Key Players of Anaerobiosis." *The Scientific World Journal*. doi:10.1155/2014/183752

Aljerf, L. 2018. "A Gateway to Metal Resistance: Bacterial Response to Heavy Metal Toxicity in the Biological Environment." *Annals of Advances in Chemistry*. doi:10.29328/journal.aac.1001012

Andosch, Ancuela, Margit Höftberger, Cornelius Lütz, and Ursula Lütz-Meindl. 2015. "Subcellular Sequestration and Impact of Heavy Metals on the Ultrastructure and Physiology of the Multicellular Freshwater Alga Desmidium Swartzii." *International Journal of Molecular Sciences*. doi:10.3390/ijms160510389

Anekwe, Ifeanyi Michael Smarte, and Yusuf Makarfi Isa. 2022. "Bioremediation of Crude Oil-Contaminated Soils – A Review." *Petroleum and Coal*. doi: 10.1016/j.promfg.2016.12.037

Arias, Milton Senen Barcos, Juan José Peña-Cabriales, Alejandro Alarcón, and María Maldonado Vega. 2015. "Enhanced Pb Absorption by Hordeum Vulgare L. and Helianthus Annuus L. Plants Inoculated with an Arbuscular Mycorrhizal Fungi Consortium." *International Journal of Phytoremediation*. doi:10.1080/15226514.2014.898023

Armendariz, Ana L., Melina A. Talano, María Florencia Olmos Nicotra, Leticia Escudero, María Laura Breser, Carina Porporatto, and Elizabeth Agostini. 2019. "Impact of Double Inoculation with Bradyrhizobium Japonicum E109 and Azospirillum Brasilense Az39 on Soybean Plants Grown under Arsenic Stress." *Plant Physiology and Biochemistry*. doi:10.1016/j.plaphy.2019.02.018

Ashraf, Sana, Qasim Ali, Zahir Ahmad Zahir, Sobia Ashraf, and Hafiz Naeem Asghar. 2019. "Phytoremediation: Environmentally Sustainable Way for Reclamation of Heavy Metal Polluted Soils." *Ecotoxicology and Environmental Safety*. doi:10.1016/j.ecoenv.2019.02.068

Bakshi, Santanu, Chumki Banik, and Zhenli He. 2018. "The Impact of Heavy Metal Contamination on Soil Health." In *Managing Soil Health for Sustainable Agriculture*. doi:10.19103/as.2017.0033.20

Beardsley, Chloe A., Kai Z. Fuller, Thomas H. Reilly, and Charles S. Henry. 2021. "Method for Analysis of Environmental Lead Contamination in Soils." *Analyst*. doi:10.1039/d1an01744f

Bhat, Shakeel Ahmad, Omar Bashir, Syed Anam Ul Haq, Tawheed Amin, Asif Rafiq, Mudasir Ali, Juliana Heloisa Pinê Américo-Pinheiro, and Farooq Sher. 2022. "Phytoremediation of Heavy Metals in Soil and Water: An Eco-Friendly, Sustainable and Multidisciplinary Approach." *Chemosphere*. doi:10.1016/j.chemosphere.2022.134788

Bortoloti, Gabriel Antonio, and Daniel Baron. 2022. "Phytoremediation of Toxic Heavy Metals by Brassica Plants: A Biochemical and Physiological Approach." *Environmental Advances*. doi:10.1016/j.envadv.2022.100204

Budania, Ravindra, and Sanyam Dangayach. 2023. "A Comprehensive Review on Permeable Reactive Barrier for the Remediation of Groundwater Contamination." *Journal of Environmental Management*. doi:10.1016/j.jenvman.2023.117343

Chandra, Krishno, Ram Proshad, Hridoy Chandra Dey, and Abubakr M. Idris. 2023. "A Review on Radionuclide Pollution in Global Soils with Environmental and Health Hazards Evaluation." *Environmental Geochemistry and Health*. doi:10.1007/s10653-023-01725-2

Chaturvedi, Amit Kumar, Manish Kumar Patel, Avinash Mishra, Vivekanand Tiwari, and Bhavanath Jha. 2014. "The SbMT-2 Gene from a Halophyte Confers Abiotic Stress Tolerance and Modulates ROS Scavenging in Transgenic Tobacco." *PLoS One*. doi:10.1371/journal.pone.0111379

Chen, Shuangshuang, Xiaojiao Han, Jie Fang, Zhuchou Lu, Wenmin Qiu, Mingying Liu, Jian Sang, Jing Jiang, and Renying Zhuo. 2017. "Sedum Alfredii SaNramp6 Metal Transporter Contributes to Cadmium Accumulation in Transgenic Arabidopsis Thaliana." *Scientific Reports*. doi:10.1038/s41598-017-13463-4

Cheng, Lang, Lei Pu, Ai Li, Xiaping Zhu, Ping Zhao, Xiangning Xu, Ningfei Lei, and Jinsong Chen. 2022. "Implication of Exogenous Abscisic Acid (ABA) Application on Phytoremediation: Plants Grown in Co-Contaminated Soil." *Environmental Science and Pollution Research*. doi:10.1007/s11356-021-16241-y

Choden, Deki, Prayad Pokethitiyook, Toemthip Poolpak, and Maleeya Kruatrachue. 2021. "Phytoremediation of Soil Co-Contaminated with Zinc and Crude Oil Using Ocimum Gratissimum (L.) in Association with Pseudomonas Putida MU02." *International Journal of Phytoremediation*. doi:10.1080/15226514.2020.1803205

Chon, Hyo Taek, Jin Soo Lee, and Jong Un Lee. 2011. "Heavy Metal Contamination of Soil, Its Risk Assessment and Bioremediation." *Geosystem Engineering*. doi:10.1080/12269328.2011.10541350

Cobbett, Christopher, and Peter Goldsbrough. 2002. "Phytochelatins and Metallothioneins: Roles in Heavy Metal Detoxification and Homeostasis." *Annual Review of Plant Biology*. doi:10.1146/annurev.arplant.53.100301.135154

Cooper, Andrew M., Didra Felix, Fatima Alcantara, Ilya Zaslavsky, Amy Work, Paul L. Watson, Keith Pezzoli, et al. 2020. "Monitoring and Mitigation of Toxic Heavy Metals and Arsenic Accumulation in Food Crops: A Case Study of an Urban Community Garden." *Plant Direct*. doi:10.1002/pld3.198

Corzo Remigio, Amelia, Rufus L. Chaney, Alan J. M. Baker, Mansour Edraki, Peter D. Erskine, Guillaume Echevarria, and Antony van der Ent. 2020. "Phytoextraction of High Value Elements and Contaminants from Mining and Mineral Wastes: Opportunities and Limitations." *Plant and Soil*. doi:10.1007/s11104-020-04487-3

Garbisu, Carlos, and Itziar Alkorta. 2001. "Phytoextraction: A Cost-Effective Plant-Based Technology for the Removal of Metals from the Environment." *Bioresource Technology*. doi:10.1016/S0960–8524(00)00108-5

Gavrilescu, Maria. 2022. "Enhancing Phytoremediation of Soils Polluted with Heavy Metals." *Current Opinion in Biotechnology*. doi:10.1016/j.copbio.2021.10.024

Gunarathne, Viraj, Sonia Mayakaduwa, Ahamed Ashiq, Shyama Ranjani Weerakoon, Jayanta Kumar Biswas, and Meththika Vithanage. 2018. "Transgenic Plants: Benefits, Applications, and Potential Risks in Phytoremediation. Benefits, Applications, and Potential Risks in Phytoremediation." In *Transgenic Plant Technology for Remediation of Toxic Metals and Metalloids*. doi:10.1016/B978-0-12–814389-6.00005-5

Guo, Junkang, and Jie Chi. 2014. "Effect of Cd-Tolerant Plant Growth-Promoting Rhizobium on Plant Growth and Cd Uptake by Lolium Multiflorum Lam. and Glycine Max (L.) Merr. in Cd-Contaminated Soil." *Plant and Soil*. doi:10.1007/s11104-013-1952-1

Guo, Zhaolai, Xinqi Yuan, Linyang Li, Ming Zeng, Jie Yang, Hong Tang, and Changqun Duan. 2022. "Genome-Wide Analysis of the ATP-Binding Cassette (ABC) Transporter Family in Zea Mays L. and Its Response to Heavy Metal Stresses." *International Journal of Molecular Sciences*. doi:10.3390/ijms23042109

Gupta, Rishil, Faryad Khan, Fatmah M. Alqahtani, Mohamed Hashem, and Faheem Ahmad. 2023. "Plant Growth–Promoting Rhizobacteria (PGPR) Assisted Bioremediation of Heavy Metal Toxicity." *Applied Biochemistry and Biotechnology*. doi:10.1007/s12010-023-04545-3

Han, Ran, Huiping Dai, Lidia Skuza, and Shuhe Wei. 2021. "Comparative Study on Different Organic Acids for Promoting Solanum Nigrum L. Hyperaccumulation of Cd and Pb from the Contaminated Soil." *Chemosphere*. doi:10.1016/j.chemosphere.2021.130446

Hawari, Jalal, Sylvie Beaudet, Annamaria Halasz, S. Thiboutot, and G. Ampleman. 2000. "Microbial Degradation of Explosives: Biotransformation versus Mineralization." *Applied Microbiology and Biotechnology*. doi:10.1007/s002530000445

Kafle, Arjun, Anil Timilsina, Asmita Gautam, Kaushik Adhikari, Anukul Bhattarai, and Niroj Aryal. 2022. "Phytoremediation: Mechanisms, Plant Selection and Enhancement by Natural and Synthetic Agents." *Environmental Advances*. doi:10.1016/j.envadv.2022.100203

Kaur, Perminder Jit, Umesh Chandra, Chaudhery Mustansar Hussain, and Geetanjali Kaushik. 2020. "Chapter 13. Landfarming: A Green Remediation Technique." In *The Handbook of Environmental Remediation*. doi:10.1039/9781788016261-00357

Kaur, Ravneet, Susmita Das, Sakshi Bansal, Gurbir Singh, Shaswati Sardar, Hena Dhar, and Hasthi Ram. 2021. "Heavy Metal Stress in Rice: Uptake, Transport, Signaling, and Tolerance Mechanisms." *Physiologia Plantarum*. doi:10.1111/ppl.13491

Kaushik, Shruti, Poonam Sharma, Gurvarinder Kaur, Anil Kumar Singh, Fahad A. Al-Misned, Hesham M. Shafik, and Geetika Sirhindi. 2022. "Seed Priming with Methyl Jasmonate Mitigates Copper and Cadmium Toxicity by Modifying Biochemical Attributes and Antioxidants in Cajanus Cajan." *Saudi Journal of Biological Sciences*. doi:10.1016/j.sjbs.2021.12.014

Khouja, Hassine Radhouane, Simona Abbà, Laurence Lacercat-Didier, Stefania Daghino, Didier Doillon, Pierre Richaud, Elena Martino, et al. 2013. "OmZnT1 and OmFET, Two Metal Transporters from the Metal-Tolerant Strain Zn of the Ericoid Mycorrhizal Fungus Oidiodendron Maius, Confer Zinc Tolerance in Yeast." *Fungal Genetics and Biology*. doi:10.1016/j.fgb.2012.11.004

Kim, Seungjin, Rosa Krajmalnik-Brown, Jong Oh Kim, and Jinwook Chung. 2014. "Remediation of Petroleum Hydrocarbon-Contaminated Sites by DNA Diagnosis-Based Bioslurping Technology." *Science of the Total Environment*. doi:10.1016/j.scitotenv.2014.08.002

Kong, Zhaoyu, and Bernard R. Glick. 2017. "The Role of Plant Growth-Promoting Bacteria in Metal Phytoremediation." In *Advances in Microbial Physiology*. doi:10.1016/bs.ampbs.2017.04.001

Kotp, Yousra H., Mohamed E. A. Ali, Saad A. Mohallal, and Mostafa M. S. Aboelfadl. 2019. "Synthesis of a Novel Inorganic Cation Exchanger Based on Molybdate: Applications for Removal of Pb2+, Fe3+ and Mn2+ Ions from Polluted Water." *Separation Science and Technology (Philadelphia)*. doi:10.1080/01496395.2018.1465079

Kristanti, Risky Ayu, Wei Jie Ngu, Adhi Yuniarto, and Tony Hadibarata. 2021. "Rhizofiltration for Removal of Inorganic and Organic Pollutants in Groundwater: A Review." *Biointerface Research in Applied Chemistry*. doi:10.33263/BRIAC114.1232612347

Krout, Ian N., Thomas Scrimale, Daria Vorojeikina, Eric S. Boyd, and Matthew D. Rand. 2022. "Organomercurial Lyase (MerB)-Mediated Demethylation Decreases Bacterial Methylmercury Resistance in the Absence of Mercuric Reductase (MerA)." *Applied and Environmental Microbiology*. doi:10.1128/aem.00010–22

Laishram, Ranu Jajo, Tensubam Basanta Singh, and Wazir Alam. 2023. "A Comprehensive Health Risk Assessment Associated with Bioaccumulation of Heavy Metals and Nutrients in Selected Macrophytes of Loktak Lake, Manipur, India." *Environmental Science and Pollution Research*. doi:10.1007/s11356-023-29606-2

Macci, Cristina, Eleonora Peruzzi, Serena Doni, Francesca Vannucchi, and Grazia Masciandaro. 2021. "Landfarming as a Sustainable Management Strategy for Fresh and Phytoremediated Sediment." *Environmental Science and Pollution Research*. doi:10.1007/s11356-021-13134-y

Machackova, Jirina, Zdena Wittlingerova, Kvetoslav Vlk, and Jaroslav Zima. 2012. "Major Factors Affecting *in Situ* Biodegradation Rates of Jet-Fuel during Large-Scale Biosparging Project in Sedimentary Bedrock." *Journal of Environmental Science and Health - Part A Toxic/Hazardous Substances and Environmental Engineering*. doi:10.1080/10934529.2012.668379

Mandaresu, Melinda, Ludovica Dessì, Andrea Lallai, Marco Porceddu, Maria Enrica Boi, Gianluigi Bacchetta, Tiziana Pivetta, et al. 2023. "Helichrysum Microphyllum Subsp. Tyrrhenicum, Its Root-Associated Microorganisms, and Wood Chips Represent an Integrated Green Technology for the Remediation of Petroleum Hydrocarbon-Contaminated Soils." *Agronomy*. doi:10.3390/agronomy13030812

Mendez, Monica O., and Raina M. Maier. 2008. "Phytostabilization of Mine Tailings in Arid and Semiarid Environments - An Emerging Remediation Technology." *Environmental Health Perspectives*. doi:10.1289/ehp.10608

Michael-Igolima, Uloaku, Samuel J. Abbey, and Augustine O. Ifelebuegu. 2022. "A Systematic Review on the Effectiveness of Remediation Methods for Oil Contaminated Soils." *Environmental Advances*. doi:10.1016/j.envadv.2022.100319

Miller, Efrat, Ofir Menashe, and Carlos G. Dosoretz. 2022. "A Tailored Permeable Reactive Bio-Barrier for *in Situ* Groundwater Remediation: Removal of 3-Chlorophenol as a Case Study." *Environmental Technology (United Kingdom)*. doi:10.1080/09593330.2020.1822922

Moore, Rebekah E. T., Ihsan Ullah, Vinicius H. de Oliveira, Samantha J. Hammond, Stanislav Strekopytov, Mark Tibbett, Jim M. Dunwell, and Mark Rehkämper. 2020. "Cadmium Isotope Fractionation Reveals Genetic Variation in Cd Uptake and Translocation by Theobroma Cacao and Role of Natural Resistance-Associated Macrophage Protein 5 and Heavy Metal ATPase-Family Transporters." *Horticulture Research*. doi:10.1038/s41438-020-0292–6

Moreira, Helena, Sofia I. A. Pereira, Michel Mench, Carlos Garbisu, Petra Kidd, and Paula M. L. Castro. 2021. "Phytomanagement of Metal(Loid)-Contaminated Soils: Options, Efficiency and Value." *Frontiers in Environmental Science*. doi:10.3389/fenvs.2021.661423

Nahar, Noor, Aminur Rahman, Neelu N. Nawani, Sibdas Ghosh, and Abul Mandal. 2017. "Phytoremediation of Arsenic from the Contaminated Soil Using Transgenic Tobacco Plants Expressing ACR2 Gene of Arabidopsis Thaliana." *Journal of Plant Physiology*. doi:10.1016/j.jplph.2017.08.001

Nedjimi, Bouzid, and Youcef Daoud. 2009. "Cadmium Accumulation in Atriplex Halimus Subsp. Schweinfurthii and Its Influence on Growth, Proline, Root Hydraulic Conductivity and Nutrient Uptake." Flora: Morphology, *Distribution, Functional Ecology of Plants*. doi:10.1016/j.flora.2008.03.004

Oustriere, Nadège, Lilian Marchand, Eli Roulet, and Michel Mench. 2017. "Rhizofiltration of a Bordeaux Mixture Effluent in Pilot-Scale Constructed Wetland Using Arundo Donax L. Coupled with Potential Cu-Ecocatalyst Production." *Ecological Engineering*. doi:10.1016/j.ecoleng.2017.04.047

Papadopoulos, Nikolaos, and Georgios Zalidis. 2019. "The Use of Typha Latifolia L. in Constructed Wetland Microcosms for the Remediation of Herbicide Terbuthylazine." *Environmental Processes*. doi:10.1007/s40710-019-00398-3

Poggere, Giovana, Amanda Gasparin, Julierme Zimmer Barbosa, George Wellington Melo, Rodrigo Studart Corrêa, and Antonio Carlos Vargas Motta. 2023. "Soil Contamination by Copper: Sources, Ecological Risks, and Mitigation Strategies in Brazil." *Journal of Trace Elements and Minerals*. doi:10.1016/j.jtemin.2023.100059

Prabakaran, K., Jian Li, A. Anandkumar, Zhanrui Leng, Chris B. Zou, and Daolin Du. 2019. "Managing Environmental Contamination through Phytoremediation by Invasive Plants: A Review." *Ecological Engineering*. doi:10.1016/j.ecoleng.2019.07.002

Rai, Prabhat Kumar, Sang Soo Lee, Ming Zhang, Yiu Fai Tsang, and Ki Hyun Kim. 2019. "Heavy Metals in Food Crops: Health Risks, Fate, Mechanisms, and Management." *Environment International*. doi:10.1016/j.envint.2019.01.067

Rajtor, Monika, and Zofia Piotrowska-Seget. 2016. "Prospects for Arbuscular Mycorrhizal Fungi (AMF) to Assist in Phytoremediation of Soil Hydrocarbon Contaminants." *Chemosphere*. doi:10.1016/j.chemosphere.2016.07.071

Raklami, Anas, Abdelilah Meddich, Khalid Oufdou, and Marouane Baslam. 2022. "Plants—Microorganisms-Based Bioremediation for Heavy Metal Cleanup: Recent Developments, Phytoremediation Techniques, Regulation Mechanisms, and Molecular Responses." *International Journal of Molecular Sciences*. doi:10.3390/ijms23095031

Ranjan, Alok, Ragini Sinha, Shambhu Krishan Lal, Sujit Kumar Bishi, and Anil Kumar Singh. 2021. "Phytohormone Signalling and Cross-Talk to Alleviate Aluminium Toxicity in Plants." *Plant Cell Reports*. doi:10.1007/s00299-021-02724-2

Ranjan, Alok, Ragini Sinha, Tilak Raj Sharma, Arunava Pattanayak, and Anil Kumar Singh. 2021. "Alleviating Aluminum Toxicity in Plants: Implications of Reactive Oxygen Species Signaling and Crosstalk with Other Signaling Pathways." *Physiologia Plantarum*. doi:10.1111/ppl.13382

Raza, Ali, Madiha Habib, Shiva Najafi Kakavand, Zainab Zahid, Noreen Zahra, Rahat Sharif, and Mirza Hasanuzzaman. 2020. "Phytoremediation of Cadmium: Physiological, Biochemical, and Molecular Mechanisms." *Biology*. doi:10.3390/biology9070177

Robles-González, Ireri V., Fabio Fava, and Héctor M. Poggi-Varaldo. 2008. "A Review on Slurry Bioreactors for Bioremediation of Soils and Sediments." *Microbial Cell Factories*. doi:10.1186/1475–2859-7-5

Roychowdhury, Abhishek, Pallabi Mukherjee, Saumik Panja, Rupali Datta, Christos Christodoulatos, and Dibyendu Sarkar. 2021. "Evidence for Phytoremediation and Phytoexcretion of NTO from Industrial Wastewater by Vetiver Grass." *Molecules*. doi:10.3390/MOLECULES26010074

Rugh, Clayton L., Scott A. Merkle, and Richard B. Meagher. 2019. "Toxic Mercury Reduction and Remediation Using a Modified Bacterial Gene in Transgenic Plants." *HortScience*. doi:10.21273/hortsci.31.4.699a

Sardar, Rehana, Shakil Ahmed, and Nasim Ahmad Yasin. 2022. "Titanium Dioxide Nanoparticles Mitigate Cadmium Toxicity in Coriandrum Sativum L. through Modulating Antioxidant System, Stress Markers and Reducing Cadmium Uptake." *Environmental Pollution*. doi:10.1016/j.envpol.2021.118373

Sayqal, Ali, and Omar B. Ahmed. 2021. "Advances in Heavy Metal Bioremediation: An Overview." *Applied Bionics and Biomechanics*. doi:10.1155/2021/1609149

Sharma, Ashita, Jatinder Kaur Katnoria, and Avinash Kaur Nagpal. 2016. "Heavy Metals in Vegetables: Screening Health Risks Involved in Cultivation along Wastewater Drain and Irrigating with Wastewater." *SpringerPlus*. doi:10.1186/s40064-016-2129-1

Sharma, Pooja, Ashutosh Kumar Pandey, Sang Hyoun Kim, Surendra Pratap Singh, Preeti Chaturvedi, and Sunita Varjani. 2021. "Critical Review on Microbial Community during In-Situ Bioremediation of Heavy Metals from Industrial Wastewater." *Environmental Technology and Innovation*. doi:10.1016/j.eti.2021.101826

Sharma, Sakshi, Inderpreet Kaur, and Avinash Kaur Nagpal. 2021. "Contamination of Rice Crop with Potentially Toxic Elements and Associated Human Health Risks—a Review." *Environmental Science and Pollution Research*. doi:10.1007/s11356-020-11696-x

Sharma, Sakshi, Avinash Kaur Nagpal, and Inderpreet Kaur. 2018. "Heavy Metal Contamination in Soil, Food Crops and Associated Health Risks for Residents of Ropar Wetland, Punjab, India and Its Environs." *Food Chemistry*. doi:10.1016/j.foodchem.2018.02.037

Singh, Nandita, Lena Q. Ma, Mrittunjai Srivastava, and Bala Rathinasabapathi. 2006. "Metabolic Adaptations to Arsenic-Induced Oxidative Stress in Pteris Vittata L and Pteris Ensiformis L." *Plant Science*. doi:10.1016/j.plantsci.2005.08.013

Singh, O. V., S. Labana, G. Pandey, R. Budhiraja, and R. K. Jain. 2003. "Phytoremediation: An Overview of Metallic Ion Decontamination from Soil." *Applied Microbiology and Biotechnology*. doi:10.1007/s00253-003-1244-4

Song, Yu, Liang Jin, and Xiaojuan Wang. 2017. "Cadmium Absorption and Transportation Pathways in Plants." *International Journal of Phytoremediation*. doi:10.1080/15226514.2016.1207598

Souri, Zahra, Naser Karimi, Marzeyh Sarmadi, and Elham Rostami. 2017. "Salicylic Acid Nanoparticles (SANPs) Improve Growth and Phytoremediation Efficiency of Isatis Cappadocica Desv., under As Stress." *IET Nanobiotechnology*. doi:10.1049/iet-nbt.2016.0202

Szuba, Agnieszka, Leszek Karliński, Magdalena Krzesłowska, and Teresa Hazubska-Przybył. 2017. "Inoculation with a Pb-Tolerant Strain of Paxillus Involutus Improves Growth and Pb Tolerance of Populus × Canescens under in Vitro Conditions." *Plant and Soil*. doi:10.1007/s11104-016-3062–3

Vries, Wim de, Paul F. Römkens, and Gudrun Schütze. 2007. "Critical Soil Concentrations of Cadmium, Lead, and Mercury in View of Health Effects on Humans and Animals." *Reviews of Environmental Contamination and Toxicology*. doi:10.1007/978-0-387–69163-3_4

Wang, Li, Dongguang Yang, Rongjian Chen, Fang Ma, and Gen Wang. 2022. "How a Functional Soil Animal-Earthworm Affect Arbuscular Mycorrhizae-Assisted Phytoremediation in Metals Contaminated Soil?" *Journal of Hazardous Materials*. doi:10.1016/j.jhazmat.2022.128991

Werkneh, Adhena Ayaliew, Gebretsadkan Gebremedhin Gebretsadik, and Shifare Berhe Gebru. 2023. "Review on Environmental Selenium: Occurrence, Public Health Implications and Biological Treatment Strategies." *Environmental Challenges*. doi:10.1016/j.envc.2023.100698

Xiao, Ran, Amjad Ali, Yaqiong Xu, Hamada Abdelrahman, Ronghua Li, Yanbing Lin, Nanthi Bolan, Sabry M. Shaheen, Jörg Rinklebe, and Zengqiang Zhang. 2022. "Earthworms as Candidates for Remediation of Potentially Toxic Elements Contaminated Soils and Mitigating the Environmental and Human Health Risks: A Review." *Environment International*. doi:10.1016/j.envint.2021.106924

Yan, An, Yamin Wang, Swee Ngin Tan, Mohamed Lokman Mohd Yusof, Subhadip Ghosh, and Zhong Chen. 2020. "Phytoremediation: A Promising Approach for Revegetation of Heavy Metal-Polluted Land." *Frontiers in Plant Science*. doi:10.3389/fpls.2020.00359

Yang, Minjune, James W. Jawitz, and Minhee Lee. 2015. "Uranium and Cesium Accumulation in Bean (Phaseolus Vulgaris L. Var. Vulgaris) and Its Potential for Uranium Rhizofiltration." *Journal of Environmental Radioactivity*. doi:10.1016/j.jenvrad.2014.10.015

Zhao, Haixia, Liangsheng Wang, Fang Jie Zhao, Longhua Wu, Anna Liu, and Wenzhong Xu. 2019. "SpHMA1 Is a Chloroplast Cadmium Exporter Protecting Photochemical Reactions in the Cd Hyperaccumulator Sedum Plumbizincicola." *Plant Cell and Environment*. doi:10.1111/pce.13456

Zhou, Pingfan, Muhammad Adeel, Noman Shakoor, Manlin Guo, Yi Hao, Imran Azeem, Mingshu Li, Mengyuan Liu, and Yukui Rui. 2021. "Application of Nanoparticles Alleviates Heavy Metals Stress and Promotes Plant Growth: An Overview." *Nanomaterials*. doi:10.3390/nano11010026

15 Plant-Fungal Interaction in Remediation of Metalloids from Environment

Venkataramana Chapara and Rani Chapara

15.1 INTRODUCTION

Metalloids are elements with an atomic mass higher than 20 and a density higher than 5,000 kg·m^{-3}. However, 51 elements in the periodic table are considered heavy metals (HMs)/metalloids (Ali and Khan, 2018). In simple words, metalloids are elements with intermediate chemical properties between metals and semiconductors. Metalloids are released extensively into the environment with the increased rate of industrialization, urbanization, mining, milling, fossil fuel burning, and agrochemicals (Sandeep et al., 2019). Metalloids coexist naturally on our planet Earth and were found in different layers. High concentrations of metalloids result in toxicity to the biota of soil, plants, animals, and humans. Weathering of the planet Earth is considered the main natural source of metalloids in the soil environment (Park et al., 2011). Elements that include HMs such as cadmium (Cd), lead (Pb), barium (Ba), mercury (Hg), antimony (Sb), thallium (Tl), arsenic (As), and uranium (U) are termed toxic heavy metalloids. These metalloids at increasing concentrations pose a high risk to global food safety and to the environment. Revised recent pollutants studies in countries like Europe, Africa, Latin American countries, and China indicated high levels of environmental contamination. Extensive agriculture, industrial, and mining activities are the major causes of contamination, with mining being the main source of metal(oid) pollution since many of the world's metals and metalloids are extracted in Latin American countries (Di Marzio et al., 2019; Khalid et al., 2017). Micronutrients such as iron (Fe), cobalt (Co), copper (Cu), manganese (Mn), nickel (Ni), zinc (Zn), and molybdenum (Mo), along with metals, are needed in plant growth in lower concentrations in general for various metabolic activities. In addition, the soil pH value is a very important aspect, since some elements are more bioavailable at pH $\leq$ 7. However, on exceeding their required concentrations, they become toxic to plant development (Rahman et al., 2019), while others (e.g., Cd, Pb, and Hg) are non-essential and are toxic even in trace amounts. Several techniques were developed over the years to mitigate the concentration of HMs and metalloids in the environment: physical remediation (e.g., washing, thermal desorption, and solidification), chemical remediation (e.g., adsorption, catalysis, precipitation/solubilization, and electrokinetic methods), biological remediation (e.g., biodegradation, phytoremediation, and bioventing), and combined remediation (e.g., electrokinetic–microbial remediation and washing–microbial degradation) (Raffa et al., 2021).

The harmful effect of the metalloids in plants and other soil biota is conspicuous when metalloids bond with normal occurring elements that are vital in plants primary and secondary metabolism. Protein imbalance is one such side effect seen occurring when metalloids bond with sulfhydryl groups. Likewise, metalloids can cause stress in the root cortex of plants by interfering in the metabolic activities of endodermal cells, deposition of suberin and lignin, cell wall thickening (exodermal and endodermal cells), formation of air spaces and alterations in intercellular spaces, and finally, modifications to the root vasculature (Siddiquee et al., 2015). Plants when exposed to Cu, As, or Pb not only have abnormalities in the root system but were also found to extend to the

DOI: 10.1201/9781032673592-15

stem tissues by disrupting cell division and enabling enlarged cortical cells with a loss of turgor in sclerenchyma cells of phloem cells. Metalloid's uptake that happens through roots interferes with the photosynthetic tissues by limiting their entrance or translocation of food materials through them to the foliar tissues. The common side effects of metalloids of Cd, As, and Mn seen on the foliar tissues (Jiang, 2000). General reduction in leaf thickness, alterations to the epidermal cell structure and a reduction in intercellular spaces (mesophyll), an increase in callose deposition, and alterations to the stomatal structure, density, and aperture frequency are the common anomalies observed in many studies. In the coevolution process, plants have evolved with complex mechanisms to overcome the side effects of metalloids if present above their threshold concentrations; thus, they are classified as hyperaccumulators (>1,000 $\mu g\ g^{-1}$) and non-hyperaccumulators (<500 $\mu g\ g^{-1}$). To name some of these complex mechanisms of plants to avoid damage control of metalloids include sequestration, exclusion, chelation, and speciation. Simply, the vital toxic metalloids that came in contact with the root tip and are internalized via symplast or apoplast, based on their chemical nature. The most common evidence of metalloid toxicity is stunting in plants, and the plant growth reduction increases with the metalloid concentration increases (Angulo-Bejarano et al., 2021).

Metalloids in all can cause damage by hyperaccumulating in plants and mushrooms used for food, which pose a direct threat to human neurological function and lead to an indirect threat by modulating the concentration of phytochemical compounds such as hydrazine, dencichine, and cyanogens, which are known to pose a neurotoxic hazard themselves to the consumer. Examples of the latter include certain plants (grass pea and cassava) and mushrooms (false morel), food use of which can precipitate neurotoxic disease. The metal-contaminated environments, such as soils, can be restored with the help of the interaction establishment of functional microbial community with plant community, leading to better soil development and biogeochemical cycling (Olayinka and Babalola, 2001).

15.1.1 Metalloid Uptake and Transport

Metalloid uptake by the plants is influenced by many factors, primarily the type of plant species, pH of the surrounding environment, root zone, cation exchange capacity (Sarma, 2011), the addition of chelators (Kinnersley, 1993), and temperature (Liao et al., 2004). Metalloids have diverse sites of action within a plant; the visual toxic response varies among the metalloid composition and plant species. Chemical valency, mostly divalent cations (Ca^{2+} and Mg^{2+} act as transporters), are the immediate attractors of metalloids in the plant´s active transport system that enable them to make an easy way into the plant's root tip. Several non-selective channels facilitate the absorption of essential and non-essential metals and create competition between cations at the surface of the root. As a result, elevated levels of metalloids interfere and compete for the transport sites, leading to the absorption of HMs over macronutrients, thus inducing or exacerbating deficiencies of nutrient cations. Trivalent ions, such as Al^{3+}, are reported to be abundant in low pH (> 6.6) soils known to release into soils due to human activity. A transporting system known as NRAMP Aluminum Transporter 1 (NRAT1) is widely related to the Al^{3+} uptake in plants (Angulo-Bejarano et al., 2021). Once the metallic ions cross the plant´s first barrier found in roots, they can be translocated using long distance transport through the phloem, resulting in hyperaccumulation and tolerance mechanisms (in non-accumulator plants). Likewise, an electrochemical gradient that influences the transport of specific nutrient metalloids that are formed in combination of Fe, Cu, Mn, Zn, and Ni is absorbed passively by plant roots via ion channels. Later, several selective transport pathways for metalloids into plants are now discovered. For instance, Fe bound to phytosiderophores can be transported by Yellow stripel (YS1) across the plasma membrane. YS1 gene synthesizes Fe(III)-Phytosiderophore (Fe-III-PS) transporter in maize (*Zea mays*). Also, the YS1-like (YSL) family of transporters can mobilize Zn and Cu from plant leaves and seeds (Roberts et al., 2004).

15.2 REMEDIATION OF METALLOID TOXICITY

Remediation of soils from metalloid contamination had been studied thoroughly around the world with success by implementing several processes, viz., physical processes, chemical processes, nanoparticles, sodification, soil washing, electrochemical remediation, and biological processes (Raffa et al., 2021). The remediation of metalloids from various sources using plants, fungi, and other microorganisms is known as a biological process. If plants alone are used for remediation known as phytoremediation. In general, metalloids are not degraded by microorganisms or plants, but they are accumulated, stabilized, or bonded in less toxic volatile compounds by reacting to contaminated soils or industrial wastes. These microorganisms or plants activate their defense mechanisms of enzyme secretion or cellular morphological changes to remediate the contamination.

15.2.1 Metalloid Toxicity Remediation by Plants

Phytoremediation is a widely adopted technique that was regarded as an efficient method. However, its validity widely depends on plant species selection. Phytoremediation follows various other mechanisms such as phytoextraction, rhizofiltration, rhizodegradation, phytostabilization, phytodegradation, and phytovolatilization (Li et al., 2016). Plant species that can accumulate metalloids are called metallophytes. The main characteristics that define the phytoextraction perspective of metallophytes are their capacity to accumulate metalloids and above-ground biomass; these are planted for phytoextraction (Table 15.1). For successful phytoremediation of metalloids, finding effective metallophytes are important, and more than 450 plant species have currently been identified as potential metal hyperaccumulators (Suman et al., 2018). Currently, research studies around the world to expand the effectiveness of phytoextraction, where novel metallophytes are studied rigorously to improve understanding of their mechanisms. Plant families, such as Asteraceae, Brassicaceae, Euphorbiaceae, Fabaceae, Flacourticeae, and Violaceae, have been proven to absorb larger concentrations of metalloids. Among these, species of the Brassicaceae family have shown larger potential to remediate and scavenge metalloids that have Pb, Cd, Zn, and Ni. Among the Brassica species, *Brassica juncea* L., *Brassica oleracea* L., *Brassica campestris* L., and *Brassica napus* L. (Kumar et al., 1995) are widely used for phytoremediation. *Brassica juncea* L. has shown tremendous potential to remediate metalloids with Cd, Cr, Cs, Cu, Ni, Pb, U, and Zn. A plant species *Pteris vittata* (Chinese brake fern) found to be a potential

TABLE 15.1
Remediation of Metalloids by Various Plants

Plant	Biosorption	Citations
Arabidopsis thaliana	Cd, Pb, and Mn	Zhang et al. (2018)
Thlaspi caerulescens	Zn, Pb, and Cd	El Kheir et al. (2008)
Oryza sativa	As, Cd, Cr, Hg, Pb, and Se	Navarathna et al. (2021)
Zea mays	Cr and Pb	Atta et al. (2023)
Lactuca sativa	Zn, Cu, and Pb	Calabró et al. (2022)
Triticum aestivum	Br and polycyclic aromatic hydrocarbons (PAHs)	Shtangeeva I. et al. (2018)
Glycine max	Cd, Pb, and Zn	Delil et al. (2020)
Brassica napus	Cr	Adiloglu et al. (2015)
Fagopyrum esculentum	Cd and Pb	Domańska et al. (2021)
Beta vulgaris	Mg, Fe, and Ba	Arthur et al. (2022)
Vetiver zizanioides	Cd, Cu, Mn, Pb, Zn, Ni, and Cr	Abaga et al. (2021)
Helianthus annus	Cd, Cr, and Ni	Chen and Cutright (2001)

accumulator of As (3,280–4,980 ppm). Sunflower (*Helianthus annus*) was proved as a feasible hyperaccumulator plant to remediate soil contaminated with cesium-137, strontium-90, and U (Sakakibara et al., 2010).

Rhizofiltration, where both aquatic as well as terrestrial plants can be utilized as metallophytes. Water hyacinth, Azolla, duckweed, cattail, and poplar are frequently utilized for the remediation of wetland water with high metalloid contamination (Liao et al., 2004).

Phytovolatilization is defined as a transformation of pollutants into different volatile compounds into the atmosphere via transpiration. Plants such as *Nicotiana tabacum*, *Crinum americanum*, *Triticum aestivum*, *Arabidopsis thaliana*, *Bacopa monnieri*, and *Trifolium repens* are commonly used plants for phytovolatilization (Singh et al., 2018). Soils with metalloids of Silver (Hg) and selenium (Se) were remediated with the phytovolatilization process. Likewise, phytovolatilization of radioactive isotopes of hydrogen (tritium) to stable helium studies were successful in remediation.

Phytodesalination, an emerging technique where halophytic plants were used to remediate the saline soils. However, not for soils polluted with metalloids. Halophytes are well-adapted in nature to accumulate metalloids in comparison to glycophytic plants. Along with rainfall, the soil properties, such as salinity, sodicity, and porosity influence phytodesalination. The halophytic plants, *Suaeda maritime* and *Sesavium partulacastrum*, found to have the capacity to remove 504 and 474 kg of NaCl, respectively, from 1 hectare of saline soil in a 4-month period (Sabreena et al., 2022).

15.2.2 Metalloid Toxicity Remediation by Fungi

Microorganisms facilitate the decay of organic compounds in the rhizosphere. Besides decay, metallophytes, fungi, and algae can also be put to use for metalloid removal (Table 15.2). Algae absorb by taking the HMs by adsorption and into the cytoplasm by chemisorption (Madejón et al., 2022). Numerous fungi with significant potential adsorption capacity have been utilized for the treatment of metalloid-contaminated soils. Fungi constitute a large variety of extracellular proteins, organic acids, and valuable metabolites that can be used to remediate the metalloid-contaminated soils. Fungi have wider adaptability in any ecosystem and any environmental conditions (Cochrane, 1958).

The filamentous fungi produce a larger quantity of biomasses that are in use for adsorption of metalloid elements of Pb, Zn, Cd, Cu, Cr, As, and Ni (Srivastava, 2011). In addition, fungi have the capacity to conjugate heavy metals in various organic molecules, such as glutathione or organic acids (Siddiquee, 2015). These mechanisms reduce the bioavailability of HMs in the soil, therefore, reducing their uptake by plants, resulting in better plant growth.

Research on the capability of *Trichoderma* sp. in removing HMs is considered one of the crucial agents in the bioremediation process (Siddiquee, 2015). Among the other fungi, *Trichoderma* was found with highest As effectively removal capacity from As-contaminated agricultural soils (Srivastava, 2011). Removal of metalloids with the use of microbes such as fungi can also be called rhizodegradation. This process was utilized in the biodegradation of the organic pollutants in the

TABLE 15.2
Remediation of Metalloids by Various Fungi

Fungus	Metalloids Biosorped	Citations
Penicillium chrysogenum	Cu, Ni, U, Cr, Th, Zn, Cd, and Pb	Wang and Cheng (2009)
Trichoderma ghanense	As, Cu, Cd, Pb, and Fe	Singh et al. (2015)
Aspergillus niger	Cd, Pb, Zn, Cu, Ni, and Cr	Pal et al. (2006)
Sachharomyces cerevisiae	Cr (VI) and Se (VI)	Cherest et al. (1997)
Phellis weirii	Al, Ca, Cu, Fe, K, Mg, Mn, Na, Ni, Pb, and Zn	McDougall and Blanchette (1996)

soil accompanied by rhizospheric microbes that secrete specific enzymes that degrade or transform exceedingly contaminated organic pollutants into less detrimental forms. The process of rhizodegradation has enhanced the soils as these organisms accelerated the extraction and amputation of pollutants by the plant (Li et al., 2016). The significant features of rhizodegradation include the disruption of the pollutant at its origin and helps in mineralization of the organic pollutant (Fiorentino et al., 2018).

15.2.3 Plant-Fungal Interaction in Remediation of Metalloids

Vascular plants host a great variety of fungi. In addition to being susceptible to soil-borne pathogens, plant roots are also colonized by non-pathogenic or mutualistic fungi, such as arbuscular mycorrhizae and ectomycorrhizae (ECM) (Table 15.3). Fungi are known to amass metalloids in their higher parts in concentrations between 100 and 500 times in conjunction with plants without affecting their development and functioning (Sheoran et al., 2016). Application of plant-fungal interactions as soil amendments is a successful technique in the restoration of contaminated and degraded mining soils (Kumpiene et al., 2019). Soil amendment by using fungi, specifically arbuscular mycorrhizal fungi (AMF), is efficient in accelerating the phytoremediation process (Kumpiene et al., 2019). AMF inoculation is regarded as a promising tool in biotechnology for the sustainable remediation of hazardous contaminants (Schneider et al., 2013). The efficiency of metalloid extraction through phytoremediation depends upon the response of the plant to AMF species and also the soil parameters.

The endophytic fungi generally colonize the tissues of the plant internally, and no external sign of infection or a negative effect were visible on the host. Further, the variables such as plant growth, productivity, carbon sequestration, and phytoremediation are helped by the colonization of the endophytic fungi in the rhizosphere (Weyens et al., 2009; Suryanarayanan, 2013). The ability of fungi in forming mutualistic symbiotic associations with photosynthetic plants, algae, and cyanobacteria (mycorrhizas and lichens), is one of the foremost reasons of their success in biogeochemical processes in terrestrial ecosystems. Likewise, needed for redistribution of inorganic nutrients and major transformations of metals and phosphate, as well as carbon flow (Gadd, 2007). In soil contaminated with metals, AMF may be a significant factor in plant growth by functioning as a bioalleviator and/or biofertilizer by increasing bioavailability and soil water holding capacity (Figure 15.1) (Karthikeyan et al., 2022). Also, reactive oxygen species are reported to generate during the long-term effect of metalloid stress in plants, which in turn demotes the plant growth and development. The creation of reactive oxygen species (ROS) and scavenging of them need to be balanced in order to reduce any potentially harmful effects of metalloid stress. Stress signals in

TABLE 15.3
Remediation of Metalloids by Plant-Fungal Interactions

Plant	Fungi	Metalloids Biosorped	Citation(s)
Helianthus annus	*Claraideoglomus claroideum*	Ni	Ma et al. (2022)
Solanum nigrum	*Rhizophagus irregularis* and *Glomus intraradices*	Cd	Ma et al. (2022)
Zea mays	*Glomus aggregatum, G. intraradices*, and *Funneliformis mosseae*	Pb, Hg, Cd, Zn, and As	Ma et al. (2022)
Medicago sativa	*Diversipora spurcum* and *Rhizophagus irregularis*	Cd and Ni	Ma et al. (2022)
Triticum aestivum	*Rhizoglomus intraradius*	As	Ma et al. (2022)
Cynodon dactylon	*Funneliformis mosseae* and *Diversispora spurcum*	Sb, Pb, and Zn	Ma et al. (2022)
Oryza sativa	*Rhizophagus irregularis*	Cd	Ma et al. (2022)
Lactuca sativa	*Funneliformis mosseae and Rhizoglomus intraradius*	Zn	Ma et al. (2022)
Sorghum bicolor	*Claroideoglomus etunicat*	Mo	Ma et al. (2022)

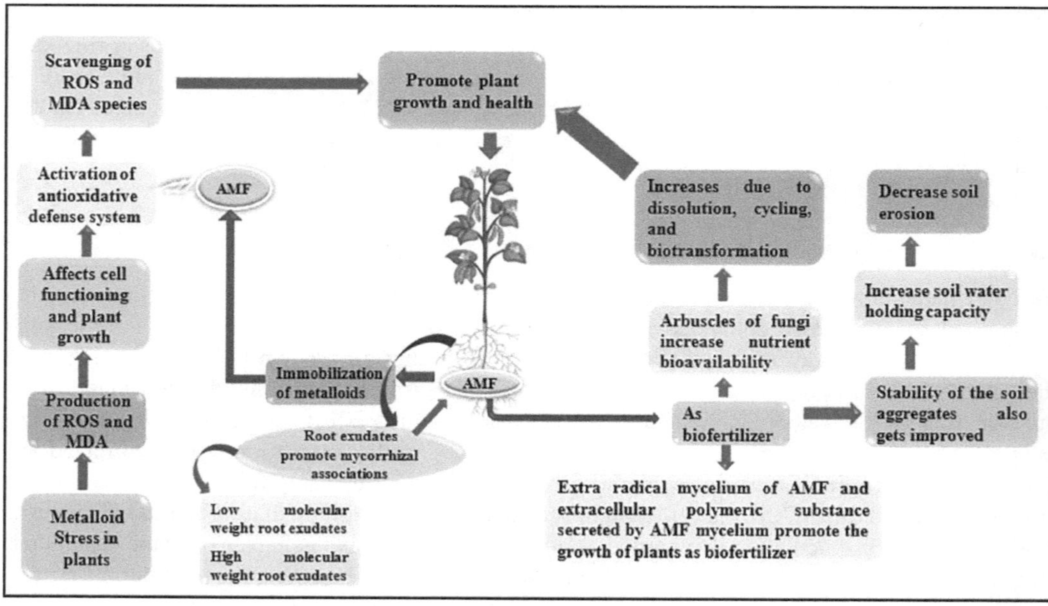

FIGURE 15.1 Mutualistic relationship of AMF and plants in the environment.

plants include antioxidant signaling (enzymes such as catalase, superoxide dismutase, ascorbate peroxidase, etc.), maintaining redox equilibrium, and continuously producing or eliminating ROS at the cellular level. According to research, the early stages of symbiotic relationships between AMF and host plant roots are when ROS formation takes place (Cuypers et al., 2013). For instance, recent meta-analysis revealed that inoculating crop plants with AMF decreases As concentration by 19% while increasing phosphorus (P) levels and biomass by 64% and 53%, respectively, under As stress. This approach offers potential to mitigate As-induced crop losses and food chain contamination, especially in As-affected regions (Neidhardt, 2021).

The most commonly endophytic fungi that interact with the plant roots in the soil are the AMF. The AMF comprise approximately 150 species from the phylum zygomycetous fungi and 6,000 species of ECM fungi that belong to the phylum basidiomycetes, ascomycetes, and zygomycetes as well. The herbaceous and the woody plant families form the mutualistic associations readily with the AMF, not the case with the ECM fungi in their association woody plant families, which is very limited (Mandyam and Jumpponen, 2005). The phyto-fungal association process in fungi, especially by mycorrhizal fungi, occurs through the absorption process of non-mobile nutrients from the soil and translocate them to host plants; further, sequester potentially harmful HM ions/metalloids that are present in higher concentrations than required, facilitate interplant transfer of nutrients, and beneficially modify plant water relations (Fomina et al., 2005). Several studies indicated that certain members of the Cruciferae family and are well known commercial crops such as Chinese cabbage, broccoli, and rapeseed, are known as nonmycorrhizal plants. These crops are known to host some dark septate endophytes (DSE), which are later found to have symbiotic associations with fungi (Usuki and Narisawa, 2007). These DSE produce conidial and sterile septate fungal endophytes, with characteristic melanized structures, known as sclerotia or microsclerotia. These are formed by amassing inter- and intracellular hyphae in the plant roots. The DSE fungi are ubiquitous and often coexist with various mycorrhizal fungi (Upson et al., 2009). Such kinds of DSE were reported from over 600 plant species, including non-mycorrhizal (Mandyam and Jumpponen, 2005). Another plant-fungal association that was well studied is ectomycorrhiza (ECM). ECM fungi show a mechanism where the fungal-plant nutrient exchange takes place through organs; this mechanism through external organs benefits the host plant in forming a resistant barrier against HM toxicity (Gadd, 2007).

As exemplary, members of the endophytic fungi of the genera *Phomopsis*, *Phoma*, *Colletotrichum*, and *Phyllosticta* colonize a wide range of plant hosts that are unrelated taxonomically. A unique character of organ and tissue specificity is exhibited by the endophytes while colonizing and in adaptation to different physiological conditions in plants (Aly et al., 2011). The resistance in canola plant roots to metals was improved by inoculation of certain endophytic yeasts. Likewise, endophytic Cryptococcus sp. Strain CBSB78 on inoculation on to *B. alboglabra* increased the extraction amount of Cd, Pb, and Zn from the multi-metal-contaminated soils (Deng et al., 2021).

15.3 CONCLUSION

It is imperative that plants and fungi play a vital role in remediation of metalloids from the biosphere through important biogeochemical cycles, primarily in the cycling of elements (metalloids) and organic and inorganic substrate transformations by their mutualistic association. Plant root systems and the ubiquitous nature of fungi in subsoil environments often play a major role in remediation of metalloid-polluted habitats. Characteristics of certain fungal species, such as their oligotrophic nature, filamentous growth habit, flexible growth strategies, and resistance to extreme environmental factors, make them successful in forming successful mutualistic symbiotic associations with plants; likewise, with algae and cyanobacteria. These associations in metalloid-rich habitats were well studied by the scientific community and were put into use in the remediation of metal-polluted habitats. It is imperative to draw attention to the 'phytoremediation process' and to the interdisciplinary approach to further understand the vital roles that fungi play in the biogeochemical cycling of metalloids in association with plants. Future studies should involve crucial objectives of the chemical and biological mechanisms of plant-fungi interactions that are involved in remediating the metalloid pollutant soils and their environmental and biotechnological significance.

REFERENCES

Abaga, Norbert Ondo Zue, Sylvie Dousset, and Colette Munier-Lamy. Phytoremediation potential of vetiver grass (Vetiveria zizanioides) in two mixed heavy metal contaminated soils from the Zoundweogo and Boulkiemde regions of Burkina Faso (West Africa). *Journal of Geoscience and Environment Protection* 9, no. 11 (2021): 73–88.

Abiodun, Aransiola Sesan, Ijah Udeme Joshua Josiah, and Abioye Olabisi Peter. Phytoremediation of Lead Polluted Soil by Glycine max L. *Applied and Environmental Soil Science* 2013 (2013): 1–7.

Adiloglu, Sevinç, and Mehmet Turgut Saglam Ali Süme. Chrome (Cr) pollution in agricultural areas improvement by phytoremediation method with Canola (Brassica napus L.) plant growing. *Journal of Essential Oil Bearing Plants* 18, no. 5 (2015): 1180–1186.

Ali, Hazrat, and Ezzat Khan. What are heavy metals? Long-standing controversy over the scientific use of the term 'heavy metals'–proposal of a comprehensive definition. *Toxicological & Environmental Chemistry* 100, no. 1 (2018): 6–19.

Aly, Amal Hassan, Abdessamad Debbab, and Peter Proksch. Fungal endophytes: unique plant inhabitants with great promises. *Applied Microbiology and Biotechnology* 90 (2011): 1829–1845.

Angulo-Bejarano, Paola I., Jonathan Puente-Rivera, and Rocío Cruz-Ortega. Metal and metalloid toxicity in plants: an overview on molecular aspects. *Plants* 10, no. 4 (2021): 635.

Arthur, Georgina D., Adeyemi O. Aremu, Manoj G. Kulkarni, Wendy A. Stirk, Theophilus C. Davies, Joseph Penaye, and Johannes Van Staden. Phytoremediation potential of beta vulgaris L. (Swiss chard) using soil from the vicinity of Kette-Batouri Goldmine (Eastern Cameroon). *South African Journal of Botany* 151 (2022): 713–719.

Atta, Muhammad Imran, Syeda Sadaf Zehra, Habib Ali, Basharat Ali, Syed Naveed Abbas, Sara Aimen, Sadia Sarwar et al. Assessing the effect of heavy metals on maize (Zea mays L.) growth and soil characteristics: plants-implications for phytoremediation. *PeerJ* 11 (2023): e16067.

Calabró, María Rosario, Gonzalo Roqueiro, Raúl Tapia, Diana Cristina Crespo, Martha Fidela Bargiela, and Brian Jonathan Young. Chronic toxicity, bioavailability and bioaccumulation of Zn, Cu and Pb in Lactuca sativa exposed to waste from an abandoned gold mine. *Chemosphere* 307 (2022): 135855.

Chen, Hong, and Teresa Cutright. EDTA and HEDTA effects on Cd, Cr, and Ni uptake by Helianthus annuus. *Chemosphere* 45, no. 1 (2001): 21–28.

Cherest, Helene, Jean-Claude Davidian, Dominique Thomas, Vladimir Benes, Wilhelm Ansorge, and Yolande Surdin-Kerjan. Molecular characterization of two high affinity sulfate transporters in Saccharomyces cerevisiae. *Genetics* 145, no. 3 (1997): 627–635.

Cochrane, Willard W. *Farm Prices: Myth and Reality*. University of Minnesota Press, 1958.

Cuypers, Ann, Tony Remans, Nele Weyens, Jan Colpaert, Andon Vassilev, and Jaco Vangronsveld. Soil-plant relationships of heavy metals and metalloids. In Alloway, Brian J. ed., *Heavy Metals in Soils: Trace Metals and Metalloids in Soils and Their Bioavailability*, pp. 161–193. Springer, 2013.

Delil, Aydeniz Demir, Nurcan Köleli, Hatice Dağhan, and Gizem Bahçeci. Recovery of heavy metals from canola (Brassica napus) and soybean (Glycine max) biomasses using electrochemical process. *Environmental Technology & Innovation* 17 (2020): 100559.

Deng, Fenglin, Fanrong Zeng, Guang Chen, Xue Feng, Adeel Riaz, Xiaojian Wu, Wei Gao, Feibo Wu, Paul Holford, and Zhong-Hua Chen. Metalloid hazards: from plant molecular evolution to mitigation strategies. *Journal of Hazardous Materials* 409 (2021): 124495.

Di Marzio, A., S. A. Lambertucci, A. J. Garcia Fernandez, and E. Martínez-López. From Mexico to the Beagle Channel: a review of metal and metalloid pollution studies on wildlife species in Latin America. *Environmental Research* 176 (2019): 108462.

Domańska, Jolanta, Danuta Leszczyńska, and Aleksandra Badora. The possibilities of using common buckwheat in phytoremediation of mineral and organic soils contaminated with Cd or Pb. *Agriculture* 11, no. 6 (2021): 562.

El Kheir, Safae Berrah, Jamal Oubbih, Nadia Saidi, and Abdelhak Bouabdli. Uptake and fixation of Zn, Pb, and Cd by Thlaspi caerulescens: application in the cases of old mines of Mibladen and Zaida (West of Morocco). *Arabian Journal of Geosciences* 1 (2008): 87–95.

Fiorentino, Nunzio, Mauro Mori, Vincenzo Cenvinzo, Luigi Giuseppe Duri, Laura Gioia, Donato Visconti, and Massimo Fagnano. Assisted phytoremediation for restoring soil fertility in contaminated and degraded land. *Italian Journal of Agronomy* 13, no. S1 (2018): 34–44.

Fomina, M. A., Ian James Alexander, J. V. Colpaert, and G. M. Gadd. Solubilization of toxic metal minerals and metal tolerance of mycorrhizal fungi. *Soil Biology and Biochemistry* 37, no. 5 (2005): 851–866.

Gadd, Geoffrey M. Geomycology: biogeochemical transformations of rocks, minerals, metals and radionuclides by fungi, bioweathering and bioremediation. *Mycological Research* 111, no. 1 (2007): 3–49.

Hemen Sarma, and Hemen Sarma. Metal hyperaccumulation in plants: a review focusing on phytoremediation technology. *Journal of Environmental Science and Technolog* 4 (2011): 118–138.

Jiang, W., D. Liu, and W. Hou. Hyperaccumulation of lead by roots, hypocotyls, and shoots of Brassica juncea. *Biologia Plantarum* 43 (2000): 603–606.

Karthikeyan, Obulisamy Parthiba, Jerusa Schneider, Chetan Keswani, Kuldeep Bauddh, and Ying Ma. Plant-mycorrhizal fungi interactions in phytoremediation of geogenic contaminated soils. In *Biotechnological Advances in Bioremediation of Heavy Metal Polluted Soils*. 2022.

Khalid, Sana, Muhammad Shahid, Nabeel Khan Niazi, Behzad Murtaza, Irshad Bibi, and Camille Dumat. A comparison of technologies for remediation of heavy metal contaminated soils. *Journal of Geochemical Exploration* 182 (2017): 247–268.

Kinnersley, A. M. The role of phytochelates in plant growth and productivity. *Plant Growth Regulation* 12 (1993): 207–218.

Kumar, P. B. A. Nanda, Viatcheslav Dushenkov, Harry Motto, and Ilya Raskin. Phytoextraction: the use of plants to remove heavy metals from soils. *Environmental Science & Technology* 29, no. 5 (1995): 1232–1238.

Kumpiene, Jurate, Juan Antelo, Evelina Brännvall, Ivan Carabante, Kristina Ek, Michael Komárek, Charlotta Söderberg, and Linda Wårell. In situ chemical stabilization of trace element-contaminated soil–field demonstrations and barriers to transition from laboratory to the field–a review. *Applied Geochemistry* 100 (2019): 335–351.

Li, Yifei, Jiefeng Zhang, Guibing Zhu, Yu Liu, Bing Wu, Wun Jern Ng, Adhityan Appan, and Soon Keat Tan. Phytoextraction, phytotransformation and rhizodegradation of ibuprofen associated with Typha angustifolia in a horizontal subsurface flow constructed wetland. *Water Research* 102 (2016): 294–304.

Liao ShaoWei, Liao ShaoWei, and Chang WenLian Chang WenLian. Heavy metal phytoremediation by water hyacinth at constructed wetlands in Taiwan. *Journal of Aquatic Plant Management* 42 (2004): 60–68.

Ma, Ying, Ankit, Jaya Tiwari, and Kuldeep Bauddh. Plant-mycorrhizal fungi interactions in phytoremediation of geogenic contaminated soils. *Frontiers in Microbiology* 13 (2022): 843415.

Madejón, Paula, María Teresa Domínguez, Ignacio Girón, Pilar Burgos, María Teresa López-Fernández, Oscar Garcia Porras, and Engracia Madejón. Assessment of the phytoremediation effectiveness in the restoration of uranium mine tailings. *Ecological Engineering* 180 (2022): 106669.

Mandyam, Keerthi, and Ari Jumpponen. Seeking the elusive function of the root-colonising dark septate endophytic fungi. *Studies in Mycology* 53, no. 1 (2005): 173–189.

McDougall, Dennis N., and Robert A. Blanchette. Metal ion adsorption by pseudosclerotial plates of Phellinus weirii. *Mycologia* 88, no. 1 (1996): 98–103.

Navarathna, Chanaka, Saman Pathiratne, D. S. M. De Silva, Jörg Rinklebe, Dinesh Mohan, and Todd Mlsna. Intrusion of heavy metals/metalloids into rice (Oryza sativa L.) in relation to their status in two different agricultural management systems in Sri Lanka. *Groundwater for Sustainable Development* 14 (2021): 100619.

Neidhardt, Harald. Arbuscular mycorrhizal fungi alleviate negative effects of arsenic-induced stress on crop plants: a meta-analysis. *Plants, People, Planet* 3, no. 5 (2021): 523–535.

Olayinka, Akinyemi, and Gbolahan Ola Babalola. Effects of copper sulphate application on microbial numbers and respiration, nitrifier and urease activities, and nitrogen and phosphorus mineralization in an alfisol. *Biological Agriculture & Horticulture* 19, no. 1 (2001): 1–8.

Pal, Arundhati, Suchhanda Ghosh, and A. K. Paul. Biosorption of cobalt by fungi from serpentine soil of Andaman. *Bioresource Technology* 97, no. 10 (2006): 1253–1258.

Park, Jin Hee, Dane Lamb, Periyasamy Paneerselvam, Girish Choppala, Nanthi Bolan, and Jae-Woo Chung. Role of organic amendments on enhanced bioremediation of heavy metal (loid) contaminated soils. *Journal of Hazardous Materials* 185, no. 2–3 (2011): 549–574.

Raffa, Carla Maria, Fulvia Chiampo, and Subramanian Shanthakumar. Remediation of metal/metalloid-polluted soils: a short review. *Applied Sciences* 11, no. 9 (2021): 4134.

Rahman, Zeeshanur, and Ved Pal Singh. The relative impact of toxic heavy metals (THMs) (arsenic (As), cadmium (Cd), chromium (Cr) (VI), mercury (Hg), and lead (Pb)) on the total environment: an overview. *Environmental Monitoring and Assessment* 191 (2019): 1–21.

Roberts, Louis A., Abbey J. Pierson, Zivile Panaviene, and Elsbeth L. Walker. Yellow stripe1. Expanded roles for the maize iron-phytosiderophore transporter. *Plant Physiology* 135, no. 1 (2004): 112–120.

Sabreena, Shahnawaz Hassan, Sartaj Ahmad Bhat, Vineet Kumar, Bashir Ahmad Ganai, and Fuad Ameen. Phytoremediation of heavy metals: an indispensable contrivance in green remediation technology. *Plants* 11, no. 9 (2022): 1255.

Sakakibara, Masayuki, Aya Watanabe, Masahiro Inoue, Sakae Sano, and Toshikazu Kaise. Phytoextraction and phytovolatilization of arsenic from As-contaminated soils by Pteris vittata. *Proceedings of the Annual International Conference on Soils, Sediments, Water and Energy* 12, no. 1 (2010): 26.

Sandeep, G., K. R. Vijayalatha, and T. Anitha. Heavy metals and its impact in vegetable crops. *International Journal of Chemical Studies* 7, no. 1 (2019): 1612–21.

Schneider, Jerusa, Claudia Regina Gontijo Labory, Wesley Melo Rangel, Eduardo Alves, and Luiz Roberto Guimarães Guilherme. Anatomy and ultrastructure alterations of Leucaena leucocephala (Lam.) inoculated with mycorrhizal fungi in response to arsenic-contaminated soil. *Journal of Hazardous Materials* 262 (2013): 1245–1258.

Sheoran, Vimla, Attar Singh Sheoran, and Poonam Poonia. Factors affecting phytoextraction: a review. *Pedosphere* 26, no. 2 (2016): 148–166.

Shtangeeva, Irina, Paavo Perämäki, Matti Niemelä, Evgeny Kurashov, and Yulia Krylova. Potential of wheat (Triticum aestivum L.) and pea (Pisum sativum) for remediation of soils contaminated with bromides and PAHs. *International Journal of Phytoremediation* 20, no. 6 (2018): 560–566.

Siddiquee, Shafiquzzaman, Kobun Rovina, Sujjat Al Azad, L. Naher, Saallah Suryani, and P. J. J. M. B. T. Chaikaew. Heavy metal contaminants removal from wastewater using the potential filamentous fungi biomass: a review. *Journal of Microbial & Biochemical Technology* 7, no. 6 (2015): 384–395.

Singh, M., P. K. Srivastava, P. C. Verma, R. N. Kharwar, N. Singh, and R. D. Tripathi. Soil fungi for mycoremediation of arsenic pollution in agriculture soils. *Journal of Applied Microbiology* 119, no. 5 (2015): 1278–1290.

Singh, Ravindra, Narendra Kumar Ahirwar, Jagrati Tiwari, and Jyotsana Pathak. Review on sources and effect of heavy metal in soil: its bioremediation. *International Journal of Research in Applied, Natural and Social Sciences* 2018 (2018): 1–22.

Srivastava, Pankaj Kumar, Aradhana Vaish, Sanjay Dwivedi, Debasis Chakrabarty, Nandita Singh, and Rudra Deo Tripathi. Biological removal of arsenic pollution by soil fungi. *Science of the Total Environment* 409, no. 12 (2011): 2430–2442.

Suman, J., Uhlik, O., Viktorova, J. and Macek, T., 2018. Phytoextraction of heavy metals: a promising tool for clean-up of polluted environment?. *Frontiers in plant science*, *9*, p.1476.

Suryanarayanan, Trichur S. Endophyte research: going beyond isolation and metabolite documentation. *Fungal Ecology* 6, no. 6 (2013): 561–568.

Upson, Rebecca, Kevin K. Newsham, P. D. Bridge, David A. Pearce, and D. J. Read. Taxonomic affinities of dark septate root endophytes of Colobanthus quitensis and Deschampsia antarctica, the two native Antarctic vascular plant species. *Fungal Ecology* 2, no. 4 (2009): 184–196.

Usuki, Fumiaki, and Kazuhiko Narisawa. A mutualistic symbiosis between a dark septate endophytic fungus, Heteroconium chaetospira, and a nonmycorrhizal plant, Chinese cabbage. *Mycologia* 99, no. 2 (2007): 175–184.

Wang, Jianlong, and Can Chen. Biosorbents for heavy metals removal and their future. *Biotechnology Advances* 27, no. 2 (2009): 195–226.

Weyens, Nele, Daniel van der Lelie, Safiyh Taghavi, Lee Newman, and Jaco Vangronsveld. Exploiting plant–microbe partnerships to improve biomass production and remediation. *Trends in Biotechnology* 27, no. 10 (2009): 591–598.

Zhang, Xu, Min Li, Huanhuan Yang, Xinxin Li, and Zhaojie Cui. Physiological responses of Suaeda glauca and Arabidopsis thaliana in phytoremediation of heavy metals. *Journal of Environmental Management* 223 (2018): 132–139.

16 Algal-Metalloid Dynamics
Unraveling the Impact on Plant Ecophysiology

Kushi Yadav, Swati Mangla, Gaurav Pathak, Pranshu Kalson, Veer Bhan, and Yogesh K. Ahlawat

16.1 INTRODUCTION

Due to their significance to the environment and impact on living things, metalloids, a large group of elements having both metallic and non-metallic properties that include arsenic (As), antimony (Sb), and selenium (Se), have drawn a lot of attention (Gupta, 2020). It is critical to comprehend the mechanisms behind metalloid absorption, transport, and accumulation in both terrestrial and aquatic environments. This multidisciplinary investigation covers a broad range of subjects, from the absorption and control of metals and metalloids by plants to their bioavailability in soil and water (Yadav et al., 2018). Furthermore, this research explores the potential of plants and algae for bioremediation in metalloid-contaminated habitats, as well as symbiotic partnerships, where plants and algae that are tolerant of metalloids collaborate to repair ecosystems (Kumar et al., 2018). In addition, this study explores the potential of plants and algae for bioremediation in metalloid-contaminated environments, as well as symbiotic relationships, where plants and algae that are tolerant of metalloids work together to restore ecosystems (Kumar et al., 2018). This study is focused on the fundamental mechanisms of metalloid absorption in terrestrial and aquatic environments. Algae, which are frequently hailed as the "sentinels" of environmental contamination, considerably contribute in the building of metalloids. Through complex absorption processes, algae may actively absorb and accumulate metalloids (Manahan, 2022). Understanding the potential bioremediation uses of algae depends critically on their capacity to absorb metalloids with diversity and adaptation. In addition, there are significant ramifications for plant health and ecosystem dynamics from plants absorbing metalloids, especially from the transport and regulatory functions of membrane proteins. This study delves into the bioavailability of metals and metalloids in soil and water, including the methods of absorption and regulatory mechanisms that preserve metal homeostasis in plants (Ulhassan et al., 2022). Comprehending the complex interactions between metalloids that plants take up from their immediate surroundings might help develop techniques for bioremediation and sustainable agriculture.

Metalloid toxicity to algae and plants is a substantial challenge, as excessive amounts can disrupt critical cellular processes in algae, disrupting photosynthesis and respiration, while in plants, they cause growth suppression and developmental abnormalities. The buildup of metalloids in plant tissues raises concerns about possible hazards to human health from consuming contaminated plants (Singh et al., 2022). This highlights the need of studying plant detoxification mechanisms, including specific transporters and metal-binding proteins that extract, modify, or sequester metalloids from essential biological processes. These processes are relevant to the interaction between plants, algae, and metalloids as well as the regulatory pathways governing metal homeostasis in plants. They are also necessary to understand plant resistance to metalloid stress (Ghuge et al., 2023). This association extends to bioremediation potential, with algae particular aptitude for metalloid bioaccumulation and biomagnification promising in metalloid-contaminated water. Additionally, the interaction

DOI: 10.1201/9781032673592-16

between metalloid-tolerant algae and plants in ecosystem restoration demonstrates environmental conservation potential, including the employment of hyperaccumulators with developed mechanisms to tolerate and sequester metalloids (Bhalla et al., 2022).

Novel solutions for the rehabilitation of soil and water contaminated by metalloids are offered by phytoremediation, the environmentally safe and sustainable use of plants to cleanse contaminated settings. The complex connections between algae and plants in phytoremediation tactics have piqued the interest of environmental researchers, since plants and algae collaborate to absorb and detoxify metalloids, placing them as efficient metalloid cleaning agents (Deng et al., 2021). Metalloid biogeochemical cycling in ecosystems is strongly influenced by the dynamic functions of algae and plants, especially the ability of algae to absorb and alter metalloids. These processes have a major impact on the cycling of these elements in both terrestrial and aquatic ecosystems, which in turn affects the sustainability and health of the ecosystem (Hussain et al., 2019). To improve crop growth and health, it is crucial to comprehend the way plants react to soil contaminated with metalloids. Moreover, research on metalloid-plant interactions in agricultural settings reveals a complex interaction between essential nutrients and potentially hazardous elements. Likewise, new possibilities for more robust and sustainable agriculture practices are presented by the ability of genetic engineering and adaptations to impart metalloid tolerance in plants (Ramezani et al., 2021). It emphasizes the importance of algae as environmental indicators because their presence in aquatic ecosystems is a useful tool for determining the level of metalloid pollution and because the complex interactions that occur between algae and metalloids in aquatic environments help us understand the dynamics and health of ecosystems (Hauser-Davis & Wosnick, 2022). To completely recognize the wider ecological effects of metalloid pollution, it is also essential to investigate whether metalloids affect plant-microbe interactions along with the mycorrhizal fungi assist plants form partnerships with metalloids. The intricacy and potential of these interactions are emphasized by the insights gleaned from multiple case studies demonstrating the varied and frequently synergistic effects of organisms and metalloids on plant growth, offering insightful guidance to researchers, environmental scientists, and industry practitioners (Deb et al., 2020). In order to commence a thorough investigation of the complicated links that exist between plants, metalloids, and algae. This exploration explores the basic processes that control the absorption and regulation of metalloids, the possibility of bioremediation, increased plant resistance, and the function of algae as environmental indicators. It is a captivating voyage through basic and applied science that broadens our understanding of the natural world and addresses pressing concerns like pollution, the health of ecosystems, and sustainable agriculture. Through a rigorous, interdisciplinary approach, this work reveals new possibilities for a more peaceful coexistence between living things and the environment by dissecting the complex web of interactions between plants, metalloids, and algae.

Thorough investigation of metalloid-induced stress responses is crucial, as evidenced by the impact of metalloids on algal cells, plant development, and detoxification processes. This investigation covers the complex interactions between metalloids and plants in agriculture, the biogeochemical cycle of metalloids, genetic engineering for metalloid tolerance, and the function of algae as environmental indicators. The complex and significant interactions among plants, metalloids, and algae are shown through case studies, which also show how these interactions may be used to restore the environment and have favorable impacts on plant development.

16.2 METALLOID UPTAKE MECHANISMS

16.2.1 Algal Accumulation of Metalloids

Algae, with their remarkable ability to adapt to a wide range of environmental circumstances, play a critical role in the accumulation and amelioration of metalloids such as As and Se within aquatic ecosystems (Labudda et al., 2022). Through a variety of physiological and biochemical processes, these diminutive organisms are able to absorb, transform, and accumulate metalloids. *Chlorella sp.*, *Spirulina sp.*, and diatoms, for illustration, exhibit an unparalleled affinity for metalloids, acting as natural inherent biofilters that facilitate the elimination of these potentially hazardous elements from aqueous domains (Chandra et al., 2020; Greeshma et al., 2022). With its many facets and growing significance, the study of algal accumulation of metalloids combines ecological and biotechnological components that provide promise for environmentally friendly resource management. Larger amounts of metals like zinc (Zn) and As were accumulated by *Sargassum natans*, causing metal trace elements, particularly As, to alter more quickly in the mangrove zone (Cipolloni et al., 2023). *Botryococcus braunii, Chlorodesmis* sp., *Cladophora* sp., *Chlamydomonas reinhardtii, Chara braunii, Chlorella pyrenoidosa, Ostreococcus tauri, Ulva fasciata*, and other algal species exhibit remarkable efficacy in As sequestration (Ghosh et al., 2023).

16.2.2 Metalloid Uptake Mechanisms in Algae

Algae exhibit a wide range of metalloid absorption mechanisms and can survive in environments with varying amounts of elements (Arora et al., 2018). These tactics mostly deal with algal cell transport mechanisms, both active and passive. Under passive transport, metalloids diffuse into algal cells and take advantage of natural concentration gradients (Parveen et al., 2022). On the other hand, active transport mechanisms are more widely used and are characterized by the efficient assistance of specific transporters, such as aquaglyceroporins and other transport proteins, in the acquisition of metalloids against concentration gradients. These transporters enable algae to reject other ions and specifically absorb metalloids due to their high substrate specificity (Kumar et al., 2022a). This deliberate choice is critical as it ensures the effective uptake of essential nutrients and the retention of potentially harmful substances, hence preserving the survival and growth of algal species in ecosystems enriched in metalloids (Garg et al., 2020). Algae can also transform and immobilize metalloids internally. Certain species, for instance, have the potential to convert inorganic Se into organic seleno compounds, such as seleno-amino acids, which have lower toxicity and can act as critical nutrients for higher trophic levels in aquatic environments (Gupta & Seth, 2019; Bhat et al., 2022). The current study focuses on the complex molecular processes that underpin these transitions. Understanding the intricate processes involved in algal-metalloid uptake and transformation offers valuable insights into the physiological mechanisms of these microbes and can be applied to the remediation of metalloid-contaminated waterways and the sustainable production of bioactive substances with industrial and nutritional value (Fagorzi, 2021).

16.2.3 Metalloid Uptake and Transport in Plants

Plant physiology relies on the essential contributions of metalloids, which are controlled in their absorption and movement by a variety of membrane proteins. Many metabolic processes require trace levels of metalloids such as As, Sb, and Se, but they might become poisonous when accumulated in excess (Uchimiya et al., 2020). For plants to effectively regulate their intake of metalloids, they utilize specific transporters and channels in their cell membranes. These allow for the targeted uptake of essential metalloids while protecting against harmful versions. Transporters like the aquaporin family, which includes nodulin 26-like intrinsic proteins (NIPs) and others, are important participants in the transport of essential metalloids like silicon and boron (Karle et al., 2022). These transporters control the uptake of metalloids and preserve tissue homeostasis in plant tissues. Moreover, research on membrane proteins involved in metalloid absorption and transport has important applications in restoration of the environment and agriculture. Researchers might possibly improve plant health and agricultural yields by improving plant health and crop accumulation of beneficial metalloids in crops by understanding these mechanisms (Tang et al., 2023). On the other hand, phytoremediation approaches, which employ plants to eliminate harmful metalloids from contaminated soils and water bodies, are informed by knowledge of metalloid exclusion mechanisms (Angulo-Bejarano et al., 2021). Thus, the discovery of the function of membrane proteins in plant metalloid uptake and transport contributes to our understanding of plant biology and provides resources for environmentally conscious and sustainable agriculture.

A crucial factor influencing plant absorption and, consequently, the potential environmental effect of metals and metalloids is their bioavailability in soil and water. Several characteristics of soil and water, including pH, organic matter content, redox conditions, and the presence of competing ions, affect the bioavailability of these elements. The pH of soil has a major impact on the solubility of metals and metalloids. Metals like iron and aluminum, for instance, become more accessible and soluble at lower pH values. Conversely, under alkaline circumstances, some metalloids, such as Se, can become more soluble (Martínez-Alcalá et al., 2020). By creating compounds with metals and metalloids, organic matter in soil might decrease their bioavailability. Metal and metalloid speciation is influenced by environmental redox potential alongside. As an illustration, some metals become more soluble in reduced settings while forming less soluble precipitates in oxidized surroundings. Plants absorb these components from the soil and water in turn. Plants absorb metals and metalloids through a well-controlled mechanism (Shahid et al., 2017). To preserve essential metal homeostasis and prevent detrimental accumulation, plants employ a range of membrane transporters and metal-binding proteins (Tang et al., 2023). Because these transporters are selective, only necessary components can be absorbed while toxic or non-essential components are rejected or detoxified. For instance, plants use transporters like IRT1 and YSL family members to guarantee an adequate supply of important metals like iron (Fe) (Kumari, 2021). Plants contain transporters like HMA and NRAMP to restrict their absorption of potentially hazardous metals like lead (Pb) and cadmium (Cd). Signaling networks and transcription factors that react to metal stress have a major role in controlling metal homeostasis (Kaur et al., 2021; Yadav et al., 2021, 2022). Plants are able to defend themselves against metal toxicity and adjust to changes in the concentration of metals in their surroundings because of these systems.

16.2.4 Metalloid Accumulation in Plant Tissues

The intricate procedure of metalloid accumulation in plant tissues is controlled by several variables, including the variety of metalloid, the species of plant, the environment, and its physiological processes. Due to their necessity for metabolic processes, some metalloids, such as As and Se, can be advantageous in trace concentrations (Abou Seeda et al., 2022). They become poisonous to plants, nevertheless, when present in excess. Metalloid absorption is often accomplished through the roots, where specialized transporters and channels play an important role in controlling the entry of these elements into plant cells. Metalloids may be found in a variety of tissues throughout the plant, including the roots, stems, leaves, and seeds. Some plant species exhibit differing abilities to collect and endure metalloids. Among them, hyperaccumulators are very skilled at capturing and retaining these substances inside their tissues (Masarovičová & Kráľová, 2012). The ability of plants to accumulate metalloids has implications for human health, ecological dynamics, and plant health. Some plants serve as organic filters, removing metalloids from the soil or water and helping to lower pollution levels in the environment (Kumar et al., 2022b). Furthermore, the buildup of advantageous metalloids in edible portions of food crops may provide nutritional advantages, such as higher Se levels in some plant-based diets. On the other hand, hazardous metalloids can be harmful to human health when they are present in food crops. Developing ways to minimize metalloid contamination, increase crop tolerance, and improve the nutritional content of food crops requires an understanding of the processes driving metalloid accumulation in plant tissues (Deng et al., 2021). Ensuring the sustainable coexistence of individuals, metalloids, and plants requires research in both agriculture and ecology.

16.3 METALLOID TOLERANCE AND TOXICITY

16.3.1 Toxic Effects of Metalloids on Algal Cells

Even though algae can adapt to a wide range of environmental stresses, excessive quantities of metalloids such as As, Sb, and Se can have negative effects on algae. These metalloids have a variety of harmful effects and have the ability to interfere with basic cellular functions in algae species (Nazir et al., 2020). Inhibiting photosystem II, which is essential for the creation of molecules high in oxygen and energy, is a means that As hinders photosynthesis. An excess of Sb can disrupt cellular respiration, a crucial process for algae cells to produce energy (Patel et al., 2018). In addition to causing oxidative stress and impairing essential cellular functions, toxic forms of Se can also have an adverse effect on the sustainability and overall health of algal communities (Wang et al., 2021). Assessing the overall health of aquatic ecosystems and potential risks associated with metalloid contamination in water bodies, including the effects of metalloids on these essential microorganisms, requires a thorough understanding of the toxic effects of metalloids on algal cells (Ramakrishnan et al., 2010).

16.3.2 Toxic Effects of Metalloids on Plant Growth

Despite minute levels of metalloids are necessary for some biochemical activities in plants, an excess of these substances can be harmful to plant development. The growth and development of plants are impacted by these components since they interact with several physiological and metabolic processes (Abbas et al., 2018; Hasanuzzaman et al., 2020; Vidya et al., 2022). For example, the metalloid As, commonly found in contaminated soils, can obstruct vital enzymes and pigments required for light absorption and energy conversion, hence impeding plant photosynthesis (Abbas et al., 2018). In a similar vein, Sb can hinder plant respiration, which lowers energy output and slows down development. Concerning Se, its hazardous forms can cause oxidative stress, which damages plant tissues and prevents development (Hasanuzzaman et al., 2020). Another aspect to be concerned about is excessive metalloid accumulation inside plant tissues. Comprehending the detrimental impacts of metalloids on plant development is essential in agricultural and environmental settings, which directs approaches to managing soils polluted with metalloids, enhance crop resilience, and reduce hazards to ecosystems and public health (Mishra et al., 2023).

16.3.3 Detoxification Mechanisms in Plants vs. the Toxicity of Metalloids

Plants have evolved a variety of detoxifying strategies to combat the damaging impacts of metalloids, including Se, As, and Sb. These mechanisms are designed to sequester, convert, or exclude metalloids from important cellular processes, allowing plants to continue functioning and even flourish in metalloid-contaminated environments (Vidya et al., 2022). ATP-binding cassette (ABC) transporters and aquaglyceroporins are two examples of the specific transporters and channels that some plant species use to actively absorb essential components while eliminating or detoxifying potentially harmful metalloids. These transporters can actively pump metalloids out of the plant or compartmentalize them inside vacuoles, limiting their detrimental impact on important cellular activities (Zhang et al., 2022).

Moreover, several plants possess the ability to convert hazardous metalloids into less dangerous or non-toxic substances. For example, Se may be bio-transformed by enzymes into organic seleno compounds that are not harmful. In the context of biofortification initiatives to raise the Se content of crops for human nutrition, this detoxification mechanism is especially pertinent as it aids plants in coping with Se stress (Hawrylak-Nowak, 2022). Crop tolerance, phytoremediation methods, and plant physiology may all be improved by comprehending these detoxification processes, which also provide management solutions for soils polluted with metalloids. Di Martino et al. (2016) suggest an area of investigation that seeks to capitalize on plant resilience to decrease the consequences of metalloid toxicity by bridging the gaps between plant biology, environmental science, and human health.

16.4 BIOREMEDIATION POTENTIAL

The process of bioremediation involves removing or neutralizing pollutants from the environment by using living organisms such as microorganisms, fungi, and plants. It can be divided into in situ or ex situ. In the case of in situ bioremediation, it treats the polluted material in the location, whereas

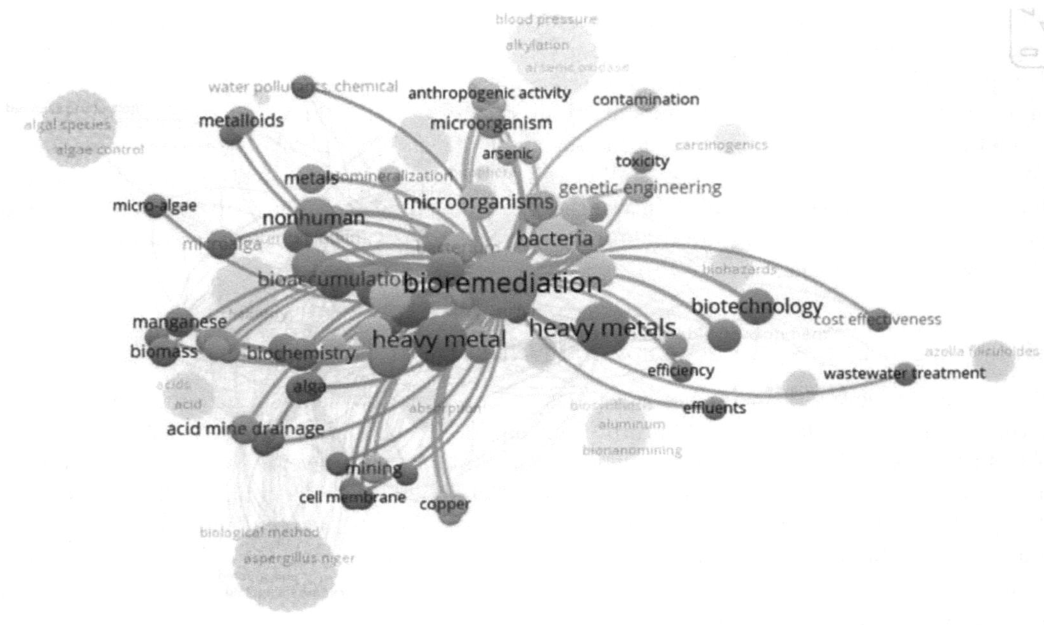

FIGURE 16.1 Mapping bioremediation: bubble sizes show keyword citations; colors indicate journal origins. Explore the interconnected dance of microorganisms, HMs, toxicity, bioaccumulation, and biomass in SCOPUS publications from 2019 to 2023. The bubble sizes reflect the citation count for each keyword, offering insights into their relative importance in scholarly publications from SCOPUS between 2019 and 2023. The varied colors of the bubbles signify the diverse journals contributing to the discourse on these topics. Additionally, the interactions between bubbles visually convey the interconnectedness of these thematic elements, highlighting the complex and interrelated nature of research in this field.

ex situ bioremediation involves treatment outside their main location (Kensa, 2011; Adams et al., 2015). It has significant potential for addressing various environmental problems, including soil and water pollution. The effectiveness of bioremediation depends on several factors, including the type of contaminant, the specific environment, and the choice of remediation organisms (Sharma, 2012). Figure 16.1 provides a comprehensive overview of the bioremediation bubble map, illustrating the intricate relationships among key concepts such as microorganisms, heavy metals (HMs), toxicity, bioaccumulation, and biomass.

16.4.1 Algal Bioaccumulation and Biomagnification of Metalloids

In order to remediate metalloids, phycoremediation involves the use of various algae and cyanobacteria that are effective at either extracting or degrading contaminants (Chabukdhara et al., 2017). Effectiveness of the microalga *Scenedesmus incrassatulus* in eliminating Cd (II), copper (Cu) (II), and chromium (Cr) (VI) was investigated. The findings demonstrated that *S. incrassatulus* successfully eliminated 25%–78% of these metalloids, with a particularly notable reduction in the concentration of Cr (VI) achieved by the microalga (Pena-Castro et al., 2004). Additionally, a review was conducted on various *Picochlorum* strains to assess their capacity for removing HMs from a range of wastewater sources. According to their research, the kinds of HMs that are present in the wastewater or growth medium have an impact on how successful different microalgae species are at lowering HM concentrations conducted (Goswami et al., 2022). Moreover, both deceased microalgae cells and materials derived from microalgae were found to have the ability to remove metalloids through adsorption and desorption processes. In a study, Pradhan et al. (2019) examined the capacity of *Scenedesmus* sp. biomass for adsorbing Cr (VI) from a solution. The findings demonstrated

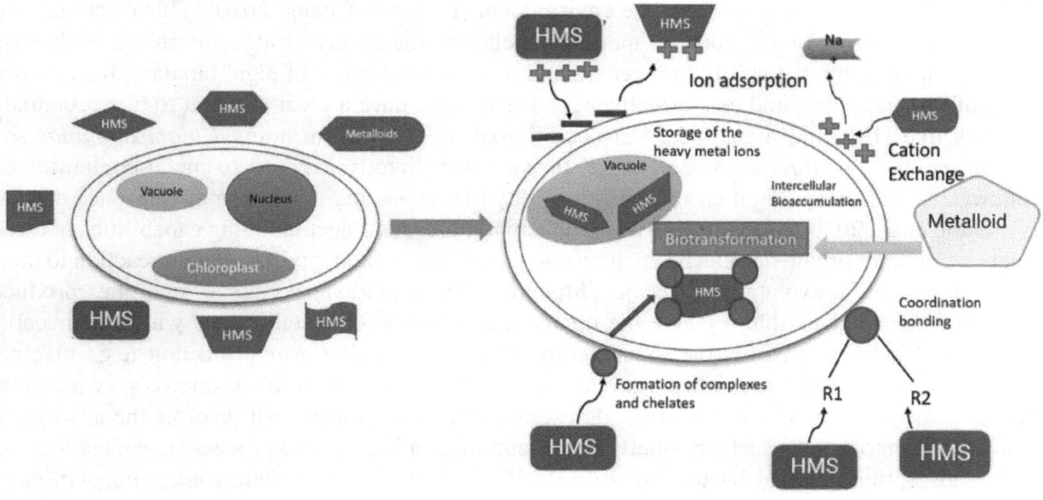

FIGURE 16.2 Possible mechanisms for the detoxification of HMs and metalloids by microalgae.

that *Scenedesmus* sp. biomass exhibited remarkable biosorption capability, accumulating approximately 92%–89% of the Cr. Analysis through Fourier-transform infrared spectroscopy (FTIR) further revealed the presence of diverse functional groups within the microalgae cells, including amides, aldehydes, carboxylic acids, halides, and phosphates. These functional groups carried positive charges and played a crucial role in facilitating the biosorption process. The diatoms *Staurosira construens* and *Pseudostaurosira brevistriata* have proven to be successful in the phytoremediation of silver (Ag). The adsorption of Ag is influenced by the photosynthetic activity rate of these diatoms (Asiandu & Wahyudi, 2021).

Various possible mechanisms for detoxification of metalloids and HMs by microalgae is listed in Figure 16.2. Microalgae have the potential to employ a bioreactor system for the immobilization and recycling of Cd ions (Shen et al., 2020). Additionally, it has been reported that a combination involving metal oxides, such as Fe_2O_3, can achieve the goal of immobilizing ions while enhancing the removal of HMs with greater efficiency. Microalgae consortiums have drawn notice for their capacity to remove nutrients from wastewater and are also used to remove metalloids. In a study, a combination of microalgae and cyanobacteria (specifically, *Chlorella vulgaris*, *Scenedesmus quadricuda*, and *Spirulina platensis*) was employed to remediate HMs like Cd, nickel (Ni), and Pb, along with the organophosphate pesticide malathion from a mixture of urban and agricultural wastewater (Abdel-razek et al., 2019). The results demonstrated that the microalgae consortia achieved impressive removal rates: 89% for Pb, 95% for Ni, 88% for Cd, and an outstanding 99% for malathion from the water sample.

16.4.2 Algae-Based Bioremediation of Metalloid-Contaminated Waters

Numerous sectors, such as electroplating, dyes, batteries, mining, autos, and textiles release metalloids into the environment. By which these toxic radicals pollute air and discharge into water bodies. Like some HMs are mercury (Hg), Ni, titanium (Ti), Cu, Ag, gold (Au), etc., and some metalloids are As, Sb, Cr, Se, etc., (Babula et al., 2010; Bienert & Tamás, 2018). Various traditional techniques like adsorption, precipitation of chemicals, exchange of ions, and membrane purification are used (Chen et al., 2002). Many contemporary strategies, in addition to more conventional ones, are employed to lessen these contaminations. For example, biological methods are utilized to remove metalloids in an environmentally acceptable manner (Kumar et al., 2022a). Biological agents, including fungus, algae, and bacteria, are employed in bioremediation to reduce the levels of HMs. Statistics show that, in contrast to bacteria or fungus, algae can remove between 15.3% and

84.6% of HMs and metalloids from the environment (Leong & Chang, 2020). This offers several benefits over existing bioremediation methods, including the ability of algae to absorb HMs with a significant capacity, the ability to regenerate, the year-round usage of algal biomass, the absence of harmful metabolites, and cost-effectiveness. Microalgae have a great efficient to bioaccumulate toxicants like HMs and metalloids. Various microalgae species, including *Chlamydomonas* sp., *Scenedesmus* sp., *Chlorella sp.*, and more, thrive when directly exposed to metal-contaminated wastewaters, demonstrating their remarkable ability to tolerate and flourish in the presence of metals (Shahid et al., 2020; Arora et al., 2019). The remarkable metal-accumulating capabilities of these hyperaccumulator organisms can be attributed to adaptation to their environment in reaction to their surroundings or the evolution of genes. Throughout the evolutionary change, microbes produce a range of mechanisms that regulate the uptake and retention of metallic ions within their cells. A biological process involves the manufacture of substances like osmoprotectant (e.g., glycine, betaine, and proline) and free radical scavenging enzymes (ascorbate peroxidase) to cope with metal exposure. Biosorption is an energy-free, quick, and reversible process that involves the adsorption of pollutants onto a cell's surface, whereas bioaccumulation is an active process (Javanbakht et al., 2014). Biosorption involves bacteria such as Bacillus subtilis, algae include *Chaetomorphalinum*, marine microalga, Rhizopus fungi, and Saccharomyces cerevisiae yeast that helps in the extraction of harmful metalloids or metals by binding to the cell surfaces (Nilanjana et al., 2008).

Bioaccumulation happens when a contaminant's rate of absorption outpaces its rate of loss. Commonly utilized organisms to research bioaccumulation and as an indicator of pollutants include bacteria (Diepens et al., 2015), fungi, plants (Somdee et al., 2016), algae (Jaiswar et al., 2015), fish, and oysters (Mok et al., 2015). Genes that are involved in bioaccumulation will be analyzed using transcriptomics analysis (Shi et al., 2015). Utilizing the metal-resistant siderophore-producing bacteria (SPB) that are close to the rhizosphere allows plants to get nutrients, especially Fe, which may lessen the detrimental impacts of metal pollution. The siderophore may form permanent interactions with metals that are of concern to the environment, including radionuclides and the elements Al, Cd, Cu, Ga, In, Pb, and Zn, in addition to solubilizing iron from minerals and biological molecules (Kumar & Seth, 2022; Rajkumar et al., 2010).

Algae are one of the renewable natural biomasses that show varying affinities for various metals, making them a promising option for use as a biosorbent material. These are less costly methods of eliminating pollution. There have been reports of using river and marine algae for the adsorption and elution of Ag, Au, and Co (Lin et al., 2020; Martínez et al. 2018). Various algal species have good bioaccumulation capacity against HMs and metalloids like *Oscillatoria* sp., *Chlorella* sp., *Cladophora* sp., *Ankistrodesmus* sp., *Spirogyra* sp., and *Oedogonium* sp. (Jaiswar et al., 2015). They can gather a sizable number of metals from the polluted surroundings (Saranya & Shanthakumar, 2019). Research by Yoon et al. (2006) stated that native species growing on contaminated sites respond to stress better than plants brought from other areas in terms of survival, growth, and reproduction. Moreover, it is plausible that species originating from contaminated areas may exhibit a greater propensity to collect elevated levels of metals and metalloids (Arıca et al., 2005).

16.4.3 Symbiotic Relationships

Symbiotic relationships between two organisms have a great impact on reducing metalloid toxicity. To improve soil fertility and clean up the contaminated environment of metalloids, researchers are turning to genetic engineering to create genetically modified rhizobia and encourage the symbiotic interaction. To evaluate the possible application of legumes and rhizobia in bioremediation, the impact of metalloid-contaminated soil on the symbiotic connection between legumes and Rhizobium bacteria needs to be clarified. The development of leguminous plants and the nodulation process should be included in this investigation. Throughout the various phases of legume plant growth in soils tainted with HMs, it becomes evident that the nodulation process is more susceptible to the influence of these contaminants than other aspects of vegetative growth, such as the development of root and shoot and the levels of photosynthetic chlorophyll pigments. These latter parameters are typically employed to gauge the extent of metal toxicity in mature leguminous plants (Hao et al., 2014).

16.5 UTILIZING METALLOID-TOLERANT ALGAE AND PLANTS IN ECOSYSTEM RESTORATION

Ecosystem restoration from the metalloids by plants and algae is an innovative and sustainable approach to rehabilitating areas contaminated with metalloids, such as As, Se, and Sb. Ecosystems that are impacted by metalloid contamination are frequently located close to industrial areas, mining operations, or places where metalloid concentrations are naturally occurring in the geology. The United States initiated the development of permeable reaction wall technology in the 1980s to mitigate water contamination, demonstrating a noteworthy progress in ecosystem restoration techniques. Over the years, this technology has attracted substantial attention and research efforts from engineers and scientists in the United States, Europe, and various other countries. As a result, it has steadily advanced toward the point of becoming a viable commercial product (Lin et al., 2021). The enduring and imperishable characteristics of HMs require their ongoing extraction from soil and water systems to prevent ecological disruption. Plants and algae are inherently equipped with the genetic, physiological, and biochemical attributes that make them the ideal selection for addressing soil and water pollution. Phytoremediation encompasses a vast range of plant-based techniques, utilizing either genetically modified or naturally occurring plants to purify and restore contaminated ecosystems (Flathman & Lanza, 1998). Plants engaged in the process of phytoremediation for metals are known as metallophytes (Ojha et al., 2023). Harmful metalloids can be effectively removed from soil over time through the process of phytoremediation. In metal-contaminated wetland environments with flooded conditions, the wetland plant *Cyperus alternifolius* demonstrated a remarkable capacity for the uptake of Pb, Zn, and Cd from both the rhizospheric region and its roots, indicating its strong metal accumulation capability (Yang et al., 2017; Agnihotri et al., 2018; Hassan et al., 2020). Research has indicated that, in a hydroponic environment using leachate from municipal solid waste, three varieties of grass, specifically Elephant grass, Rhodes grass, and Vetiver, exhibited notable efficiency in the uptake of metals like manganese (Mn), Cu, and Zn (Hassan et al., 2020; Phang et al., 2015a). *Hydrilla verticillate* is one of the plants, which forms a thick covering on the water surface while the plants are still in the top 2 feet of water. It has shown significant tolerance to high metal exposure concentrations. Metals react with thiol (-SH)-containing proteins in these plants, which may contribute to their metal tolerance and help in ecosystem restoration from the toxicants (Shrivastava & Srivastava, 2021). Other aquatic plants, like *Eichhornia crassipes, verticillata, Lemna,* and *Pistia stratiotes*, among others, are promising because of their quick growth, large biomass production, and strong resistance to various water pollutants (Ali et al., 2020; Rezania et al., 2016; Havel et al., 2015).

The total metal detoxification process that microalgae do consists of three basic steps. Metal ions stick to extracellular polymeric components that are released on the cell wall in the first procedure. By obstructing the metals' entry into the cell, this link lessens their availability and toxicity (Monteiro et al., 2012; Singh et al., 2017). The second process involves the internal absorption of metal ions with the assistance of cysteine-containing peptides like metallothioneins (MTs) and phytochelatins (PCs). This facilitates the sequestration of non-toxic ions within vacuoles. Meanwhile, the third mechanism encompasses the removal of surplus metal ions from the cell, releasing them back into the surrounding environment (Hanikenne et al., 2005; Monteiro et al., 2012). Among these various phenomena, MT-assisted detoxification stands out, drawing particular attention due to its consistent expression, which provides compelling evidence of its vital role in regulating metal homeostasis (Sharma et al., 2016). Macrophytes water hyacinths (E. crassipes) have demonstrated their effectiveness as robust bioremediation systems for a wide spectrum of HMs in aquatic environments (Leguizamo et al., 2017; Rai et al., 2019; Seth et al., 2012). Furthermore, these wetland plants have shown remarkable efficacy in removing As, making them excellent resources for As bioremediation in freshwater environments (Favas et al., 2012).

16.6 METALLOID-INDUCED STRESS RESPONSES: HYPERACCUMULATION AND TOLERANCE OF METALLOIDS IN A COMPLEX EVOLUTIONARY NETWORK IN PLANTS

It has been determined that more than 700 plant species are hyperaccumulators of non-metals, metalloids, and trace metals. The hyperaccumulator species are from 52 families in the pteridophyte and angiosperm kingdoms (Reeves et al., 2017). Certain species, such as the Zn hyperaccumulators *Noccaea caerulescens* and *Arabidopsis halleri*, have hyperaccumulation as a constitutive characteristic that is present in all populations. According to Clauss et al. (2002) both are near-wild relatives of *Arabidopsis thaliana*, which provides a wealth of genomic, proteomic, and transcriptomic data in addition to molecular markers and genomic tools. This makes it possible to analyze the characteristic in these species more easily and thoroughly. In contrast to inherent hyperaccumulating species like *A. halleri* and *N. caerulescens*, some species like *Senecio coronatus* and *Sedum alfredii* have populations that are both hyperaccumulating and non-accumulating (Lu et al., 2008; Mesjasz–Przybyłowicz, Barnabas & Przybyłowicz, 2007). There have been reports of additional hyperaccumulating metals, including As, Se, and Pb. Hyperaccumulation of Pb is extremely uncommon and limited to a select few species. Pb has a low solubility in soil as they may form complexes with organic materials found in nature and precipitate as phosphates, hydroxides, and carbonates cultivated in a nutrient solution contaminated with Pb, *Sesbania drummondii*, a perennial leguminous shrub native to the southern coastal areas of the United States, exhibits a remarkable ability to hyperaccumulate Pb in its shoots (Epstein et al., 1999; Sahi et al., 2002).

Fagopyrum esculentum is another example of a Pb hyperaccumulator, as it collects a significant quantity of Pb in its stems, leaves, as well as in roots. The most researched *Pteris vittata*, a hyperaccumulator plant, was found in soils polluted with chromated copper arsenate (CCA) in Central Florida. Isatis cappadocica and *Hesperis persica* are two species of *Brassicaceae* that exhibit hyperaccumulation, aside from ferns (Karimi et al., 2009a, b). All living things must adjust to and accommodate their surroundings, but those that are unable to flee unfavorable environmental circumstances must make additional attempts.

Ecologists have been interested in hypertolerance and hyperaccumulation since their discovery because of their high levels of natural selection and quick evolutionary histories. For plants to respond to a variety of abiotic stimuli, including drought, salt, intense light, nutritional shortages, and HMs, abscisic acid (ABA) is essential (Choudhary et al., 2024). ABA is a major phytohormone that helps plants become resistant to dangerous metals and metalloids like As, Cd, and Pb. The Cd-tolerant rice cultivar's leaves and roots demonstrated a higher level of Cd-induced rapid ABA synthesis than the Cd-sensitive genotype, indicating a positive correlation between endogenous ABA content and Cd tolerance (Hsu & Kao, 2003; Mariyam et al., 2023). Toxic metalloid As has characteristics halfway between those of metals and non-metals. Large areas in India, Argentina, Vietnam, Ghana, Laos, Mexico, Bangladesh, Chile, China, Canada, the United States, and numerous other countries suffer from high levels of As contamination, surpassing the threshold value of 10 $\mu g\ L^{-1}$ established by the World Health Organization (WHO) and the U.S. Environmental Protection Agency (EPA). Drinking water tainted with As or ingesting it through food chains are the two ways that As may enter the human body. The contamination of drinking water supplies by As affects about 150 million people globally. The majority of these people live in countries in South and Southeast Asia. Large concentrations of As can also be found in edible sections of foods, including grains and vegetables, which pose a major risk to human health. To combat the harmful effects of As, all living things, from bacteria to higher plants, have evolved complex resistance mechanisms. Because As (V) and inorganic phosphate have a strong structural resemblance, prokaryotic cells absorb As(V) through phosphate transporters. Two distinct forms of ArsC proteins, thioredoxin-coupled and glutaredoxin-linked, subsequently rapidly convert As(V) to As(III). As compounds resistance 2 (ACR2) or *A. thaliana* arsenate reductase QTL1 (ARQ1)/high as content (HAC) proteins mediate the same mechanism in eukaryotes. While ARFs and HACs primarily function as arsenate reductases in plants, ACR2 facilitates As (V) reduction in yeast. Subsequently,

As (III) can be either enclosed within vacuoles through the action of ABC transporters or expelled from the cell via aquaglyceroporins, which belong to the aquaporin family, or through the As(III)/ H^+ antiporter known as ACR3. Furthermore, there are other pathways for As tolerance, including As (V) outflow and As methylation.

16.7 PHYTOREMEDIATION TECHNIQUES

16.7.1 Algae-Metalloid Interactions in Phytoremediation Strategies

Phycoremediation is the employment of algae to ameliorate and purify waste materials or effluents (Phang et al., 2015b). This sustainable method employs microalgae and macroalgae (seaweeds) to efficiently eliminate pollutants, HMs, and noxious organic compounds from the environment (Koul et al., 2022). By harnessing the natural power of these aquatic plants, phycoremediation not only cleanses water resources but also promotes environmental sustainability (Shackira et al., 2021; Misra et al., 2009). Phycoremediation uses microalgae's capacity to effectively remove or degrade a range of environmental pollutants, such as HMs and organic pollutants, in addition to absorbing nitrogen and phosphorus, two nutrients necessary for their growth. Microalgae have developed defense mechanisms that increase their ability to withstand these toxins and enable them to break down or detoxify the dangerous materials. (Kumar et al., 2023). Microalgae can convert HMs like Cd and Hg into less soluble sulfides, reducing their toxicity and rendering them biologically unavailable. The *C. reinhardtii* (green alga), the *Cyanidioschyzon merolae* (red alga), and *Synechococcus leopoliensis* were observed to facilitate the detoxification of Cd through a process that converts it into CdS (Mandal et al., 2016). In these investigations, the microorganisms were provided with additional sulfate, thereby enhancing their ability to tolerate Cd. Other study has demonstrated that algae serve as efficient and cost-effective bioabsorbents for the efficient removal of HMs from industrial wastewater (Mohammadi & Mahmoudnia, 2023).

Because of the significant attraction of their cell membranes to bond with Cu, algae have a stronger capacity to remove Cu than cobalt, which results in increased bioaccumulation of Cu (Dwivedi, 2012). *Codium tomentosum*, a marine green macroalgae, demonstrates an adsorption capacity of 5.032 ± 0.644 for the Cr(VI) compound at a pH of 2 (Anandaraj et al., 2018). In the endeavor to cleanse dumpsite leachate, a consortium comprising the vibrant *Ankistrodesmus* sp. and the cyanobacterial marvels *Nostoc* sp. and *Anabaena* sp. was harnessed. The outcomes elegantly revealed the hierarchy of removal to be Pb (II) > Cu (II) > Zn (II) > Fe (II) > Cr (VI) (Iqbal et al., 2022). In a study, *G. sulphuria* can detoxify HMs by changing them into less harmful forms. *G. sulphuria* is an extremophilic red alga that can grow in a range of diverse habitats, including those having thermophilic, acidophilic, halophilic, and toxic metal conditions (Thangaraj et al., 2011). Microalgae possess the capability to synthesize MTs or PCs, which aid in the detoxification of HMs (Perales-Vela et al., 2006). These peptides play an essential role in microalgae's heavy metal detoxification process by binding to HMs and isolating them within vacuoles (Tripathi & Poluri, 2021). *C. vulgaris*, a freshwater microalga strain, has shown the capacity to expedite the agglomeration of NiO nanoparticles and convert them into zero-valence Ni (Gong et al., 2011). In a study conducted by (Saunders et al., 2012), it was observed that three species of green algae, *Hydrodictyon sp., Oedogonium sp.*, and *Rhizoclonium sp.*, exhibited the capacity to accumulate HMs, including vanadium (V) and As, from wastewater originating from coal-fired power generation. The bioconcentration of these HMs was notably associated with the growth rate of the algae. However, it's worth noting that certain elements, like Cd and Zn, displayed higher concentrations when the growth rates of the algae were lower.

By attaching to both the cell surface and internal ligands, HMs can be bound by microalgae and then absorbed from wastewater (Pradhan et al., 2022). Metals can be bound by the following functional sites: amino, phosphate, sulfate, thioether, carbonyl, amide, hydroxyl, carboxyl, sulphydryl, and phenol. The carboxyl group is the most important of these chemical groups for metal-binding (Sahoo & Seckbach, 2015).

Therefore, the future holds substantial potential for algae in HM remediation, with opportunities including refining immobilization methods, advancing algal bioreactors, and exploring synergistic integration with other remediation technologies to create more effective and sustainable solutions for addressing HM contamination.

16.7.2 Phytoremediation by Plants as Metalloid Cleanup Agents

The innovative and environmentally friendly approach offers a versatile solution for remediating a wide range of pollutants, including HMs, radionuclides, and organic substances like polynuclear aromatic hydrocarbons, polychlorinated biphenyls, and pesticides (Kumar et al., 2020; Novikau & Lujaniene, 2022). This approach minimizes installation and maintenance expenses and can cost as little as 5% of other cleanup methods. Additionally, vegetation on polluted soils helps prevent erosion and metal leaching, contributing to long-term soil quality improvement, as shown in Table 16.1 (Suresh & Ravishankar, 2004).

TABLE 16.1
Overview of Phytoremediation Process - A Compilation of Diverse Plants and Their Effective Mitigation of Specific Metals

Technique	Plant Species	Metals	References
Phytoextraction/ Phytoaccumulation	*P. hortorum*	Cd	Gul et al. (2023)
	Calotropis procera	Cu, Zn, Pb, and Cd	Siraj et al. (2022)
	Sedum alfredii	Zn	Feng et al. (2019)
	Pteris vittata	As	Danh et al. (2014), Kalve et al. (2011)which is reported elsewhere as an arsenic hyperaccumulator. This is the first report of characterization of arsenic (As
	Hyparrhenia. Hirta	Pb, Zn, and Cu	Conesa et al. (2007), Poschenrieder et al. (2001)
Phytostabilization	*Cynara cardunculus, Brassica juncea*	Cd, Pb, and Zn	Lacalle et al. (2023)
	Myriophylhum aquaticum, Ludwigina palustris, Mentha aquatic	Fe, Zn, Cu, and Hg	Kumar et al. (2017), Kamal et al. (2004) transportation, translocation and storage of heavy metals (HMs
	Amaranthus spinosus L.	Cu, Pb, and Cd	Huang et al., (2019), Chinmayee et al. (2012)
	Solanum tuberosum L.	As, Ag, Cr, and Sb	Baghour et al. (2001)
Phytovolatilization	*Brassica sp.*	Hg, Se, and As	Bortoloti et al. (2022)
	Quercus ilex	Cd	Ramezani et al. (2021)
	Agrostis castellana	Cu, Pb, and Zn	Gonzaga et al. (2006)
	Lolium perenne	Cu, Pb, and Zn	Alvarenga et al. (2008)
Phytodegradation	*Brassica juncea and Lupinus sp.*	Hg	Martínez-Alcalá et al. (2020)this technique requires the use of plants that are able to accumulate high concentrations of the target elements in their aerial part, while producing high plant biomass. Here, pot experiments were carried out in order to determine the interaction between a hyperaccumulator (Nocaea caerulescens
	Arundo donax L.	As	Mirza et al. (2011)

Phytoremediation of HMs from contaminated sites typically occurs via one or a combination of the following mechanisms or processes:

i. **Phytoaccumulation:** This technique involves the uptake and accumulation of various pollutants, such as HMs, in the harvestable biomass of plants, primarily in the aboveground parts such as shoots. This method is valuable for removing contaminants from soil or water, with a focus on the concentration of pollutants in easily harvestable plant components (Chen et al., 2004). The availability of metals in the root zone (rhizosphere), the rate at which roots absorb metals, the percentage of metals bound within the roots, the rate at which metals are transported through the xylem to the aboveground shoots, and the plant's tolerance of toxic metals is some of the important factors that determine the extent of metal accumulation by hyperaccumulators (Huang et al., 2020).

ii. **Phytostabilization:** Plants are used in this remediation technique to restrict the bioavailability and mobility of contaminants in the soil. Plant roots play a crucial role in immobilizing contaminants, preventing their movement through the soil, and reducing their potential to harm ecosystems and human health (Bolan et al., 2011).

iii. **Phytovolatilization:** It is a phytoremediation process in which plants take up pollutants and convert them into volatile forms. Subsequently, these volatile compounds are released into the atmosphere. This technique can be employed to address contaminants that are amenable to transformation into gaseous forms (Limmer & Burken, 2016). It offers a distinct advantage over other phytoremediation strategies, as it involves the removal of HM and metalloid contaminants from the site, dispersing them as gaseous compounds, eliminating the necessity for plant harvesting and disposal (Yan et al., 2020).

iv. **Phytodegradation:** It is the process of using plant enzymes within plant tissues to break down and degrade organic xenobiotics. Plants actively participate in the decomposition of organic contaminants, contributing to their detoxification and removal from the environment (Muthusaravanan et al., 2018). The breakdown and modification of HMs within plants through metabolic processes or the use of root-generated enzymes to break down HMs outside the plant is known as phytodegradation of HMs (Newman & Reynolds, 2004). This method offers a promising solution for eliminating HMs from contaminated regions, particularly when employing hyperaccumulator plant species.

16.8 ALGAE AS METALLOID INDICATORS IN ENVIRONMENTAL MONITORING

Algal species have been employed as metal biomonitor agents all over the world. The types of algae can be used as a real-time method to track the hazardous substances found in saltwater and coastal sediments. In the Aegean Sea, brown algae of the *Cystoseira* species and green algae of the Ulva and Enteromorpha species serve as metal biomonitoring agents. It has been stated that brown and green algae may be used to detect metalloids in rivers, gulfs, and seas (García-Seoane et al., 2018). In seawater and sediment samples, the *C. myrica* species was shown to be the most effective biomonitoring instrument for Cd. Among the algae under study, *R. riparium* was the most effective biomonitor of V in the sediments and a suitable agent for monitoring V, Cd, and Ni in the saltwater. It was also recognized that the *S. boveanum* species was an effective biomonitor for Cd in seawater, Ni in the sediments, and V in both seawater and sediments. As a result, the investigated algae species appear to be effective biomonitoring agents for reflecting the contamination of the marine environment with Ni, V, Cd, and As.

Due to the rapid increase in urbanization, the introduction of agricultural practices and industrialization cause contamination in water bodies. Industrialization includes many kinds of factories, which cause seepage of harmful, non-biodegradable substances in the water bodies and hence cause water pollution. The WHO and EPA have defined permissible limits for contaminants in water bodies, and exceeding these limits can have devastating consequences (Nishikawa et al., 2018). It is believed that sea algae are a practical and effective source of soil conditioner for wastewater treatment. It considers how renewable resources are distributed throughout the world. It can be categorized by its color and colloid content. Both macro and microalgae were utilized to remove metalloids from aquatic systems and as suitable candidates for different organic dyes (Chen et al., 2019). Macroalgae in particular are frequently employed as bioindicators in the investigation of marine radionuclide and metalloid pollution. As their capacity to collect a variety of contaminants, macroalgae, or marine sediments are frequently utilized for environmental management. Macroalgae has numerous benefits over other agents, including significant biomass, simple identification, long life, availability, poor mobility (which makes it a valuable agent for monitoring the contaminants in saltwater), and the capacity to collect metals (Aydın-Önen & Öztürk, 2017). Alginate, carrageenan, and polycolloid are the primary polysaccharides that make up the algal cell wall. These ingredients have the capability to eliminate various types of water pollutants. In the maritime environment, the concentration of radionuclide and the capability for accumulation of several algae species were also ascertained for three distinct algal phyla. The three primary kinds of algae that are often found in seas are rhodophyta, phaeophyta, and chlorophyta (Mokhtar et al., 2017). The primary component of green algae is cellulose, which is very proportionately linked to proteins and polysaccharides to produce glycoproteins. A wide variety of metabolites, including fucoidans, mannitol, alginates, fucoxanthin, halogenated chemicals, polyphenols, and laminarins, are characteristics of brown algae. Alginates are a combination of salts, including Na, P, and Ca. In the black sea ecosystems, chlorophyta (green algae) has been widely found to remove environmental contaminants. The use of biochar produced from the pyrolysis of microalgae as an additional sorbent for the removal of HMs or metalloids from aqueous solutions (Amin & Chetpattananondh, 2019; Misra et al., 2009).

In the natural environment, coexistence between plants and microorganisms gives rise to interactions that can be either mutually beneficial or deleterious. Even in minimal concentrations, HMs such as Cd, Pb, As, and Hg pose significant hazards to plants, affecting their physiological systems. The pathways of these metals are influenced by various factors, including soil characteristics, plant developmental stage, environmental conditions, root exudates, and compounds emanating from rhizospheric bacteria. In plant metal homeostasis, key ligands binding to HMs are PCs and MTs. The principal focus of heavy-metal-contaminated site restoration strategies is strengthening plant resilience via improved growth and decreased metal transfer. Plant growth-promoting rhizobacteria produce a range of compounds that aid in plant growth.

Interactions between plants and microbes play a crucial role in combating metal toxicity and improving the efficiency of restoration processes, as shown in Table 16.2. Symbiotic relationships, exemplified by *Frankia* sp. and alder plants, demonstrate potential for the re-cultivation of metal-polluted areas. Specific microbial strains, such as *Pseudomonas putida* CZ1 and *Variovorax paradoxus*, have been identified to enhance Cu and Ni phytoextraction (Xu et al., 2015; Durand et al., 2016). *Pseudomonas koreensis* AGB-1 has been observed to aid in plant growth in HM-contaminated soil. Rhizosphere bacteria are shown to alter soil pH, thereby enhancing Ni absorption in *Alyssum murale*. The contribution of rhizospheric bacteria to phytoremediation involves affecting redox processes and releasing chelators (Weyens et al., 2009).

Three mechanisms by which microorganisms actively remove metalloids are uptake, bacterial cell absorption, and remediation via siderophores and biosurfactants. In interactions between plants and microbes, siderophores help take iron, and biosurfactants help remove metal from soil (Zarei et al., 2010; Rajkumar et al., 2012; Chen et al., 2017). Organic acids produced by microbes enhance Zn accumulation, and certain bacteria are identified to oxidize As and convert Cr into a less toxic form (Ryan et al., 2001). Recent research on *Hydrilla* plants has uncovered bacteria capable of oxidizing As(III), reducing As(V), and methylating As(III). When combined with *Hydrilla*, these bacteria have demonstrated the oxidation of As(III) and reduction of As(V). Microorganisms thriving in biofilms around *Hydrilla* exert influence on metal uptake and deposition by altering metal availability or chemical forms (Zhen et al., 2020; Tavanappanavar et al., 2024).

TABLE 16.2
Interplay of Microbes, Plants, and Metals: A Compilation of Biological Remediation Partnerships

Microbes	Plants	Metals	References
Chaetomium cupreum	*Miscanthus sinensis*	Al	Haruma et al. (2019)
Claroideoglomus claroideum, Pseudomonas libanensis	*Helianthus annuus*	Ni	Ma et al. (2019)
Simplicillium chinense	*Phragmites communis*	Pb and Cd	Jin et al. (2019)
Bacillus cereus	*Vetiveria zizanioides*	Fe, Zn, Cr (VI), Cd, Mn, Cu, and Ni	Nayak et al. (2018)
Rhizophagus irregularis	*Medicago sativa*	Cd and Ni	Mnasri et al. (2017)
Fungal strains RSF-4L and RSF-6L	*Solanum nigrum*	Cd	Khan et al. (2017)
Arbuscular mycorrhiza fungi	*Maize*	Hg	Kodre et al. (2017)
Streptomyces sp.	*Salix dasyclados*	Zn, Cd, and Pb	Złoch et al. (2017)
Rhizobacterial strains	*Alyssum murale and Noccaea tymphaea*	Ni	Durand et al. (2016)
Agrobacterium tumefaciens, Rahnella aquatilis, Pseudomonas sp.	*Lens culinaris*	Pb	Jebara et al. (2015)
Bacillus megaterium 1Y31	*Pennisetum*	Mn	Zhang et al. (2015)

16.9 CASE STUDIES

16.9.1 Synergistic Effects of Algae and Metalloids on Plant Growth

Coexistence of these microorganisms can have complex impacts on plant development through interactions between algae and metalloids in aquatic environments. Metalloids may be bioaccumulated and transformed by algae, which reduces their toxicity and may even have positive effects on plants (Bilal et al., 2018). One noteworthy characteristic of some algae species is their capacity to absorb and hold onto metalloids from their surroundings, which reduces the number of metalloids in the water. Therefore, by using higher-quality irrigation water, plants may benefit from lower metalloid exposure and toxicity (Martínez-Alcalá et al., 2020). Moreover, the biomass of algae that accumulates metalloids can be utilized as a source of nutrients and organic matter to improve the soil and encourage the development of plants.

Moreover, some types of algae could transform metalloids like Se into less harmful forms like organic seleno compounds. For instance, *C. reinhardtii, C. vulgaris, Scenedesmus* sp., and *Selenastrum capricornutum* are reported to transform Se into organic seleno compounds (Bogstie, 2022). These algae species can reduce metalloid toxicity while increasing the availability of vital nutrients when they cohabit with plants (Tan et al., 2018). The mutually beneficial impacts of metalloids and algae on plant development highlight the intricate and interdependent dynamics of aquatic environments. Gaining knowledge of these relationships can help with ecological preservation, sustainable agricultural methods, and the possibility of utilizing the advantages of algal-metalloid linkages for improved plant development and general ecosystem health (Papry et al., 2022).

16.9.2 Case Studies Based on Algae and Plant Interactions with Metalloids

Research investigates the relationship between waterways contaminated with As and the ubiquitous green microalga *C. vulgaris*. It was revealed that *C. vulgaris* can actively absorb and change high concentrations of As into less harmful compounds like dimethylarsenate and arsenobetaine in a controlled setting (De Francisco et al., 2021). The interaction between algae and As has great significance for phytoremediation initiatives in soils and streams polluted by As. Ranjan et al. (2021) claim that chlorella's ability to reduce As bioavailability increased rice growth. This case study demonstrated the potential of plant-algae interactions in sustainable remediation and agricultural operations by lowering As toxicity and improving soil quality (Rizvi et al., 2020). The contribution of Se-accumulating algae to improving Se biofortification in crops is examined in a different case study. According to research, Se may be absorbed by the green microalga *C. reinhardtii* and converted into selenomethionine, an organic form that is less hazardous and higher in nutrients (Schiavon et al., 2017). The Se content of plant tissues rose when these Se-accumulating algae were added to the growth medium of crops such as lettuce and wheat. Through sustainable agriculture techniques, this case study demonstrated how algae-plant interactions might improve the nutritional content of food crops, alleviate Se deficiency concerns, and improve human health (Terry et al., 2000). These case studies showcase the variety of ways that plants and algae might interact with metalloids, providing novel remedies for problems related to food security and restoration of the environment.

16.10 CONCLUSION

In conclusion, the exploration of algal-metalloid dynamics presents a critical avenue for understanding the complex interactions within plant ecophysiology. This field, which intersects the domains of plant biology, ecology, and environmental chemistry, provides invaluable insights into how metalloids, often viewed as contaminants, interact with algal systems, and affect broader ecological networks.

First, the study of algal responses to metalloids has revealed a fascinating array of adaptive mechanisms. Algae, as resilient and versatile organisms, have evolved various strategies to tolerate, accumulate, and even detoxify metalloid compounds. These mechanisms not only highlight the resilience of algal species but also point to potential bioremediation applications where algae could be used to mitigate metalloid pollution in aquatic ecosystems. Moreover, the impact of metalloids on algal physiology and biochemistry has far-reaching implications for the health and functioning of aquatic ecosystems. Metalloids can influence algal growth, photosynthesis, and nutrient cycling, thereby altering food web dynamics and ecosystem productivity. Understanding these effects is crucial for predicting and managing the ecological consequences of metalloid pollution, particularly in the face of increasing industrial and agricultural activities. Furthermore, the study of algal-metalloid dynamics is instrumental in informing environmental policy and management practices. By enhancing our understanding of how metalloids behave in aquatic environments and their impact on primary producers like algae, we can develop more effective strategies for monitoring, regulating, and remediating metalloid pollution.

REFERENCES

Abbas, G., Murtaza, B., Bibi, I., Shahid, M., Niazi, N. K., Khan, M. I., Amjad, M., Hussain, M., and Natasha. Arsenic uptake, toxicity, detoxification, and speciation in plants: physiological, biochemical, and molecular aspects. *International Journal of Environmental Research and Public Health* 15, no. 1 (2018): 59.

Abdel-Razek, M. A., Abozeid, A. M., Eltholth, M. M., Abouelenien, F. A., El-Midany, S. A., Moustafa, N. Y., and Mohamed, R. A. Bioremediation of a pesticide and selected heavy metals in wastewater from various sources using a consortium of microalgae and cyanobacteria. *Slovakian Veterinary Journal* 56, no. Suppl 22 (2019): 61–73.

Abou Seeda, M., Abou El-Nour, E. Z., Abdallah, M., and El-Bassiouny, H. Impacts of metal, metalloid and their effects in plant physiology: a review. *Middle East Journal of Agriculture Research* 11, no. 03 (2022): 838–931.

Adams, G. O., Fufeyin, P. T., Okoro, S. E., and Ehinomen, I. Bioremediation, biostimulation and bioaugmentation: a review. *International Journal of Environmental Bioremediation & Biodegradation* 3, no. 1 (2015): 28–39.

Agnihotri, A., Gupta, P., Dwivedi, A., and Seth, C. S. Counteractive mechanism (s) of salicylic acid in response to lead toxicity in Brassica juncea (L.) Czern. cv. Varuna. *Planta* 248 (2018): 49–68.

Ali, S., Abbas, Z., Rizwan, M., Zaheer, I. E., Yavaş, İ., Ünay, A., Abdel-Daim, M. M., Bin-Jumah, M., Hasanuzzaman, M., and Kalderis, D. Application of floating aquatic plants in phytoremediation of heavy metals polluted water: a review. *Sustainability* 12, no. 5 (2020): 1927.

Alvarenga, P., Gonçalves, A. P., Fernandes, R. M., de Varennes, A., Vallini, G., Duarte, E., and Cunha-Queda, A. C. Evaluation of composts and liming materials in the phytostabilization of a mine soil using perennial ryegrass. *Science of the Total Environment* 406 (2008): 43–56.

Amin, M., and Chetpattananondh, P. Biochar from extracted marine Chlorella sp. residue for high efficiency adsorption with ultrasonication to remove Cr (VI), Zn (II) and Ni (II). *Bioresource Technology* 289 (2019): 121578.

Anandaraj, B., Eswaramoorthi, S., Rajesh, T. P., Aravind, J., and Suresh Babu, P. Chromium(VI) adsorption by Codium tomentosum: evidence for adsorption by porous media from sigmoidal dose–response curve. *International Journal of Environmental Science and Technology* 15 (2018): 2595–2606.

Angulo -Bejarano, P. I., Puente-Rivera, J., and Cruz-Ortega, R. Metal and metalloid toxicity in plants: an overview on molecular aspects. *Plants* 10, no. 4 (2021): 635.

Arıca, M. Y., Tüzün, İ., Yalçın, E., İnce, Ö., and Bayramoğlu, G. Utilisation of native, heat and acid-treated microalgae Chlamydomonas reinhardtii preparations for biosorption of Cr(VI) ions. *Process Biochemistry* 40, no. 7 (2005): 2351–2358.

Arora, N., Gulati, K., Tripathi, S., and Pruthi, V. Algae as a budding tool for mitigation of arsenic from aquatic systems. In: Hasanuzzaman, M., Nahar, K., and Fujita, M. (Eds.), *Mechanisms of Arsenic Toxicity and Tolerance in Plants*, 269–297. Springer, 2018.

Arora, N., Tripathi, S., Pruthi, V., and Poluri, K. M. An integrated approach of wastewater mitigation and biomass production for biodiesel using Scenedesmus sp. In: Gupta, S. K., and Bux, F. (Eds.), *Application of Microalgae in Wastewater Treatment: Volume 2: Biorefinery Approaches of Wastewater Treatment*, 467–494. Springer Nature, 2019.

Asiandu, A. P., and Wahyudi, A. Phycoremediation: heavy metals green-removal by microalgae and its application in biofuel production. *Journal of Environmental Treatment Techniques* 9, no. 3 (2021): 647–656.

Aydın-Önen, S., and Öztürk, M. Investigation of heavy metal pollution in Eastern Aegean Sea coastal waters by using Cystoseira barbata, Patella caerulea, and Liza aurata as biological indicators. *Environmental Science and Pollution Research* 24 (2017): 7310–7334.

Babula, P., Adam, V., Opatrilova, R., Zehnalek, J., Havel, L., and Kizek, R. Uncommon heavy metals, metalloids and their plant toxicity: a review. In: Lichtfouse, E. (Ed.), *Organic Farming, Pest Control and Remediation of Soil Pollutants: Organic Farming, Pest Control and Remediation of Soil Pollutants*, 275–317. Springer, 2010.

Baghour, M., Moreno, D. A., Hernández, J., Castilla, N., and Romero, L. Influence of root temperature on phytoaccumulation of As, Ag, Cr, and Sb in potato plants (Solanum tuberosum L. var. Spunta). *Journal of Environmental Science and Health, Part A* 36 (2001): 1389–1401.

Bhalla, G., Bhalla, B., Kumar, V., and Sharma, A. Bioremediation and phytoremediation of pesticides residues from contaminated water: a novel approach. In: Dehghani, M. H., Karri, R. R., and Anastopoulos, I. (Eds.), *Pesticides Remediation Technologies from Water and Wastewater*, 339–363. Elsevier, 2022.

Bhat, R. A., Mir, M. Y., Dar, G. H., and Dervash, M. A. *Aquatic Environmental Bioengineering: Monitoring and Remediation of Contamination*. John Wiley & Sons, 2022.

Bienert, G. P., and Tamás, M. J. Molecular mechanisms of metalloid transport, toxicity and tolerance. *Frontiers in Cell and Developmental Biology* 6 (2018): 99.

Bilal, M., Rasheed, T., Sosa-Hernández, J. E., Raza, A., Nabeel, F., and Iqbal, H. M. Biosorption: an interplay between marine algae and potentially toxic elements—a review. *Marine Drugs* 16, no. 2 (2018): 65.

Bogstie, C. *The Relationship between Cellular Protein Content and Selenium Accumulation in Freshwater Micro-Algae*. Simon Fraser University, 2022.

Bolan, N. S., Park, J. H., Robinson, B., Naidu, R., and Huh, K. Y. Phytostabilization: a green approach to contaminant containment. In: Sparks, D. L. (Ed.), *Advances in Agronomy*. Academic Press, 145–204.

Bortoloti, G. A., and D. Baron. Phytoremediation of toxic heavy metals by Brassica plants: a biochemical and physiological approach. *Environmental Advances* 8 (2022): 100204.

Chabukdhara, M., Gupta, S. K., and Gogoi, M. (2017). Phycoremediation of heavy metals coupled with generation of bioenergy. In: Gupta, S. K., Malik, A., and Bux, F. (Eds.), *Algal Biofuels: Recent Advances and Future Prospects*, 163–188. Springer, 2017.

Chandra Shekharaiah, P. S., Sanyal, D., Dasgupta, S., Sapre, A., and Banik, A. Heavy metal mitigation with special reference to bioremediation by mixotrophic algae-bacterial protocooperation. In: Faisal, M., Saquib, Q., Alatar, A. A., and Al-Khedhairy, A. A. (Eds.), *Cellular and Molecular Phytotoxicity of Heavy Metals*, 305–334. Springer, 2020.

Chen, B., Zhao, H., Chen, S., Long, F., Huang, B., Yang, B., and Pan, X. A magnetically recyclable chitosan composite adsorbent functionalized with EDTA for simultaneous capture of anionic dye and heavy metals in complex wastewater. *Chemical Engineering Journal* 356 (2019): 69–80.

Chen, J. P., Hong, L., Wu, S., and Wang, L. Elucidation of interactions between metal ions and ca alginate-based ion-exchange resin by spectroscopic analysis and modeling simulation. *Langmuir* 18, no. 24 (2002): 9413–9421.

Chen, Y., Li, X., and Shen, Z. Leaching and uptake of heavy metals by ten different species of plants during an EDTA-assisted phytoextraction process. *Chemosphere* 57 (2004): 187–196.

Chen, Y., Yang, W., Chao, Y., Wang, S., Tang, Y. T., and Qiu, R. L. Metal-tolerant Enterobacter sp. strain EG16 enhanced phytoremediation using hibiscus cannabinus via siderophore-mediated plant growth promotion under metal contamination. *Plant and Soil* 413 (2017): 203–216.

Chinmayee, M. D., Mahesh, B., Pradesh, S., Mini, I., and Swapna, T. S. The assessment of phytoremediation potential of invasive weed Amaranthus spinosus L. *Applied Biochemistry and Biotechnology* 167 (2012): 1550–1559.

Choudhary, R., Rajput, V. D., Ghodake, G., and Seth, C. S. Comprehensive journey from past to present to future about seed priming with hydrogen peroxide and hydrogen sulfide concerning drought, temperature, UV and ozone stresses-a review. *Plant and Soil* 500 (2024): 351–373.

Cipolloni, O. A., Baudrimont, M., Simon-Bouhet, B., Dassié, É. P., Gigault, J., Connan, S., and Pascal, P. Y. Kinetics of metal and metalloid concentrations in holopelagic Sargassum reaching coastal environments. *Environmental Science and Pollution Research* 30, no. 47 (2023): 104779–104790.

Clauss, M.J., Cobban, H. and Mitchell-Olds, T. Cross-species microsatellite markers for elucidating population genetic structure in Arabidopsis and Arabis (Brassicaeae). *Molecular Ecology* 11, no. 3 (2002): 591–601.

Conesa, H. M., Faz, Á., and Arnaldos, R. Initial studies for the phytostabilization of a mine tailing from the Cartagena-La Union Mining District (SE Spain). *Chemosphere* 66 (2007): 38–44.

Danh, L. T., Truong, P., Mammucari, R., and Foster, N. A critical review of the arsenic uptake mechanisms and phytoremediation potential of Pteris vittata. *International Journal of Phytoremediation* 16 (2014): 429–453.

Deb, V. K., Rabbani, A., Upadhyay, S., Bharti, P., Sharma, H., Rawat, D. S., and Saxena, G. Microbe-assisted phytoremediation in reinstating heavy metal-contaminated sites: concepts, mechanisms, challenges, and future perspectives. In: Arora, P. K. (Eds), *Microbial Technology for Health and Environment*, 161–189. Springer Nature, 2020.

Deng, F., Zeng, F., Chen, G., Feng, X., Riaz, A., Wu, X., Gao, W., Wu, F., Holford, P., Chen, Z. H. Metalloid hazards: from plant molecular evolution to mitigation strategies. *Journal of Hazardous Materials* 409 (2021): 124495.

Di Martino, C., Palumbo, G., Di Iorio, E., Di Santo, P., and Fuggi, A. Nitrogen assimilation in mycorrhized durum wheat plants. In *Abstracts of Plant Biology Europe EPSO/FESPB 2016 Congress* 2016 (No. ID 895). CS.

Diepens, N. J., Dimitrov, M. R., Koelmans, A. A., and Smidt, H. Molecular assessment of bacterial community dynamics and functional end points during sediment bioaccumulation tests. *Environmental Science & Technology* 49, no. 22 (2015): 13586–13595.

Durand, A., Piutti, S., Rue, M., Morel, J. L., Echevarria, G., and Benizri, E. Improving nickel phytoextraction by co-cropping hyperaccumulator plants inoculated by plant growth promoting rhizobacteria. *Plant and Soil* 399 (2016): 179–192.

Dwivedi, S. Bioremediation of heavy metal by algae: current and future perspective. *Journal of Advanced Laboratory Research in Biology* 3 (2012): 195–199.

Epstein, A. L., Gussman, C. D., Blaylock, M. J., Yermiyahu, U., Huang, J. W., Kapulnik, Y., Orser, C. S. EDTA and Pb-EDTA accumulation in Brassica juncea grown in Pb-amended soil. *Plant and Soil* 208 (1999), 87–94.

Favas, P. J., Pratas, J., and Prasad, M. N. V. Accumulation of arsenic by aquatic plants in large-scale field conditions: opportunities for phytoremediation and bioindication. *Science of the Total Environment* 433 (2012): 390–397.

Feng, Y., Wang, Q., Meng, Q., Liu, Y., Pan, F., Luo, S., Wu, Y., and Ma, L., Yang, X. Chromosome doubling of Sedum alfredii hance: a novel approach for improving phytoremediation efficiency. *Journal of Environmental Sciences* 86 (2019): 87–96.

Flathman, P. E., and Lanza, G. R. Phytoremediation: current views on an emerging green technology. *Journal of Soil Contamination* 7, no. 4 (1998): 415–432.

García-Seoane, R., Fernández, J. A., Villares, R., and Aboal, J. R. Use of macroalgae to biomonitor pollutants in coastal waters: optimization of the methodology. *Ecological Indicators* 84 (2018): 710–726.

Garg, N., Bharti, A., Sharma, A., Saroy, K., Cheema, A., and Bisht, A. Prokaryotic and eukaryotic microbes: potential tools for detoxification and bioavailability of metalloids. In: Deshmukh, R., Guerriero, G., and Tripathi, D. K. (Eds.), *Metalloids in Plants: Advances and Future Prospects*, 149–183. John Wiley & Sons, Incorporated, Newark, 2020.

Ghosh, S., Nayak, J., Islam, M. A., and Thongmee, S. Arsenic bioremoval using algae: a sustainable process. In: Adetunji, C. O., Oloke, J. K., Dwivedi, N., Ummalyma, S. B., Dwivedi, S., Hefft, D. I., and Adetunji, J. B. (Eds.), *Next-Generation Algae: Volume I: Applications in Agriculture, Food and Environment*, 91–108. 2023. John Wiley & Sons.

Ghuge, S. A., Nikalje, G. C., Kadam, U. S., Suprasanna, P., and Hong, J. C. Comprehensive mechanisms of heavy metal toxicity in plants, detoxification, and remediation. *Journal of Hazardous Materials* 450 (2023): 131039.

Gong, N., Shao, K., Feng, W., Lin, Z., Liang, C., and Sun, Y. Biotoxicity of nickel oxide nanoparticles and bio-remediation by microalgae Chlorella vulgaris. *Chemosphere* 83 (2011): 510–516.

Gonzaga, M. I. S., Santos, J. A. G., and Ma, L. Q. Arsenic phytoextraction and hyperaccumulation by Fern species. *Scientia Agricola* 63 (2006): 90–101.

Goswami, R. K., Agrawal, K., and Verma, P. Phycoremediation of nitrogen and phosphate from wastewater using Picochlorum sp.: a tenable approach. *Journal of Basic Microbiology* 62, no. 3–4 (2022): 279–295.

Greeshma, K., Kim, H. S., and Ramanan, R. The emerging potential of natural and synthetic algae-based microbiomes for heavy metal removal and recovery from wastewaters. *Environmental Research* 215 (2022): 114238.

Gul, I., Manzoor, M., Ahmad, I., Kallerhoff, J., and Arshad, M. Phytoaccumulation of cadmium by Pelargonium×hortorum—tolerance and metal recovery. *Environmental Science and Pollution Research* 30, no. 12 (2023): 32673–32682.

Gupta, A. *Heavy Metal and Metalloid Contamination of Surface and Underground Water: Environmental, Policy and Ethical Issues*. CRC Press, 2020.

Gupta, P., and Seth, C. S. Nitrate supplementation attenuates As (V) toxicity in Solanum lycopersicum L. cv Pusa Rohini: insights into As (V) sub-cellular distribution, photosynthesis, nitrogen assimilation, and DNA damage. *Plant Physiology and Biochemistry* 139 (2019): 44–55.

Hanikenne, M., Krämer, U., Demoulin, V., and Baurain, D. A comparative inventory of metal transporters in the green alga Chlamydomonas reinhardtii and the red alga Cyanidioschizon merolae. *Plant Physiology* 137, no. 2 (2005): 428–446.

Hao, X., Taghavi, S., Xie, P., Orbach, M. J., Alwathnani, H. A., Rensing, C., and Wei, G. Phytoremediation of heavy and transition metals aided by legume-rhizobia symbiosis. *International Journal of Phytoremediation* 16, no. 2 (2014): 179–202.

Hartmann, A., Schmid, M., Tuinen, D. V., Berg, G. Plant-driven selection of microbes. Plant and Soil 321, (2009): 235-257.

Haruma, T., Yamaji, K., Ogawa, K., Masuya, H., Sekine, Y., and Kozai, N. Root-endophytic Chaetomium cupreum chemically enhances aluminium tolerance in Miscanthus sinensis via increasing the aluminium detoxicants, chlorogenic acid and oosporein. *PLoS One* 14, no. 2 (2019): e0212644.

Hasanuzzaman, M., Bhuyan, M. B., Raza, A., Hawrylak-Nowak, B., Matraszek-Gawron, R., Nahar, K., and Fujita, M. Selenium toxicity in plants and environment: biogeochemistry and remediation possibilities. *Plants* 9, no. 12 (2020): 1711.

Hassan, M. M., Haleem, N., Baig, M. A., and Jamal, Y. Phytoaccumulation of heavy metals from municipal solid waste leachate using different grasses under hydroponic condition. *Scientific Reports* 10, no. 1 (2020): 15802.

Hauser -Davis, R. A., and Wosnick, N. Climate change implications for metal and metalloid dynamics in aquatic ecosystems and its context within the decade of ocean sciences. *Water* 14, no. 15 (2022): 2415.

Havel, J. E., Kovalenko, K. E., Thomaz, S. M., Amalfitano, S., and Kats, L. B. Aquatic invasive species: challenges for the future. *Hydrobiologia* 750 (2015): 147–170.

Hawrylak-Nowak, B. Biological activity of selenium in plants: physiological and biochemical mechanisms of phytotoxicity and tolerance. In: Hossain, M. A. (Ed.), *Selenium and Nano-Selenium in Environmental Stress Management and Crop Quality Improvement*, 341–363. Cham: Springer International Publishing, 2022.

Hsu, Y.T. and Kao, C.H. Role of abscisic acid in cadmium tolerance of rice (Oryza sativa L.) seedlings. *Plant, Cell & Environment*, 26, no. 6 (2003): 867–874.

Huang, R., Dong, M., Mao, P., Zhuang, P., Paz-Ferreiro, J., Li, Y., Li, Y., Hu, X., Netherway, P., Li, Z. Evaluation of phytoremediation potential of five Cd (hyper)accumulators in two Cd contaminated soils. *Science of the Total Environment* 721 (2020): 137581.

Huang, Y., Xi, Y., Gan, L., Johnson, D., Wu, Y., Ren, D., and Liu, H. Effects of lead and cadmium on photosynthesis in Amaranthus spinosus and assessment of phytoremediation potential. *International Journal of Phytoremediation* 21 (2019): 1041–1049.

Hussain, M. M., Bibi, I., Shahid, M., Shaheen, S. M., Shakoor, M. B., Bashir, S., Younas, F., Rinklebe, J., and Niazi, N. K. Biogeochemical cycling, speciation and transformation pathways of arsenic in aquatic environments with the emphasis on algae. In: Wilson, C. L., Wilson, D. W., and Svehla G. (Eds.), *Comprehensive Analytical Chemistry*, 85, 15–51. Elsevier, 2019.

Iqbal, J., Javed, A., and Baig, M. A. Heavy metals removal from dumpsite leachate by algae and cyanobacteria. *Bioremediation Journal* 26 (2022): 31–40.

Jaiswar, S., Kazi, M. A., and Mehta, S. Bioaccumulation of heavy metals by freshwater algal species of Bhavnagar, Gujarat, India. *Journal of Environmental Biology* 36, no. 6 (2015): 1361.

Javanbakht, V., Alavi, S. A., and Zilouei, H. Mechanisms of heavy metal removal using microorganisms as biosorbent. *Water Science and Technology* 69, no. 9 (2014): 1775–1787.

Jebara, S. H., Saadani, O., Fatnassi, I. C., Chiboub, M., Abdelkrim, S., and Jebara, M. Inoculation of Lens culinaris with Pb-resistant bacteria shows potential for phytostabilization. *Environmental Science and Pollution Research* 22 (2015): 2537–2545.

Jin, Z., Deng, S., Wen, Y., Jin, Y., Pan, L., Zhang, Y., Black, T., Jones, K. C., Zhang, H., and Zhang, D. Application of Simplicillium chinense for Cd and Pb biosorption and enhancing heavy metal phytoremediation of soils. *Science of the Total Environment* 697 (2019): 134148.

Kalve, S., Sarangi, B. K., Pandey, R. A., and Chakrabarti, T. Arsenic and chromium hyperaccumulation by an ecotype of Pteris vittata – prospective for phytoextraction from contaminated water and soil. *Current Science* 100 (2011): 888–894.

Kamal, M., Ghaly, A. E., Mahmoud, N., and Côté, R. Phytoaccumulation of heavy metals by aquatic plants. *Environment International* 29 (2004): 1029–1039.

Karle, S. B., Kumar, K., and Dhankher, O. P. The versatile role of plant aquaglyceroporins in metalloid transport. In: Kumar, K., and Srivastava, S. (Eds.), *Plant Metal and Metalloid Transporters*, 133–150. Springer International Publishing, 2022.

Kaur, R., Das, S., Bansal, S., Singh, G., Sardar, S., Dhar, H., and Ram, H. Heavy metal stress in rice: uptake, transport, signaling, and tolerance mechanisms. *Physiologia Plantarum* 173, no. 1 (2021): 430–448.

Kensa, V. M. Bioremediation-an overview. *Journal of Industrial Pollution Control* 27, no. 2 (2011): 161–168.

Khan, A. R., Ullah, I., Waqas, M., Park, G. S., Khan, A. L., Hong, S. J., Ullah, R., Jung, B. K., Park, C. E., Ur-Rehman, S., Lee, I.-J., and Shin, J. H. Host plant growth promotion and cadmium detoxification in Solanum nigrum, mediated by endophytic fungi. *Ecotoxicology and Environmental Safety* 136 (2017): 180–188.

Kodre, A., Arčon, I., Debeljak, M., Potisek, M., Likar, M., and Vogel-Mikuš, K. Arbuscular mycorrhizal fungi alter Hg root uptake and ligand environment as studied by X-ray absorption fine structure. *Environmental and Experimental Botany* 133 (2017): 12–23.

Koul, B., Sharma, K., and Shah, M. P. Phycoremediation: a sustainable alternative in wastewater treatment (WWT) regime. *Environmental Technology & Innovation* 25 (2022): 102040.

Kumar, A., Tripti, M. N., Maiti, S. K., and Favas, P. J. Mycoremediation site rehabilitation for mine. In: Prasad, M. N. V., de Campos Favas, P. J., Maiti, S. K., and de Campos Favas, P. J. (Eds.), *Bio-Geotechnologies for Mine Site Rehabilitation*, 233. Springer, 2018.

Kumar, D., Dhankher, O. P., Tripathi, R. D., and Seth, C. S. Titanium dioxide nanoparticles potentially regulate the mechanism(s) for photosynthetic attributes, genotoxicity, antioxidants defense machinery, and phytochelatins synthesis in relation to hexavalent chromium toxicity in Helianthus annuus L. *Journal of Hazardous Materials* 454 (2023): 131418.

Kumar, D., and Seth, C. S. Photosynthesis, lipid peroxidation, and antioxidative responses of Helianthus annuus L. against chromium (VI) accumulation. *International Journal of Phytoremediation* 24, no. 6 (2022): 590–599.

Kumar, S. S., Kadier, A., Malyan, S. K., Ahmad, A., and Bishnoi, N. R. Phytoremediation and rhizoremediation: uptake, mobilization and sequestration of heavy metals by plants. In: Singh, D. P., Singh, H. B., and Prabha, R. (Eds.), *Plant-Microbe Interactions in Agro-Ecological Perspectives: Volume 2: Microbial Interactions and Agro-Ecological Impacts*, 367–394. Springer, Singapore, 2017.

Kumar, V., Katyal, D., and Nayak, S. Removal of heavy metals and radionuclides from water using nanomaterials: current scenario and future prospects. *Environmental Science and Pollution Research* 27 (2020): 41199–41224.

Kumari, P. Metals homeostasis and chelators. In: Kumar, P., and Dey, S. R. (Eds.), *Metal Toxicity in Agriculture Crops: Emerging Trends*, 1–20. Scientific Publishers, 2021.

Labudda, M., Dziurka, K., Fidler, J., Gietler, M., Rybarczyk-Płońska, A., Nykiel, M., Prabucka, B., Morkunas, I., and Muszyńska, E. The alleviation of metal stress nuisance for plants—a review of promising solutions in the face of environmental challenges. *Plants* 11, no. 19 (2022): 2544.

Lacalle, R. G., Bernal, M. P., Álvarez-Robles, M. J., and Clemente, R. Phytostabilization of soils contaminated with As, Cd, Cu, Pb and Zn: physicochemical, toxicological and biological evaluations. *Soil & Environmental Health* 1, no. 2 (2023): 100014.

Leguizamo, M. A. O., Gómez, W. D. F., and Sarmiento, M. C. G. Native herbaceous plant species with potential use in phytoremediation of heavy metals, spotlight on wetlands—a review. *Chemosphere* 168 (2017): 1230–1247.

Leong, Y. K., and Chang, J. S. Bioremediation of heavy metals using microalgae: recent advances and mechanisms. *Bioresource Technology* 303 (2020): 122886.

Limmer, M., and Burken, J. Phytovolatilization of organic contaminants. *Environmental Science & Technology* 50 (2016): 6632–6643.

Lin, Y., Jiao, Y., Zhao, M., Wang, G., Wang, D., Xiao, W., Li, H., Xu, Z., and Jiang, Y. Ecological restoration of wetland polluted by heavy metals in Xiangtan manganese mine area. *Processes* 9, no. 10 (2021): 1702.

Lin, Z., Li, J., Luan, Y., and Dai, W. Application of algae for heavy metal adsorption: a 20-year meta-analysis. *Ecotoxicology and Environmental Safety* 190 (2020): 110089.

Lu, L.L., Tian, S.K., Yang, X.E., Wang, X.C., Brown, P., Li, T.Q. and He, Z.L.. Enhanced root-to-shoot translocation of cadmium in the hyperaccumulating ecotype of Sedum alfredii. *Journal of Experimental Botany* 59, no. 11 (2008): 3203–3213.

Ma, Y., Rajkumar, M., Oliveira, R. S., Zhang, C., and Freitas, H. Potential of plant beneficial bacteria and arbuscular mycorrhizal fungi in phytoremediation of metal-contaminated saline soils. *Journal of Hazardous Materials* 379 (2019): 120813.

Manahan, S. E. *Environmental Chemistry*. CRC Press, 2022.

Mandal, R. P., Sekh, S., Sen Sarkar, N., Chattopadhyay, D., and De, S. Algae mediated synthesis of cadmium sulphide nanoparticles and their application in bioremediation. *Materials Research Express* 3 (2016): 055007.

Mariyam, S., Upadhyay, S. K., Chakraborty, K., Verma, K. K., Duhan, J. S., Muneer, S., and Seth, C. S. Nanotechnology, a frontier in agricultural science, a novel approach in abiotic stress management and convergence with new age medicine-a review. *Science of the Total Environment*, 912 (2023): 169097.

Martínez -Alcalá, I., Clemente, R., and Bernal, M. P. Interactions between the hyperaccumulator Noccaea caerulescens and Brassica juncea or Lupinus albus for phytoextraction. *Agronomy* 10 (2020): 1367.

Masarovičová, E., and Kráľová, K. Plant-heavy metal interaction: phytoremediation, biofortification and nanoparticles. In: Montanaro, G., and Dichio, B. (Eds.), *Advances in Selected Plant Physiology Aspects*. InTech, Rijeka, 2012.

Mesjasz -Przybyłowicz, J., Barnabas, A. and Przybyłowicz, W. Comparison of cytology and distribution of nickel in roots of Ni-hyperaccumulating and non-hyperaccumulating genotypes of Senecio coronatus. *Plant and Soil* 293 (2007): 61–78.

Mirza, N., Pervez, A., Mahmood, Q., Shah, M. M., and Shafqat, M. N. Ecological restoration of arsenic contaminated soil by Arundo donax L. *Ecological Engineering* 37 (2011): 1949–1956.

Mishra, S., Singh, G., Gupta, A., and Tiwari, R. K. Heavy metal/metalloid contamination: their sources in environment and accumulation in food chain. In: Singh, R. P., Singh, P., and Srivastava, A. (Eds.), *Heavy Metal Toxicity: Environmental Concerns, Remediation and Opportunities* 19–47. Springer Nature Singapore, 2023.

Misra, V., Tiwari, A., Shukla, B., and Seth, C. S. Effects of soil amendments on the bioavailability of heavy metals from zinc mine tailings. *Environmental Monitoring and Assessment* 155 (2009): 467–475.

Mnasri, M., Janoušková, M., Rydlová, J., Abdelly, C., and Ghnaya, T. Comparison of arbuscular mycorrhizal fungal effects on the heavy metal uptake of a host and a non-host plant species in contact with extraradical mycelial network. *Chemosphere* 171 (2017): 476–484.

Mohammadi, A., and Mahmoudnia, F. Biological treatment of heavy metals with algae. In: Almayyahi, B. (Eds.), *Heavy Metals - Recent Advances*. IntechOpen, 2023.

Mok, J. S., Yoo, H. D., Kim, P. H., Yoon, H. D., Park, Y. C., Lee, T. S., Kwon, J. Y., Son, K. T., Lee, H. J., Ha, K. S., Shim, K. B., and Kim, J. H. Bioaccumulation of heavy metals in oysters from the southern coast of Korea: assessment of potential risk to human health. *Bulletin of Environmental Contamination and Toxicology* 94 (2015): 749–755.

Mokhtar, N., Aziz, E. A., Aris, A., Ishak, W. F. W., and Ali, N. S. M. Biosorption of azo-dye using marine macro-alga of Euchema spinosum. *Journal of Environmental Chemical Engineering* 5, no. 6 (2017): 5721–5731.

Monteiro, C. M., Castro, P. M., and Malcata, F. X. Metal uptake by microalgae: underlying mechanisms and practical applications. *Biotechnology Progress* 28, no. 2 (2012): 299–311.

Mustafa, S., Bhatti, H. N., Maqbool, M., and Iqbal, M. Microalgae biosorption, bioaccumulation and biodegradation efficiency for the remediation of wastewater and carbon dioxide mitigation: prospects, challenges and opportunities. *Journal of Water Process Engineering* 41 (2021): 102009.

Muthusaravanan, S., Sivarajasekar, N., Vivek, J. S., Paramasivan, T., Naushad, M., Prakashmaran, J., Gayathri, V., and Al-Duaij, O. K. Phytoremediation of heavy metals: mechanisms, methods and enhancements. *Environmental Chemistry Letters* 16 (2018): 1339–1359.

Nayak, A. K., Panda, S. S., Basu, A., and Dhal, N. K. Enhancement of toxic Cr(VI), Fe, and other heavy metals phytoremediation by the synergistic combination of native Bacillus cereus strain and Vetiveria zizanioides L. *International Journal of Phytoremediation* 20, no. 7 (2018): 682–691.

Nazir, M. M., Ulhassan, Z., Zeeshan, M., Ali, S., and Gill, M. B. *Toxic metals/metalloids accumulation, tolerance, and homeostasis in Brassica oilseed species*. In: Hasanuzzaman, M. (Ed.), *The Plant Family Brassicaceae: Biology and Physiological Responses to Environmental Stresses*, 379–408. Springer, 2020.

Newman, L. A., and Reynolds, C. M. Phytodegradation of organic compounds. *Current Opinion in Biotechnology* 15 (2004): 225–230.

Nilanjana, D., Vimla, R. and Karthika, K. Biosorption of heavy metal - An overview. *Indian Journal of Biotechnology* 7 (2008): 159–169.

Nishikawa, E., da Silva, M. G. C., and Vieira, M. G. A. Cadmium biosorption by alginate extraction waste and process overview in life cycle assessment context. *Journal of Cleaner Production* 178 (2018): 166–175.

Novikau, R., and Lujaniene, G. Adsorption behaviour of pollutants: heavy metals, radionuclides, organic pollutants, on clays and their minerals (raw, modified and treated): a review. *Journal of Environmental Management* 309 (2022): 114685.

Ojha, A., Jaiswal, S., Thakur, P., and Mishra, S. K. Bioremediation techniques for heavy metal and metalloid removal from polluted lands: a review. *International Journal of Environmental Science and Technology* 20, no. 9 (2023): 10591–10612.

Papry, R. I., Miah, S., and Hasegawa, H. Integrated environmental factor-dependent growth and arsenic biotransformation by aquatic microalgae: a review. *Chemosphere* 303 (2022): 135164.

Parveen, N., Berni, R., Sudhakaran, S., Bhat, J. A., Shinde, S., Ramawat, N., Singh, V. P., Sahi, S., Deshmukh, R., Chauhan, D. K., and Tripathi, D. K. Metalloids in plants: a systematic discussion beyond description. *Annals of Applied Biology* 180, no. 1 (2022): 7–25.

Patel, A., Tiwari, S., and Prasad, S. M. Toxicity assessment of arsenate and arsenite on growth, chlorophyll a fluorescence and antioxidant machinery in Nostoc muscorum. *Ecotoxicology and Environmental Safety* 157 (2018): 369–379.

Pena -Castro, J. M., Martınez-Jerónimo, F., Esparza-Garcıa, F., and Canizares-Villanueva, R. O. Heavy metals removal by the microalga Scenedesmus incrassatulus in continuous cultures. *Bioresource Technology* 94, no. 2 (2004): 219–222.

Perales -Vela, H. V., Peña-Castro, J. M., and Cañizares-Villanueva, R. O. Heavy metal detoxification in eukaryotic microalgae. *Chemosphere* 64 (2006): 1–10.

Phang, S. M., Chu, W. L., and Rabiei, R. P. The algae world. In *Cellular Origin, Life in Extreme Habitats and Astrobiology,* 10, 978–94. Springer, 2015a.

Phang, S.-M., Chu, W.-L., and Rabiei, R. P. Phycoremediation. In: Sahoo, D., and Seckbach, J. (Eds.), *The Algae World, Cellular Origin, Life in Extreme Habitats and Astrobiology*, 357–389. Springer Netherlands, Dordrecht, 2015b.

Pilon -Smits, E. Phytoremediation. *Annual Review of Plant Biology* 56 (2005): 15–39.

Poschenrieder, C., Bech, J., Llugany, M., Pace, A., Fenés, E., and Barceló, J. Copper in plant species in a copper gradient in Catalonia (North East Spain) and their potential for phytoremediation. *Plant and Soil* 230 (2001): 247–256.

Pradhan, B., Bhuyan, P. P., Nayak, R., Patra, S., Behera, C., Ki, J.-S., Ragusa, A., Lukatkin, A. S., and Jena, M. Microalgal phycoremediation: a glimpse into a sustainable environment. *Toxics* 10 (2022): 525.

Pradhan, D., Sukla, L. B., Mishra, B. B., and Devi, N. Biosorption for removal of hexavalent chromium using microalgae Scenedesmus sp. *Journal of Cleaner Production* 209 (2019): 617–629.

Rai, P. K., Lee, S. S., Zhang, M., Tsang, Y. F., and Kim, K. H. Heavy metals in food crops: health risks, fate, mechanisms, and management. *Environment International* 125 (2019): 365–385.

Rajkumar, M., Ae, N., Prasad, M. N. V., and Freitas, H. Potential of siderophore-producing bacteria for improving heavy metal phytoextraction. *Trends in Biotechnology* 28, no. 3 (2010): 142–149.

Rajkumar, M., Sandhya, S., Prasad, M. N. V., and Freitas, H. Perspectives of plant-associated microbes in heavy metal phytoremediation. *Biotechnology Advances* 30, no. 6 (2012): 1562–1574.

Ramachandra, T. V., Sudarshan, P. B., Mahesh, M. K., and Vinay, S. Spatial patterns of heavy metal accumulation in sediments and macrophytes of Bellandur wetland, Bangalore. *Journal of Environmental Management* 206 (2018): 1204–1210.

Ramakrishnan, B., Megharaj, M., Venkateswarlu, K., Naidu, R., and Sethunathan, N. The impacts of environmental pollutants on microalgae and cyanobacteria. *Critical Reviews in Environmental Science and Technology* 40, no. 8 (2010): 699–821.

Ramezani, M., Enayati, M., Ramezani, M., and Ghorbani, A. A study of different strategical views into heavy metal(oid) removal in the environment. *Arabian Journal of Geosciences* 14 (2021): 1–6.

Rani, S., Gunjyal, N., Ojha, C. S. P., and Singh, R. P. Review of challenges for algae-based wastewater treatment: strain selection, wastewater characteristics, abiotic, and biotic factors. *Journal of Hazardous, Toxic, and Radioactive Waste* 25 (2021): 03120004.

Ranjan, R., Kumar, N., Gautam, A., Dubey, A. K., Pandey, S. N., and Mallick, S. Chlorella sp. modulates the glutathione mediated detoxification and S-adenosylmethionine dependent methyltransferase to counter arsenic toxicity in Oryza sativa L. *Ecotoxicology and Environmental Safety* 208 (2021): 111418.

Reeves, R.D., Baker, A.J.M., Jaffré, T., Erskine, P.D., Echevarria, G., van der Ent, A. A global database for hyperaccumulator plants of metal and metalloid trace elements. *New Phytologist* 218 (2017):407–411.

Rezania, S., Taib, S. M., Din, M. F. M., Dahalan, F. A., and Kamyab, H. Comprehensive review on phytotechnology: heavy metals removal by diverse aquatic plants species from wastewater. *Journal of Hazardous Materials* 318 (2016): 587–599.

Rizvi, A., Zaidi, A., Ameen, F., Ahmed, B., AlKahtani, M. D., and Khan, M. S. Heavy metal induced stress on wheat: phytotoxicity and microbiological management. *RSC Advances* 10, no. 63 (2020): 38379–38403.

Ryan, P. R., Delhaize, E., and Jones, D. L. Function and mechanism of organic anion exudation from plant roots. *Annual Review of Plant Biology* 52, no. 1 (2001): 527–560.

Sahi, S.V., Bryant, N.L., Sharma, N.C. and Singh, S.R. Characterization of a lead hyperaccumulator shrub, Sesbania drummondii. *Environmental Science & Technology*, 36, no. 21 (2002), 4676–4680.

Sahoo, D., and Seckbach, J. (Eds.). *The Algae World, Cellular Origin, Life in Extreme Habitats and Astrobiology*. Springer Netherlands, Dordrecht, 2015.

Saranya, D., and Shanthakumar, S. Green microalgae for combined sewage and tannery effluent treatment: performance and lipid accumulation potential. *Journal of Environmental Management* 241 (2019): 167–178.

Saunders, R. J., Paul, N. A., Hu, Y., and Nys, R. de. Sustainable sources of biomass for bioremediation of heavy metals in waste water derived from coal-fired power generation. *PLoS One* 7 (2012): e36470.

Schiavon, M., Ertani, A., Parrasia, S., and Dalla Vecchia, F. Selenium accumulation and metabolism in algae. *Aquatic Toxicology* 189 (2017): 1–8.

Schreck, E., Foucault, Y., Geret, F., Pradère, P., and Dumat, C. Influence of soil ageing on bioavailability and ecotoxicity of lead carried by process waste metallic ultrafine particles. *Chemosphere*, 85, no. 10 (2011): 1555–1562.

Seth, C. S., Misra, V., and Chauhan, L. K. S. Accumulation, detoxification, and genotoxicity of heavy metals in Indian mustard (Brassica juncea L.). *International Journal of Phytoremediation* 14, no. 1 (2012): 1–13.

Shackira, A. M., Jazeel, K., and Puthur, J. T. Phycoremediation and phytoremediation: promising tools of green remediation. In: Kumar Mishra, V., and Kumar, A. (Eds.), *Sustainable Environmental Clean-Up,* 273–293. Elsevier, 2021.

Shahid, A., Malik, S., Zhu, H., Xu, J., Nawaz, M. Z., Nawaz, S., Alam, M. A., and Mehmood, M. A. Cultivating microalgae in wastewater for biomass production, pollutant removal, and atmospheric carbon mitigation: a review. *Science of the Total Environment* 704 (2020): 135303.

Shahid, M., Dumat, C., Khalid, S., Schreck, E., Xiong, T., and Niazi, N. K. Foliar heavy metal uptake, toxicity and detoxification in plants: a comparison of foliar and root metal uptake. *Journal of Hazardous Materials* 325 (2017): 36–58.

Shahid, M., Shamshad, S., Rafiq, M., Khalid, S., Bibi, I., Niazi, N. K., Dumat, C., and Rashid, M. I. Chromium speciation, bioavailability, uptake, toxicity and detoxification in soil-plant system: a review. *Chemosphere* 178 (2017): 513–533.

Sharma, R., Bhardwaj, R., Handa, N., Gautam, V., Kohli, S. K., Bali, S., Kaur, P., Thukral, A. K., Arora, S., Ohri, P., and Vig, A. P. Responses of phytochelatins and metallothioneins in alleviation of heavy metal stress in plants: an overview. In: Ahmad, P. (Ed.), *Plant Metal Interaction*, 263–283. Elsevier, 2016.

Sharma, S. Bioremediation: features, strategies and applications. *Asian Journal of Pharmacy and Life Science* 2231 (2012): 4423.

Shen, L., Wang, J., Li, Z., Fan, L., Chen, R., Wu, X., Li, J., and Zeng, W. A high-efficiency Fe_2O_3@ microalgae composite for heavy metal removal from aqueous solution. *Journal of Water Process Engineering* 33 (2020): 101026.

Shi, B., Huang, Z., Xiang, X., Huang, M., Wang, W. X., and Ke, C. Transcriptome analysis of the key role of GAT2 gene in the hyper-accumulation of copper in the oyster Crassostrea angulata. *Scientific Reports* 5, no. 1 (2015): 17751.

Shrivastava, M., and Srivastava, S. Application and research progress of Hydrilla verticillata in ecological restoration of water contaminated with metals and metalloids. *Environmental Challenges* 4 (2021): 100177.

Singh, D., Agnihotri, A., and Seth, C. S. Interactive effects of EDTA and oxalic acid on chromium uptake, translocation and photosynthetic attributes in Indian mustard (*Brassica juncea* L. var. Varuna). *Current Science* 112 (2017): 2034–2042.

Singh, N., Afzal, S., Singh, N. K., Sohrab, S., Mishra, S. K., and Agrawal, S. C. Physiological, morphological, and biochemical responses of metals and metalloids on algae. In: Aftab, T., and Hakeem, K. (Eds.), *Metals Metalloids Soil Plant Water Systems*, 271–286. Academic Press, 2022.

Siraj, N. K., Ali, K., Khan, M. E. H., and Jones, D. A. Phytoaccumulation of heavy metals by sodom apple (Calotropis procera (Aiton) WT Aiton) along an urban–rural gradient. *Applied Sciences* 12, no. 3 (2022): 1003.

Somdee, T., Thathong, B., and Somdee, A. The removal of cyanobacterial hepatotoxin [Dha 7] microcystin-LR via bioaccumulation in water lettuce (*Pistia stratiotes* L.). *Bulletin of Environmental Contamination and Toxicology* 96 (2016): 388–394.

Suresh, B., and Ravishankar, G. A. Phytoremediation—a novel and promising approach for environmental clean-up. *Critical Reviews in Biotechnology* 24 (2004): 97–124.

Tan, L. C., Nancharaiah, Y. V., van Hullebusch, E. D., and Lens, P. N. Selenium: environmental significance, pollution, and biological treatment technologies. In Lea Tan (Ed.), *Anaerobic Treatment of Mine Wastewater for the Removal of Selenate and Its Co-Contaminants*, 9–71, CRC Press, 2018.

Tang, Z., Wang, H. Q., Chen, J., Chang, J. D., and Zhao, F. J. Molecular mechanisms underlying the toxicity and detoxification of trace metals and metalloids in plants. *Journal of Integrative Plant Biology* 65, no. 2 (2023): 570–593.

Tavanappanavar, A. N., Mulla, S. I., Seth, C. S., Bagewadi, Z. K., Rahamathulla, M., Ahmed, M. M., and Farhana, S. A. Phytochemical analysis, GC–MS profile and determination of antibacterial, antifungal, anti-inflammatory, antioxidant activities of peel and seeds extracts (chloroform and ethyl acetate) of *Tamarindus indica* L. *Saudi Journal of Biological Sciences* 31, no. 1 (2024): 103878.

Terry, N., Zayed, A. M., De Souza, M. P., and Tarun, A. S. Selenium in higher plants. *Annual Review of Plant Biology* 51, no. 1 (2000): 401–432.

Thangaraj, B., Jolley, C. C., Sarrou, I., Bultema, J. B., Greyslak, J., Whitelegge, J. P., Lin, S., Kouřil, R., Subramanyam, R., Boekema, E. J., and Fromme, P. Efficient light harvesting in a dark, hot, acidic environment: the structure and function of PSI-LHCI from *Galdieria sulphuraria. Biophysical Journal* 100 (2011): 135–143.

Tripathi, S., and Poluri, K. M. Metallothionein-and phytochelatin-assisted mechanism of heavy metal detoxification in microalgae. In: Hasanuzzaman, M. (Ed.), *Approaches to the Remediation of Inorganic Pollutants*, 323–344. Springer, Singapore, 2021.

Uchimiya, M., Bannon, D., Nakanishi, H., McBride, M. B., Williams, M. A., and Yoshihara, T. Chemical speciation, plant uptake, and toxicity of heavy metals in agricultural soils. *Journal of Agricultural and Food Chemistry* 68, no. 46 (2020): 12856–12869.

Ulhassan, Z., Farooq, M. U., Basit, F., Nazir, M. M., Zhu, J., Ishaaq, I., Maqbool, R., Hakeem, K. R., and Zhou, W. Uptake and translocation mechanisms of metals/metalloids in plants through soil and water. In: Aftab, T., and Hakeem, K. (Eds.), *Metals Metalloids Soil Plant Water Systems*, 1–28. Academic Press, 2022.

Vidya, C. S., Shetty, R., Vaculíková, M., and Vaculik, M. Antimony toxicity in soils and plants, and mechanisms of its alleviation. *Environmental and Experimental Botany* 202 (2022): 104996.

Wang, Y., Wang, Z. J., Huang, J. C., Zhou, C., Zou, H., He, S., and Chen, V. Y. Feasibility of using *Chlorella vulgaris* for the removal of selenium and chromium in water: competitive interactions with sulfur, physiological effects on algal cells and its resilience after treatment. *Journal of Cleaner Production* 313 (2021): 127939.

Weyens, N., van der Lelie, D., Taghavi, S., and Vangronsveld, J. Phytoremediation: plant–endophyte partnerships take the challenge. *Current Opinion in Biotechnology* 20, no. 2 (2009): 248–254.

Xu, C., Chen, X., Duan, D., Peng, C., Le, T., and Shi, J. Effect of heavy-metal-resistant bacteria on enhanced metal uptake and translocation of the Cu-tolerant plant, *Elsholtzia splendens. Environmental Science and Pollution Research* 22 (2015): 5070–5081.

Yadav, B., Dubey, R., Gnanasekaran, P. and Narayan, O.P. OMICS approaches towards understanding plant's responses to counterattack heavy metal stress: an insight into molecular mechanisms of plant defense. *Plant Gene* 28 (2021): 100333.

Yadav, K. K., Gupta, N., Kumar, A., Reece, L. M., Singh, N., Rezania, S., and Khan, S. A. Mechanistic understanding and holistic approach of phytoremediation: a review on application and future prospects. *Ecological Engineering* 120 (2018): 274–298.

Yadav, M., Gupta, P., and Seth, C. S. Foliar application of α-lipoic acid attenuates cadmium toxicity on photosynthetic pigments and nitrogen metabolism in *Solanum lycopersicum* L. *Acta Physiologiae Plantarum* 44, no. 11 (2022): 112.

Yan, A., Wang, Y., Tan, S. N., Mohd Yusof, M. L., Ghosh, S., and Chen, Z. Phytoremediation: a promising approach for revegetation of heavy metal-polluted land. *Frontiers in Plant Science* 11 (2020): 359.

Yang, J., Zheng, G., Yang, J., Wan, X., Song, B., Cai, W., and Guo, J. Phytoaccumulation of heavy metals (Pb, Zn, and Cd) by 10 wetland plant species under different hydrological regimes. *Ecological Engineering* 107 (2017): 56–64.

Yang, Y., Guo, Y., O'Brien, A. M., Lins, T. F., Rochman, C. M., and Sinton, D. Biological responses to climate change and nanoplastics are altered in concert: full-factor screening reveals effects of multiple stressors on primary producers. *Environmental Science & Technology* 54, no. 4 (2020): 2401–2410.

Yoon, J., Cao, X., Zhou, Q., and Ma, L. Q. Accumulation of Pb, Cu, and Zn in native plants growing on a contaminated Florida site. *Science of the Total Environment* 368, no. 2–3 (2006): 456–464.

Zarei, M., Hempel, S., Wubet, T., Schäfer, T., Savaghebi, G., Jouzani, G. S., ... and Buscot, F. Molecular diversity of arbuscular mycorrhizal fungi in relation to soil chemical properties and heavy metal contamination. *Environmental Pollution* 158, no. 8 (2010): 2757–2765.

Zhang, J., Liu, J., Zheng, F., Yu, M., Shabala, S., and Song, W. Y. Comparative analysis of arsenic transport and tolerance mechanisms: evolution from prokaryote to higher plants. *Cells* 11, no. 17 (2022): 2741.

Zhang, W. H., He, L. Y., Wang, Q., and Sheng, X. F. Inoculation with endophytic *Bacillus megaterium* 1Y31 increases Mn accumulation and induces the growth and energy metabolism-related differentially-expressed proteome in Mn hyperaccumulator hybrid *pennisetum. Journal of Hazardous Materials* 300 (2015): 513–521.

Zhen, Z., Yan, C., and Zhao, Y. Influence of epiphytic bacteria on arsenic metabolism in *Hydrilla verticillata. Environmental Pollution* 261 (2020): 114232.

Złoch, M., Kowalkowski, T., Tyburski, J., and Hrynkiewicz, K. Modeling of phytoextraction efficiency of microbially stimulated *Salix dasyclados* L. in the soils with different speciation of heavy metals. *International Journal of Phytoremediation* 19, no. 12 (2017): 1150–1164.

17 Siderophores and their Action against Metalloid Toxicity

Roohi Sharma, Deepak Kumar, Parkirti, Nahida Rehman Mir, Manu Sharma, Amandeep Singh, Bilal Ahmad Mir, Puja Ohri, Renu Bhardwaj, and Nitika Kapoor

17.1 INTRODUCTION

The last century has witnessed a significant surge in industrialization, resulting in substantial environmental contamination. Various pollutants, ranging from heavy metals and metalloids such as cadmium (Cd), vanadium (V), copper (Cu), nickel (Ni), cobalt (Co), zinc (Zn), lead (Pb), selenium (Se), antimony (Sb), and arsenic (As) to petroleum hydrocarbons, halogenated hydrocarbons, pesticides, and solvents, have become pervasive (Arihilam and Arihilam, 2019). The persistent nature of metals and metalloids poses a considerable challenge for remediation efforts (Ali et al., 2019). Traditional *in situ* physico-chemical remediation methods are often prohibitively expensive and can have adverse effects on the subsurface environment (Zhang et al., 2019). In recent times, there has been a shift toward developing biological approaches to efficiently restore polluted environments. Understanding this interplay is crucial for assessing metalloid contamination. Strategies manipulating siderophore production or designing novel chelators hold promise for environmental and therapeutic applications for metal homeostasis, which offers an eco-friendly alternative that is also cost-effective (Deb et al., 2020).

Siderophores, which mean "iron carriers," are composites of the ancient Greek terms *sidero* (meaning "iron") and *phore* (meaning "carriers"); thus, these are compounds produced by different organisms to survive under iron-depleted conditions by capturing the insoluble ferric ion (Fe^{3+}), which is generally not available (Nagoba and Vedpathak, 2011; Ahmed and Holmström, 2014). These are "rhizospheric bacteria" in iron-limited conditions that create low-molecular-weight (<10 kDa) iron-chelating compounds to boost plant development by scavenging iron from the environment and making the mineral available to the cell near the root (Albelda et al., 2019). For a number of essential plant processes, including respiration, nucleoside and amino acid (AA) synthesis, enzymatic cofactors, the photosynthesis procedure, and redox reagents, iron (Fe) is a key component (Sah and Singh, 2015; Ansari et al., 2017; Ashraf et al., 2021). It has been demonstrated that siderophores are organic substances generated by plants and microorganisms that are crucial to the mechanisms involved in metal detoxification. Siderophores primarily improve the bioavailability of Fe(III), but usually their structure and source vary, which indicates that they are not able to selectively bind to a single metal ion (Hofmann et al., 2020). Because of Fe deficiency, many plants appear to rely on siderophore secreted by a group of microbes and phytosiderophore excreted by the roots (chelate compounds, common in grasses that sequester Fe) to aid in the uptake of the Fe complex, which binds to iron with a high affinity (Kumar et al., 2016; Molnár et al., 2023).

Siderophores dissolve metals and metalloids and complicate soil bioremediation by forming stable complexes. The key elements that affect the binding affinity of siderophores toward various metals are the characteristics and the coordinative environment of the attached ligands, as well as the kind of central metal ion (e.g., ligand configuration or size and charge of the central metal ion) (Crumbliss et al., 2009; Bhardwaj et al., 2021). Siderophores are known to provide microbial resistance in plants by selectively binding with metals and aiding their separation from certain

DOI: 10.1201/9781032673592-17

nutrient solutions (Saha et al., 2016). In this process, diffusion is the primary method by which metals and metalloids enter bacterial cells (Sreedevi et al., 2022). By binding to siderophores, there is a reduction in the environmental concentration of free metals and metalloids and stops them from diffusing across the membranes of bacteria, which further lowers the toxicity of metals and metalloids (Saha et al., 2016). Metals and metalloids can also be sequestered via a variety of extracellular processes, including biosorption and bioaccumulation after being chelated by siderophores (Bramhachari and Nagaraju, 2017). Siderophores also function in stimulation of plant development (Pahari et al., 2017), act as drivers for bioremediation (de Carvalho et al., 2010; Rawat et al., 2022), and aid in biocontrol purposes (Ahmed and Holmström, 2014). Siderophores are also proposed to play an important role in bioremediation of organic pollutants as they facilitate the generation of reactive oxygen species (ROS) by controlling redox cycling that furthers oxide organic compounds (Gu et al., 2018; Dickson et al., 2020). Different research has been performed in recent decades to look into the function of siderophores in environmental cleanup of heavy metals and metalloids. Therefore, this chapter's objective is to provide an overview of siderophore structure and identification along with the processes behind siderophore-assisted bioremediation.

17.2 CHEMICAL COMPOSITION AND CLASSIFICATION OF SIDEROPHORES

A wide range of differences is seen in siderophore structures that are produced by numerous microbes (bacteria and fungi) and plants. Generally, siderophores are categorized according to the synchorinizing groups that chelate the Fe(III) ion. On the basis of their chemical composition, siderophores can be categorized as catecholates, hydroxamates, carboxylates, "mixed type," or "mixed ligands," or those with two or more classes that function (Holden and Bachman, 2015; Ustiatik et al., 2021). Pyoverdines are categorized as mixed ligands since they have coordinating groups that belong to distinct chemical classifications. Furthermore, all siderophores have a higher attraction for Fe(III) as compared to Fe(II), besides any exception. They have the potential to participate in the absorption of several heavy metals from polluted soils, perhaps having applications in bioremediation. Fe(III) can be connected with hydroxamates, hydroxycarboxylates, and catecholates (bidentate functional groups) and make a joint structure of three chemical compounds, including polydentate phenolates, nitrogen heterocycles, and carboxylates in the siderophore structure. Salicylic acid and oxazoline are among the handful of siderophores with Fe(III) ion groups that differ from other groups or nitrogen from thizoline. Electrophoretic mobility, spectrophotometric titration, mass spectrometry, proton nuclear magnetic resonance spectroscopy, acid hydrolysis, and biological processes were used to identify different forms of siderophores. Figure 17.1 illustrates the general classification and types of siderophores.

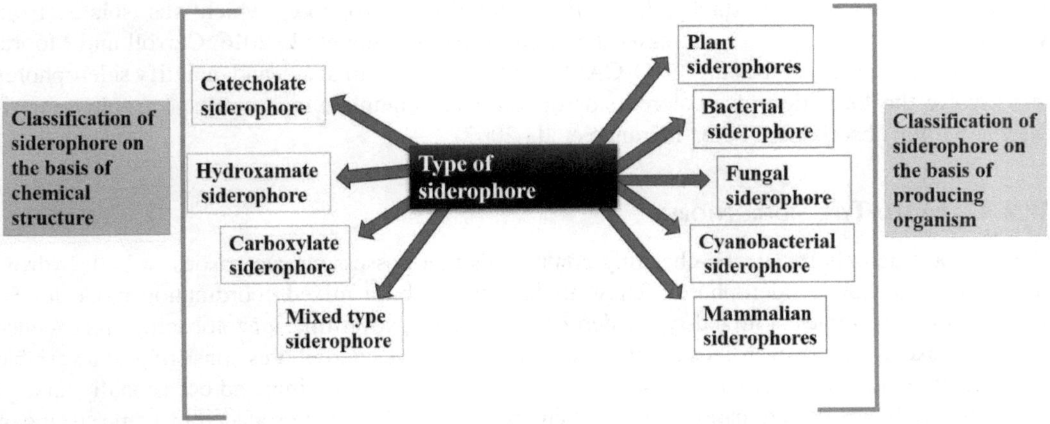

FIGURE 17.1 General classification and types of siderophores.

17.2.1 Catecholate (Phenolate) Siderophores

Chemical interaction occurs between Fe(III) ions and hydroxyl or catecholate groups in catecholate-type siderophores derived from salicylic/2,3-dihydroxybenzoic acid derivatives (Mishra et al., 2021). Chelation with Fe(III) results in the synthesis of a complex compound (hexadentate and octahedral complex), with the involvement two oxygen molecules from each catecholate group (Ghosh et al., 2020). All catecholate siderophores are salicylic or 2,3-dihydroxybenzoic acid derivatives (2,3-DHBA). *Azospirillum lipoferum* synthesizes 2,3-DHBA and 3,5-DHBA, covalently linked to threonine and lysine; both of them have exhibited siderophore activity. Similarly, *Azotobacter vinelandii*, a bacterial strain, produced major types of four catecholate siderophores, including 2,3-dihydroxybenzoic acid, protochelin, nitrocholine, and aminocholine (Ustiatik et al., 2021; Kraepiel et al., 2009). Other catecholate siderophores include agrobactin, aminochelin, bacillibactin, and protochilin (Timofeeva et al., 2022).

17.2.2 Hydroxamate Siderophores

Hydroxamate siderophores have the structure C(=O) N–(OH)R, where R represents that AA or an oxygen-modified derivative that combine to form a bidentate ligand that interacts with Fe ions. With Fe(III) ions, each siderophore can form hexadentate ligands and octahedral complex compounds. When these types of siderophores make complex with the Fe(III) ion and further loses the proton ion from the hydroxylamine group (–NH_2OH) by its functional group, resulting in the formation of a bidentate ligand. The hydroxamate siderophores are the most commonly found type of siderophores secreted by both bacteria and fungi (Saha et al., 2016).

According to another study, under Fe-limited parameters, the bacterial strain (*Bacillus megaterium)* is processed to generate hydroxamate siderophores, including shizokinen, rhizobactin, N-deoxyshizokinen, and vicibactin (Timofeeva et al., 2022). A cyclic trihydroxamate siderophore known as vibicin has been found in *Rhizobium phaseoli* and *Rhizobium leguminosarum* (Aguirre-Noyola et al., 2021).

17.2.3 Carboxylate Siderophores

In carboxylate-type siderophores, Fe binds to the carboxyl/hydroxyl group, which is present in them (Ghosh et al., 2020). These are primarily synthesized by microbes such as *Staphylococcus* spp. and *Rhizobium* spp. and fungi with the order Mucorales. Carboxylate siderophores include staphyloferrin A, staphyloferrin B, and rhizobactin. Furthermore, *Rhizobium meliloti* strain rhizobactin is a carboxylate siderophores with a composition of amino poly carboxylic acid with ethylene diamine (dicarboxyl, hydroxyl, and carboxyl moieties), which act as Fe-chelating groups. Similarly, according to another study, staphyloferrin B (hydrophilic siderophore), which was isolated from *Staphylococcus aureus*, under decreased Fe circumstances (Saha et al., 2016; Carroll and Moore, 2018). The overlaid-chrome azurol S (O-CAS) assay can be used to detect and quantify siderophores that involve the formation of a colored siderophore-CAS complex, measured spectrophotometrically at 630nm absorbance (Pérez-Miranda et al., 2007).

17.2.4 Mixed-Type Siderophores

Mixed-type siderophores are Fe-chelating compounds that possess characteristics of both hydroxamate and catecholate siderophores. These molecules exhibit a mixed coordination mode for Fe binding. Some examples of mixed-type siderophores include pyoverdine, y-azotobactin, and pseudobactin. As exemplary, pyoverdines are the bacterial fluorescent siderophores consisting of a variable fluorophore, a peptide chain with AAs, a chelating moiety for Fe binding, and occasionally an acyl chain that facilitates Fe acquisition in diverse environments. These are produced by some strains of bacteria, especially *Pseudomonas* species. The composition of pyoverdines can vary among different bacterial species and even among strains. Nonetheless, in alkaline environments, *Pantoea eucalypti* M91 can create siderophores that resemble pyoverdine and pyochelin (Campestre et al., 2016).

17.3 SOURCES OF SIDEROPHORES

Siderophores are molecules unique to Fe that are secreted when there is a low concentration of Fe and that absorb Fe from the surrounding environment. During the course of evolution, the increasing oxygen concentration caused the parallel oxidation of the soluble ferrous ions to the insoluble ferric ions, which in turn forced the microorganisms to synthesize particular Fe-binding compounds known as siderophores (Cain et al., 2021).

As already mentioned above, various organisms, including fungi, bacteria, plants, and mammals that are living in low Fe environments, are capable of producing siderophores. These compounds are able to remove Fe from a variety of organic substrates (Correnti and Strong, 2012). The synthesis of siderophores takes place inside the cells by the nonribosomal peptide synthetase (NRPS) pathway or by the NRPS-independent pathway, and then they are released into the environment (Gulick, 2017; Carroll and Moore, 2018) (Table 17.1).

TABLE 17.1
Sources of Siderophores and Their Biological Function

Siderophore	Source	Function	References
FsC, Triacetylfusarinine C, Ferricrocin, Hydroxyferricrocin	*Aspergillus fumigatus*	Iron acquisition, intracellular iron storage and handling	Misslinger et al. (2021)
Ferrioxamines	*Streptomyces ciscaucasicus* strain GS2	Reduced apple replant disease, promoted plant growth and production	Armin et al. (2021)
Bacillibactin	*Bacillus subtilis* (LSBS2)	Boosted peanut plant immunity	Lalitha and Nithyapriya (2021)
Pyoverdine	*Pseudomonads*	Lengthened the roots and shoots of pea plants	Lurthy et al. (2020)
Hydroxamate	*Pseudomonas citronellolis* strain SLP6	Improved chlorophyll content, synthesis of antioxidant enzyme, and growth in sunflower	Silambarasan et al. (2020)
Hydroxymate and catecholate	*Pseudomonas* sp.	Increased plant growth and provided defense against plant pathogens	Parveen and Latha (2019)
Catecholate	*Rhizobium* sp. Strain R1	Improved the germination of soyabean seeds	Igiehon et al. (2019)
Fimsbactin	*Acinetobacter*	Hindered the growth of phytopathogen *Pseudomonas syringae* pv. tomato DC3000 in *Arabidopsis thaliana*	Bohac et al. (2019)
Rhodopetrobactin A and B	*Rhodopseudomonas palustris* strain CGA009	Supported iron acquisition under variable oxygen and light conditions	Baars et al. (2018)
Hydroxymate	*Pseudomonas* sp.	Served as an organic fertilizer	Joshi et al. (2018)
Catecholate and salicylate	*Bacillus* spp.	Stimulated plant growth in potatoes and suppressed wilting disease infection in bananas	Kesaulya et al. (2018)

(Continued)

TABLE 17.1 (*Continued*)
Sources of Siderophores and Their Biological Function

Siderophore	Source	Function	References
Catecholate	*Pantoeadispersa* MPJ9 *and Pseudomonas putida* MPJ2	MPJ9 and MPJ2 both boosted siderophore synthesis as well as iron, protein, and carbohydrate content in the *Vigna radiata* plant	Patel et al. (2018)
Hydroxymate and Catecholate	*Enterobacter species, Azotobacter species, and Pseudomonas species*	Potentially function as biocontrol agents against detrimental plant pathogens in *Gossypium hirsutum*	Patel and Minocheherhomji (2018)
Yersinibactin	*Yersinia pestis* *E. coli*	Acted as zincophore	Bobrov et al. (2014)
Dimerum acid, Fusigen, CPG, and Ferricrocin	*Trichoderma* spp.	Increased iron availability to plants	Lehner et al. (2013)
Fusigen	*Laccaria laccata* *Laccaria bicolor*	Increased mycorrhizal potential of the producing fungi	Haselwandter et al. (2013)
Pyoverdine	*Pseudomonas fluorescens*	Improved plant growth	Gamalero and Glick (2011)
Enterobactin and Salmochelin	*E. coli*	Bind with silicon and formed silicon complexes	Schmiederer et al. (2011)
Frrioxamines	*Azadirachta indica*	Encouraged root and shoot development and assisted in the nourishment of plants with iron	Verma et al. (2011)
FsC	*Fusarium roseum*	Improved fungi uptake of iron	Saharan and Nehra (2011)
FRC	*Ustilago sphaerogena*	Helped fungi to grow in Fe-deficient conditions	Saharan and Nehra (2011)
Staphyloferrin A, Staphyloferrin B	*Staphylococcus aureus*	Removed Fe from human holotransferrin to promote the development of *S. aureus*	Beasley et al. (2011)
Azotochelin Azotobactin	*Azotobacter vinelandii*	Helped in the uptake of molybednum and V	Wichard et al. (2009)
Pyochelin	*Pseudomonas aeruginosa*	Can chelates a wide range of metals	Braud et al. (2009a)

17.3.1 Microbial Siderophores

Microorganisms like bacteria and fungi live in low Fe conditions that produce siderophores. These microbes enhance the availability of Fe by producing and excreting siderophores, which are Fe transporters and chelators that promote plant growth (Pattnaik et al., 2021). The growth of plants may be stimulated either directly or indirectly by microbial siderophores by either boosting Fe accessibility in the soil around the roots or by actively preventing the growth of plant pathogens that have less effective Fe absorption mechanisms (Ghosh et al., 2021). In a study, siderophore produced by plant growth-promoting rhizobacteria *Rhizobiaceae* within the root area promoted functioning and overall productivity of the plant *Agrostis capillaris* with the increase in Fe uptake (Grobelak et al., 2015). Similarly, siderophore of plant growth-promoting fungi *Aspergillus tubingensis* strain HZ123 also promoted the growth of *Solanum melongena* (Li et al., 2021).

17.3.1.1 Bacterial Siderophore

Bacteria mostly produce extracellular (ecto-rhizospheric) siderophores, which are primarily synthesized by both distinct types of bacteria (gram-negative bacteria and gram-positive bacteria) under an Fe-scarce scenario (Saharan and Nehra, 2011). Facultative, gram-negative anaerobe *Escherichia coli*, which is often found in the gut, is among the most extensively researched bacteria for siderophore synthesis in addition to other rhizoplane bacteria in the soil (Grobelak and Hiller, 2017). *E. coli* is the primary producer of enterobactin, a catechol siderophore that has the greatest affinity for the Fe(III) ion of any known siderophore. Additionally, *Pestis Yersinia*, which causes lethal plague distinguished by a highly pathogenic island, also encodes for a siderophore named yersiniabactin (Ybt) siderophore (Paauw et al., 2009). In another study, gram-negative, nosocomial, multidrug resistant aerobic bacterial strain 6A *Pseudomonas aeruginosa* synthesizes two siderophores, mainly pyoverdine and pyochelin (Marathe et al., 2015). While chelating Fe^{3+} is siderophores' principal activity, they may also form complex molecules with heavy metals such as Cd^{2+}, Cu^{2+}, and Zn^{2+} (Johnstone and Nolan, 2015). Siderophores are important players in the biogeochemical cycling of Fe in the ocean, where most of such organic Fe chelators in seawater are produced by marine bacteria (Sinha et al., 2019). Bacteria and phytoplanktons compete for Fe via siderophores, influencing the substance's solubility in the ocean (Sanchez et al., 2018). Mawji et al. (2008) reported that marine siderophores enhance Fe availability in the Atlantic Ocean that played a crucial role in the Fe biogeochemical cycle. Siderophores enhanced the plant growth by inhibiting the pathogen growth. For instance, siderophore synthesized by a bacterium named *Bacillus subtilis* CAS15 controlled the wilt of pepper induced by *Fusarium oxysporum* (Yu et al., 2011). Similarly, siderophores synthesized by *Bacillus siamensis* Gxun-6 inhibited the growth of *F. oxysporum* in *Musa paradisiaca* (Shen et al., 2022).

17.3.1.2 Fungal Siderophores

Fungi colonize a variety of habitats and this colonization requires the regular uptake of nutrients, one of which is Fe. To maintain Fe uptake, fungi have evolved different strategies, like the production of siderophores. Fungi usually generate hydroxamate and carboxylate siderophores. The majority of fungi have hydroxamate siderophores, which belong to the ferrichrome (FRC) family and have three hydroxamate units connected by ester or peptide linkages, forming a hexadentate configuration. This structure greatly improves the affinity for ferric ions (Weissman, 2015). *Aspergillus fumigatus* and *Aspergillus nidulans* are two intensively researched fungi for siderophore production with 55 related types of siderophores. *A. fumigatus* and *A. nidulans* are saprophytic fungal species that play a vital part in the carbon and nitrogen recycling in the environment (Figure 17.2). *Aspergillus fumigatus* produces extracellular as well as intracellular siderophores for Fe acquisition and Fe storage, respectively (Misslinger et al., 2021). Other fungi, such as *U. maydis, Aspergillus niger,* etc., produce fusaririnine, coprogen (CPG)-type, and FRC-type siderophores (Franken et al., 2014; Gressler et al., 2015). Yeast taxa like *Schizosaccharomyces pombe*, *Rhodotorula* spp., and *Aureobasidium melanogenum* also produce siderophores (Lu et al., 2019; Misslinger et al., 2021). Most of the fungi species from Ascomycota and Basidiomycota produce hydroxamate-type siderophores except Mucoromycota, which produces a carboxylate-type siderophore called rhizoferrin, which was initially obtained from *Rhizopus* microspores (Brandenburger et al., 2017; Haselwandter et al., 2020). Ester containing siderophore has been reported in two mycorrhiza fungi, basidiomycetes, *Laccaria laccata* (a typical woodland fungus) and *L. bicolor* (a model microorganism used for research purposes). Siderophore linear fusigen occurs in small amounts in addition to CPG, ferricrocin, and triacetylfusarinine C (Haselwandter et al., 2013).

17.3.1.3 Siderophore Produced by Cyanobacteria

The majority of cyanobacterial species such as schizokinen and anachelin H are known to produce dihydroxymate-type siderophores. Furthermore, frequent observations have been made of the electronic structures of metals with siderophores including Cu, aluminum, Ni, Zn, Co, and gallium

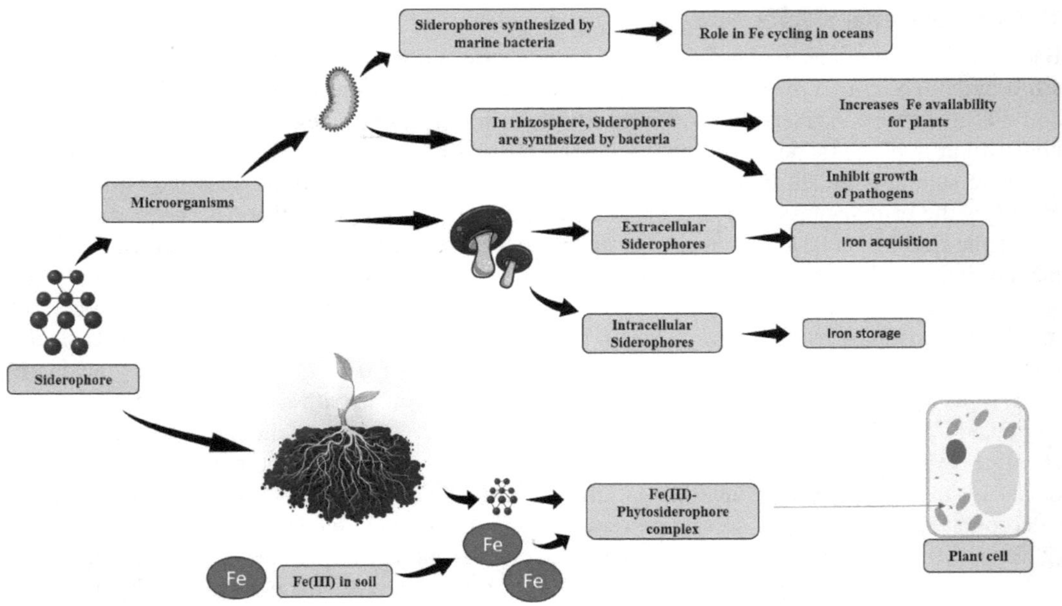

FIGURE 17.2 Production of siderophores in plants and microbes.

(Duckworth et al., 2009). The first cyanobacterial siderophore was produced from cyanobacteria *Bacillus megaterium* (Řezanka et al., 2018). In 2005, synechobactins (4, 5, and 6) were identified as siderophores after they were recovered from a strain of *Synechococcus* that lived in coastal saltwater (Ito and Butler, 2005). For instance, common paddy field cyanobacterium *Anabaena oryza* may produce siderophore, a biological solubilizing agent that reduces the amount of Cd metal ions in the environment and increases crop yield (Singh et al., 2016).

17.3.2 Plant Siderophores

Phytosiderophores are Fe(III)-chelating molecules synthesized by plants to form complexes with Fe(III) under Fe-deficient conditions (Molnár et al., 2023). Phytosiderophores are plant-secreted compounds that enhance Fe uptake from the soil, particularly in calcium-rich soil with limited availability of Fe (Kobayashi et al., 2019). These organic molecules form complexes with Fe, aiding its absorption and utilization by plants. When released into the rhizosphere, the phytosiderophore acts as hexadentate ligands by complexing Fe(III) with their amino and carboxyl groups, producing a complex named Fe(III)-phytosiderophore complexes that are capable of transversing the root plasma membrane (Ahmed and Holmström, 2014).

Phytosiderophores are mostly produced by the poaceae family that includes crops like wheat and barley. The first plant siderophore to be discovered was mugineic acid (MA), which is the most common phytosiderophore (Takemoto et al., 1978). There are other phytosiderophores as well, such as 2′-deoxymugineic acid (DMA), derivatives of MA (e.g., 3-epi-hydroxymugineic acid), Avenic acid from *Avenic sativa*, nicotinamine, and asteropeptin from legume *Desmodium uncinatum* (Davis and Stocksdale, 2013). MA is a precursor to DMA and is also produced by grass species. Each of these compounds plays a role in improving the capability of plants to acquire important minerals from the environment. The molecular mass of phytosiderophore also ranges between 500 and 1,000 Da as compared to microbial siderophores, which have molecular masses ranging from 200 to 2,000 Da (Ahmed et al., 2014). In plants, siderophores mostly act as a biological control against phytopathogens by decreasing the Fe content that is available to plant pathogens, aiding their elimination (Beneduzi et al., 2012; Ahmed and Holmström, 2014).

17.3.3 Siderophores Produced by Mammals

Mammalian siderophores are molecules facilitating Fe transport, crucial for various physiological functions in animals, particularly in Fe uptake and utilization (Pecoraro et al., 2021). Mammalian cells may include certain siderophores that are structurally and functionally comparable to those found in bacteria, as demonstrated in the *FL5.12/EC-24p3* murine interleukin-3 (*IL-3*) reliant pro-B lymphocytes. A knockdown cell known as hydroxybutyrate dehydrogenase type 2 (*BDH2*) results in enhanced intracellular iron concentration, iron regulatory protein (*IRP1*) breakdown, excessive level of aconitase, increased generation of ROS, and programmed cell death induction by oxidative stress, suggesting that *BDH2* plays an essential part in maintaining Fe homeostasis (Devireddy et al., 2010). According to Liu et al. (2014), enterobactin, which binds Fe in mammalian cells, is bound by lipocalin *24p3*. Mammals are thought to possess a siderophore known as 2,3-dihydroxybenzoic acid (2,5-DHBA), connected to the Fe-binding moiety of bacterial enterobactin (2,3-DHBA). 2,5-DHBA binds to Fe(III) and promotes mitochondrial Fe absorption. Plants under heavy metal stress may benefit from mammalian siderophores, which chelate metals to reduce toxicity. For the purpose of developing sustainable phytoremediation techniques, research on their function in plant-metal interactions is still underway.

17.4 BIOSYNTHESIS OF SIDEROPHORES

The siderophore biosynthesis in microorganisms is activated by intracellular Fe deficiency. These are secreted by microbes to sequester Fe from the environment. Its biosynthesis involves two pathways. The first one depends on NRPS, and the second is independent of NRPS, also called nonribosomal-peptide-synthetase-independent siderophore (NIS) (Soares, 2022; Saha et al., 2013). In the NIS mechanism, mostly carboxylate and hydroxamate siderophores are synthesized. The enzymes that synthesize siderophore in the NIS pathway are aldolase, AA ligases, amino and acetyltransferases, decarboxylases, and monooxygenases and in the final step, the enzymes *IucC*- and *IucA*-type (Fe uptake chelate) NIS synthetases are required that catalyze the condensation of citric acid with an alcohol or amine group (Puja et al., 2023). The NRPS pathway, on the other hand, involves enzyme NRPSs, which are multi-enzyme complexes capable of making peptide scaffolds without RNA, and each of its components participates in incorporating a single AA into a peptide chain already in place. Thus, several important peptide molecules are synthesized by NRPS. Generally, NRPS has three main domains, namely, the adenylation domain for recognition of prime AA, the peptidyl carrier protein domain (PCP) for binding AA and growing peptide chains covalently to NRPS, and the condensation domain to catalyze peptide bond formation. Other domains include all accessory domains and are involved in other processes like acylation, glycosylation, etc., while the thioesters domain is involved in the final step of hydrolysis of the peptide chain, eventually releasing it from NRPS (Swayambhu et al., 2021; Schalk et al., 2020). In NRPS, the adenylation domain activates and recognizes AA and binds it to a cofactor present in the thiolation domain. Then the condensation domain forms a peptide chain and incorporates this AA to the growing polypeptide chain. In the end, the C-terminal thioester domain catalyzes the cyclization process and ultimately releases the polypeptide chain from synthetase (Iacovelli et al., 2021).

17.4.1 Biosynthesis of Bacterial Siderophores

Siderophores are synthesized in many genera of bacteria, including *Serratia, Pseudomonas, Pantoea, Bascillus, Azotobacter, Enterobacter, Azospirillum, Streptomyces*, etc. There are three methods of biosynthesis of siderophore in bacteria, i.e., biosynthesis by NRPS, biosynthesis by polyketide synthases (PKS), and biosynthesis by NIS synthetase (Mishra and Baek, 2021). Biosynthesis by NRPS involves three stages, i.e., initiation, condensation, and termination, as illustrated in Figure 17.3. A typical NRPS structure comprises A (adenylation), C (condensation), P (PCP), and T (thioesterase) sequences. Beside these, other functional domains are also present, like cyclization (Cy), oxidation (Ox),

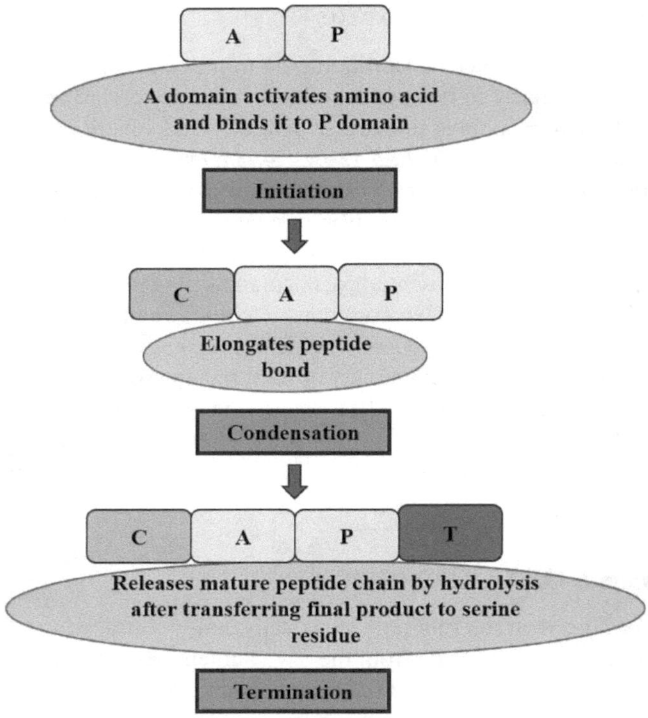

FIGURE 17.3 NPRS biosynthesis involves three stages: The figure is bound to the P domain by domain A after being selectively activated. The peptide bond is extended via the condensation process. The termination step releases the mature peptide chain by hydrolysis after transferring the final product to serine residue.

epimerization (E), and methylation (M) (Chen et al., 2022). The AA is covalently bound to the P domain by the A domain, which preferentially turns on it. AA and primed modules bonded to create bonds made of peptides during the condensation process. In the termination step, the mature peptide is released after hydrolysis in the thioester domain. In PKS biosynthesis of siderophore, the main domains are ketosynthase, an acyltransferase, and a carrier domain. The acyltransferase domain covalently binds the initiation module to the carrier domain, releasing CoA. The acyl chain is transferred to the cysteine in the ketosynthase domain via the loading module with the P domain. Growing chains in condensation reactions can latch to the first module with the carrier domain and then move to the subsequent module with the ketosynthase domain (Timofeeva et al., 2022; Pecoraro et al., 2021). In NIS synthetase siderophore biosynthesis, an acyladenylation domain of NIS synthetase forms citrate–AMP, which supplies an energetic bond for the condensation process with a polyamine or AA. Usually, the NIS pathway consists of three NIS synthetases as type A that uses citrate as a substrate, type B catalyze the C5 carboxyl group of α-ketoglutarate, and type C NIS synthetases citric and succinic acid derivatives. Thus, NIS synthetase forms siderophore consisting of α-ketoglutarate, citric acid, or succinic acid (Tang et al., 2020).

17.4.2 Biosynthesis of Fungal Siderophores

There are different types of siderophores as discussed above, of which fungi mainly synthesize hydroxymates, which are further of three types: fusarinines (FSR), CPGs, and FRC, with rhizoferrin as few exceptions. Rhizoferrin is produced by some zygomycetes as a carboxylate-type siderophore. In fungi, hydroxamate siderophores contain alkylated ornithine AA and hydroxylated, while in bacteria, they contain hydroxylated and acylated alkylamines (Happacher et al., 2023; Pecoraro et al., 2021). The general biosynthetic pathway of fungal siderophore initiates with

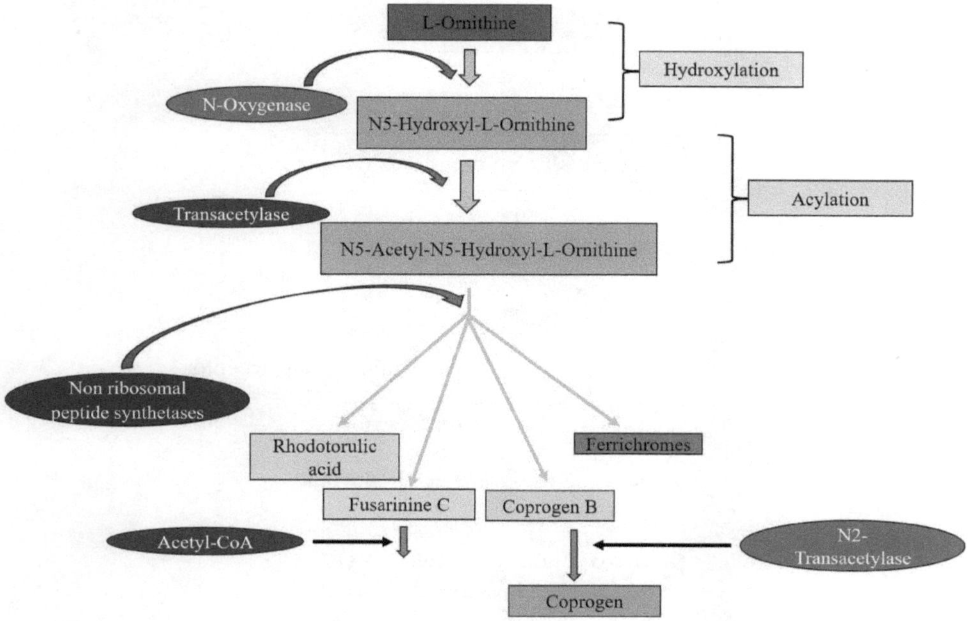

FIGURE 17.4 The precursor L-Ornithine initiates the biosynthetic pathway of fungal siderophore. The subsequent steps involve hydroxylation, acylation, and acetylation. The final product form is rhodotorulic acid, FsC, and CPG. (Modified and adapted from Pecoraro et al., 2021.)

hydroxylation of L-ornithine as first step. The enzymes L-ornithine N^5-oxygenase and L-ornithine catalyze the reaction of converting this precursor into *N*-hydroxy-L-ornithine. The second step involves acylation, catalyzed by the enzyme transacetylase, and converts *N*-hydroxy L-ornithine into *N*-acyl-*N*-hydroxy-l-ornithine, as shown in Figure 17.4. In this reaction, the acyl-CoA derivative acts as an acyl donor in the presence of the enzyme acyl-CoA: *N*-hydroxy-L-ornithine *N*-acyl transferase. The reaction is then followed by the condensation process. Many *N*-acyl-*N*-hydroxy-L-ornithine units combine and form FRC, FSR C (FsC), CPG B, and rhodotorulic acid-like dipeptides and triesters (Pecoraro et al., 2021; Tripathi et al., 2020; Perez-Cuesta et al., 2021).

17.4.3 Biosynthesis of Plant Siderophores

Graminaceous plants (like barley and rice), experiencing Fe deficiency, employ a strategy wherein they secrete phytosiderophores, such as the MA family, from their roots to chelate external insoluble Fe (ferric) and form soluble ferric-PS complexes, facilitating Fe uptake. In MA synthesis, methionine serves as the precursor and undergoes conversion to *S*-adenosyl-L-methionine (SAM) through the action of SAM synthetase; the synthesis of SAM involves ATP and occurs in both chloroplasts and the cytoplasm, where *S*-adenosylmethionine synthase (SAMS) is localized. This is followed by the conversion of SAM into nicotianamine (NA). NA plays the role of precursor of DMA and also acts as potent Fe chelator that helps in suppression of toxicity and the translocation of the Fe. NA biosynthesis is a single step reaction that involves condensation reactions among three SAM molecules in the presence of nicotianamine synthase (NAS). NA is converted into 3″-keto acid and DMA in the presence of NA aminotransferase (NAAT) and enzyme DMA synthase (DMAS) (Seregin and Kozhevnikova, 2023; Roy et al., 2022; Dey et al., 2020). Different enzymes utilized are encoded by different genes (Suzuki et al., 2016). In all plant species, the synthesis of NA is conserved. Also, in plants like rice, DMA and NA are found in xylem and phloem exudates, suggesting their role as chelators in xylem and phloem iron transport (Kobayashi et al., 2019). Plants undergo two stages

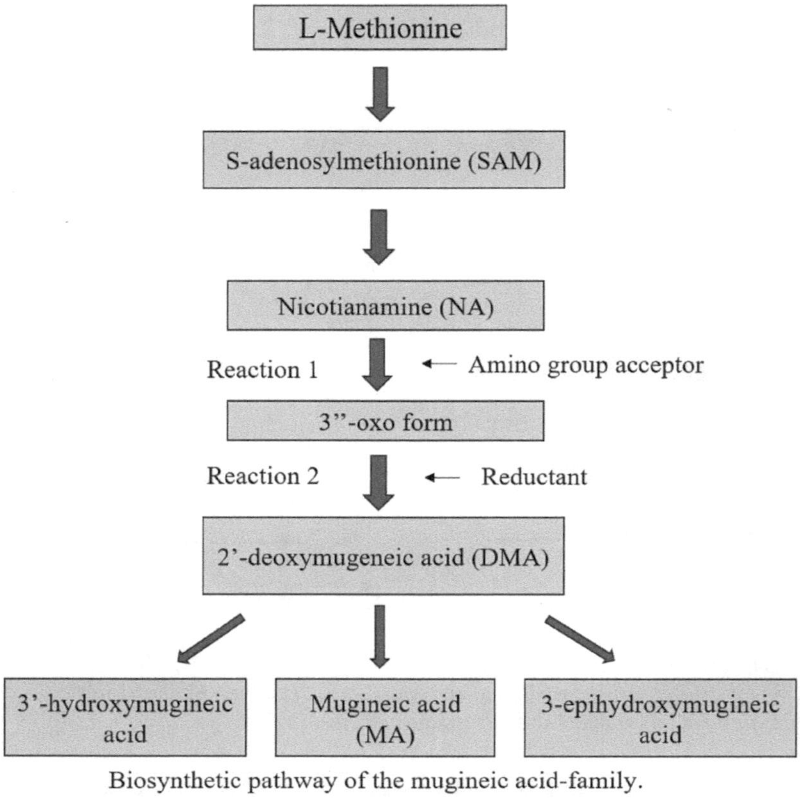

FIGURE 17.5 Biosynthetic pathway of the MA family.

to transform NA to DMA. The first step is the transfer of the amino group from NA's 3″ carbon to the amino group acceptors pyruvate, oxalacetate, and 2-oxoglutarate. The second reaction is the reduction of 3″-oxoform by NADH and NADPH. The enzyme that catalyzes the transamination of aminoacid at 3″-carbon of NA is NA aminotransferase in the presence of $NaBH_4$ as a chemical reductant (Hall et al., 2011). The so obtained product, i.e., 3″-keto intermediate form, is the result of the action of enzyme NA amino synthase (NAAT). While the enzyme DMAS functions to convert 3″-keto intermediate to DMA. DMA is only synthesized by rice plants, while other graminaceous plants also secrete other MAs (Figure 17.5), such as 3-epihydroxymugineic acid (epiHMA) and 3-hydroxymugineic acid (HMA), along with MA (Bashir and Ishimaru, 2022). The Fe uptake transporters and enzymes responsible for Fe uptake are present in vascular bundles and cortex. The Fe-DMA complex is also present in phloem sap, indicating its importance in uptake and internal translocation of Fe. In a study, DMA was detected in sap of xylem in olive (*Olea europaea* L.) (Dey et al., 2020; Suzuki et al., 2016).

17.5 METAL/METALLOID RESISTANCE MECHANISM BY BACTERIA

Bacteria have evolved various resistance mechanisms that enable them to immobilize, mobilize, or transform metals, thereby decreasing their toxicity and enhancing their ability to tolerate the uptake of heavy metal ions (Nanda et al., 2019). This process encompasses physical sequestration, exclusion, complexation, and detoxification. Metals can become immobile and can't enter cells when they bond to extracellular components. For instance, usually metals bind with the anionic functional groups present on the outermost layer of cells, like amine, sulfhydryl, carboxyl, hydroxyl, sulfonate, and amide groups (Priyadarshanee et al., 2021). In a comparable manner, heavy metals can be extensively bound by microbial extracellular polymers like polysaccharides, proteins, and humic materials, which can then eliminate toxins from the metals by forming complexes or by erecting a reliable barrier surrounding the cell (Mahapatra et al., 2020). Since microbial siderophores attach to metal ions with biochemistry similar to Fe's, they help reduce the bioavailability and toxicity of metals (Hofmann et al., 2021). It was noted that siderophore having citrate and catecholate forms complex with boron primarily because of the siderophore's vicinal dianionic oxygen-binding group (Hofmann et al., 2020). For example, in recent research, it was found that *Marinobacter* sp. strain 893 produces siderophore vibrioferrin capable of boron binding (de Carvalho et al., 2010). In fact, additional research supported that vibrioferrin is not the only substance that binds boron. Enterobactin and salmochelin, two extensively researched siderophores produced by *E. coli*, have the potential to bind to Si(IV) under physiological conditions (Schmiederer et al., 2011). Similarly, the use of siderophore preparations from *Aspergillus nidulans* significantly reduced the toxicity associated with As in *Triticum aestivum* (Kumari et al., 2019). Furthermore, a lot of bacteria promote interactions or create metabolites that cause metals to precipitate and crystallize (Robles-Fernández et al., 2022). Numerous bacteria are equipped with efflux transporters designed to expel toxic metals or those present in excessive concentrations from the cell through ATPase pumps or chemiosmotic ion/proton pumps (Etesami, 2018). In consequence of the high affinity that such transporters have for their substrates, the cytosol's metal concentration is effectively reduced. Cytosolic sequestration is a defense mechanism used by some bacteria against metals. Metal ions can be segregated or changed into less toxic forms once they are within the cell, providing an effective detoxification method (Dubey et al., 2018). Certain microbes produce metal-binding proteins like metallothioneins (a low molecular mass and cysteine-rich proteins) with high affinities for metals such as Cd, Cu, Ag, and Hg. These proteins are induced in response to the presence of metals and primarily function in metal detoxification. Although it is a metal-dependent mechanism, some metals use bacteria for methylation as a metal resistance or detoxification mechanism. Methyl groups are transferred to metals and metalloids during this process, resulting in reduced toxicity, as seen with elements like Se under methylated state (Sun et al., 2014; Anju, 2017).

17.5.1 Iron-Siderophore Complex Transport Mechanism

The accessible iron-siderophore complex can be internalized through one of the two pathways, i.e., either the Fe is released from the complex and enters the cell as a single ion or the cell absorbs the entire iron-siderophore complex (Saha et al., 2013). The receptor proteins at the outer membrane mediate the uptake of the Fe-Si complex (Josts et al., 2021). Recently, various studies highlighted the structure and action mechanisms of these receptors. Figure 17.6 explains the iron-siderophore complex mechanism. It is now understood that a conformational change in a specific domain of these receptors facilitates the substrate's passage (Chakraborty et al., 2007). After entering the periplasm, the Fe-Si complex passes through the inner membrane into the cytosol with the help of the ABC transporter or Fe-acquisition process (Josts et al., 2021). The transportation process is energy-dependent, which is provided by the *TonB* protein complex. The *TonB* protein complex has two inner membrane proteins, *viz., ExbD* and *ExbB*. Following the transport of the Fe-Si complex into the cell, the Fe is released from the complex by several mechanisms (Saha et al., 2013). While this mechanism typically involves chemical modification of the siderophore, released soluble Fe can be directly utilized in the metabolic and physiological processes. Recent findings have revealed that Fe is liberated from the Fe-Si complex by a simple reduction process, and the siderophore is subsequently again released outside the cell membrane without undergoing any chemical alteration (Schalk et al., 2011; Leventhal et al., 2019).

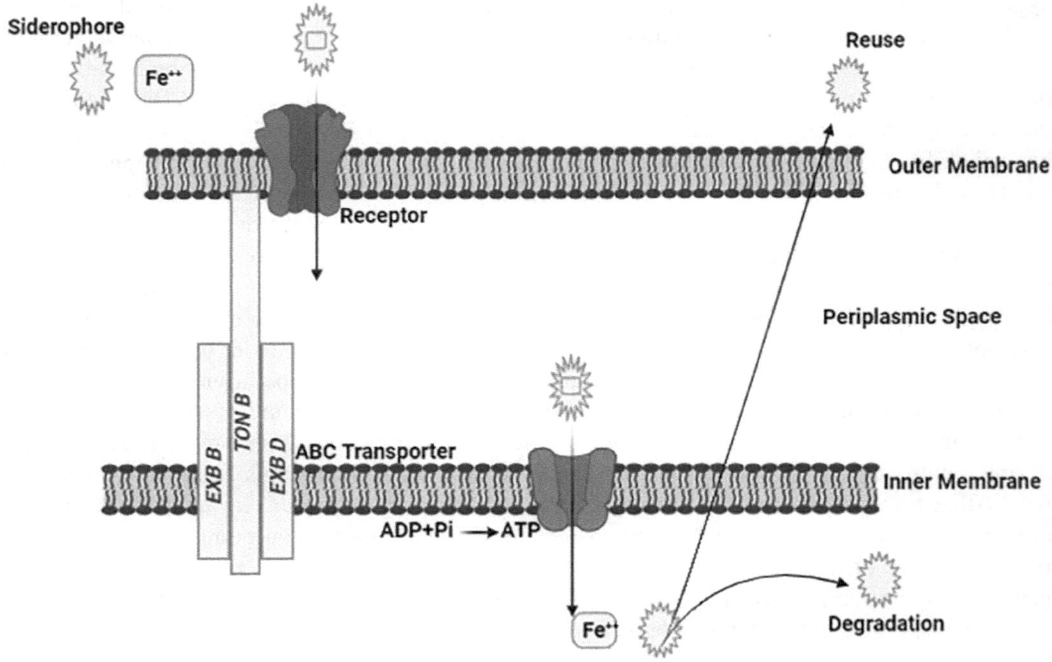

FIGURE 17.6 Transport of the Fe-siderophore complex across membranes. (Modified after Sah and Singh, 2015.)

17.5.2 Siderophores: Expanding Their Metal-Capturing Abilities beyond Iron

Siderophores have the potential to form stable complexes with metal cations other than Fe (Albelda-Berenguer et al., 2019). According to a study, *Pseudomonas aeruginosa* produces the siderophores pyoverdine and pyochelin, which have the potential to chelate 16 different metals (Ag^{+}, Al^{3+}, Cd^{2+}, Co^{2+}, Cr^{2+}, Cu^{2+}, Eu^{3+}, Ga^{3+}, Hg^{2+}, Mn^{2+}, Ni^{2+}, Pb^{2+}, Sn^{2+}, Tb^{3+}, Tl^{+}, and Zn^{2+}) (Braud et al., 2009a, b). V(IV) (vanadyl ion) and V(V) (vanadate ion) have the capability to create complexes with pyochelin and putrebactin. However, only the vanadyl ion forms complexes with pyoverdine and enterobactin, while the vanadate ion forms a complex with desferrioxamine B (Johnstone and Nolan, 2015). The interaction between pyoverdine, produced by *Pseudomonas fluorescens* strain CCUG 32456, was also found in rock aquifers and curium(III) (Moll et al., 2008). It is noteworthy to observe that the binding strength of curium(III) with this siderophore is higher compared to its complexation with substances like EDTA, hydroxide, or carbonate. Additionally, studies have shown that desferrioxamine can coordinate plutonium(V) having a structurally diverse metal center than Fe (Boukhalfa et al., 2007). These findings highlight the role of microorganisms producing siderophores in transporting toxic metals in our environment. This underscores their importance in assessing the risks associated with waste disposal sites. Furthermore, dedicated *TonB*-dependent transporters (TBDTs) have been extensively characterized for their role in transporting various metals extending beyond the external membrane. These include well-established examples for Fe complexed with siderophore, as well as transporters for vitamin B12, which has a corrin ring containing a Co ion. More lately, similar transporters have been identified for Ni transport (Tang et al., 2012). There is proof in the literature that *Heliobacter pylori* transports Ni across its outer membrane via TBDT called *FrpB4* (Li et al., 2009). Accordingly, vitamin B12 is a scarce and necessary substance for numerous living things. It is imported by the particular TBDT *BtuB* in *E. coli*, which has the universally preserved structure of ferrisiderophore TBDTs (Nijland et al., 2022).

17.6 ROLE OF SIDEROPHORE IN ALLEVIATION OF METAL/METALLOID TOXICITY

Elements possessing physical and chemical characteristics halfway between those of metals and non-metals are called metalloids or semimetals (Praveen Kumar et al., 2012). While As, Ge, Sb, and Te are non-essential and hazardous to plants even at extremely low concentrations, whereas Br, Si, and Se are necessary to plants in tiny amounts (Pandey et al., 2019; Behera et al., 2023). When there is an excess of these metalloids in the soil, the roots of developing plants absorb them and exhibit their toxic impacts in plants in terms of reduced growth and levels of photosynthetic pigments, increased oxidative stress, and modulated levels of enzymatic and non-enzymatic antioxidants, including superoxide dismutase (SOD), ascorbate peroxidase (APOX), peroxidase (POD), catalase (CAT), glutathione-*S*-transferase (GST), glutathione, etc. (Feng et al., 2021; Hasanuzzaman et al., 2022). Global worry over metalloid contamination of soil environments has resulted in a global drop in agricultural production. When metalloids are present in soil, water, or the air at excessive concentrations, they enter the food chain. Animals and humans who consume grains or plant parts high in metalloids, run the risk of developing major health problems (Deng et al., 2021). Through the use of transporters, plant roots absorb metalloids from the earth, which they then move to various plant sections (Sharma et al., 2021). After being absorbed by the cells of various organs, metalloids are eliminated from the plant system via sequestration or active efflux (Podar and Maathuis, 2022). In addition to the sequestration of metalloids within the cells, efforts have been made by different groups of workers to reduce the toxicity of metalloids to plants by using plant- and microbial-produced metal chelators.

Microbial and plant metal chelators known as siderophores have the ability to bind the metalloids in addition to Fe (Johnstone and Nolan, 2015). The ability of other metalloids, such as boron and silicon, to form a complex with siderophores has not been reported as frequently. Enterobactin and salmochelin, a twofold glycosylated enterobactin generated by *E. coli*, are two extremely well-studied siderophores that have demonstrated silicon (IV)-binding capacity under physiological settings (Schmiederer et al., 2011). The ability of siderophores to bind Ge is comparable to that of Si. It has been demonstrated that a chemically generated counterpart of the siderophore enterobactin, which is produced by *E. coli*, can bind both Si(IV) and Ge(IV) (Baramov et al., 2013). Further, siderophores have been reported to form complexes with other metalloids as well as to remediate metalloid toxicity (Table 17.2) (Gusain and Singh, 2016; Singh et al., 2021). For instance, the administration of a hydroxamate-type siderophore that was obtained from *Aspergillus nidulans* under hazardous conditions of As stress greatly restored the wheat genotype's growth and assisted in modifying the enzymatic activities of SOD, CAT, and POD. Injury resulting from lipid peroxidation was also markedly decreased at the same time. Further, improved binding affinity of the ferricrocin–arsenate complex, resulting in a thermodynamically stable compound, was demonstrated by in silico simulations (Kumari et al., 2019). Similarly, siderophore-producing bacteria, *Brevundimonas intermedia* and *Pseudomonas gessardii*, increased wheat tolerance against As. Treated plants exhibited increased expression of genes related to metal stress defense, expression of SOD and APOX, and dry biomass (Soto et al., 2019). In another study, bacteria *Delftia* sp. P2III5 and *Pseudomonas* sp. P1III2 that reduce arsenate and produce siderophores have been reported to increase biomass of *Pteris vittate* grown in As-contaminated soil (Lampis et al., 2015). Similarly, when compared to an uninoculated set, the positive effects of the siderophore-producing bacterium *Bacillus aryabhattai* AS6 strain on the growth promotion of rice seedlings under As stress were reported, considering several morphological and biochemical parameters, including antioxidant activities. This strain was discovered to have an enhanced rate of As(V) reduction, which made it useful for cleaning up As in contaminated agricultural fields (Ghosh et al., 2018). Additionally, six As-resistant bacteria that were isolated from the rhizosphere of the As-hyperaccumulator *Pteris vittata* were used to study the association between bacterial abilities in As transformation and siderophore synthesis. Results revealed that greater As(III) oxidation and resistance to As(III) were seen in bacteria that produced more siderophores than in bacteria that produced less siderophores. Results from mutants of

TABLE 17.2
Role of Plant and Microbial Siderophores/Siderophore-Producing Microbes in Remediation of Metal/Metalloid Toxicity

Siderophore	Source	Metal/Metalloid	Plant Species/Experiment	Impact	References
Bacillibactin	*Bacillus amyloliquefaciens* BM3	As	*Solanum melongena*	Exhibited As chelating ability	Pradhan et al. (2023)
Siderophore	*Enterobacter* sp. YG-14	As	Soil and *Robinia pseudoacacia*	Promoted growth and assisted host plant in As phytoremediation	Chen et al. (2023)
Chlorogenic acid	Exogenous application	As	*Zea mays*	Improved activity of antioxidant enzymes, maintained membrane integrity, and improved plant growth	Arikan et al. (2022)
Siderophore	Fungi	As (Sodium arsenite)	Invitro experiment	Fungal species tolerated high As concentration by forming As-siderophore complex	Ceci et al. (2020)
Siderophore	*B. pumilus*	As	*Oryza sativa*	Reduced As accumulation in rice grains	Dolphen and Paitip (2019)
Oosporein	*Chaetomium cupreum*	Al	*Miscanthus sinensis*	Enhanced Al tolerance and might promote plant growth	Haruma et al. (2019)
Hydroxamate-type	*Aspergillus nidulans*	As	*Triticum aestivum*	Reduced lipid peroxidation by amending activity of antioxidant enzymes, including SOD, CAT, and POD, leading to the growth improvement	Kumari et al. (2019)
Siderophore	*Staphylococcus sp.* TA6	As (Arsenate and arsenite)	In vitro experiment	Siderophore was found directly related with As resistance and bacterium transformed As(V) to As(III)	Das and Barooah (2018)
Siderophore	*Rhodopseudomonas palustris* C1 and *Rubrivivax benzoatilyticus* C31	As	*Oryza sativa*	Reduced As uptake and increased chlorophyll a, chlorophyll b, and carotenoid content	Nookongbut et al. (2018)
Siderophore	Actinobacteria	As	In vitro experiment	Exhibited As bound capacity and can be used as a possible candidate for environmental cleanup	Retamal-Morales et al. (2018)
Siderophore	Bacteria	As	–	Increased bacterial tolerance to As	Ghosh et al. (2015)
Catecholate or Desferrioxamine B	*Pseudomonas* sp. PG12 or fungus	As ($FeAsO_4$)	*Pteris vittata*	Increased plant biomass	Liu et al. (2015)
Siderophore	*P. aeruginosa*	As	*P. cretica*	Increased As accumulation in plant aerial parts in the form of As-siderohpre complex	Jeong et al. (2014)
Siderophore	*P. azotoformans*	As	Soil	Form complex with As and removed 92.8% As	Nair et al. (2007)
Pyridine-2,6-bis (thiocarboxylic acid)	*P. stutzeri KC*	As	–	Reduces As toxicity	Zawadzka et al. (2007)
Pyridine-2,6-bis (thiocarboxylic acid)	*P. stutzeri KC*	Se and Te	Bacterial cultures	Precipitated and detoxified Se and tellurium oxyanions	Zawadzka et al. (2006)

P. fluorescens with decreased siderophore synthesis further corroborated this observation, as they showed 23%–25% less tolerance to As(III) than the wild-type (Ghosh et al., 2015). In another experiment, Samreen et al. (2019) reported that siderophore-producing *Bacillus* spp. MN-54 mitigated Br toxicity in canola plants and enhanced growth in terms of increased shoot length, root length, shoot weight, root weight, and chlorophyll content.

17.7 CONCLUSION

Siderophores are Fe-chelating compounds produced by microorganisms to acquire Fe. Types include catecholates, hydroxamates, and carboxylates, each with distinct chemical structures facilitating Fe binding and uptake. Siderophores play a pivotal role in combating metalloid toxicity by forming complexes with various metals. These specialized molecules sequester and transport essential metals, preventing harmful interactions. Their unique ability to selectively bind with specific elements makes siderophores valuable in promoting microbial resistance and safeguarding against metalloid-induced stress. Thus, understanding their action and role provides insights into potential applications for environmental remediation and biotechnological advancements.

REFERENCES

Aguirre-Noyola, José Luis, Mónica Rosenblueth, Michel Geovanni Santiago-Martínez, and Esperanza Martínez-Romero. Transcriptomic responses of Rhizobium phaseoli to root exudates reflect its capacity to colonize maize and common bean in an intercropping system. *Frontiers in Microbiology* 12 (2021): 740818.

Ahmed, Engy, and Sara JM Holmström. Siderophores in environmental research: roles and applications. *Microbial biotechnology* 7, no. 3 (2014): 196–208.

Albelda-Berenguer, Magdalena, Mathilde Monachon, and Edith Joseph. Siderophores: from natural roles to potential applications. *Advances in Applied Microbiology* 106 (2019): 193–225.

Ali, Hazrat, Ezzat Khan, and Ikram Ilahi. Environmental chemistry and ecotoxicology of hazardous heavy metals: environmental persistence, toxicity, and bioaccumulation. *Journal of Chemistry* 253 (2019): 65–113.

Anju, M. Biotechnological strategies for remediation of toxic metal(loid)s from environment. In: Gahlawat, Suresh Kumar, Raj Kumar Salar, Priyanka Siwach, Joginder Singh Duhan, Suresh Kumar, and Pawan Kaur (Eds.) *Plant Biotechnology: Recent Advancements and Developments*, 315–359. Springer Nature, 2017.

Ansari, Rizwan Ali, Irshad Mahmood, Rose Rizvi, Aisha Sumbul, and Safiuddin. Siderophores: augmentation of soil health and crop productivity. In: Kumar, Vivek, Shivesh Sharma, Manoj Kumar, and Ram Prasad (Eds.) *Probiotics in Agroecosystem*: 291–312. Springer Nature, 2017.

Arihilam, Ngozi H., and E. C. Arihilam. Impact and control of anthropogenic pollution on the ecosystem–a review. *Journal of Bioscience and Biotechnology Discovery* 4, no. 3 (2019): 54–59.

Arikan, Busra, Ceyda Ozfidan-Konakci, Evren Yildiztugay, Gokhan Zengin, Fatma Nur Alp, and Fevzi Elbasan. Exogenous hesperidin and chlorogenic acid alleviate oxidative damage induced by arsenic toxicity in *Zea mays* through regulating the water status, antioxidant capacity, redox balance and fatty acid composition. *Environmental Pollution* 292 (2022): 118389.

Armin, Reyhaneh, Sebastian Zühlke, Gisela Grunewaldt-Stöcker, Felix Mahnkopp-Dirks, and Souvik Kusari. Production of siderophores by an apple root-associated Streptomyces ciscaucasicus strain GS2 using chemical and biological OSMAC approaches. *Molecules* 26, no. 12 (2021): 3517.

Ashraf, Sadia, Kalyani Dhusia, and Susheel Verma. Siderophores mediated iron acquisition and virulence of brown rot disease in stone fruits caused by *Monilinia fructicola* in Jammu and Kashmir. In: Dhusia, Kalyani, Kalpana Raja, and Pramod Ramteke (Eds.) *Fungal Siderophores: From Mineral—Microbe Interactions to Anti-Pathogenicity*, 51–68. Springer, 2021.

Baars, Oliver, François M. M. Morel, and Xinning Zhang. The purple non-sulfur bacterium Rhodopseudomonas palustris produces novel petrobactin-related siderophores under aerobic and anaerobic conditions. *Environmental Microbiology* 20, no. 5 (2018): 1667–1676.

Baramov, Todor, Karlijn Keijzer, Elisabeth Irran, Eva Mösker, Mu-Hyun Baik, and Roderich Süssmuth. Synthesis and structural characterization of hexacoordinate silicon, germanium and titanium complexes of the *E. coli* siderophore enterobactin. *Chemistry–A European Journal* 19, no. 32 (2013): 10536–10542.

Bashir, Khurram, and Yasuhiro Ishimaru. Challenges and opportunities to regulate mineral transport in rice. *Bioscience, Biotechnology, and Biochemistry* 86, no. 1 (2022): 12–22.

Beasley, Federico C., Cristina L. Marolda, Johnson Cheung, Suzana Buac, and David E. Heinrichs. Staphylococcus aureus transporters Hts, Sir, and Sst capture iron liberated from human transferrin by Staphyloferrin A, Staphyloferrin B, and catecholamine stress hormones, respectively, and contribute to virulence. *Infection and Immunity* 79, no. 6 (2011): 2345–2355.

Behera, Biswaranjan, Mrunalini Kancheti, Md Basit Raza, Aalok Shiv, Vikas Mangal, Gajendra Rathod, Muhammad Ahsan Altaf, Awadhesh Kumar, Tariq Aftab, Ravinder Kumar, Rahul Kumar Tiwari, Milan Kumar Lal, and Brajesh Singh. Mechanistic insight on boron-mediated toxicity in plant vis-a-vis its mitigation strategies: a review. *International Journal of Phytoremediation* 25, no. 1 (2023): 9–26.

Beneduzi, Anelise, Adriana Ambrosini, and Luciana M. P. Passaglia. Plant growth-promoting rhizobacteria (PGPR): their potential as antagonists and biocontrol agents. *Genetics and Molecular Biology* 35 (2012): 1044–1051.

Bhardwaj, Snigdha, Shaminder Singh, and Sonam Bhatia. Contrasting role of fungal siderophore in metal ion complex formation. In: Dhusia, Kalyani, Kalpana Raja, and Pramod Ramteke (Eds.) *Fungal Siderophores: From Mineral—Microbe Interactions to Anti-Pathogenicity*, 99–117. Springer, 2021.

Bobrov, Alexander G., Olga Kirillina, Jacqueline D. Fetherston, M. Clarke Miller, Joseph A. Burlison, and Robert D. Perry. The Y ersinia pestis siderophore, yersiniabactin, and the ZnuABC system both contribute to zinc acquisition and the development of lethal septicaemic plague in mice. *Molecular Microbiology* 93, no. 4 (2014): 759–775.

Bohac, Tabbetha J., Luting Fang, Daryl E. Giblin, and Timothy A. Wencewicz. Fimsbactin and acinetobactin compete for the periplasmic siderophore binding protein BauB in pathogenic Acinetobacter baumannii. *ACS Chemical Biology* 14, no. 4 (2019): 674–687.

Boukhalfa, Hakim, Sean D. Reilly, and Mary P. Neu. Complexation of Pu (IV) with the natural siderophore desferrioxamine B and the redox properties of Pu (IV)(siderophore) complexes. *Inorganic Chemistry* 46, no. 3 (2007): 1018–1026.

Bramhachari, P. V., and Ganji Purnachandra Nagaraju. Extracellular polysaccharide production by bacteria as a mechanism of toxic heavy metal biosorption and biosequestration in the marine environment. In: Naik, Mohan, M., and D. Santosh Kumar (Eds.) *Marine Pollution and Microbial Remediation*, 67–85. Springer, 2017.

Brandenburger, Eileen, Markus Gressler, Robin Leonhardt, Gerald Lackner, Andreas Habel, Christian Hertweck, Matthias Brock, and Dirk Hoffmeister. A highly conserved basidiomycete peptide synthetase produces a trimeric hydroxamate siderophore. *Applied and Environmental Microbiology* 83, no. 21 (2017): e01478–17.

Braud, Armelle, Françoise Hoegy, Karine Jezequel, Thierry Lebeau, and Isabelle J. Schalk. New insights into the metal specificity of the Pseudomonas aeruginosa pyoverdine–iron uptake pathway. *Environmental Microbiology* 11, no. 5 (2009a): 1079–1091.

Braud, Armelle, Mélissa Hannauer, Gaëtan L. A. Mislin, and Isabelle J. Schalk. The Pseudomonas aeruginosa pyochelin-iron uptake pathway and its metal specificity. *Journal of Bacteriology* 191, no. 11 (2009b): 3517–3525.

Cain, Timothy J., and Aaron T. Smith. Ferric iron reductases and their contribution to unicellular ferrous iron uptake. *Journal of Inorganic Biochemistry* 218 (2021): 111407.

Campestre, María Paula, Luis Nazareno Castagno, María Julia Estrella, and Oscar Adolfo Ruiz. Lotus japonicus plants of the Gifu B-129 ecotype subjected to alkaline stress improve their Fe^{2+} bio-availability through inoculation with Pantoea eucalypti M91. *Journal of Plant Physiology* 192 (2016): 47–55.

Carroll, Cassandra S., and Margo M. Moore. Ironing out siderophore biosynthesis: a review of non-ribosomal peptide synthetase (NRPS)-independent siderophore synthetases. *Critical Reviews in Biochemistry and Molecular Biology* 53, no. 4 (2018): 356–381.

Ceci, Andrea, Veronica Spinelli, Lorenzo Massimi, Silvia Canepari and Anna Maria Persiani. Fungi and arsenic: tolerance and bioaccumulation by soil saprotrophic species. *Applied Sciences* 10, no. 9 (2020): 3218.

Chakraborty, Ranjan, Erin Storey, and Dick Van Der Helm. Protection against myocardial ischemia/reperfusion injury in tlr4-deficient mice is mediated through a phosmolecular mechanism of ferricsiderophore passage through the outer membrane receptor proteins of Escherichia coliphoinositide 3-KINASE-DEPENDENT MECHANISM. *BioMetals* 20, no. 3–4 (2007): 263–274.

Chen, Hanna, Lin Zhong, Haibo Zhou, Tao Sun, Guannan Zhong, Qiang Tu, Yan Zhuang et al. Biosynthesis of glidomides and elucidation of different mechanisms for formation of β-OH amino acid building blocks. *Angewandte Chemie* 134, no. 35 (2022): e202203591.

Chen, Jiawei, Xuan Zhang, Min Kuang, Kunpeng Cui, Ting Xu, Xuanming Liu, Rui Zhuo et al. Endophytic Enterobacter sp. YG-14 mediated Arsenic mobilization through siderophore and its role in enhancing phytostabilization. *Journal of Hazardous Materials* (2023): 133206.

Correnti, Colin, and Roland K. Strong. Mammalian siderophores, siderophore-binding lipocalins, and the labile iron pool. *Journal of Biological Chemistry* 287, no. 17 (2012): 13524–13531.

Crumbliss, Alvin L., and James M. Harrington. Iron sequestration by small molecules: thermodynamic and kinetic studies of natural siderophores and synthetic model compounds. *Advances in Inorganic Chemistry* 61 (2009): 179–250.

Das, Saurav, and Madhumita Barooah. Characterization of siderophore producing arsenic-resistant *Staphylococcus* sp. strain TA6 isolated from contaminated groundwater of Jorhat, Assam and its possible role in arsenic geocycle. *BMC Microbiology* 18, no. 1 (2018): 1–11.

Davis, George Thomas, and Mark Stocksdale. A putative Fe^{+3}/phytosiderophore (PS) transporter isolated from oats (*Avena sativa*). *FASEB Journal* (2013): 1017.2.

De Carvalho, Carla C. C. R., and Pedro Fernandes. Production of metabolites as bacterial responses to the marine environment. *Marine Drugs* 8, no. 3 (2010): 705–727.

Deb, Vishal Kumar, Ahmad Rabbani, Shashi Upadhyay, Priyam Bharti, Hitesh Sharma, Devendra Singh Rawat, and Gaurav Saxena. Microbe-assisted phytoremediation in reinstating heavy metal-contaminated sites: concepts, mechanisms, challenges, and future perspectives. In: Arora, Pankaj Kumar (Eds.) *Microbial Technology for Health and Environment*, 161–189. Springer, 2020.

Deng, Fenglin, Fanrong Zeng, Guang Chen, Xue Feng, Adeel Riaz, Xiaojian Wu, Wei Gao, Feibo Wu, Paul Holford and Zhong-Hua Chen. Metalloid hazards: from plant molecular evolution to mitigation strategies. *Journal of Hazardous Materials* 409 (2021): 124495.

Devireddy, Laxminarayana R., Daniel O. Hart, David H. Goetz, and Michael R. Green. A mammalian siderophore synthesized by an enzyme with a bacterial homolog involved in enterobactin production. *Cell* 141, no. 6 (2010): 1006–1017.

Dey, Sangita, Preetom Regon, Saradia Kar, and Sanjib Kumar Panda. Chelators of iron and their role in plant's iron management. *Physiology and Molecular Biology of Plants* 26, no. 8 (2020): 1541–1549.

Dickson, Kayle B., and Juan Zhou. Role of reactive oxygen species and iron in host defense against infection. *Frontiers in Bioscience-Landmark* 25, no. 8 (2020): 1600–1616.

Dolphen, Rujira, and Paitip Thiravetyan. Reducing arsenic in rice grains by leonardite and arsenic-resistant endophytic bacteria. *Chemosphere* 223 (2019): 448–454.

Dubey, Sonali, Manju Shri, Anubhuti Gupta, Vibha Rani, and Debasis Chakrabarty. Toxicity and detoxification of heavy metals during plant growth and metabolism. *Environmental Chemistry Letters* 16 (2018): 1169–1192.

Duckworth, Owen W., John R. Bargar, and Garrison Sposito. Quantitative structure–activity relationships for aqueous metal–siderophore complexes. *Environmental Science & Technology* 43, no. 2 (2009): 343–349.

Etesami, Hassan. Bacterial mediated alleviation of heavy metal stress and decreased accumulation of metals in plant tissues: mechanisms and future prospects. *Ecotoxicology and Environmental Safety* 147 (2018): 175–191.

Feng, RenWei, PingPing Zhao, YanMing Zhu, JiGang Yang, XinQi Wei, Li Yang, Hong Liu, Christopher Rensing, and YongZhen Ding. Application of inorganic selenium to reduce accumulation and toxicity of heavy metals (metalloids) in plants: the main mechanisms, concerns, and risks. *Science of the Total Environment* 771 (2021): 144776.

Franken, Angelique C. W., Beatrix E. Lechner, Ernst R. Werner, Hubertus Haas, B. Christien Lokman, Arthur F. J. Ram, Cees A. M. J. J. van den Hondel, Sandra de Weert, and Peter J. Punt. Genome mining and functional genomics for siderophore production in *Aspergillus niger*. *Briefing in Functional Genomics* 13 (2014): 482–492.

Franken, Angelique C. W., Beatrix E. Lechner, Ernst R. Werner, Hubertus Haas, B. Christien Lokman, Arthur F. J. Ram, Cees A. M. J. J. van den Hondel, Sandra de Weert, and Peter J. Punt. Genome mining and functional genomics for siderophore production in Aspergillus niger. *Briefings in Functional Genomics* 13, no. 6 (2014): 482–492.

Gamalero, Elisa, and Bernard R. Glick. Mechanisms used by plant growth-promoting bacteria. In: Maheshwari, D. K. (Eds.) *Bacteria in Agrobiology: Plant Nutrient Management*, 17–46. Springer, 2011.

Ghosh, Pallab Kumar, Tushar Kanti Maiti, Krishnendu Pramanik, Sudip Kumar Ghosh, Soumik Mitra, and Tarun Kumar De. The role of arsenic resistant *Bacillus aryabhattai* MCC3374 in promotion of rice seedlings growth and alleviation of arsenic phytotoxicity. *Chemosphere* 211 (2018): 407–419.

Ghosh, Piyasa, Bala Rathinasabapathi, Max Teplitski, and Lena Q. Ma. Bacterial ability in AsIII oxidation and AsV reduction: relation to arsenic tolerance, P uptake, and siderophore production. *Chemosphere* 138 (2015): 995–1000.

Ghosh, Sougata, Tanay Bhagwat, and Thomas J. Webster. Endophytic microbiomes and their plant growth-promoting attributes for plant health. In: Yadav, Ajar Nath, Joginder Singh, Chhatarpal Singh, and Neelam Yadav (Eds.) *Current Trends in Microbial Biotechnology for Sustainable Agriculture* (2021): 245–278. Springer, 2021.

Ghosh, Swapan Kr, Tanmay Bera, and Ananda M. Chakrabarty. Microbial siderophore–A boon to agricultural sciences. *Biological Control* 144 (2020): 104214.

Gressler, Markus, Florian Meyer, Daniel Heine, Peter Hortschansky, Christian Hertweck, and Matthias Brock. Phytotoxin production in Aspergillus terreus is regulated by independent environmental signals. *Elife* 4 (2015): e07861.

Grobelak, A., A. Napora, and M. J. E. E. Kacprzak. Using plant growth-promoting rhizobacteria (PGPR) to improve plant growth. *Ecological Engineering* 84 (2015): 22–28.

Grobelak, Anna, and Joanna Hiller. Bacterial siderophores promote plant growth: screening of catechol and hydroxamate siderophores. *International Journal of Phytoremediation* 19, no. 9 (2017): 825–833.

Gu, Chen, Jing Wang, Mengfan Guo, Meng Sui, Hong Lu, and Guangfei Liu. Extracellular degradation of tetrabromobisphenol A via biogenic reactive oxygen species by a marine Pseudoalteromonas sp. *Water Research* 142 (2018): 354–362.

Gulick, Andrew M. Nonribosomal peptide synthetase biosynthetic clusters of ESKAPE pathogens. *Natural Product Reports* 34, no. 8 (2017): 981–1009.

Gusain, P., and V. Singh. Hydroxamate and catecholate siderophore synthesizing arsenic resistant PGPR isolated from west Bengal. *Journal of Environmental and Applied Bioresearch* 4 (2016): 1–4.

Hall, Mélanie, and Andreas S. Bommarius. Enantioenriched compounds via enzyme-catalyzed redox reactions. *Chemical Reviews* 111, no. 7 (2011): 4088–4110.

Happacher, Isidor, Mario Aguiar, Annie Yap, Clemens Decristoforo, and Hubertus Haas. Fungal siderophore metabolism with a focus on Aspergillus fumigatus: impact on biotic interactions and potential translational applications. *Essays in Biochemistry* 67 (2023): 824–842.

Haruma, Toshikatsu, Keiko Yamaji, Kazuyoshi Ogawa, Hayato Masuya, Yurina Sekine, and Naofumi Kozai. Root-endophytic *Chaetomium cupreum* chemically enhances aluminium tolerance in *Miscanthus sinensis* via increasing the aluminium detoxicants, chlorogenic acid and oosporein. *PLoS One* 14, no. 2 (2019): e0212644.

Hasanuzzaman, Mirza, Kamrun Nahar, Pedro García-Caparrós, Khursheda Parvin, Faisal Zulfiqar, Naznin Ahmed, and Masayuki Fujita. Selenium supplementation and crop plant tolerance to metal/metalloid toxicity. *Frontiers in Plant Science* 12 (2022): 792770.

Haselwandter, Kurt, Gerlinde Häninger, Markus Ganzera, Hubertus Haas, Graeme Nicholson, and Günther Winkelmann. Linear fusigen as the major hydroxamate siderophore of the ectomycorrhizal Basidiomycota Laccaria laccata and Laccaria bicolor. *Biometals* 26 (2013): 969–979.

Haselwandter, Kurt, Hubertus Haas, Gerlinde Haeninger, and Günther Winkelmann. Siderophores in plant root tissue: tagetes patula nana colonized by the arbuscular mycorrhizal fungus Gigaspora margarita. *Biometals* 33 (2020): 137–146.

Hofmann, Marika, Gerardo Retamal-Morales, and Dirk Tischler. Metal binding ability of microbial natural metal chelators and potential applications. *Natural Product Reports* 37, no. 9 (2020): 1262–1283.

Hofmann, Marika, Thomas Heine, Luise Malik, Sarah Hofmann, Kristin Joffroy, Christoph Helmut Rudi Senges, Julia Elisabeth Bandow, and Dirk Tischler. Screening for microbial metal-chelating siderophores for the removal of metal ions from solutions. *Microorganisms* 9, no. 1 (2021): 111.

Holden, Victoria I., and Michael A. Bachman. Diverging roles of bacterial siderophores during infection. *Metallomics* 7, no. 6 (2015): 986–995.

Iacovelli, Riccardo, Roel A. L. Bovenberg, and Arnold J. M. Driessen. Nonribosomal peptide synthetases and their biotechnological potential in Penicillium rubens. *Journal of Industrial Microbiology and Biotechnology* 48, no. 7–8 (2021): kuab045.

Igiehon, Nicholas O., Olubukola O. Babalola, and Bukola R. Aremu. Genomic insights into plant growth promoting rhizobia capable of enhancing soybean germination under drought stress. *BMC Microbiology* 19, no. 1 (2019): 1–22.

Igiehon, Nicholas O., Olubukola O. Babalola, and Bukola R. Aremu. Genomic insights into plant growth promoting rhizobia capable of enhancing soybean germination under drought stress. *BMC Microbiology* 19, no. 1 (2019): 1–22.

Ito, Yusai, and Alison Butler. Structure of synechobactins, new siderophores of the marine cyanobacterium Synechococcus sp. PCC 7002. *Limnology and Oceanography* 50, no. 6 (2005): 1918–1923.

Jeong, Seulki, Hee Sun Moon, and Kyoungphile Nam. Enhanced uptake and translocation of arsenic in Cretan brake fern (*Pteris cretica* L.) through siderophorearsenic complex formation with an aid of rhizospheric bacterial activity. *Journal of Hazardous Materials* 280 (2014): 536–543.

Johnstone, Timothy C., and Elizabeth M. Nolan. Beyond iron: non-classical biological functions of bacterial siderophores. *Dalton Transactions* 44, no. 14 (2015): 6320–6339.

Joshi, R. S., S. H. Shaikh, and S. S. Joshi. Optimization and partial characterization of siderophore produced by Pseudomonas species isolated from agricultural soil. *Journal of Global Biosciences* 7 (2018): 5342–5349.

Josts, Inokentijs, Katharina Veith, Vincent Normant, Isabelle J. Schalk, and Henning Tidow. Structural insights into a novel family of integral membrane siderophore reductases. *Proceedings of the National Academy of Sciences* 118, no. 34 (2021): e2101952118.

Kesaulya, Henry, J. V. Hasinu, and G. NC Tuhumury. Potential of Bacillus spp produces siderophores insuppressing thewilt disease of banana plants. In *IOP Conference Series: Earth and Environmental Science* 102 (2018): 012016. IOP Publishing.

Kobayashi, Takanori, Tomoko Nozoye, and Naoko K. Nishizawa. Iron transport and its regulation in plants. *Free Radical Biology and Medicine* 133 (2019): 11–20.

Kraepiel, A. M. L., J. P. Bellenger, T. Wichard, and Francois M. M. Morel. Multiple roles of siderophores in free-living nitrogen-fixing bacteria. *Biometals* 22 (2009): 573–581.

Kumar, Lalit, Nand Lal Meena, and Ummed Singh. Role of phytosiderophores in acquisition of iron and other micronutrients in food legumes. In: Singh, Ummed, C. S. Praharaj, S. S. Singh, and N. P. Singh (Eds.) *Biofortification of Food Crops*, 291–302. Springer, 2016.

Kumari, Sneha, Azmi Khan, Pratika Singh, Sharad Kumar Dwivedi, Krishna Kumar Ojha, and Amrita Srivastava. Mitigation of As toxicity in wheat by exogenous application of hydroxamate siderophore of Aspergillus origin. *Acta Physiologiae Plantarum* 41 (2019): 1–11.

Lalitha, S., and S. Nithyapriya. Production of bacillibactin siderophore from soil bacteria, Bacillus subtilis: a bioinoculant enhances plant growth in Arachis hypogaea L. through elevated uptake of nutrients. In *International Seminar on Promoting Local Resources for Sustainable Agriculture and Development (ISPLRSAD* 2020), 71–82. Atlantis Press, 2021.

Lampis, Silvia, Chiara Santi, Adriana Ciurli, Marco Andreolli, and Giovanni Vallini. Promotion of arsenic phytoextraction efficiency in the fern *Pteris vittata* by the inoculation of As-resistant bacteria: a soil bioremediation perspective. *Frontiers in Plant Science* 6 (2015): 80.

Lehner, Sylvia M., Lea Atanasova, Nora K. N. Neumann, Rudolf Krska, Marc Lemmens, Irina S. Druzhinina, and Rainer Schuhmacher. Isotope-assisted screening for iron-containing metabolites reveals a high degree of diversity among known and unknown siderophores produced by Trichoderma spp. *Applied and Environmental Microbiology* 79, no. 1 (2013): 18–31.

Leventhal, Gabriel E., Martin Ackermann, and Konstanze T. Schiessl. Why microbes secrete molecules to modify their environment: the case of iron-chelating siderophores. *Journal of the Royal Society Interface* 16, no. 150 (2019): 20180674.

Li, Xuqing, Dingyi Li, Jianli Yan, Ya Zhang, Hong Wang, Jingze Zhang, Temoor Ahmed, and Bin Li. Effect of plant-growth-promoting fungi on eggplant (Solanum melongena L.) in new reclamation land. *Agriculture* 11, no. 11 (2021): 1036.

Li, Yanjie, and Deborah B. Zamble. Nickel homeostasis and nickel regulation: an overview. *Chemical Reviews* 109, no. 10 (2009): 4617–4643.

Liu, Xue, Guang-Mei Yang, Dong-Xing Guan, Piyasa Ghosh, and Lena Q. Ma. Catecholate-siderophore produced by As-resistant bacterium effectively dissolved $FeAsO_4$ and promoted Pteris vittata growth. *Environmental Pollution* 206 (2015): 376–381.

Liu, Zhuoming, Scott Reba, Wei-Dong Chen, Suheel Kumar Porwal, W. Henry Boom, Robert B. Petersen, Roxana Rojas, Rajesh Viswanathan, and L. Devireddy. Regulation of mammalian siderophore 2,5-DHBA in the innate immune response to infection. *Journal of Experimental Medicine* 211, no. 6 (2014): 1197–1213.

Lu, Yi, Hongying Wang, Zhuangzhuang Wang, Ying Cong, Peng Zhang, Guanglei Liu, Chenguang Liu, Zhenming Chi, and Zhe Chi. (2019). Metabolic rewiring improves the production of the fungal active targeting molecule Fusarinine C. *ACS Synthetic Biology*, *8*(8), 1755–1765.

Lurthy, Tristan, Cécile Cantat, Christian Jeudy, Philippe Declerck, Karine Gallardo, Catherine Barraud, Fanny Leroy, Alain Ourry, Philippe Lemanceau, Christophe Salon, and Sylvie Mazurier. Impact of bacterial siderophores on iron status and ionome in pea. *Frontiers in Plant Science* 11 (2020): 730.

Mahapatra, Biswajita, Nabin Kumar Dhal, Abanti Pradhan, and Bibhu Prasad Panda. Application of bacterial extracellular polymeric substances for detoxification of heavy metals from contaminated environment: a mini-review. *Materials Today: Proceedings* 30 (2020): 283–288.

Marathe, R., Y. Phatake, and A. Sonawane. Bioprospecting of Pseudomonas aeruginosa for their potential to produce siderophore: Process optimization and evaluation of its bioactivity. *International Journal of Bioassays* 4 (2015): 3667–3675.

Mawji, Edward, Martha Gledhill, James A. Milton, Glen A. Tarran, Simon Ussher, Anu Thompson, George A. Wolff, Paul J. Worsfold, and Eric P. Achterberg. Hydroxamate siderophores: occurrence and importance in the Atlantic Ocean. *Environmental Science & Technology* 42, no. 23 (2008): 8675–8680.

Mishra, Awdhesh Kumar, and Kwang-Hyun Baek. Salicylic acid biosynthesis and metabolism: a divergent pathway for plants and bacteria. *Biomolecules* 11, no. 5 (2021): 705.

Misslinger, Matthias, Peter Hortschansky, Axel A. Brakhage, and Hubertus Haas. Fungal iron homeostasis with a focus on Aspergillus fumigatus. *Biochimica et Biophysica Acta (BBA)-Molecular Cell Research* 1868, no. 1 (2021): 118885.

Moll, Henry, Anna Johnsson, Mathias Schäfer, Karsten Pedersen, Herbert Budzikiewicz, and Gert Bernhard. Curium (III) complexation with pyoverdins secreted by a groundwater strain of Pseudomonas fluorescens. *Biometals* 21 (2008): 219–228.

Molnár, Zoltán, Wogene Solomon, Lamnganbi Mutum, and Tibor Janda. Understanding the mechanisms of Fe deficiency in the rhizosphere to promote plant resilience. *Plants* 12, no. 10 (2023): 1945.

Nagoba, Basavraj, and Deepak V. Vedpathak. Medical applications of siderophores–a review. *European Journal of General Medicine* 8, no. 3 (2011): 230–233.

Nair, Anupa, Asha A. Juwarkar, and Sanjeev K. Singh. Production and characterization of siderophores and its application in arsenic removal from contaminated soil. *Water, Air, and Soil Pollution* 180 (2007): 199–212.

Nanda, Manisha, Vinod Kumar, and D. K. Sharma. Multimetal tolerance mechanisms in bacteria: the resistance strategies acquired by bacteria that can be exploited to 'clean-up' heavy metal contaminants from water. *Aquatic Toxicology* 212 (2019): 1–10.

Nijland, Mark, Jose M. Martínez Felices, Dirk J. Slotboom, and Chancievan Thangaratnarajah. Membrane transport of cobalamin. In: Thimann, Kenneth Vivian (Eds.) *Vitamins and Hormones*, 119, 121–148. Academic Press, 2022.

Nookongbut, Phitthaya, Duangporn Kantachote, Mallavarapu Megharaj, and Ravi Naidu. Reduction in arsenic toxicity and uptake in rice (*Oryza sativa* L.) by As-resistant purple nonsulfur bacteria. *Environmental Science and Pollution Research* 25 (2018): 36530–36544.

Paauw, Armand, Maurine A. Leverstein-van Hall, Kok P. M. van Kessel, Jan Verhoef, and Ad C. Fluit. Yersiniabactin reduces the respiratory oxidative stress response of innate immune cells. *PLoS One* 4, no. 12 (2009): e8240.

Pahari, Avishek, Alisha Pradhan, Suraja Kumar Nayak, and Bhusan B. Mishra. Bacterial siderophore as a plant growth promoter. In: Patra, Jayanta Kumar, Chethala N. Vishnuprasad, and Gitishree Das (Eds.) *Microbial Biotechnology: Volume 1. Applications in Agriculture and Environment*, 163–180. Springer, 2017.

Pandey, Akhilesh Kumar, Arti Gautam, and Rama Shanker Dubey. Transport and detoxification of metalloids in plants in relation to plant-metalloid tolerance. *Plant Gene* 17 (2019): 100171.

Parveen, Nishat, Roberto Berni, Sreeja Sudhakaran, Javaid A. Bhat, Suhas Shinde, Naleeni Ramawat, Vijay P. Singh, Shivendra Sahi, Rupesh Deshmukh, Devendra K. Chauhan, and Durgesh Kumar Tripathi. Metalloids in plants: a systematic discussion beyond description. *Annals of Applied Biology* 180, no. 1 (2022): 7–25.

Parveen, Rashitha S., and Latha D. Characterization of siderophore producing *Pseudomonas Sp* for its plant growth promoting properties. *Bioscience Biotechnology Research Communications* 12, no. 4 (2019): 1031–1037.

Patel, Twisha M., and Farida P. Minocheherhomji. Isolation and characterization of several siderophore producing bacteria from cotton plant. *International Journal of Innovative Research in Science, Engineering and Technology* 7 (2018): 159–166.

Pattnaik, Swati, Balaram Mohapatra, and Abhishek Gupta. Plant growth-promoting microbe mediated uptake of essential nutrients (Fe, P, K) for crop stress management: microbe–soil–plant continuum. *Frontiers in Agronomy* 3 (2021): 689972.

Pecoraro, Lorenzo, Xiao Wang, Dawood Shah, Xiaoxuan Song, Vishal Kumar, Abdul Shakoor, Keshawanand Tripathi, Pramod W. Ramteke, and Rupa Rani. Biosynthesis pathways, transport mechanisms and biotechnological applications of fungal siderophores. *Journal of Fungi* 8, no. 1 (2021): 21.

Perez-Cuesta, Uxue, Xabier Guruceaga, Saioa Cendon-Sanchez, Eduardo Pelegri-Martinez, Fernando L. Hernando, Andoni Ramirez-Garcia, Ana Abad-Diaz-de-Cerio, and Aitor Rementeria. Nitrogen, iron, and zinc acquisition: key nutrients to Aspergillus fumigatus virulence. *Journal of Fungi* 7, no. 7 (2021): 518.

Pérez-Miranda, Sandra, Nathalie Cabirol, Rocío George-Téllez, Luís S. Zamudio-Rivera, and Francisco J. Fernández. O-CAS, a fast and universal method for siderophore detection. *Journal of Microbiological Methods* 70, no. 1 (2007): 127–131.

Podar, Dorina, and Frans J. M. Maathuis. The role of roots and rhizosphere in providing tolerance to toxic metals and metalloids. *Plant, Cell & Environment* 45, no. 3 (2022): 719–736.

Pradhan, Smriti, Abhinandan Choudhury, Sovan Dey, Md Firoj Hossain, Aniruddha Saha, and Dipanwita Saha. Siderophore-producing *Bacillus amyloliquefaciens* BM3 mitigate arsenic contamination and suppress Fusarium wilt in brinjal plants. *Journal of Applied Microbiology* 134, no. 10 (2023): lxad217.

Praveen Kumar, G., N. Kishore, E. Leo Daniel Amalraj, S. K. Mir Hassan Ahmed, Abdul Rasul, and Suseelendra Desai. Evaluation of fluorescent Pseudomonas spp. with single and multiple PGPR traits for plant growth promotion of sorghum in combination with AM fungi. *Plant Growth Regulation* 67 (2012): 133–140.

Priyadarshanee, Monika, and Surajit Das. Biosorption and removal of toxic heavy metals by metal tolerating bacteria for bioremediation of metal contamination: a comprehensive review. *Journal of Environmental Chemical Engineering* 9, no. 1 (2021): 104686.

Puja, Hélène, Gaëtan L. A. Mislin, and Coraline Rigouin. Engineering siderophore biosynthesis and regulation pathways to increase diversity and availability. *Biomolecules* 13, no. 6 (2023): 959.

Rawat, Deepak, Udita Sharma, Pankaj Poria, Arran Finlan, Brenda Parker, Radhey Shyam Sharma, and Vandana Mishra. Iron-dependent mutualism between *Chlorella sorokiniana* and *Ralstonia pickettii* forms the basis for a sustainable bioremediation system. *ISME Communications* 2, no. 1 (2022): 83.

Retamal-Morales, Gerardo, Marika Mehnert, Ringo Schwabe, Dirk Tischler, Claudia Zapata, Renato Chávez, Michael Schlömann, and Gloria Levicán. Detection of arsenic-binding siderophores in arsenic-tolerating Actinobacteria by a modified CAS assay. *Ecotoxicology and Environmental Safety* 157 (2018): 176–181.

Řezanka, Tomáš, Andrea Palyzová, and Karel Sigler. Isolation and identification of siderophores produced by cyanobacteria. *Folia Microbiologica* 63 (2018): 569–579.

Robles-Fernández, Ana, Camila Areias, Daniele Daffonchio, Volker C. Vahrenkamp, and Mónica Sánchez-Román. The role of microorganisms in the nucleation of carbonates, environmental implications and applications. *Minerals* 12, no. 12 (2022): 1562.

Roy Chandan, and Sudhir Kumar, Rakesh Deo Ranjan, Sita Ram Kumhar, and Velu Govindan. (2022). Genomic approaches for improving grain zinc and iron content in wheat. *Frontiers in Genetics*, *13*, 1045955.

Sah, Stuti, and Rajni Singh. Siderophore: structural and functional characterisation-a comprehensive review. *Agriculture* 61, no. 3 (2015): 97.

Saha, Maumita, Subhasis Sarkar, Biplab Sarkar, Bipin Kumar Sharma, Surajit Bhattacharjee, and Prosun Tribedi. Microbial siderophores and their potential applications: a review. *Environmental Science and Pollution Research* 23 (2016): 3984–3999.

Saha, Ratul, Nabaneeta Saha, Robert S. Donofrio, and Lorelle L. Bestervelt. Microbial siderophores: a mini review. *Journal of Basic Microbiology* 53, no. 4 (2013): 303–317.

Saharan, Baljeet Singh, and Vibha Nehra. Plant growth promoting rhizobacteria: a critical review. *Life Science and Medical Research* 21, no. 1 (2011): 30.

Samreen, Tayyaba, Zahir Ahmad Zahir, Muhammad Naveed, and Muhammad Asghar. Boron tolerant phosphorus solubilizing *Bacillus* spp. MN-54 improved canola growth in alkaline calcareous soils. *International Journal of Agriculture and Biology* 21, no. 3 (2019): 538–546.

Sanchez, Nicolas, Eunice A. Brown, Yngvar Olsen, Olav Vadstein, Jose L. Iriarte, Humberto E. Gonzalez, and Murat Van Ardelan. Effect of siderophore on iron availability in a diatom and a dinoflagellate species: contrasting response in associated bacteria. *Frontiers in Marine Science* 5 (2018): 118.

Schalk, Isabelle J., Coraline Rigouin, and Julien Godet. An overview of siderophore biosynthesis among fluorescent Pseudomonads and new insights into their complex cellular organization. *Environmental Microbiology* 22, no. 4 (2020): 1447–1466.

Schalk, Isabelle J., Mélissa Hannauer, and Armelle Braud. New roles for bacterial siderophores in metal transport and tolerance. *Environmental Microbiology* 13, no. 11 (2011): 2844–2854.

Schmiederer, Timo, Saskia Rausch, Marianne Valdebenito, Yogita Mantri, Eva Mösker, Todor Baramov, Kamil Stelmaszyk, Peter Schmieder, Diane Butz, Silke I. Müller, Kathrin Schneider, Mu-Hyun Baik, Klaus Hantke, and Roderich D. Süssmuth. The *E. coli* siderophores enterobactin and salmochelin form six-coordinate silicon complexes at physiological pH. *Angewandte Chemie International Edition* 18, no. 50 (2011): 4230–4233.

Seregin, Ilya V., and Anna D. Kozhevnikova Nicotianamine: a key player in metal homeostasis and hyperaccumulation in plants. *International Journal of Molecular Sciences* 24, no. 13 (2023): 10822.

Sharma, Shanti S., Vijay Kumar, and Karl-Josef Dietz. Emerging trends in metalloid-dependent signaling in plants. *Trends in Plant Science* 26, no. 5 (2021): 452–471.

Shen, Naikun, Shuyan Li, ShiYong Li, Hongyan Zhang, and Mingguo Jiang. The siderophore-producing bacterium, Bacillus siamensis Gxun-6, has an antifungal activity against Fusarium oxysporum and promotes the growth of banana. *Egyptian Journal of Biological Pest Control* 32, no. 1 (2022): 34.

Silambarasan, Sivagnanam, Peter Logeswari, Alexander Valentine, Pablo Cornejo, and Velu Rajesh Kannan. Pseudomonas citronellolis strain SLP6 enhances the phytoremediation efficiency of Helianthus annuus in copper contaminated soils under salinity stress. *Plant and Soil* 457 (2020): 241–253.

Singh, Anumeha, Manish Singh Kaushik, Meenakshi Srivastava, Devendra Nath Tiwari, and Arun Kumar Mishra. Siderophore mediated attenuation of cadmium toxicity by paddy field cyanobacterium *Anabaena oryzae*. *Algal Research* 16 (2016): 63–68.

Singh, Pratika, Azmi Khan, and Amrita Srivastava. Biological means of arsenic minimization with special reference to siderophore. In: Kumar, Nitish (Eds.) *Arsenic Toxicity: Challenges and Solutions*, 253–278. Springer, 2021.

Sinha, Alok K., Bhaskar Parli Venkateswaran, Sarat C. Tripathy, Amit Sarkar, and Sabu Prabhakaran. Effects of growth conditions on siderophore producing bacteria and siderophore production from Indian Ocean sector of Southern Ocean. *Journal of Basic Microbiology* 59, no. 4 (2019): 412–424.

Soares, Eduardo V. Perspective on the biotechnological production of bacterial siderophores and their use. *Applied Microbiology and Biotechnology* 106, no. 11 (2022): 3985–4004.

Soto, Javiera, Javier Ortiz, Hector Herrera, Alejandra Fuentes, Leonardo Almonacid, Trevor C. Charles, and César Arriagada. Enhanced arsenic tolerance in *Triticum aestivum* inoculated with arsenic-resistant and plant growth promoter microorganisms from a heavy metal-polluted soil. *Microorganisms* 7, no. 9 (2019): 348.

Sreedevi, P. R., K. Suresh, and Guangming Jiang. Bacterial bioremediation of heavy metals in wastewater: a review of processes and applications. *Journal of Water Process Engineering* 48 (2022): 102884.

Sun, Hong-Jie, Bala Rathinasabapathi, Bing Wu, Jun Luo, Li-Ping Pu, and Lena Q. Ma. Arsenic and selenium toxicity and their interactive effects in humans. *Environment International* 69 (2014): 148–158.

Suzuki, Motofumi, Tomoko Nozoye, Seiji Nagasaka, Hiromi Nakanishi, Naoko K. Nishizawa, and Satoshi Mori. The detection of endogenous 2′-deoxymugineic acid in olives (*Olea europaea* L.) indicates the biosynthesis of mugineic acid family phytosiderophores in non-graminaceous plants. *Soil Science and Plant Nutrition* 62, no. 5–6 (2016): 481–488.

Swayambhu, Girish, Michael Bruno, Andrew M. Gulick, and Blaine A. Pfeifer. Siderophore natural products as pharmaceutical agents. *Current Opinion in Biotechnology* 69 (2021): 242–251.

Takemoto, Tsunematsu, Kyosuke Nomoto, Shinji Fushiya, Reiko Ouchi, Genjiro Kusano, Hiroshi Hikino, Sei-Ichi Takagi, Yoshiki Matsuura, and Masao Kakudo. Structure of mugineic acid, a new amino acid possessing an iron-chelating activity from roots washings of water-cultured Hordeum vulgare L. *Proceedings of the Japan Academy, Series B* 54, no. 8 (1978): 469–473.

Tang, Jieyu, Yingchen Ju, Jingwei Zhou, Junsong Guo, Qiong Gu, Jun Xu, and Huihao Zhou. Structural and biochemical characterization of SbnC as a representative type B siderophore synthetase. *ACS Chemical Biology* 15, no. 10 (2020): 2731–2740.

Tang, Kai, Nianzhi Jiao, Keshao Liu, Yao Zhang, and Shuhui Li. Distribution and functions of TonB-dependent transporters in marine bacteria and environments: implications for dissolved organic matter utilization. *PLoS One* 7, no. 7 (2012): e41204.

Timofeeva, Anna M., Maria R. Galyamova, and Sergey E. Sedykh. Bacterial siderophores: classification, biosynthesis, perspectives of use in agriculture. *Plants* 11, no. 22 (2022): 3065.

Tripathi, Keshawanand, Narendra Kumar, Meenakshi Singh, and Ravi Kant Singh. Fungal siderophore: biosynthesis, transport, regulation, and potential applications. In: Sushil Kumar Sharma, Udai B.Singh, Pramod Kumar Sahu, Harsh Vardhan Singh, Pawan Kumar Sharma (Eds.), *Rhizosphere Microbes: Soil and Plant Functions*, 387–408. Springer, 2020.

Ustiatik, Reni, Yulia Nuraini, and Eko Handayanto. Siderophore production of the Hg-resistant endophytic bacteria isolated from local grass in the Hg-contaminated soil. *Journal of Ecological Engineering* 22, no. 5 (2021): 129–138.

Verma, V. C., S. K. Singh, and Satya Prakash. Bio-control and plant growth promotion potential of siderophore producing endophytic Streptomyces from Azadirachta indica A. Juss. *Journal of Basic Microbiology* 51, no. 5 (2011): 550–556.

Weissman, Kira J. The structural biology of biosynthetic megaenzymes. *Nature Chemical Biology* 11, no. 9 (2015): 660–670.

Wichard, Thomas, Jean-Philippe Bellenger, François M. M. Morel, and Anne M. L. Kraepiel. Role of the siderophore azotobactin in the bacterial acquisition of nitrogenase metal cofactors. *Environmental Science & Technology* 43, no. 19 (2009): 7218–7224.

Yu, Xianmei, Chengxiang Ai, Li Xin, and Guangfang Zhou. The siderophore-producing bacterium, Bacillus subtilis CAS15, has a biocontrol effect on Fusarium wilt and promotes the growth of pepper. *European Journal of Soil Biology* 47, no. 2 (2011): 138–145.

Zawadzka, Anna M., Ronald L. Crawford, and Andrzej J. Paszczynski. Pyridine-2,6-bis (thiocarboxylic acid) produced by Pseudomonas stutzeri KC reduces chromium (VI) and precipitates mercury, cadmium, lead and arsenic. *Biometals* 20 (2007): 145–158.

Zawadzka, Anna M., Ronald L. Crawford, and Andrzej J. Paszczynski. Pyridine-2,6-bis (thiocarboxylic acid) produced by Pseudomonas stutzeri KC reduces and precipitates selenium and tellurium oxyanions. *Applied and Environmental Microbiology* 72, no. 5 (2006): 3119–3129.

Zhang, Tong, Gregory V. Lowry, Natalie L. Capiro, Jianmin Chen, Wei Chen, Yongsheng Chen, Dionysios D. Dionysiou, Daniel W. Elliott, Subhasis Ghoshal, Thilo Hofmann, Heileen Hsu-Kim, Joseph Hughes, Chuanjia Jiang, Guibin Jiang, Chuanyong Jing, Michael Kavanaugh, Qilin Li, Sijin Liu, Jie Ma, Bingcai Pan, Tanapon Phenrat, Xiaolei Qu, Xie Quan, Navid Saleh, Peter J. Vikesland, Qiuquan Wang, Paul Westerhoff, Michael S. Wong, Tian Xia, Baoshan Xing, Bing Yan, Lunliang Zhang, Dongmei Zhou, and Pedro J. J. Alvarez. In situ remediation of subsurface contamination: opportunities and challenges for nanotechnology and advanced materials. *Environmental Science: Nano* 6, no. 5 (2019): 1283–1302.

18 Perspective of Biotechnological Mechanisms of Metalloid Detoxification in Plants

Gagan Jain, Gurvarinder Kaur, Rashmi Mishra, Aarti, Babita Dhiman, Palampreet Singh, and Geetika Sirhindi

18.1 INTRODUCTION

Metalloids are also known as semimetals; these elements exhibit physical and chemical properties between non-metals and metals (Borges, 2023). This group consists of boron (B), arsenic (As), antimony (Sb), silicon (Si), selenium (Se), germanium (Ge), and tellurium (Te). Si, As, and Sb are abundant in Earth's crust and are widely used in modern day industrial operations such as electronics, semiconductors, timber, and the construction industry. These elements are ubiquitous environmental contaminants from natural sources, but enhanced use of these elements in the industry has caused a soil contamination problem around the world; metalloids can be found in high concentrations in soil in certain areas, which leads to adverse effects on plants, animals, and human health (Parveen et al., 2022; Deng et al., 2021).

From a plant physiological perspective, some metalloids can induce a positive response; for example, B is an important growth regulator in plants and is essential for cell wall formation, nutrient uptake, and stress tolerance. Its deficiency can cause a significant reduction in root growth, leaf expansion, and reproductivity (Pereira et al., 2021). Similarly, metalloid-like Si is important for maintaining cell wall integrity and expansion and controlling transpiration by reducing stomatal conductance, and Se plays a crucial role in antioxidant defense mechanisms in plants (Verma et al., 2021; Sharma et al., 2018). On the other hand, some metalloids can be quite detrimental for the physiological functioning of plants, such as Ge, As, Sb, and Te, which are primarily known for their toxic effects even at very low concentrations (Angulo-Bejarano et al., 2021). Accumulation of these toxic metalloids in plants can be caused by contaminated soil, polluted water, agrichemicals, insecticides, or herbicides. Even though plants have selective transporters for nutrient uptake, certain metalloids are chemically similar to essential nutrients and can easily enter the plant system and accumulate in various plant tissue. This accumulation of toxic metalloids can disrupt plant growth and development on cellular and physiological levels; therefore, it is crucial for plants to detoxify these metalloids. This book chapter lays particular emphasis on biotechnological advancements in strengthening the strategies for metalloid detoxification in plants.

18.2 NECESSITY FOR DETOXIFICATION OF METALLOIDS IN PLANTS

Metalloids can enter plant roots through passive diffusion or *via* active uptake facilitated by transporters. Passive uptake occurs when metalloids move along a concentration gradient, while active uptake involves transporter proteins that facilitate specific uptake mechanisms (Angulo-Bejarano et al., 2021). A higher concentration of metalloids in soil leads to a quicker uptake in plants.

DOI: 10.1201/9781032673592-18

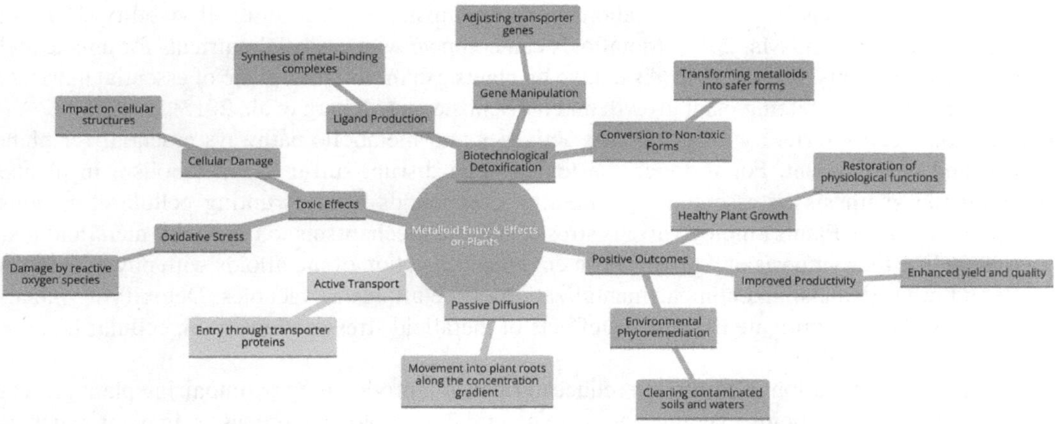

FIGURE 18.1 Entry of metalloids in plants and its various impacts.

This is also affected by other factors like fungal or microbial population in soil, soil type and structure, and the genetic profile of plants. Metalloid speciation and their chemical form can also influence their mobility and bioavailability to plants (Kumpiene et al., 2006). A diagrammatic representation of metalloid entry into plant systems and its interaction is explained in Figure 18.1.

In response to soil contaminants, plants have developed a specialized and diverse transporter system in their root systems. Various transporter proteins, such as aquaporins and phosphate transporters, play crucial roles in metalloid uptake. These transporters are selective and can discriminate between different metalloid species based on their physicochemical properties (Yamaji and Ma, 2021).

Properties of metalloids, such as charge, size, and chemical form, influence metalloid uptake by plants. For example, arsenate (As(V)) and selenate (Se(VI)) are chemically similar to phosphate, allowing them to be taken up by phosphate transporters (Angulo-Bejarano et al., 2021; Sharma et al., 2023). These include influx channels of the nodulin 26-like intrinsic protein (NIP) subfamily (Wallace et al., 2006), C-type ATP-binding cassette (ABC) transporters (Do et al., 2021), solute carrier 4 (SLC4) anion exchanger (Saier et al., 2014), and H+ antiporter (Ye at al., 2013). These transporter channels are the main entry gate for metalloid uptake in plants. For example, Si is mainly uptaken by Si-transporters present in the root system (Wallace et al., 2006). Some metalloids, like As can also be uptaken through leaves, which is also called 'Atmospheric Deposition'. This is more common in industrial areas (Saha et al., 2021).

Once absorbed by roots, metalloids can be transported within the plant through the xylem and phloem. Xylem tissues move metalloids from roots to shoots, while phloem tissues redistribute them to different plant parts (Zhao et al., 2022). Efficiency of metalloid translocation within plant tissues depends on factors such as the plant species involved, the chemical form of the metalloid, and its mobility within the plant. Some metalloids, like As, are more readily translocated within the plant compared to others (Allevato et al., 2019).

Metalloids can be localized within specific organelles or compartments within plant cells, such as vacuoles, cell walls, or chloroplasts. Subcellular localization influences the toxicity and bioavailability of metalloids within plant tissues. Plants try to sequester metalloids in specific tissues or cellular compartments to reduce their toxicity. Metalloid exposure induces various morphological and biochemical changes in plants, including alterations in root architecture, leaf morphology, and chlorophyll content. Plants may also activate antioxidant systems and stress-responsive genes to mitigate metalloid-induced stress (Angulo-Bejarano et al., 2021).

On a cellular and physiological level, it is important for plants to detoxify the metalloids, as they can severely disrupt cellular homeostasis by interfering with ionic balance and essential biochemical reactions within plant cells. High concentrations of metalloids can lead to oxidative stress, membrane damage, and alterations in cellular metabolism, compromising plant health and viability (Fryzova et al., 2018; Clemens and Ma, 2016). Metalloids can compete with essential nutrients for uptake and transport within plants. For example, As uptake by plants can inhibit the uptake of essential nutrients such as phosphorus, affecting plant growth and nutrient status (Meharg et al., 2012).

Metalloids can interfere with enzymatic activities and metabolic pathways essential for plant growth and development. For instance, Se toxicity can disrupt sulfur (S) metabolism in plants, affecting the synthesis of essential S-containing compounds and disrupting cellular functions (Terry et al., 2000). Plants employ various stress response mechanisms to cope with metalloid toxicity, including the synthesis of detoxification enzymes, chelation of metalloids with phytochelatins (PCs) and other ligands, and compartmentalization of metalloids in vacuoles. Detoxifying metalloids allows plants to mitigate the adverse effects of metalloid stress and maintain cellular homeostasis (Chen et al., 2015).

Metalloid accumulation in plants can reduce agricultural productivity by impairing plant growth and development. Metalloids may interfere with essential physiological processes in plants, such as photosynthesis, nutrient uptake, and enzyme activity, leading to stunted growth, reduced yields, and crop failure. Detoxifying metalloids in plants can help maintain crop productivity and ensure food security (Deng et al., 2021; Parveen et al., 2022).

Metalloids can induce genotoxic effects in plant cells by causing DNA strand breaks, DNA adduct formation, and chromosomal aberrations. This genotoxicity can lead to mutations, cell death, and impaired reproductive capacity in plants. As, in particular, has been shown to induce DNA damage and chromosomal abnormalities in plant cells, leading to genetic instability and reduced plant fitness (Kocadal et al., 2020). Metalloids are capable to modulating gene expression patterns in plant cells, leading to changes in the expression of stress-responsive genes, detoxification enzymes, and signaling pathways involved in stress perception and adaptation. Se, for instance, can alter the expression of genes encoding antioxidant enzymes, heat shock proteins, and stress-responsive transcription factors, enabling plants to cope with Se-induced stress.

From an agriculture perspective, detoxifying metalloids in plants is also crucial for several reasons, primarily related to human health, environmental sustainability, and agricultural productivity. Metalloids such as As, cadmium, and Se, when accumulated in edible parts of plants, can pose serious health risks to humans upon consumption. Chronic exposure to elevated levels of these metalloids through contaminated food can lead to various health problems, including cancer, neurological disorders, and developmental abnormalities (Meharg et al., 2012). Metalloid contamination in plants can contribute to environmental pollution, affecting soil and water quality. Runoff from agricultural fields containing metalloid-contaminated plants can contaminate water bodies, leading to ecosystem disruption and harm to aquatic organisms. Phytoremediation, which involves using plants to remove contaminants from soil and water, can help mitigate environmental contamination by detoxifying metalloids (Chen et al., 2015).

18.3 MANAGEMENT OF THE UPTAKE SYSTEM AND RENDERING TOXIC METALLOIDS NONTOXIC AND FORMATION OF VOLATILE METALLOIDS

A highly toxic metalloid is As, which is classified as a group I carcinogen for humans by the International Agency for Research on Cancer (IARC) (Prakash and Verma, 2021). Arsenate (AsV) and Arsenite (AsIII) are inorganic As forms, which are common in the environment and more toxic than their organic forms (Mosa et al., 2012). Arsenate is a phosphate analog and is taken up by the phosphate transporters (Zhao et al., 2009). Arsenite is transported by aquaporins in plants (Mosa et al., 2012). Genetically modified plants for As include enhancing plant uptake for phytoextraction, improving the plants tolerance to As contamination, decreasing plant uptake, and increasing methylation for enhanced food safety (Zhao et al., 2022). Other metalloids such as B, antimonite (SbIII), Si, and antimonite (SbIII) have been reported to be transported through aquaporin channel proteins (Kumar et al., 2014).

Many research groups have focused their efforts on developing phytoremediation strategies for Se, Sb, and As using biotechnological approaches applying for the conversion of these metalloids to less toxic and volatile form.

Se is an essential micronutrient for many organisms but leads to toxicity when taken in excessive amounts and is a universal environmental pollutant (Banerjee et al., 2022). It occurs naturally in soil and is chemically similar to S. Plants uptake the inorganic and organic forms of Se via S transporters and metabolize them to volatile forms through S assimilation pathways to nontoxic dimethylselenide (DMSe) gas (Jiang et al., 2021). A constitutive overexpression of *A. thaliana* ATP sulfurylase (APS), converting selenate to selenite, in *Brassica juncea* showed enhanced reduction of selenate to organic Se forms in the APS overexpressed plants, whereas wild-type (WT) plants accumulated mainly selenate. The APS transgenic plants express high tolerance to selenate as compared to WT plants. Selenocysteine lyase (Scly) gene showed by Arabidopsis plants, which enhance shoot Se concentrations (up to 1.5-fold) compared to the WT (Mosa et al., 2016). The overexpression of the selenocysteine methyltransferase (SMT) gene from the Se hyperaccumulator *Astragalus bisulcatusin A. thaliana* and *B. juncea* improved the tolerance of transgenic plants to Se and increased Se accumulation in shoots (Usmani et al., 2019). The genetic engineered plants also increased Se volatilization rates (Zayed et al., 2020). Small RNA and degradome sequencing analyses are used to identify many miRNAs induced by Se treatment in the Se hyperaccumulator *Astragalus chyrsochlorus* plant callus (Cakir et al., 2016).

Sb is metalloid, which is present in two different chemical forms: non-metallic form and metallic form (He et al., 2019). Due to human activities, Sb emissions into the living environment have increased. Sb is emitted in the form of Sb trioxide as a result of the smelting of Sb-containing ores or coal burning (Nishad and Bhaskarapillai, 2021). In environment Sb is present mainly as Sb(III) and Sb(V) (Zhang et al., 2021). Toxicity of Sb is not well known, but Sb(V) species are usually less toxic than Sb(III) species, which is comparable in its biochemical behavior with As. It shows that algae and plants with high ability to accumulate As are also able to accumulate Sb. Microorganism *Chlorella vulgaris* that express better growth parameters in medium with potassium tartrate in comparison with Sb-poor medium. Its ability to accumulate Sb is 12 mg Sb to 1 g dry matter. It is observed that toxic Sb(III) in living cells is converted to less toxic Sb(V). Sb(V) attached to low molecular-mass proteins and accumulates in the vacuoles of the plant cells (Pandey et al., 2019).

As detoxification is its chelation as AsIII with glutathione (GSH) and PCs (Abbas et al., 2018). The synthesis of metalloid-binding peptides like PCs and GSH leads to detoxification of a specific ion. Plants treated with As increase the synthesis of GSH and PCs, and PC synthesis has been seen in non-accumulator, non-tolerant, and as well as in hyperaccumulator and hypertolerant plants (Shri et al., 2020). The combination of As with PCs and GSH has been experimentally observed in various plants, such as *Rauvolfia serpentina, Holcus Lanatus, Cicer arietinum, Pteris cretica, Brassica juncea, and Helianthus annuus* (Roychoudhury et al., 2020). Tolerance to As is improved by increased thiol synthesis in genetically engineered plants overexpressing the genes of the thiol biosynthesis pathway. An As hypertolerant *Isatis capadocica,* Brassica (Souri et al., 2021) had PC-based tolerance (>50% As complexed with PCs). These findings resulted that thiols, especially PCs, play an important role in As detoxification. In the case of As, PCs play a crucial role in both adaptive tolerance and constitutive tolerance (Upadhyaya and Roychoudhury, 2022), PCs do not show the As tolerance phenotype in hyperaccumulator plants (*P. cretica and P. vittata*), hypertolerant (*Silene paradoxa* Souri et al., 2017), *and Holcus lanatus*, and the amount of PCs synthesized is too low in these plants. Other studies observed that non-hyperaccumulator plants like *Ceratophyllumdemersum, Helianthus annuus, Hydrilla verticillata, and Ceratophyllumdemersum* (Mitra et al., 2017) show more on PC-based detoxification mechanisms.

Metalloids Hyperaccumulator and Hypertolerant Plants

Metalloids	Name of the Plants	References
	Hyperaccumulator	
Se	*Astragalus bisulcatus*	Usmani et al. (2019)
	Astragalus chyrsochlorus	Cakir et al. (2016)
As	*Pteris cretica*	Wang et al. (2021)
	Pteris vittata	Yan et al. (2019)
	Hypertolerant	
As	*Isatis capadocica*	Souri et al. (2021)
As	*Silene paradoxa*	Souri et al. (2017)
As	*Holcus lanatus*	Dradrachet al. (2020)

18.4 SCOPE OF BIOTECHNOLOGY IN PROMOTING PHYTOREMEDIATION

Phytoremediation is a successful bioremediation technique, which refers to the use of plants and associated microbes to reduce the toxic effects of contaminants in the environment. Phytoremediation is a cost-effective environmental restoration technique. Phytoremediation has different processes, such as phytoextraction or phytoaccumulation, which uses plants to transfer inorganic pollutants and accumulate them in plant tissues (Banerjee et al., 2011).

Phytodegradation or phytotransformation is mostly suitable for organic pollutants since plants can break down or degrade the contaminants by metabolic processes with enzymatic modification within the plants (Doty et al., 2007; James et al.,2008). In phytoevaporation, metalloids are attached to volatile organic substances and then removed from the plants into the atmosphere through the process of transpiration (Sakakibara et al., 2010). Some studies show that the use of the plant *Pterisvittata* is efficient of metabolizing As into volatile form and helps in decreasing As content in the soil (Zhao et al., 2023).

Various biotechnological tools, including genetic modification of plant, can be used to enhance the phytoremediation capacity. For the detoxification, resistance, and accumulation of metalloids, a transgenic approach has been adopted. Advancement in genetically modified plants has been adopted to change specific properties for resistance, accumulation, or detoxification (Meyers et al., 2010). Formation of transgenic plants for phytoremediation can be an effective approach to utilize conceivable genes involved in metal uptake, reduction translocation, vacuolar sequestration, and volatilization (Kumar et al., 2023).

Se is present naturally in the soil, and it is harmful at high levels. It is chemically identical to S. It is assumed that misplacement of S by the Se in proteins caused its toxicity. It is evident that methylation of secys and semet averts the misuse of these amino acid derivatives in proteins and their transformation to volatile forms to eliminate the Se from the soil. It also helps in the breakdown of Se compounds into nontoxic forms. These all strategies of Se are done by the expression of key enzymes in transgenic plants (Pilon-Smitset and LeDuc, 2009).

Developments in genome editing technology (CRISPR-Cas9) that allow the specific editing of targeted genomes accompanied with the haploid induction technology are assuring prospects in plants for phytoremediation. A proof of concept for gene pyramiding using haploid genetics has been revealed in *Aradiopsis thaliana* by pyramiding 6 genes of the TRF-like family, which is a great achievement in itself (Thondehaalmath et al., 2021).

Many plants genomes hold a key in discovering the genes behind the phytoextraction of metalloids without interfering with plant development. Plant microbe interaction has been observed to be more effective in carrying out phytoremediation than the plant alone itself (Mesa-Marín et al., 2020). A transgenic approach for the decontamination of As from the soil was demonstrated in A.thaliana by inserting two genes, namely, γ-ECS (γ-glutamycysteine synthase) and arsc (AesenareReductase), which are encoded by E.coli and catalyze the GSH-coupled electrochemical reduction of arsenate to the more toxic arsenate. The genetically engineered plant showed a 4–17 times rise in the accumulation of arsenate than the WTs (Dhankher et al., 2002).

PvACR3, an arsenite antiporter in the As-hyperaccumulating plant *Pterisvittate*, was expressed in *A. thaliana* as part of a transgenic strategy to address As hyperaccumulation by genetic engineering. Compared to the control, the transgenic plants expressing PvACR3 via the CAMV promoter demonstrated tolerance to 80 µm arsenite and 1,200 µm arsenate, which is known to be a lethal dosage for A. thaliana WT. When compared to the wild kinds, the GM plants exhibited a 7.5-fold increase in As uptake (Chen et al., 2013). The health of humans and ecosystems is seriously threatened by the release of toxic Sb into the environment. It has recently been demonstrated that the Sb-mineralizing bacterium *Ochrobactrum* anthropic reduces Sb absorption in rice by precipitating Sb_2O_3 in the soil (Long et al., 2023).

18.5 CONCLUSION: FUTURE PERSPECTIVES OF EXPLOITING BIOTECHNOLOGICAL TOOLS FOR METALLOID DETOXIFICATION

The potential and future prospects of using biotechnological tools to improve phytoremediation processes on the biotechnological mechanisms of metalloid detoxification in plants are gaining emerging attention. The development of transgenic plants in conjunction with genetic modification and genome editing technologies, such as CRISPR-Cas9, can greatly enhance the effectiveness of metalloid detoxification in plants. These technologies present viable avenues for genetically modifying plants to enhance their resistance, accumulation, and detoxifying capacities, thereby aiding in environmental restoration and pollution mitigation.

REFERENCES

Abbas, G., Murtaza, B., Bibi, I., Shahid, M., Niazi, N.K., Khan, M.I., Amjad, M., Hussain, M. and Natasha, H., 2018. Arsenic uptake, toxicity, detoxification, and speciation in plants: physiological, biochemical, and molecular aspects. *International Journal of Environmental Research and Public Health*, 15(1), 59.

Allevato, E., Stazi, S.R., Marabottini, R. and D'Annibale, A., 2019. Mechanisms of arsenic assimilation by plants and countermeasures to attenuate its accumulation in crops other than rice. *Ecotoxicology and Environmental Safety*, 185, 109701.

Angulo-Bejarano, P.I., Puente-Rivera, J. and Cruz-Ortega, R., 2021. Metal and metalloid toxicity in plants: An overview on molecular aspects. *Plants*, 10(4), 635.

Banerjee, M., Chakravarty, D., Kalwani, P. and Ballal, A., 2022. Voyage of selenium from environment to life: Beneficial or toxic? *Journal of Biochemical and Molecular Toxicology*, 36(11), e23195.

Banerjee, S., Datta, S., Chattyopadhyay, D. and Sarkar, P., 2011. Arsenic accumulating and transforming bacteria isolated from contaminated soil for potential use in bioremediation. *Journal of Environmental Science and Health, Part A*, 46(14), 1736–1747.

Borges, E.M., 2023. Data visualization using boxplots: Comparison of metalloid, metal, and nonmetal chemical and physical properties. *Journal of Chemical Education*.

Cakir, O., Candar-Cakir, B. and Zhang, B., 2016. Small RNA and degradome sequencing reveals important micro RNA function in A stragaluschrysochlorus response to selenium stimuli. *Plant Biotechnology Journal*, 14(2), 543–556.

Chen, J., Chen, Y., Shi, Z.Q., Su, Y. and Han, F.X., 2015. Phytoremediation to remove metals/metalloids from soils. *Phytoremediation: Management of Environmental Contaminants*, 2, 297–304.

Chen, Y., Xu, W., Shen, H., Yan, H., Xu, W., He, Z. and Ma, M., 2013. Engineering arsenic tolerance and hyperaccumulation in plants for phytoremediation by a PvACR3 transgenic approach. *Environmental Science & Technology*, 47(16), 9355–9362

Clemens, S. and Ma, J.F., 2016. Toxic heavy metal and metalloid accumulation in crop plants and foods. *Annual Review of Plant Biology*, 67, 489–512.

Deng, F., Zeng, F., Chen, G., Feng, X., Riaz, A., Wu, X., Gao, W., Wu, F., Holford, P. and Chen, Z.H., 2021. Metalloid hazards: From plant molecular evolution to mitigation strategies. *Journal of Hazardous Materials*, 409, 124495.

Dhankher, O.P., Li, Y., Rosen, B.P., Shi, J., Salt, D., Senecoff, J.F., Sashti, N.A. and Meagher, R.B., 2002. Engineering tolerance and hyperaccumulation of arsenic in plants by combining arsenate reductase and γ-glutamylcysteinesynthetase expression. *Nature Biotechnology*, 20(11), 1140–1145.

Do, T.H.T., Martinoia, E., Lee, Y. and Hwang, J.-U., 2021. Update on ATP-binding cassette (ABC) transporters: How they meet the needs of plants. *Plant Physiology*, 187(4): 1876–1892. https://doi.org/10.1093/plphys/kiab193

Doty, S.L., James, C.A., Moore, A.L., Vajzovic, A., Singleton, G.L., Ma, C., Khan, Z., Xin, G., Kang, J.W., Park, J.Y. and Meilan, R., 2007. Enhanced phytoremediation of volatile environmental pollutants with transgenic trees. *Proceedings of the National Academy of Sciences*, 104(43), 16816–16821.

Dradrach, A., Karczewska, A. and Szopka, K., 2020. Arsenic uptake by two tolerant grass species: Holcuslanatus and Agrostis capillaris growing in soils contaminated by historical mining. *Plants*, 9(8), 980.

Fryzova, R., Pohanka, M., Martinkova, P., Cihlarova, H., Brtnicky, M., Hladky, J. and Kynicky, J., 2018. Oxidative stress and heavy metals in plants. *Reviews of Environmental Contamination and Toxicology*, 245, 129–156.

He, M., Wang, N., Long, X., Zhang, C., Ma, C., Zhong, Q., Wang, A., Wang, Y., Pervaiz, A. and Shan, J., 2019. Antimony speciation in the environment: Recent advances in understanding the biogeochemical processes and ecological effects. *Journal of Environmental Sciences*, 75, 14–39.

James, C.A., Xin, G., Doty, S.L. and Strand, S.E., 2008. Degradation of low molecular weight volatile organic compounds by plants genetically modified with mammalian cytochrome P450 2E1. *Environmental Science & Technology*, 42(1), 289–293.

Jiang, H., Lin, W., Jiao, H., Liu, J., Chan, L., Liu, X., Wang, R. and Chen, T., 2021. Uptake, transport, and metabolism of selenium and its protective effects against toxic metals in plants: A review. *Metallomics*, 13(7), 040.

Kocadal, K., Alkas, F., Battal, D. and Saygi, S., 2020. Cellular pathologies and genotoxic effects arising secondary to heavy metal exposure: A review. *Human & Experimental Toxicology*, 39(1), 3–13.

Kumar, K., Mosa, K.A., Chhikara, S., Musante, C., White, J.C. and Dhankher, O.P., 2014. Two rice plasma membrane intrinsic proteins, OsPIP2; 4 and OsPIP2; 7, are involved in transport and providing tolerance to boron toxicity. *Planta*, 239, 187–198.

Kumar, K., Shinde, A., Aeron, V., Verma, A. and Arif, N.S., 2023. Genetic engineering of plants for phytoremediation: Advances and challenges. *Journal of Plant Biochemistry and Biotechnology*, 32(1), 12–30.

Kumpiene, J., Ore, S., Renella, G., Mench, M., Lagerkvist, A. and Maurice, C., 2006. Assessment of zerovalent iron for stabilization of chromium, copper, and arsenic in soil. *Environmental Pollution*, 144(1), 62–69.

Long, J., Zhou, D., Wang, J., Huang, B., Luo, Y., Zhang, G., Liu, Z. and Lei, M., 2023. Repeated inoculation of antimony resistant bacterium reduces antimony accumulation in rice plants. *Chemosphere*, 327, 138335.

Meharg, A.A., Zhao, F.J., Meharg, A.A. and Zhao, F.J., 2012. *Arsenic in rice grain.* Springer, Netherlands, pp. 11–30.

Mesa-Marín, J., Pérez-Romero, J.A., Redondo-Gómez, S., Pajuelo, E., Rodríguez-Llorente, I.D. and Mateos-Naranjo, E., 2020. Impact of plant growth promoting bacteria on Salicorniaramosissimaecophysiology and heavy metal phytoremediation capacity in estuarine soils. *Frontiers in Microbiology*, 11, 553018.

Meyers, B., Zaltsman, A., Lacroix, B., Kozlovsky, S.V. and Krichevsky, A., 2010. Nuclear and plastid genetic engineering of plants: Comparison of opportunities and challenges. *Biotechnology Advances*, 28(6), 747–756.

Mitra, A., Chatterjee, S. and Gupta, D.K., 2017. Uptake, transport, and remediation of arsenic by algae and higher plants. In: Gupta, D.K. and Chatterjee, S. (Eds.) *Arsenic contamination in the environment: The issues and solutions*, pp.145–169. Springer.

Mosa, K.A., Kumar, K., Chhikara, S., Mcdermott, J., Liu, Z., Musante, C., White, J.C. and Dhankher, O.P., 2012. Members of rice plasma membrane intrinsic proteins subfamily are involved in arsenite permeability and tolerance in plants. *Transgenic Research*, 21, 1265–1277.

Mosa, K.A., Saadoun, I., Kumar, K., Helmy, M. and Dhankher, O.P., 2016. Potential biotechnological strategies for the cleanup of heavy metals and metalloids. *Frontiers in Plant Science*, 7, 303.

Nishad, P.A. and Bhaskarapillai, A., 2021. Antimony, a pollutant of emerging concern: A review on industrial sources and remediation technologies. *Chemosphere*, 277, 130252.

Pandey, A.K., Gautam, A. and Dubey, R.S., 2019. Transport and detoxification of metalloids in plants in relation to plant-metalloid tolerance. *Plant Gene*, 17, 100171.

Parveen, N., Berni, R., Sudhakaran, S., Bhat, J.A., Shinde, S., Ramawat, N., Singh, V.P., Sahi, S., Deshmukh, R., Chauhan, D.K. and Tripathi, D.K., 2022. Metalloids in plants: A systematic discussion beyond description. *Annals of Applied Biology*, 180(1), 7–25.

Pereira, G.L., Siqueira, J.A., Batista-Silva, W., Cardoso, F.B., Nunes-Nesi, A. and Araújo, W.L., 2021. Boron: More than an essential element for land plants? *Frontiers in Plant Science*, 11, 610307.

Pilon-Smits, E.A. and LeDuc, D.L., 2009. Phytoremediation of selenium using transgenic plants. *Current Opinion in Biotechnology*, 20(2), 207–212.

Prakash, S. and Verma, A.K., 2021. Arsenic: It's toxicity and impact on human health. *International Journal of Biological Innovations, IJBI*, 3(1), 38–47.

Roychoudhury, A., Krishnamoorthi, S. and Paul, R., 2020. Arsenic toxicity and molecular mechanism of arsenic tolerance in different members of Brassicaceae. In: Wani, S.H., Thakur, A. K. and Khan, Y.J. (Eds.) *Brassica improvement: Molecular, genetics and genomic perspectives*, pp. 159–186. Springer Nature.

Saha, I., Hasanuzzaman, M. and Adak, M.K., 2021. Abscisic acid priming regulates arsenite toxicity in two contrasting rice (Oryza sativa L.) genotypes through differential functioning of sub1A quantitative trait loci. *Environmental Pollution*, 287, 117586.

Saier Jr, M.H., Reddy, V.S., Tamang, D.G. and Västermark, Å., 2014. The transporter classification database. *Nucleic Acids Research*, 42(D1), D251–D258.

Sakakibara, M., Watanabe, A., Inoue, M., Sano, S. and Kaise, T., 2010, January. Phytoextraction and phytovolatilization of arsenic from As-contaminated soils by Pterisvittata. *Proceedings of the Annual International Conference on Soils, Sediments, Water and Energy*, 12(1), 26.

Sharma, P., Srivastava, V., Kumar, A.M.I.T. and Misra, A.N., 2018. Mechanisms of metalloid uptake, transport, toxicity, and tolerance in plants. In: Abbas, Z. (Ed.) *Emerging Trends of Plant Physiology for Sustainable Crop Production Apple*, pp. 167–221. Apple Academic Press.

Sharma, Y., Soni, P., Raturi, G., Mandlik, R., Rachappanavar, V.K., Kumar, M., Salvi, P., Tripathi, D.K., Ram, H. and Deshmukh, R., 2023. Regulation of metalloid uptake in plants by transporters and their solute specificity. *Environmental and Experimental Botany*, 206, 105180.

Shri, M., Chakrabarty, D. and Verma, G., 2020. Mechanisms of arsenic hyperaccumulation by plants. In: Hasanuzzaman, M. (Ed.) *Plant Ecophysiology and Adaptation under Climate Change: Mechanisms and Perspectives II: Mechanisms of Adaptation and Stress Amelioration*, pp.767–785. Springer.

Souri, Z., Karimi, N. and Ahmad, P., 2021. The effect of NADPH oxidase inhibitor diphenyleneiodonium (DPI) and glutathione (GSH) on Isatis cappadocica, under Arsenic (As) toxicity. *International Journal of Phytoremediation*, 23(9), 945–957.

Souri, Z., Karimi, N. and Sandalio, L.M., 2017. Arsenic hyperaccumulation strategies: An overview. *Frontiers in Cell and Developmental Biology*, 5, 67.

Terry, N., Zayed, A.M., De Souza, M.P. and Tarun, A.S., 2000. Selenium in higher plants. *Annual Review of Plant Biology*, 51(1), 401–432.

Thondehaalmath, T., Kulaar, D.S., Bondada, R. and Maruthachalam, R., 2021. Understanding and exploiting uniparental genome elimination in plants: Insights from Arabidopsis thaliana. *Journal of Experimental Botany*, 72(13), 4646–4662.

Upadhyaya, G. and Roychoudhury, A., 2022. Arsenic-toxicity and tolerance: Phytochelatin-mediated detoxification and genetic engineering-based remediation. In: Niazi, N.K., Bibi, I. and Aftab, T. (Eds.) *Global Arsenic Hazard: Ecotoxicology and Remediation*, pp. 481–508. Cham: Springer International Publishing.

Usmani, Z., Kumar, A., Ahirwal, J. and Prasad, M.N.V., 2019. Scope for applying transgenic plant technology for remediation and fortification of selenium. In: Prasad, M.N.V. (Ed.) *Transgenic Plant Technology for Remediation of Toxic Metals and Metalloids*, pp. 429–461. Academic Press.

Verma, K.K., Song, X.P., Tian, D.D., Guo, D.J., Chen, Z.L., Zhong, C.S., Nikpay, A., Singh, M., Rajput, V.D., Singh, R.K. and Minkina, T., 2021. Influence of silicon on biocontrol strategies to manage biotic stress for crop protection, performance, and improvement. *Plants*, 10(10), 2163.

Wallace, I.S., Choi, W.G. and Roberts, D.M., 2006. The structure, function and regulation of the nodulin 26-like intrinsic protein family of plant aquaglyceroporins. *Biochimica et Biophysica Acta (BBA)-Biomembranes*, 1758(8), 1165–1175.

Wang, H., Cui, S., Wu, D., Yang, X., Wang, H. and Wang, Z., 2021. Effects of kinetin on arsenic speciation and antioxidative enzymes in fronds of the arsenic hyperaccumulator Pteris cretica var. nervosa and non-hyperaccumulator Pteris ensiformis. *Environmental and Experimental Botany*, 191, 104622.

Yamaji, N. and Ma, J.F., 2021. Metalloid transporters and their regulation in plants. *Plant Physiology*, 187(4), 1929–1939.

Yan, H., Gao, Y., Wu, L., Wang, L., Zhang, T., Dai, C., Xu, W., Feng, L., Ma, M., Zhu, Y.G. and He, Z., 2019. Potential use of the Pteris vittata arsenic hyperaccumulation-regulation network for phytoremediation. *Journal of Hazardous Materials*, 368, 386–396.

Ye, C.Y., Yang, X., Xia, X. and Yin, W., 2013. Comparative analysis of cation/proton antiporter superfamily in plants. *Gene*, 521(2), 245–251.

Zayed, A., Pilon-Smits, E., deSouza, M., Lin, Z.Q. and Terry, N., 2020. Remediation of selenium-polluted soils and waters by phytovolatilization. In: Terry, N. and Banuelos, G.S. (Eds.) *Phytoremediation of Contaminated Soil and Water*, pp. 61–83. CRC Press.

Zhang, Y., Ding, C., Gong, D., Deng, Y., Huang, Y., Zheng, J., Xiong, S., Tang, R., Wang, Y. and Su, L., 2021. A review of the environmental chemical behavior, detection and treatment of antimony. *Environmental Technology & Innovation*, 24, 102026.

Zhao, F.J., Han, Y., Shi, H., Wang, G., Zhou, M. and Chen, Y., 2023. Arsenic in the hyperaccumulator Pteris vittata: A review of benefits, toxicity, and metabolism. *Science of the Total Environment*, 896, 165232. https://doi.org/10.1016/j.scitotenv.2023.165232

Zhao, F.J., Ma, J.F., Meharg, A.A. and McGrath, S.P., 2009. Arsenic uptake and metabolism in plants. *New Phytologist*, 181(4), 777–794.

Zhao, F.J., Tang, Z., Song, J.J., Huang, X.Y. and Wang, P., 2022. Toxic metals and metalloids: Uptake, transport, detoxification, phytoremediation, and crop improvement for safer food. *Molecular Plant*, 15(1), 27–44.

19 Biological Applications of Metalloid Nanoparticles

Isha Madaan, Gurvarinder Kaur, Shruti Kaushik, Anmol Sidhu, Kuljinder Kaur, Shamsher Singh, Akanksha Sharma, Simranpreet Kaur, Manjot Kaur, Jahanvi, Jetina, Nivedita Sharma, Tarandeep Kaur, Renu Bhardwaj, and Geetika Sirhindi

19.1 INTRODUCTION

Metalloid nanoparticles (NPs), often referred to as nanomaterials, occupy a unique and versatile position in the realm of nanotechnology due to their intermediate properties between metals and non-metals. These NPs are typically composed of elements found in the metalloid region of the periodic table, such as silicon (Si), germanium, arsenic, antimony, and tellurium (Te). Their distinctive electronic, optical, and chemical characteristics make them promising candidates for a wide range of applications, with a particular focus on biological systems. One of the key properties of metalloid NPs is their tuneable electronic structure, which arises from the combination of metallic and non-metallic elements. This property facilitates the manipulation of their conductivity and optical behavior, allowing for customization to meet specific application requirements. For instance, Si NPs exhibit semiconductor-like behavior, enabling their use in electronic and optoelectronic devices (Wang et al., 2016). The ability to modulate these properties makes metalloid NPs highly adaptable for various technological advancements.

In the context of biological applications, metalloid NPs have gained considerable attention due to their unique features and the emerging field of nanomedicine. The biocompatibility of certain metalloid NPs, such as Si and germanium, makes them suitable for biomedical applications. These NPs can be engineered to interact with biological systems at the molecular and cellular levels. Additionally, their small size facilitates efficient delivery into biological tissues, enabling targeted therapeutic and diagnostic interventions. Metalloid NPs possess remarkable optical properties that have implications for both imaging and therapeutic applications in biology. Size-dependent optical behavior, known as quantum confinement, results in fluorescence and absorbance properties that can be harnessed for imaging purposes. Si NPs, for instance, exhibit strong photoluminescence, making them valuable in bioimaging techniques (Vidya et al., 2013). This capability allows for real-time visualization of cellular processes and structures with high sensitivity and resolution. Furthermore, the surface chemistry of metalloid NPs plays a crucial role in their interactions with biological entities. The functionalization of NP surfaces can be tailored to enhance biocompatibility, stability, and specific targeting. This surface modification allows for the conjugation of biomolecules, such as proteins, peptides, or nucleic acids, onto the NP surface, enabling precise interactions with biological molecules and cells (Zhu et al., 2010). The versatility in surface engineering contributes to the development of targeted drug delivery systems and diagnostic agents. The potential of metalloid NPs in the field of drug delivery is particularly noteworthy. Their ability to encapsulate and release therapeutic agents in a controlled manner offers advantages in terms of treatment efficacy and reduced side effects. Si NPs, for example, have been explored for drug delivery due to their biodegradability and compatibility with biological systems. These NPs can be designed to carry

DOI: 10.1201/9781032673592-19

therapeutic payloads, such as drugs or nucleic acids, to targeted sites within the body, enhancing the precision and effectiveness of treatments.

In addition to drug delivery, metalloid NPs exhibit antimicrobial properties that are significant in biological applications. The unique surface chemistry and charge distribution of these NPs contribute to their interaction with microorganisms, leading to antibacterial and antifungal effects. This antimicrobial activity has potential applications in the development of coatings for medical devices, wound healing materials, and antimicrobial agents for combating infections. The use of metalloid NPs in biosensing is another area of active research. The unique electronic properties of these NPs can be harnessed for the development of highly sensitive and selective biosensors. Si NPs, for instance, have been employed in the detection of biomolecules and pathogens (Salunke et al., 2016). The functionalization of their surfaces with specific receptors enables the recognition and binding of target molecules, resulting in measurable changes in electrical or optical signals. This capability holds promise for rapid and accurate diagnostic applications in healthcare. This chapter delves into the unique properties and biological applications of a fascinating class of nanomaterials that bridge the characteristics of metals and non-metals. Their tunable electronic structure, optical behavior, and surface chemistry make them highly versatile for various applications, especially in the realm of biotechnology. From drug delivery and imaging to biosensing and antimicrobial applications, metalloid NPs hold immense promise for revolutionizing biological interventions. As research in this field progresses, addressing safety considerations will be paramount to realizing the full potential of metalloid NPs in advancing healthcare and biotechnology.

19.2 METHODS OF SYNTHESIS

19.2.1 Synthesis Approaches

I. **Top-Down Approach**: It uses initial macroscopic structures and begins with larger particles, which are further broken down into the smaller ones (Figure 19.1). It takes time and needs large setup so it is an expensive method (Siwach and Sen, 2008; Muñoz et al., 2007).
II. **Bottom-Up Approach**: Involves assembling of miniature atoms and molecules to bulk particles (Doyle and Glass, 2010; Rangarajan et al., 2014), Figure 19.2. Self-assembly of monomer/polymer molecules, chemical or electrochemical nanostructural precipitation, sol–gel processing, laser pyrolysis, chemical vapor deposition (CVD), plasma or flame spraying synthesis, and bio-assisted synthesis are some of its examples (Nichols et al., 2006).

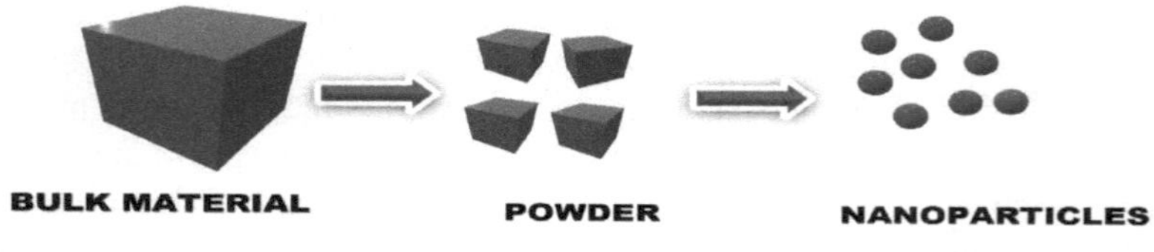

FIGURE 19.1 Top-down approach of synthesizing nanomaterials.

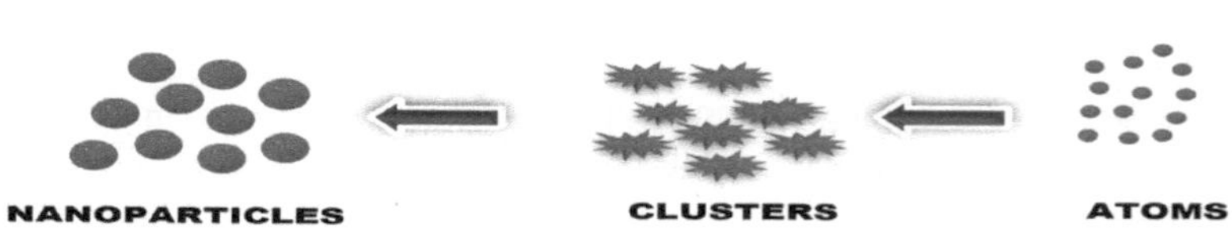

FIGURE 19.2 Bottom-up approach of synthesizing nanomaterials.

TABLE 19.1
Different Methods of Synthesis of Metalloid Nanoparticles

Methods of Synthesis		
A. Physical Methods	**B. Chemical Methods**	**C. Biological Methods**
1. Ball milling	1. Sol–gel method	1. Synthesis using microorganisms
2. Pulse laser ablation	2. Sono chemical synthesis	2. Synthesis using plant extracts
3. Melt mixing	3. Inert gas condensation method	3. Synthesis using algae
4. CVD	4. Hydrothermal synthesis	
5. Laser pyrolysis		
6. Ionized cluster beam deposition		

19.2.2 Methods of Synthesis of Nanoparticles

There are three kinds of methods for the production of NPs (Table 19.1) (Diallo et al., 2017; Liu et al., 2010). For small-scale production, physical methods are suitable. Where the cost of production is a concern, chemical methods are selected. Biological methods have significance due to their eco-friendly nature (Winterer and Hahn, 2003; Krishnappa et al., 2018).

1. **Physical Methods:**
   i. **Ball Milling:** The container contains hard balls made up of steel or carbide, planetary, vibratory, rod, and tumbler forms of mills are used. Nanocrystalline Co, Cr, W, and Ag-Fe, are synthesized using this. 2:1 ratio for balls and material is used. Inert gas is filled in the container and then rotated at high speed. The materials are pressed between the walls of the container and balls. For synthesizing NPs of optimum size, speed, and duration of milling are significant (Konrad et al., 2001).
   ii. **Pulse Laser Ablation:** Inside a vacuum chamber, the target sample is placed. On this sample, the high-pulsed laser beam is focused and plasma is generated, which is formerly transformed into a colloidal solution of NPs, as described in Figure 19.3. To formulate NPs, the second-harmonic group-type laser is frequently used (Park et al., 2010). The type of laser, some pulses, type of solvent, and pulsing time are vital for the final product (Ponon et al., 2015; Patscheider, 2003; Guler et al., 2015; Dhand et al., 2015).
   iii. **Melt Mixing:** Molten streams of metals are mixed using turbulence at a very high speed, leading to the formation of NPs. These NPs arrest in the amorphous glass led to the formation of amorphous solid-metallic glasses after cooling. Ex: A melted stream of Cu-B and a heated stream of Ti form NPs of TiB_2 (Rastogi, 2017; Dikusar et aI., 2009).
   iv. **Chemical Vapor Deposition:** At around 300–1,200°C, a thin film of gaseous reactant is deposited on the substrate. A thin film of product formed on the surface of the substrate as an outcome of a chemical reaction between the heated substrate and combining gas. Growth rate and quality of film are affected by gas pressure and the substrate temperature. This technique is advantageous for manufacturing stiff, uniform, robust, and highly pure NPs (Dikusar et al., 2009).
   v. **Ionized Cluster Beam Deposition:** To obtain high-quality single-crystalline thin films is the main aim of this method, developed in 1985. Having a source of evaporation, a nozzle through which material can expand into the chamber, an electron beam to ionize the clusters, an arrangement to accelerate the clusters, and a substrate on which NP film can be deposited, all housed in a suitable vacuum chamber where the films of nanocrystalline material using an ionized cluster beam can be produced (Karthikeyan and Loganathan, 2013).

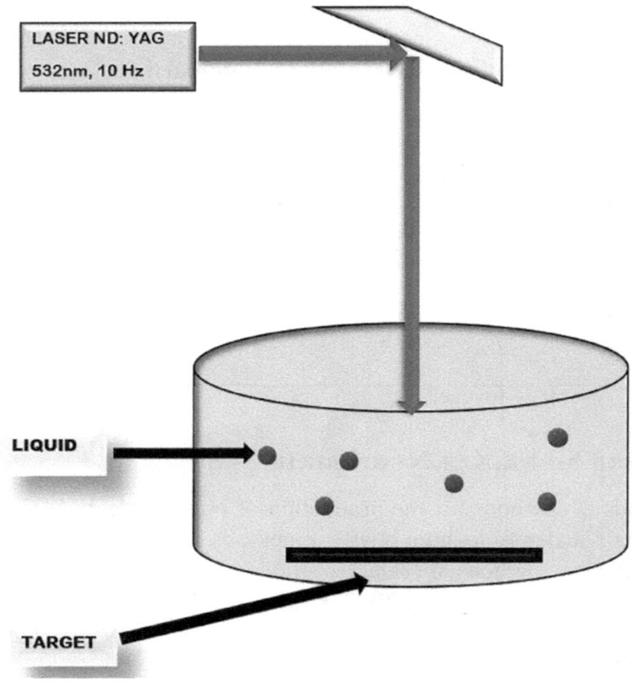

FIGURE 19.3 Pulse laser ablation technique.

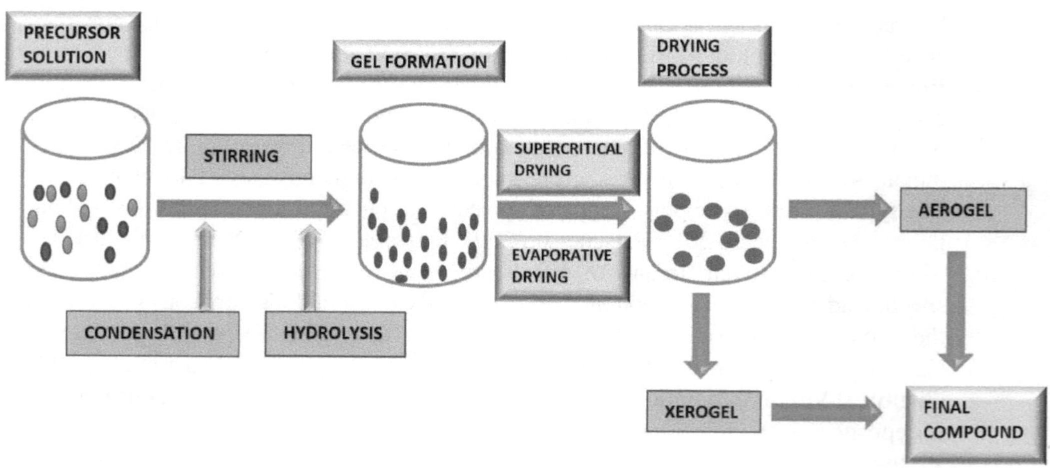

FIGURE 19.4 Steps under the sol–gel method.

2. **Chemical Methods:**
   i. **Sol–Gel Method:** It comprises the condensation, hydrolysis, and thermal decomposition of metal alkoxides or metal precursors in solution, as mentioned in Figure 19.4. The gel is formed upon hydrolysis or condensation with increased viscosity. By changing precursor concentration, temperature, and pH values, the particle size can be monitored (Li et al., 2015).

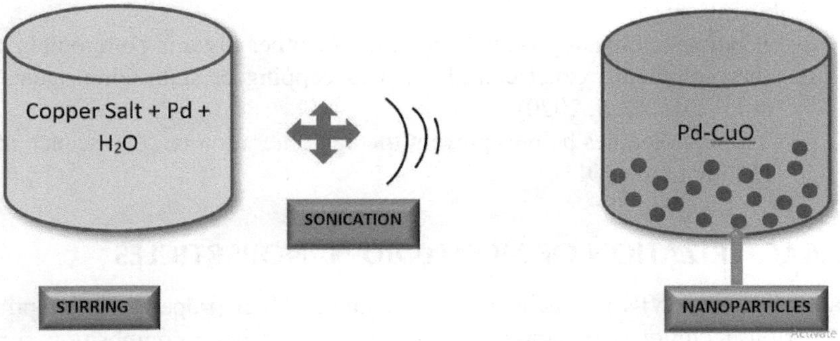

FIGURE 19.5 Sonochemical synthesis of NPs.

ii. **Sonochemical Synthesis:** In the existence of palladium (Pd) and water, Pd-CuO nano-hybrids have been effectively invented by the sonochemical fusion with copper salt, as described in Figure 19.5. Either pure metallic Pd or the Pd salts are used as sources (Sivasankaran et al., 2013; Ziylan Yavas et al., 2015).

iii. **Inert Gas Condensation Method:** This process is broadly used for the creation of metal NPs. NPs are produced via the vanishing of metallic sources in an inactive gas. Here, metals are vaporized at a reasonable rate at an attainable temperature. For example, copper metal NPs are synthesized by subjecting metal inside a chamber containing argon or helium or neon, where the metal is vaporized (Perez Tijerina et al., 2008).

iv. **Hydrothermal Synthesis:** Principally a chemical reaction-based approach that involves a broad temperature range from room temperature to very high temperatures for the synthesis of NPs. There may be unstable NPs generated at higher temperature ranges (Banerjee et al., 2008).

3. **Biological Methods:**

The biosynthesis of metal NPs refers to the process of using biological entities, such as microorganisms, plants, or their extracts, to reduce metal ions into nanoscale metallic particles (Pandey et al., 2012). This eco-friendly and cost-effective (Manjula et al., 2012) method has gained attention due to its potential applications in various fields, including medicine, catalysis, and electronics (Vidya et al., 2013).

The synthesis usually involves the following steps:

1. Selection of Biological Entity:
   - **Microorganisms:** Bacteria, fungi, and algae are commonly used for their ability to reduce metal ions (Yasmin et al., 2014).
   - **Plants:** Plant extracts contain bioactive compounds that can reduce metal ions (Konishi et al., 2007).
2. Preparation of Extracts:
   - If using plants, extracts are prepared by grinding or homogenizing the plant material and extracting it in a suitable solvent like water, ethanol, or acetone (Ghosh et al., 2021).
3. Metal Ion Reduction:
   - The biological entity or its extract is mixed with a solution containing metal ions (usually metal salts like silver nitrate, gold chloride, etc.) (Deplanche and Macaskie, 2008; Murray et al., 2017).
4. Bioreduction Process:
   - The bioactive compounds present in the biological entity or its extract act as reducing agents.
   - They facilitate the reduction of metal ions to form metal NPs (Heidelberg et al., 2002; Harris et al., 2018).

5. Stabilization:
   - Biomolecules like proteins, enzymes, or other organic compounds present in the biological extract can also act as capping or stabilizing agents (Ahmad et al., 2022; Ali, 2020).
   - These molecules help to prevent the agglomeration or coalescence of the NPs (Mittal et al., 2014).

19.3 CHARACTERIZATION OF METALLOID NANOPARTICLES

Characterization of metal NPs is crucial for understanding their properties and optimizing their applications. Various techniques are employed to study the size, shape, composition, surface charge, and other characteristics of metal NPs (SP et al., 2019). Here are some common methods for the characterization of metal NPs:

- **Transmission Electron Microscopy (TEM):** TEM is a powerful technique for imaging NPs at the atomic or molecular scale. It provides information about the size, shape, and distribution of NPs. The high resolution of TEM allows for detailed examination of the NP structure (Takenaka et al., 2006; Pulskamp et al., 2007; Beck-Speier et al., 2005).
- **Scanning Electron Microscopy (SEM):** SEM is used to obtain topographical information about the surface of NPs. It provides 3D images and helps in determining the size and shape of the NPs (Aguirre et al., 2015; Goldstein et al., 2003; Bogner et al., 2007).
- **Dynamic Light Scattering (DLS):** DLS is a technique used to measure the size distribution of NPs in suspension (Dzakpasu and Axelrod, 2004). By analyzing the fluctuations in the intensity of scattered light caused by Brownian motion, DLS provides information on the hydrodynamic diameter of NPs (Satoh et al., 1999; Digman and Gratton, 2011).
- **X-ray Diffraction (XRD):** XRD is used to determine the crystalline structure of metal NPs. It provides information about the arrangement of atoms in the NPs and can be used to identify the crystal phases present (Cantor et al., 1980; Waseda et al., 2011; Strasser et al., 2010).
- **UV-Visible Spectroscopy:** This technique is employed to study the absorption and scattering of light by metal NPs. The presence of surface plasmon resonance (SPR) peaks in the UV-visible spectrum is indicative of the formation of metal NPs (Tomaszewska et al., 2013).
- **Fourier Transform Infrared Spectroscopy (FTIR):** FTIR is used to analyze the surface chemistry of metal NPs. It provides information about functional groups and ligands on the NP surface (Cao, 2004; Gurunathan et al., 2014; Lin et al., 2014; Shang et al., 2007; Perevedentseva et al., 2010).
- **Energy Dispersive X-ray Spectroscopy (EDS or EDX):** EDS is used in conjunction with TEM or SEM to determine the elemental composition of metal NPs. It provides information about the presence and distribution of elements within the NPs (Mishra et al., 2017).
- **Zeta Potential Measurement:** Zeta potential reflects the surface charge of NPs in a colloidal solution. It is important for understanding the stability and dispersibility of NPs (Rasmussen et al., 2020).
- **Thermogravimetric Analysis (TGA):** TGA is used to determine the thermal stability and composition of metal NPs by measuring changes in mass as a function of temperature (Rami et al., 2021).
- **Magnetic Measurements:** For magnetic NPs, techniques such as vibrating sample magnetometry (VSM) can be employed to study their magnetic properties (Liebl et al., 2021).
- **Nuclear Magnetic Resonance (NMR):** NMR spectroscopy can be used to study the interaction between metal NPs and other molecules in solution, providing information on surface functionalization (Agarwal et al., 2018)

By employing a combination of these techniques, a comprehensive understanding of the physical and chemical properties of metal NPs can be attained, allowing for their effective utilization in various applications, including catalysis, sensing, imaging, and drug delivery.

19.4 APPLICATIONS OF NANOPARTICLES

19.4.1 Drug Delivery

Over the past decade, nanotechnology has revolutionized sciences across electronics, medicine, and the environment. Nanomaterials, particularly NPs that are smaller than 100 nm in size, have been instrumental in advancing diagnostics and therapeutics in the field of nanomedicine. Inorganic NPs, composed of metal and/or metalloid compounds, show potential for addressing health challenges, though concerns arise due to the energy intensity and toxic chemical use in their preparation. Biogenic NP production is cost-effective and eco-friendly, yet extraction challenges exist. Metal and metalloid NPs, like gold and silver, have diverse biomedical applications, especially in cancer diagnosis and therapy. Some, like arsenic, played a therapeutic role. Metalloid NPs, bismuth sulfide, have unique properties beneficial for imaging. Various metalloids, such as antimony and Te, also have potential therapeutic applications. Specific focus on selenium (Se) NPs reveals promising anticancer properties under *in vitro* and *in vivo* studies (Roy et al., 2022; Sarraf et al., 2022).

Metalloid NPs, including those made of Si, germanium, arsenic, and Te, have various biological applications owing to their unique properties. Due to their numerous advantages, metalloid NPs are widely used in the field of biomedical science and engineering, such as their ability to bind, absorb, or entrap guest molecules like drugs, probes, or proteins (Sekhon, 2013). These compounds exhibit various biological effects on cells, showcasing their potential as drug delivery agents.

TeNPs have applications in both diagnosis and therapeutic modalities for cancer, serving as quantum dot (QD) agents. Cd and Te are used in conjunction with the human epidermal growth factor receptor 2 (HER2). Monoclonal antibodies have recently been employed as safe and effective nanoprobes for target imaging and selective therapy of gastric cancer (Raun et al., 2012). The advantage of drug delivery using NPs lies in their ability to safely deliver drugs to the tumor site without leakage, effectively penetrate cancer cells, and release drugs inside cells. Spherical boron nitride NPs (BNNPs) with a diameter of 100–200 nm and a petal-like structure are widely used for drug delivery. Due to this structure, they can absorb a large amount of antitumor drugs, showing lower toxicity and easy uptake by cells. BNNPs loaded with doxorubicin (DOX) have been found to be toxic for IAR-6-1 cells. The BNNPs loaded with DOX penetrate neoplastic IAR-6-1 cells by endocytosis, releasing DOX into the cytoplasm and cell nuclei, leading to cell death. Another use of boron nanoparticles (BNs) involves BN-containing azolectin phospholipid-based liposomes that contain poly (maleic anhydride-alt-1-octane) and polyethylene glycol (PEG) on their surface. PEG is conjugated with the tumor-specific targeted ligand folic acid, forming folate-functionalized liposomes (FA-3PCB). Monitoring of cellular uptake is done by fluorescence microscopy.

Arsenic trioxide (ATO) manifests as white arsenic and is industrially produced as a byproduct during the roasting of arsenic-containing ores and purifying the smoke. The transformation of ATO from being the "king of poisons" to a "broad-spectrum anti-cancer drug" is one of the most exciting stories in clinical oncology (Wang et al., 2012). Porous Si has numerous pores that can accumulate drugs, making them highly potential as drug delivery agents. Porous-sized NPs with dimensions of 20–100 nm, produced by electrochemical etching of large Si, have been demonstrated to carry fluorescein isothiocyanate (FITC) insulin and sodium laurate (C12) as a permeation enhancer to cross the intestinal Caco-2 cell monolayers (Forakar et al. 2003). The ability of porous Si to cross the cell monolayer with FITC-insulin loaded is enhanced nearly 10 times compared to liquid formulations with permeation enhancers. The 126 nm porous Si particles with pore sizes of 5–10 nm exhibit near IR emission and are produced by electrochemical etching combined with ultrasonication (Park et al., 2009). These materials have the ability to self-distribute in the environment. *In vivo* imaging of mouse tumors and other organs using SiNPs has also been demonstrated. The NPs injected intravenously mainly accumulate in mononuclear phagocytic system-related organs, degrade *in vivo* within a few days, and are removed from the body. For the drug delivery process, the drug should not leak out before cellular internalization, but once internalized, the drug should be able to reach the required target. Intracellular trafficking is widely studied in drug delivery systems by labeling the particle and the cargo with different dyes. Some drug molecules, such as camptothecin and DOX, also exhibit fluorescent properties, allowing the study of uptake and routing by optical methods on time scales when the drug has not affected cellular integrity. Barui and Cauda (2020) reported cell growth inhibition by folate-functionalized mesoporous silica NPs (MSNs) containing the anticancer drug camptothecin adsorbed inside the pores. About 60% of cells died after 24 hours for PANC-1 cells. The use of silica NPs in drug delivery systems is classified as xerogel (Czarnobaj, 2008) and MSNs, such as MCM-41 and SBA-15 (Wei et al., 2013). Their easy functionalization, biocompatibility, and high porosity make them highly advantageous (Amato, 2010). Xerogels, with amorphous structures manufactured by sol–gel techniques, are loaded with commonly used drugs such as phenytoin (Fidalgo et al., 2009), DOX (Prokopowicz, 2010), cisplatin (Czarnobaj, Lukasiak, 2007), metronidazole (Czarnobaj, Sawicki, 2013), and nifedipine.

Metalloid NPs, such as Si and germanium, offer several advantages in targeted drug delivery. They possess tunable physicochemical properties, enabling precise control over size, surface charge, and functionalization for specific drug loading. Silica NPs have gained prominence in targeted drug delivery due to their unique advantages. They offer a high surface area for drug loading, ensuring efficient encapsulation of therapeutic agents (Wei et al., 2013). Their biocompatibility and low toxicity enhance their suitability for medical applications. Silica NPs also provide a stable and protective environment for drugs, preventing premature degradation. Moreover, their surface can be easily modified for targeted delivery, enabling specific interactions with biological tissues. The controlled release of drugs from silica NPs further contributes to minimizing side effects and optimizing therapeutic outcomes in targeted drug delivery systems (Barui and Cauda, 2020).

Germanium NPs exhibit several advantages in targeted drug delivery. Their unique properties make them promising candidates for this application. Germanium NPs can be synthesized by the gas-phase method. Ge nanocrystals with sizes varying from 4 to 50 nm can exist in amorphous or crystalline form depending on the plasma power (Gresback et al., 2007). In physical methods, GeNPs were constructed with reactive laser ablation, while chemical methods included metathesis reactions involving Zintl salts. Germanium NPs can be precisely engineered for size, shape, and surface characteristics, allowing for efficient drug loading and delivery. Their biocompatibility ensures minimal toxicity, facilitating safe use in biomedical contexts. Moreover, germanium NPs can be functionalized to enhance targeting specificity, directing drugs to specific cells or tissues. Additionally, these NPs may possess imaging capabilities, enabling real-time monitoring of drug delivery. Overall, germanium NPs hold great potential in advancing targeted drug delivery systems for improved therapeutic outcomes. In summary, metalloid NPs play a crucial role in the development of controlled release systems, offering a versatile platform for designing efficient and targeted drug delivery strategies.

19.4.2 Antimicrobial Potential of Metalloid Nanoparticles

Metals are used from ancient times in order to eradicate infectious disease, and the introduction of nanotechnology serves as a boon. Many metal NPs are used, including gold, zinc oxide, titanium oxide, copper oxide, magnesium oxide, etc. But silver NP is most widely used. They generate their effect by causing damage to the microbial cell membrane, providing oxidative stress, and causing damage or injury to their proteins and DNA (Bhushan et al., 2018; Labulo et al., 2016). The application of metal and metalloid NPs can be seen in many fields, such as catalysis, microbial activity, drug delivery, cancer treatment, medical diagnostics, and neurosensors (Salunke et al., 2016b). Metalloid NPs, including those composed of Si, germanium, arsenic, and Te, have emerged as promising candidates in the field of antimicrobial research due to their unique physicochemical properties. This burgeoning area of study has significant implications for combating microbial infections, with metalloid NPs exhibiting antimicrobial potential against a wide range of pathogens.

SiNPs have gained attention for their antimicrobial properties. Research by Johnson et al. (2018) demonstrated the effectiveness of SiNPs against Gram-positive and Gram-negative bacteria. The NPs exhibited notable antibacterial activity, attributed to their interaction with bacterial cell membranes and disruption of essential cellular functions. Furthermore, SiNPs have shown potential in inhibiting biofilm formation, a key factor in the persistence of bacterial infections (Smith et al., 2019).

In the case of germanium NPs, recent studies have highlighted their antimicrobial capabilities. Nguyen et al. (2020) reported the inhibitory effects of germanium NPs on various pathogenic fungi. The NPs disrupted fungal cell membranes, leading to compromised integrity and ultimately, fungal cell death. This antifungal activity positions germanium NPs as a promising tool in the fight against fungal infections.

TeNPs have also demonstrated significant antimicrobial potential. In a study by Martinez et al. (2017), TeNPs exhibited potent antibacterial activity against drug-resistant strains. The NPs were found to disrupt bacterial cell membranes and interfere with cellular processes, showcasing their potential as effective antibacterial agents. Additionally, Te NPs have been explored for their antiviral properties, with promising results against enveloped viruses (Brown et al., 2018).

The antimicrobial efficacy of arsenic NPs has been investigated in the context of both bacteria and fungi. Arsenic NPs have shown inhibitory effects against a spectrum of bacteria, including multidrug-resistant strains (Chen et al., 2016). Moreover, these NPs displayed antifungal activity by disrupting the cell walls of pathogenic fungi, suggesting their potential application in combating fungal infections (Gomez et al., 2019).

The mechanisms underlying the antimicrobial potential of metalloid NPs are multifaceted. Interaction with microbial cell membranes, disruption of cellular structures, and interference with essential cellular processes contribute to their antimicrobial effects. Additionally, the high surface area of NPs enhances their contact with microbial cells, augmenting their efficacy. In the development of metalloid NPs as antimicrobial agents, it is crucial to consider their biocompatibility and potential cytotoxicity. Understanding the balance between antimicrobial efficacy and safety is imperative for their successful translation into practical applications (Lagana et al., 2023). Further research should focus on optimizing NP size, surface modifications, and delivery methods to maximize antimicrobial effects while minimizing adverse effects on host cells.

Hence, metalloid NPs, including Si, germanium, arsenic, and Te, exhibit significant antimicrobial potential against a broad spectrum of pathogens. Their ability to disrupt microbial cell structures and interfere with essential processes positions them as promising candidates in the ongoing battle against microbial infections. As research in this field progresses, metalloid NPs may emerge as valuable additions to the arsenal of antimicrobial agents, offering innovative solutions to combat the challenges posed by drug-resistant microbes.

19.4.3 Metalloid Nanoparticles in Diagnostics

Metalloid NPs play a significant role in diagnostic assays. Cadmium telluride NPs and polymer-coated bismuth sulfide NPs (Bi_2S_3) serve as QDs, finding application in *in vivo* imaging and computed tomography (CT). For years, Si NPs, a metalloid, which are also well-known QD agents, have been widely used in various areas like biosensors, optical materials, biocatalysts, electrochemistry, and immunochemistry. Si dioxide (SiO_2), akin to iron oxide, has been employed as a carrier for targeted delivery of drugs and genes. BNPs serve as carriers for the targeted delivery of drugs and find application in neutron capture therapy (NCT) (Yazdi et al., 2016a; Wade and White, 2018).

Despite being known for their toxicity, metalloids like arsenic can play essential roles in natural processes. Recently, ATO (As_2O_3) NPs are gaining recognition as potential and promising anticancer agents specifically targeted to solid tumors. Arsenic-containing compounds are found in various organisms, including terrestrial mushrooms and marine algae. Understanding the diverse pathways through which arsenic is involved highlights its complex role in various ecosystems (Yazdi et al., 2016b; Dias et al., 2016). Te is a rare, silver-white metalloid in the chalcogen group. It does not play a significant biological role. In biological systems, both Te and its compounds exhibit toxicity. Its chemical properties relate to Se and sulfur. Biomedical applications face concerns due to acute toxicity, affecting glutathione peroxidase and various cells. TeNPs, less toxic than ions, gain interest, particularly in cancer research. CdTe QDs, combining cadmium and Te, show promise in targeted imaging and cancer therapy. TeNPs, used alongside magnetic NPs and QDs, exhibit potential in screening micrometastasis for early cancer diagnosis. Although they have the potential to be toxic, strategies such as coating with biocompatible polymers, such as gelatin, are aimed at increasing safety. Te, recognized for immunomodulation, may aid cancer treatment indirectly by enhancing the immune response. AS101, a Te IV compound, is in phase II clinical trials for cancer treatment. Considering the historical use of Se in medicine and the parallels between Te and Se, Te, especially in NP form, could offer new avenues in cancer treatment and diagnosis protocols (Dias et al., 2016; Yazdi et al., 2016).

Se is recognized as a crucial trace element for human well-being, providing protection against various chronic diseases. Its significance extends to plants and microorganisms, though certain soils may contain elevated levels, posing risks to animals like those grazing on locoweed. Se, a micronutrient ion, offers antioxidant, cancer preventive, and antiviral benefits. Two supplement types exist: salt form (sodium selenite or selenate and Se dioxide) and organoselenium (selenocysteine, selenomethionine). Se yeast, derived from fermenting *Saccharomyces cerevisiae* in a Se-rich medium, is a recognized organic food-form source. Biological effects rely on selenocysteine incorporation into selenoproteins, influencing redox balance and immune responses. Clinical trials on Se as a chemopreventive agent yield controversial results, yet its immunostimulatory effects suggest efficiency in cancer treatment when combined with conventional therapies. Se's therapeutic potential is hindered by toxicity near effective doses. Nanoselenium (Se^0) shows promise as a less toxic alternative, with various allotropic forms. Elemental Se toxicity is lower than selenate (Se^{+2}) or selenite (Se^{+4}) ions, making Se NPs viable for clinical use, offering similar biological properties at lower doses with reduced toxicity. Porous Si (PSi) is also explored for biomedical applications, particularly drug delivery and imaging (Dias et al., 2016; Yazdi et al., 2016; Lumen et al., 2021). Se NPs induce apoptosis in cancer cells; a crucial mechanism is programmed cell death, primarily death, primarily through the generation of reactive oxygen species (ROS). This apoptotic process is pivotal in inhibiting cancer cell proliferation. Combining SeNPs with the anticancer properties of secondary metabolites from medicinal plants, such as ascorbic acid and curcumin, was harnessed by synthesizing SeNPs. These nanocompounds, in conjunction with irinotecan (a medication exploited for treating small cell lung cancer and colon cancer), penetrated cancer cells via the lysosomal pathway, resulting in reduced HCT-8 tumor size and DNA breakage (Alavi et al., 2022).

Bismuth (Bi), a metalloid with atomic number 83, finds application in oncology, specifically using the 213 radioisotopes for alpha radiation treatment in melanoma, breast, prostate, and colorectal cancers. BiNPs, particularly biogenic ones, exhibit antibacterial effects against *Helicobacter pylori*, a major risk factor for stomach cancer. This suggests a potential cancer prevention strategy using metalloid NPs. In diagnostic X-ray sources, BiNPs offer higher dose enhancements compared to gold and platinum NPs. Additionally, BiNPs, when combined with a tumor-targeting antigen (LyP-1 molecule), show promise in CT scans. Safety tests demonstrate well-tolerance in mice, with efficient clearance from the blood and accumulation in tumors within 24 hours (Yazdi et al., 2016).

Inorganic NPs show potential not only in cancer diagnostics but also as effective agents in cancer treatment. The emerging field of bio-based techniques further enhances the potential, offering a sustainable and cost-effective approach for stable metalloid NP synthesis. These biogenic NPs, derived from bacteria, fungi, and plants, show multifunctional roles, from targeted drug delivery to antibacterial and fungicidal activities. As we look ahead, the anticipation is that these advancements will not only elevate diagnostics but also play a pivotal role in refining cancer treatment strategies, ultimately enhancing the duration and quality of patients' lives.

19.4.4 Cancer Treatment

Cancer is a disease in which excessive cell proliferation occurs that spreads from an initial focal point to the other body organs and ultimately causes death (Jin et al., 2020). In the modern world, NPs have transformed many areas of science and technology from the human health sector, agricultural sector, and environmental health to the electronic sector (Yazdi et al., 2016). Despite the fact that knowledge about NPs and their applications in the health sector is at a very initial stage though, NPs have opened a new and encouraging window toward solving unresolved complications such as cancer (Yazdi et al., 2016). Inorganic NPs, which include metal and/or metalloid compounds, can be prepared either by converting from parents to nanomaterials. However, this utilizes various toxic chemicals and is an energy-intensive process. Therefore, nowadays biogenic production of NPs has attracted researchers' attention (Yazdi et al., 2016). Cancer is the leading cause of death and poor quality of life, still there remain inadequacies of the cancer therapies. Among various vital steps toward the cancer treatment major milestone is the early detection of cancer cells and drug application with high specificity to reduce the toxicity (Jin et al., 2020). To do the same nowadays, nanotechnology is being used in cancer treatment. Nanotechnology has shown positive results with its applications in the diagnosis and treatment of cancer, including drug delivery (Hu et al., 2016; Jin et al., 2020), gene therapy, targeted therapy, and molecular imaging (Figure 19.6). Nanotechnology is used in the development of different nanomaterials (Huang et al., 2017a; Jin et al., 2020). For example, QDs, AuNPs that are utilized for diagnosis of cancer at molecular level by the development of biomarkers that help in precise and quick detection of cancerous cells (Jin et al., 2020). Moreover, nanoscale drug delivery systems help in accurate cancerous tissue targeting with minimal spread of toxic chemicals and minimal side effects (Jin et al., 2020). However, much more research is yet to be done in this field before it is readily accepted in the clinic for cancer management (Bharali et al., 2017; Jin et al., 2020).

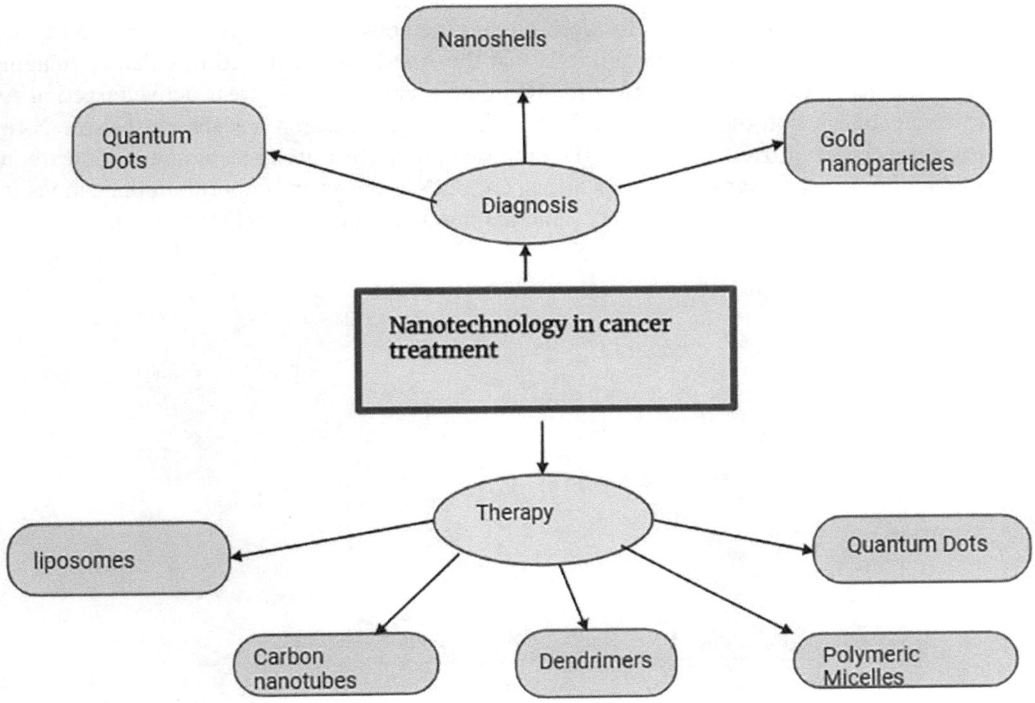

FIGURE 19.6 Different diagnostic and therapeutic tools of nanotechnology used for cancer treatment.

Various diagnostic tools for cancer cells are CT, magnetic resonance imaging (MRI), ultrasound, and positron emission tomography (PET) (Kim et al., 2010; Jin et al., 2020), but all these have limitations of relevant clinical information about different cancer types and the stage (Akhter et al., 2013;; Jin et al., 2020).

Hence, due to their small size, good biocompatibility, and high atomic number, the application of NPs, viz., semiconductors, QDs, iron oxide nanocrystals (Popescu et al., 2015; Jin et al., 2020) in cancer diagnosis and monitoring has enhanced with their usage for molecular imaging. In addition to these different antitumor drugs, biomolecules such as peptides, antibodies, etc. can be used along with NPs to label highly specific tumors (Singh et al., 2019; Jin et al., 2020). In lung cancer, detection of metastasis can be done by the use of superparamagnetic iron oxide NPs (SPIONs) in MRI (Wang et al., 2016). Furthermore, studies have shown SPIONs as suitable building blocks for aerosols in lung cancer due to their high specificity and no side effects (Jafari et al., 2015; Stocke et al., 2015; Jin et al., 2020). Also, magnetic powder imaging has also been used in tomographic imaging technology, where it has shown high accuracy and sensitivity to cancer tissues (Garrigue et al., 2018).

Visible spectral imaging has the limitation that it lacks the ability to penetrate into the objects. Therefore, QDs that emit the fluorescence in the near infrared spectrum have been used to overcome this problem, making them more efficient for imaging of colorectal cancer, pancreatic cancer, liver cancer, and lymphoma (Parungo et al., 2004; Zhu et al., 2022; Jin et al., 2020). In addition, these silver-rich Ag_2Te QDs that contain the source of sulfur have been reported to allow better visualization and a special resolution of images over a wide infrared range (Zhu et al., 2022). NanoShells are the dielectric cores that have a size between 10 and 300 nm; they are generally made of Si and are coated with a thin metal (Hirsch et al., 2003; Loo et al., 2004). Nanoshells are advantages because their imaging has no heavy metal toxicity (Nunes et al., 2019), but they still have limited uses because of their huge sizes.

AuNPs are a very good contrasting agent since they have a small size, a high atomic number, and excellent biocompatibility.

Researchers have shown that AuNPs work by two methods, *viz.*, passive targeting and active targeting. The principle of passive targeting is that the AuNPs are gathered to enhance imaging because of the permeability tension effect (EPR) in the tumor tissue, whereas active targeting Au particles are coupled with the tumor-specific targeted drugs. Research has shown that the X-ray imaging of the cluster of liver cells in AUNP-composed groups has strong resolution and clearer as compared to those in the liver cancer cells alone. The AuNPs have very important implications for the early diagnosis and treatment of cancer tumors of minute to large size (Figure 19.7).

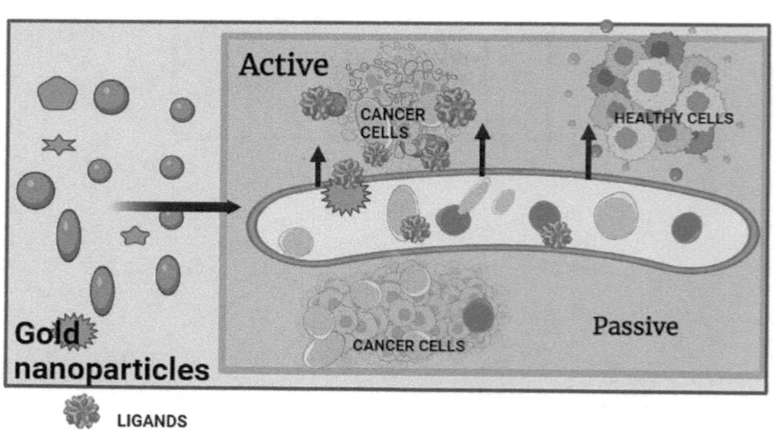

FIGURE 19.7 Active and passive targeting by AuNPs in tumor cells.

The use of nanotechnology for cancer therapy simply means the use of the smaller molecular structures and particles as tools for the detection and delivery of the drugs at the side of the cancer tumor (Jin et al., 2020). Liposomes are composed of natural or synthesized phospholipid bilayer membranes and hydrophilic nuclei (Malam et al., 2009). Due to the amphiphilicity of phospholipids (i.e., having both hydrophobic and hydrophilic parts), it can be used to incorporate the various different types of drugs into it and deliver it at the cancer cells, such as hydrophilic drugs, which preferentially stay in the monolayer liposomes while the hydrophobic drugs form multilayer liposome (Zhao et al., 2020). In the tumors, these lysosomes fuse with the cell, internalize by endocytosis, and later release the drugs in the intercellular spaces (Zhao et al., 2020).

The life of the liposome is dependent on its size; if they are up to 100 nm, then they can easily penetrate into the tumor and stay longer, but if they are of a greater size, then they can be easily recognized and cleared by the mononuclear phagocyte system of the body (Berger et al., 2014).

Carbon nanotubes are the new form of carbon similar to the rolled graphene sheets (Jin et al., 2020). Based on their diameter, the carbon and its tubes are of two types:

1. Single-walled carbon nanotubes (SWCNTs)
2. Multi-walled carbon nanotubes (MWCNTs) (Kesharwani and Iyer, 2015)

CNTs possess a property that allows them to absorb light from the near infrared region (NIR), inducing heat into them due to the thermal effects; therefore, can target tumor cells (60–62). Moreover, CNTs are very good carriers for the drug delivery (Jin et al., 2020). Polymeric micelles have a hydrophobic core and a hydrophilic corona having size between 10 and 100 nm (Aslan et al., 2020). Micelles can improve the bioavailability of the hydrophobic drugs and provide protection and prevention of the inactivation of drugs under the influence of biological surroundings. Similar to the AuNPs, polymeric micelle formulations can also be used both for passive and active targeting in anticancer therapy. Genexol-PM is a paclitaxel-loaded polymeric micelle formulation used for treating breast, lung, and pancreatic cancer through passive missile targeting (Aslan et al., 2013). NC-6004 is a carboplatin-loaded formulation that is also studied in early clinical trials for treatment of solid tumors (Wilson et al., 2008). Furthermore, polymeric micelles formulations are used for active targeting by binding them with various ligands such as folate and mAb C225. Dendrimers are the multibranched NPs having a central core and branching units along with functionalized terminal groups. The major highlight of using dendrimers for cancer therapy is that multiple anticancer agents can bind to the functionalized terminal groups or can be incorporated in the central core. Also, depolymerization of dendrimers can be controlled to modify the release profiles of the payload (Wong et al., 2020). For instance, polyamidoamine (PAMAM) dendrimers can be tailored to enhance their biocompatibility and release properties through PEGylation and acetylation and modified with anionic, neutral ligand molecules (Cai et al., 2015). Moreover, DOX was combined to PEGylated PAMAM dendrimers by acid sensitive linkages in order to trigger the release of DOX in acidic conditions (Zhu et al., 2010).

19.4.5 Metalloid Nanoparticles in Crop Production

Crop production worldwide faces many challenges, including climate and stress. Nanotechnology has become an important part of sustainable development to overcome these problems. Because they easily enter the plant, they have better adsorption and target attainment also play a role in photosynthetic efficiency, water absorption, and detoxification of reactive oxygen, thereby increasing seed germination, growth, and yield. For good results, carefully analyzing the number of different NPs can be used sustainably in agriculture. Interaction of NPs with ecosystems still needs to be evaluated (Verma et al., 2022). By increasing the antioxidant defense of plants, nanomaterials have the ability to eliminate the harmful effects of plant biotic stress. Abiotic stress is an important problem in terms of agricultural production and continuity. It includes drought, salinity, alkalinity, floods, mineral and metal toxicity, and many other factors that limits crop growth and productivity. Various NPs, such as MgNPs, TiO_2 NPs, and ZnO NPs, are used under field conditions to ensure crop growth and yield (Elmer et al., 2018a).

SiNPs with low concentrations improve the rate of seed germination in tomatoes. Also, SiNPs increase the seed germination of maize by providing better nutrients and controlling the pH and conductivity of the growing medium. Research on Si-based NPs on seedling of *Larix olgenesis* shows that NPs promote seed growth, quality, height, diameter of root collar, and accelerated chlorophyll synthesis. In cases of salinity stress in squash, it improves seed germination and promotes antioxidant effects. Exogenous application of silica NPs has shown that the germination rate of soybean seeds can be improved by increasing the concentration of nitrate reductase and the ability of seeds to absorb and utilize water and nutrients. SiNPs can also act as fertilizer (Verma et al., 2022). Similarly, SiNPs can reduce heavy metal stress in plants. SiNPs reduce the Pb, Cd, and Cr toxicity (Hussain et al., 2020). Many studies have shown that SiNPs play different roles in different plants to reduce environmental stress. The foliar application of NP of metalloid B and metal oxides affects growth and inhibits disease development, possibly by improving mineral nutrition and host defense. NP suspensions (500–1,000 mL) were sprayed on leaves of fresh watermelon plants (1–2 mL per plant) and plants in potting mix infested with *F. oxysporum* f. sp. niveum. CuO NP suppressed disease in five of eight greenhouse experiments. CuO NP increased biomass or yield compared to untreated controls and others tested NP. To suppress disease, it may serve as a highly effective delivery agent for this micronutrient. In tomato, eggplant, and watermelon, NP of CuO suppressed *Fusarium* and *Verticillium* diseases (Elmer et al., 2018b; Elmer and White, 2018).

NPs of germanium or ferrum carboxylates to inoculation suspension of rhizobia Tn5 mutant B1–20 show insignificant changes in enzyme activity during dehydration and after exposure to water stress in root nodules and soybean leaves, indicating high tolerance of these to growing conditions. Pre-treatment with GeNPS causes an increase in grain production by 15.6% and the weight of 1,000 grains by 16.4%. The foliar application of pomegranate trees with BNs under salt stress is a good tool in reducing salt stress. To increase the pomegranate plant's resistance to salinity, the ability to BNs was low but enhanced the presence of macronutrient-rich biochar (MNB) significantly. For improving pomegranate yield and quality, a combined application of MNB and BNs can be used. On pea plants, which are affected by drought after spraying them with 12.5 ppm of B_2O_3 NPs, the most sustainable positive effect was found. It also reduced the amount of hydrogen peroxide and malondialdehyde and stimulated phenol accumulation, antioxidant capacity, ascorbate peroxidase, superoxide dismutase, and enzyme activities in peas (Abd El-Wahed et al., 2023). In improving the quality of plants grown under different environmental stresses, more studies are needed on the mechanism of effect of other NPs. Foremost the phytotoxic behavior of Nanomaterials (NMs) needs to be understood at different plant growth stages. The use of NPs in agriculture before the commercial use of nanotechnology is required to be worked on as a priority.

19.5 CONCLUSION

In conclusion, the exploration of metalloid NPs in the biological realm, as outlined in this chapter, opens new vistas for transformative applications. From the foundational understanding presented in the introduction to the eco-friendly biosynthesis methods and detailed characterization techniques, the versatility of metalloid NPs is evident. Their pivotal role in drug delivery, crop production enhancement, cancer treatment, and diagnostics signifies a paradigm shift in various fields. Looking forward, the future holds exciting prospects for metalloid NPs. Ongoing research is expected to unveil novel biosynthesis approaches, refining their production, and minimizing environmental impact. Further advancements in characterization techniques will deepen our understanding, enabling more precise tailoring of these NPs for specific applications. In drug delivery, the evolution of targeted therapies and personalized medicine is anticipated, maximizing efficacy while minimizing side effects. In agriculture, the integration of metalloid NPs is poised to revolutionize crop production practices, promoting sustainability and addressing food security challenges. Continued exploration of their applications in cancer treatment and diagnostics is expected to result in breakthroughs, offering more efficient and precise solutions. As we stand at the cusp of a new era in nanotechnology, the convergence of scientific innovation and practical applications positions metalloid NPs as integral tools for addressing global challenges. The future beckons with possibilities as researchers delve deeper, unlocking the full potential of these NPs to shape a healthier and more sustainable world.

REFERENCES

Abd El-Wahed, M. H., M. A. Eissa, N. M. Almasoudi, and K. A. M. Abo-Elyousr. Macronutrient-rich biochar induces boron nanoparticles in improving the salt tolerance of pomegranate (Punica granatum L.) in arid degraded soils. *Scientia Horticulturae* 313 (2023): 111908.

Agarwal, N., M. S. Nair, A. Mazumder, and K. M. Poluri. Characterization of nanomaterials using nuclear magnetic resonance spectroscopy. In: Bhagyaraj, S. M., O. S. Oluwafemi, N. Kalarikkal, and S. Thomas (Eds.) *Characterization of Nanomaterials.* Woodhead Publishing, online 22 June 2018, Version of Record 22 June 2018.

Aguirre, F. M. A., L. Bazán-Díaz, R. Mendoza-Cruz, A. Gómez-Rodríguez, C. Zorrilla-Cangas, and R. Herrera-Becerra. Nano phase characterization by transmission electron microscopy: experimental and simulation. *Materials Sciences and Applications* 6 (2015): 935.

Ahmad, K. S., S. Yaqoob, and M. M. Gul. Dynamic green synthesis of iron oxide and manganese oxide nanoparticles and their cogent antimicrobial, environmental and electrical applications. *Reviews in Inorganic Chemistry* 42 (2022): 239–263.

Akhter, S., I. Ahmad, M. Z. Ahmad, F. Ramazani, A. Singh, Z. Rahman, F. J. Ahmad, G. Storm, and R. J. Kok. Nanomedicines as cancer therapeutics: current status. *Current Cancer Drug Targets* 13, no. 4 (2013): 362–378.

Alavi, M., M. Rai, F. Martinez, D. Kahrizi, H. Khan, I. R. A. De Menezes, H. D. M. Coutinho, and J. G. M. Costa. The efficiency of metal, metal oxide, and metalloid nanoparticles against cancer cells and bacterial pathogens: different mechanisms of action. *Cellular, Molecular and Biomedical Reports* 2, no. 1 (2022): 10–21.

Ali, M. A., Ahmed, T., Wu, W., Hossain, A., Hafeez, R., Islam Masum, M. M., Wang, Y., An, Q., Sun, G. and Li, B. Advancements in plant and microbe-based synthesis of metallic nanoparticles and their antimicrobial activity against plant pathogens. *Nanomaterials* 10, no. 6 (2020): 1146.

Amato, G. Silica-encapsulated efficient and stable si quantum dots with high biocompatibility. *Nanoscale Research Letters* 5 (2010): 1156–1160.

Aslan, N., B. Ceylan, M. M. Koç, and F. Findik. Metallic nanoparticles as X-ray computed tomography (CT) contrast agents: a review. *Journal of Molecular Structure* 1219 (2020): 128599.

Banerjee, A. N., R. Krishna, and B. Das. Size controlled deposition of Cu and Si nano-clusters by an ultra-high vacuum sputtering gas aggregation technique. *Applied Physics A* 90 (2008): 299–303.

Barui, S. and Cauda, V. Multimodal decorations of mesoporous silica nanoparticles for improved cancer therapy. *Pharmaceutics* 12, no. 6 (2020): 527.

Beck-Speier, I., N. Dayal, E. Karg, K. L. Maier, G. Schumann, H. Schulz, M. Semmler, S. Takenaka, K. Stettmaier, W. Bors, A. Ghio, J. M. Samet, and J. Heyder. Oxidative stress and lipid mediators induced in alveolar macrophages by ultrafine particles. *Free Radical Biology and Medicine* 38 (2005): 1080–1092.

Berger, J. L., A. Smith, K. K. Zorn, P. Sukumvanich, A. B. Olawaiye, J. Kelley, and T. C. Krivak. Outcomes analysis of an alternative formulation of PEGylated liposomal doxorubicin in recurrent epithelial ovarian carcinoma during the drug shortage era. *OncoTargets and Therapy* 7 (2014): 1409–1413.

Bharali, D. J., T. Sudha, H. Cui, B. M. Mian, and S. A. Mousa. Anti-CD24 nano-targeted delivery of docetaxel for the treatment of prostate cancer. *Nanomedicine: Nanotechnology, Biology and Medicine* 13, no. 1 (2017): 263–273.

Bhushan, M., Kumar, Y., Periyasamy, L. and Viswanath, A.K. Antibacterial applications of α-Fe_2O_3/Co_3O_4 nanocomposites and study of their structural, optical, magnetic and cytotoxic characteristics. *Applied Nanoscience* 8 (2018): 137–153.

Bogner, A., P.-H. Jouneau, G. Thollet, D. Basset, and C. Gauthier. A history of scanning electron microscopy developments: towards "wet-STEM" imaging. *Micron* 38 (2007): 390–401.

Brown, A. N., K. Smith, T. A. Samuels, J. Lu, S. O. Obare, and M. E. Scott. Nanoparticles functionalized with Ampicillin destroy multiple-antibiotic-resistant isolates of Pseudomonas aeruginosa and Enterobacter aerogenes and methicillin-resistant Staphylococcus aureus. *Applied and Environmental Microbiology* 84, no. 2 (2018): e01067–17.

Cai, L., G. Xu, C. Shi, D. Guo, X. Wang, and J. Luo. Telodendrimer nanocarrier for co-delivery of paclitaxel and cisplatin: a synergistic combination nanotherapy for ovarian cancer treatment. *Biomaterials* 37 (2015): 456–468.

Cantor, C. R., P. R. Schimmel, and W. H. Freeman Ed. *Techniques for the Study of Biological Structure and Function*; Standford Libraries, Standford Publishers: San Francisco, CA, 1980.

Cao, G. *Nanostructures and Nanomaterials. Synthesis, Properties, and Applications*; Imperial College Press: London, 2004.

Chen, Y., S. Li, and J. L. Brown. Rapid and sensitive detection of pathogenic bacteria in small sample volumes using light-scattering sensor. *Biosensors and Bioelectronics* 82 (2016): 233–241.

Czarnobaj, K. Preparation and characterization of silica xerogels as carriers for drugs. *Drug Delivery* 15, no. 8 (2008): 485–492.

Czarnobaj, K. and Łukasiak, J. In vitro release of cisplatin from sol–gel processed organically modified silica xerogels. *Journal of Materials Science: Materials in Medicine* 18 (2007): 2041–2044.

Czarnobaj, K. and Sawicki, W. Influence of surfactants on the release behaviour and structural properties of sol-gel derived silica xerogels embedded with metronidazole. *Pharmaceutical Development and Technology* 18, no. 2 (2013): 377–383.

Deplanche, K., and L. E. Macaskie. Biorecovery of gold by Escherichia coli and Desulfovibrio desulfuricans. *Biotechnology and Bioengineering* 99 (2008): 1055–1064.

Dhand, C., N. Dwivedi, X. J. Loh, A. Ying, N. Verma, R. W. Beuerman, R. Lakshminarayanan, and S. Ramakrishna. Methods and strategies for the synthesis of diverse nanoparticles and their applications: a comprehensive overview. *RSC Advances* 5 (2015): 105003–105037.

Diallo, A., T. B. Doyle, B. M. Mothudi, E. Manikandan, V. Rajendran, and M. Maaza. Magnetic behavior of biosynthesized Co3O4 nanoparticles. *Journal of Magnetism and Magnetic Materials*, 424 (2017), pp. 251–255

Dias, D. A., K. A. Kouremenos, D. J. Beale, D. L. Callahan, and O. A. H. Jones. Metal and metalloid containing natural products and a brief overview of their applications in biology, biotechnology and biomedicine. *BioMetals* 29 (2016): 1–13.

Digman, M. A., and E. Gratton. Lessons in fluctuation correlation spectroscopy. *Annual Review of Physical Chemistry* 62 (2011): 645–668.

Dikusar, A. I., P. G. Globa, S. S. Belevskii, and S. P. Sidel'nikova. On limiting rate of dimensional electrodeposition at meso-and nanomaterial manufacturing by template synthesis. *Surface Engineering and Applied Electrochemistry* 45 (2009): 171–179.

Doyle, M. E., and K. A. Glass. Sodium reduction and its effect on food safety, food quality, and human health. *Comprehensive Reviews in Food Science and Food Safety* 9 (2010): 44–56.

Dzakpasu, R., and D. Axelrod. Dynamic light scattering microscopy. A novel optical technique to image submicroscopic motions. I: Theory. *Biophysical Journal* 87 (2004): 1279–1287.

Elmer, W., and J. C. White. The future of nanotechnology in plant pathology. *Annual Review of Phytopathology* 56 (2018): 111–133.

Elmer, W., C. Ma, and J. White. Nanoparticles for plant disease management. *Current Opinion in Environmental Science & Health* 6 (2018a): 66–70.

Elmer, W., R. De La Torre-Roche, L. Pagano, S. Majumdar, N. Zuverza-Mena, C. Dimkpa, J. Gardea-Torresdey, and J. C. White. Effect of metalloid and metal oxide nanoparticles on Fusarium wilt of watermelon. *Plant Disease* 102, no. 7 (2018b): 1394–1401.

Fidalgo, A., Lopez, T.M. and Ilharco, L.M. Wet sol–gel silica matrices as delivery devices for phenytoin. *Journal of Sol-Gel Science and Technology* 49 (2009): 320–328.

Garrigue, P., J. Tang, L. Ding, A. Bouhlel, A. Tintaru, E. Laurini, Y. Huang, Z. Lyu, M. Zhang, S. Fernandez, L. Balasse, W. Lan, E. Mas, D. Marson, Y. Weng, X. Liu, S. Giorgio, J. Iovanna, S. Pricl, B. Guillet, and L. Peng. Self-assembling supramolecular dendrimer nanosystem for PET imaging of tumors. *Proceedings of the National Academy of Sciences* 115, no. 45 (2018): 11454–11459.

Ghosh, S., R. Ahmad, M. Zeyaullah, and S. K. Khare. Microbial nano-factories: synthesis and biomedical applications. *Frontiers in Chemistry* 9 (2021): 626834. https://doi.org/10.3389/fchem.2021.626834. PMID: 33937188; PMCID: PMC8085502

Goldstein, J. I., D. E. Newbury, P. Echlin, D. C. Joy, C. E. Lyman, E. Lifshin, L. Sawyer, and J. R. Michael. *Scanning Electron. Microscopy and X-Ray Microanalysis*, 3rd ed.; Kluwer Academic/Plenum Publishers: New York, 2003.

Gomez, M., F. Perez, G. Reyes, and T. L. Church Development of arsenic nanoparticles with antifungal activity against Aspergillus flavus. *Nanomaterials* 9, no. 10 (2019): 1437.

Gresback, R., Holman, Z. and Kortshagen, U. Nonthermal plasma synthesis of size-controlled, monodisperse, freestanding germanium nanocrystals. *Applied Physics Letters*, *91*(9) 2007.

Guler, U., V. M. Shalaev, and A. Boltasseva. Nanoparticle plasmonics: going practical with transition metal nitrides. *Materials Today* 18 (2015): 227–237.

Gurunathan, S., J. W. Han, D. N. Kwon, and J.-H. Kim. Enhanced antibacterial and anti-biofilm activities of silver nanoparticles against gram-negative and gram-positive bacteria. *Nanoscale Research Letters* 9 (2014): 373.

Harris, H. W., I. Sanchez-Andrea, J. S. Mclean, E. C. Salas, W. Tran, M. Y. El-Naggar, and K. H. Nealson. Redox sensing within the genus Shewanella. *Frontiers in Microbiology* 8 (2018): 2568. https://doi.org/10.3389/fmicb.2017.02568

Hirsch, L. R., A. M. Gobin, A. R. Lowery, F. Tam, R. A. Drezek, N. J. Halas, and J. L. West. Metal nanoshells. *Annals of Biomedical Engineering* 34 (2006): 15–22.

Hirsch, L. R., R. Jason Stafford, J. A. Bankson, S. R. Sershen, B. Rivera, R. E. Price, J. D. Hazle, N. J. Halas, and J. L. West. Nanoshell-mediated near-infrared thermal therapy of tumors under magnetic resonance guidance. *Proceedings of the National Academy of Sciences* 100, no. 23 (2003): 13549–13554.

Hu, D., Z. Sheng, G. Gao, F. Siu, C. Liu, Q. Wan, P. Gong, H. Zheng, Y. Ma, and L. Cai. Activatable albumin-photosensitizer nanoassemblies for triple-modal imaging and thermal-modulated photodynamic therapy of cancer. *Biomaterials* 93 (2016): 10–19.

Hussain, B., Lin, Q., Hamid, Y., Sanaullah, M., Di, L., Khan, M.B., He, Z. and Yang, X. Foliage application of selenium and silicon nanoparticles alleviates Cd and Pb toxicity in rice (Oryza sativa L.). *Science of the Total Environment* 712 (2020): 136497.

Jafari, S. H., Z. Saadatpour, A. Salmaninejad, F. Momeni, M. Mokhtari, J. S. Nahand, M. Rahmati, H. Mirzaei, and M. Kianmehr. Breast cancer diagnosis: imaging techniques and biochemical markers. *Journal of Cellular Physiology* 233, no. 7 (2018): 5200–5213.

Jafari, M.D., Wexner, S.D., Martz, J.E., McLemore, E.C., Margolin, D.A., Sherwinter, D.A., Lee, S.W., Senagore, A.J., Phelan, M.J. and Stamos, M.J. Perfusion assessment in laparoscopic left-sided/anterior resection (PILLAR II): a multi-institutional study. *Journal of the American College of Surgeons* 220, no. 1 (2015): 82–92.

Jin, C., K. Wang, A. Oppong-Gyebi, and J. Hu. Application of nanotechnology in cancer diagnosis and therapy-a mini-review. *International Journal of Medical Sciences* 17, no. 18 (2020): 2964.

Johnson, L., Allen, M. J., and Caswell, T. A. Evaluation of silicon nanoparticles for their inhibitory effect against Escherichia coli DH5α and Staphylococcus aureus. *Nanoscale Research Letters* 13, no. 1 (2018): 196.

Karthikeyan, B., and Loganathan, B. A close look of Au/Pt/Ag nanocomposites using SERS assisted with optical, electrochemical, spectral and theoretical methods. *Physics E (Amsterdam Netherlands)* 49 (2013): 105–110.

Kesharwani, P., and A. K. Iyer. Recent advances in dendrimer-based nanovectors for tumor-targeted drug and gene delivery. *Drug Discovery Today* 20, no. 5 (2015): 536–547.

Kim, D., Y. Y. Jeong, and S. Jon. A drug-loaded aptamer–gold nanoparticle bioconjugate for combined CT imaging and therapy of prostate cancer. *ACS Nano* 4, no. 7 (2010): 3689–3696.

Konishi, Y., K. Ohno, N. Saitoh, T. Nomura, S. Nagamine, H. Hishida, …, T. Uruga. Bioreductive deposition of platinum nanoparticles on the bacterium Shewanella algae. *Journal of Biotechnology* 128 (2007), 648–653.

Konrad, A., U. Herr, R. Tidecks, F. Kummer, and K. Samwer. Luminescence of bulk and nanocrystalline cubic yttria. *Journal of Applied Physics* 90 (2001): 3516–3523.

Krishnappa, K., J. Pandiyan, J. Paramanandham, and K. Elumalai. Antibacterial and mosquitocidal potentials of selected Indian medicinal plants extracts and synthesized silver nanoparticles. *International Journal of Zoology Studies* 3 (2018): 15–5.

Labulo, A. H., Adesuji, E.T., Dedeke, O.A., Bodede, O.S., Oseghale, C.O., Moodley, R., Nyamori, V.O., Dare, E.O. and Adegoke, O.A. A dual-purpose silver nanoparticles biosynthesized using aqueous leaf extract of Detarium microcarpum: an under-utilized species. *Talanta 160* (2016): 735–744.

Laganà, A., G. Visalli, F. Corpina, M. Ferlazzo, A. Di Pietro, and A. Facciolà. Antibacterial activity of nanoparticles and nanomaterials: a possible weapon in the fight against healthcare-associated infections. *European Review for Medical & Pharmacological Sciences* 27, no. 8 (2023): 3645–3663.

Li, J., Q. Wu, and J. Wu. Synthesis of nanoparticles via solvothermal and hydrothermal methods. In: Aliofkhazraei, M. (Ed.) *Handbook of Nanoparticles*, 1–28. Springer, 2015.

Liebl, M., D. Eberbeck, A. Coene, J. Leliaert, P. Jauch, M. Kruteva, L. Fruhner, L. Barnsley, S. Mayr, and F. Wiekhorst. 9 Magnetic measurement methods to probe nanoparticle–matrix interactions. *Physical Sciences Reviews* 8 (2021). https://doi.org/10.1515/9783110569636-009

Lin, P. C., S. Lin, P. C. Wang, and R. Sridhar. Techniques for physicochemical characterization of nanomaterials. *Biotechnology Advances* 32 (2014): 711–726.

Liu, W., Wu, Y., Wang, C., Li, H. C., Wang, T., Liao, C. Y., Cui, L., Zhou, Q. F., Yan, B. and Jiang, G. B. Impact of silver nanoparticles on human cells: effect of particle size. *Nanotoxicology* 4, no. 3 (2010): 319–330.

Loo, C., A. Lin, L. Hirsch, M.-H. Lee, J. Barton, N. Halas, J. West, and R. Drezek. Nanoshell-enabled photonics-based imaging and therapy of cancer. *Technology in Cancer Research & Treatment* 3, no. 1 (2004): 33–40.

Lumen, D., S. Wang, E. Mäkilä, S. Imlimthan, M. Sarparanta, A. Correia, C. W. Haug, J. Hirvonen, H. A. Santos, A. J. Airaksinen, W. Filtvedt, and J. Salonen. Investigation of silicon nanoparticles produced by centrifuge chemical vapor deposition for applications in therapy and diagnostics. *European Journal of Pharmaceutics and Biopharmaceutics* 158 (2021): 254–265.

Malam, Y., M. Loizidou, and A. M. Seifalian. Liposomes and nanoparticles: nanosized vehicles for drug delivery in cancer. *Trends in Pharmacological Sciences* 30, no. 11 (2009): 592–599.

Manjula, P., R. Boppella, and S. V. Manorama. A facile and green approach for the controlled synthesis of porous SnO2 nanospheres: application as an efficient photocatalyst and an excellent gas sensing material. *ACS Applied Materials Interfaces* 4 (2012): 6252–6260.

Mishra, R. K., A. K. Zachariah, and S. Thomas. Energy-dispersive X-ray spectroscopy techniques for nanomaterial. In: S. Thomas, R. Thomas, A. K. Zachariah, and R. K. Mishra (Eds.) *Microscopy Methods in Nanomaterials Characterization*, 383–405. Elsevier, 2017. https://doi.org/10.1016/B978-0-323-46141-2.00012-2

Mittal, J., B. Amla, A. Singh, and M. Sharma. Phytofabrication of nanoparticles through plant as nanofactories. *Advances in Natural Sciences: Nanoscience and Nanotechnology* 5 (2014): 153–169.

Muñoz, J. E., J. Cervantes, R. Esparza, and G. Rosas. Iron nanoparticles produced by high-energy ball milling. *Journal of Nanoparticle Research* 9 (2007): 945–950.

Murray, A. J., J. Zhu, J. Wood, and L. E. Macaskie. A novel biorefinery: biorecovery of precious metals from spent automotive catalyst leachates into new catalysts effective in metal reduction and in the hydrogenation of 2-pentyne. *Minerals Engineering* 113 (2017): 102–108. https://doi.org/10.1016/j.mineng.2017.08.011

Nguyen, D. H., M. H. Kim, and J. H. Lee. Antifungal effect of germanium nanoparticles and their application on controlling Fusarium oxysporum f. sp. cucumerinum. *Materials Science and Engineering: C* 116 (2020): 111199.

Nichols, W. T., T. Sasaki, and N. Koshizaki. Laser ablation of a platinum target in water. III. Laser-induced reactions. *Journal of Applied Physics* 100 (2006): 114911.

Nunes, T., T. Pons, X. Hou, K. Van Do, B. Caron, M. Rigal, M. Di Benedetto, B. Palpant, C. Leboeuf, A. Janin, and G. Bousquet. Pulsed-laser irradiation of multifunctional gold nanoshells to overcome trastuzumab resistance in HER2-overexpressing breast cancer. *Journal of Experimental & Clinical Cancer Research* 38 (2019): 1–13.

Pandey, S., G. Oza, A. Mewada, and M. Sharon. Green synthesis of highly stable gold nanoparticles using momordicacharantia as nano fabricator. *Archives of Applied Science Research* 4 (2012): 1135–1141.

Park, D. K., S. J. Lee, J. Lee, M. Y. Choi, and S. W. Han. Effect of polymeric stabilizers on the catalytic activity of Pt nanoparticles synthesized by laser ablation. *Chemical Physics Letters* 484 (2010): 254–257.

Park, E. J. and Park, K. Oxidative stress and pro-inflammatory responses induced by silica nanoparticles in vivo and in vitro. *Toxicology Letters*, 184, no. 1 (2009): 18–25.

Parungo, C. P., S. Ohnishi, A. M. De Grand, R. G. Laurence, E. G. Soltesz, Y. L. Colson, P. M. Kang, T. Mihaljevic, L. H. Cohn, and J. V. Frangioni. In vivo optical imaging of pleural space drainage to lymph nodes of prognostic significance. *Annals of Surgical Oncology* 11 (2004): 1085–1092.

Patscheider, J. Nanocomposite hard coatings for wear protection. *MRS Bulltion* 28 (2003): 180–183.

Perevedentseva, E. V., F. Y. Su, T. H. Su, Y. C. Lin, C. L. Cheng, A. V. Karmenyan, A. V. Priezzhev, and A. E. Lugovtsov. Laser-optical investigation of the effect of diamond nanoparticles on the structure and functional properties of proteins. *Quantum Electron* 40 (2010): 1089–1093.

Perez Tijerina, E., M. G. Pinilla, S. Mejia Rosales, U. Ortiz Mendez, A. Torres, and M. Jose Yacaman. Highly size-controlled synthesis of Au/Pd nanoparticles by inert-gas condensation. *Faraday Discussions* 138 (2008): 353–362.

Ponon, N. K., Appleby, D. J., Arac, E., King, P. J., Ganti, S., Kwa, K. S., and A. O'Neill. Effect of deposition conditions and post-deposition anneal on reactively sputtered titanium nitride thin films. *Thin Solid Films* 578 (2015): 31–37.

Popescu, T., Lupu, A.R., Raditoiu, V., Purcar, V. and Teodorescu, V.S. On the photocatalytic reduction of MTT tetrazolium salt on the surface of TiO2 nanoparticles: Formazan production kinetics and mechanism. *Journal of Colloid and Interface Science* 457 (2015): 108–120.

Prokopowicz, M. and Łukasiak, J. Synthesis and in vitro characterization of freeze-dried doxorubicin-loaded silica/PEG composite. *Journal of Non-Crystalline Solids* 356, no. 33-34 (2010): 1711–1720.

Pulskamp, K., S. Diabaté, and H. Krug. Carbon nanotubes show no sign of acute toxicity but induce intracellular reactive oxygen species in dependence on contaminants. *Toxicology Letters* 168 (2007): 58–74.

Rami, J. M., C. D. Patel, C. M. Patel, and M. V. Patel. Thermogravimetric analysis (TGA) of some synthesized metal oxide nanoparticles. *Materials Today Proceedings* 43 (2021): 655–659. https://doi.org/10.1016/j.matpr.2020.12.554

Rangarajan, M., R. Vasanthakumari, and S. Vikram. Superparamagnetic iron oxide nanoparticles from coprecipitation: composition, size, and magnetization. *Nanoscience and Nanotechnology* 14 (2014): 1–9.

Rasmussen, M. K., J. N. Pedersen, and R. Marie. Size and surface charge characterization of nanoparticles with a salt gradient. *Nature Communications* 11, 2337 (2020). https://doi.org/10.1038/s41467-020-15889-3

Rastogi, A. A mini-review practice of formulations of nanoparticles. *The International Journal of Chemical Synthesis and Chemical Reactions* 3 (2017): 1–7.

Raun, T., 2012. DIY therapy: Exploring affective self-representations in trans video blogs on YouTube. In *Digital cultures and the politics of emotion: Feelings, affect and technological change* (pp. 165–180). London: Palgrave Macmillan UK.

Roy, A., .Datta, R. Luthra, M. A. Khan, A. Gacem, M. A. Hasan, K. K. Yadav, Y. Ahn, and B.-H., Jeon. Green synthesis of metalloid nanoparticles and its biological applications: a review. *Frontiers in Chemistry* 10 (2022): 994724.

Roy, A., S. Datta, R. Luthra, M. A. Khan, A. Gacem, M. A. Hasan, K. K. Yadav, Y. Ahn, and B.-H. Jeon. Green synthesis of metalloid nanoparticles and its biological applications: a review. *Frontiers in Chemistry* 10 (2022): 994724.

Salunke, B. K., Sawant, S. S., Lee, S. I. and Kim, B. S. Microorganisms as efficient biosystem for the synthesis of metal nanoparticles: current scenario and future possibilities. *World Journal of Microbiology and Biotechnology* 32 (2016a): 1–16.

Salunke, B. K., Sawant, S. S. and Kim, B. S. Enhancement of antibacterial effect by biosynthesized silver nanoparticles with antibiotics. *Journal of Nanoscience and Nanotechnology* 16, no. 7 (2016b): 7191–7194.

Sarraf, M., K. Vishwakarma, V. Kumar, N. Arif, S. Das, R. Johnson, E. Janeeshma et al. Metal/metalloid-based nanomaterials for plant abiotic stress tolerance: an overview of the mechanisms. *Plants* 11, no. 3 (2022): 316.

Satoh, A., R. W. Chantrell, G. N. Brownian. Dynamics simulations of ferromagnetic colloidal dispersions in a simple shear flow Coverdale. *Journal of Colloid Interface Sciences* 209 (1999): 44–59.

Sekhon, B.S., 2013. Metalloid compounds as drugs. *Research in pharmaceutical sciences*, *8*(3), p.145.

Shang, L., Y. Wang, J. Jiang, and S. Dong. pH-dependent protein conformational changes in albumin: gold nanoparticle bioconjugates: a spectroscopic study. *Langmuir* 23 (2007): 2714.

Siwach, O. P., and P. Sen. Synthesis and study of fluorescence properties of Cu nanoparticles. *Journal of Nanoparticle Research* 10 (2008): 107–114.

Smith, A. M., H. Duan, A. M. Mohs, and S. Nie. Bioconjugated quantum dots for in vivo molecular and cellular imaging. *Advanced Drug Delivery Reviews* 60, no. 11 (2019): 1226–1240.

SP, V., U. Udayabhanu, and L. HS. Plant-mediated green synthesis of Ag nanoparticles using rauvolfiatetraphylla (L.) flower extracts: characterization, biological activities and screening of their catalytic activity in formylation reaction. *ScientiaIranica* (2019). https://doi.org/10.24200/SCI.2019.51275.2093

Stocke, N. A., S. A. Meenach, S. M. Arnold, H. M. Mansour, and J. Zach Hilt. Formulation and characterization of inhalable magnetic nanocomposite microparticles (MnMs) for targeted pulmonary delivery via spray drying. *International Journal of Pharmaceutics* 479, no. 2 (2015): 320–328.

Tomaszewska, E., K. Soliwoda, K. Kadziola, G. Celichowski, M. Cichomski, W. Szmaja, and J. Grobelny. Detection limits of DLS and UV-Vis spectroscopy in characterization of polydisperse nanoparticles colloids. *Journal of Nanomaterials* 2013 (2013): 60.

Verma, K.K., Zeng, Y., Song, X.P., Singh, M., Wu, K.C., Rajput, V.D. and Li, Y.R. Nanosilicon: an approach for abiotic stress mitigation and sustainable agriculture. *Frontiers in Plant Science* 13 (2022): 1025974.

Wang, K., Y. Zhang, J. Wang, A. Yuan, M. Sun, J. Wu, and Y. Hu. Self-assembled IR780-loaded transferrin nanoparticles as an imaging, targeting and PDT/PTT agent for cancer therapy. *Scientific Reports* 6, no. 1 (2016): 27421.

Wang, J. and Jiang, Y.F. Natural compounds as anticancer agents: Experimental evidence. *World Journal of Experimental Medicine*, 2, no. 3 (2012): 45.

Waseda, Y., E. Matsubara, K. Shinoda, and V. Springer. *X-Ray Diffraction Crystallography*; Springer: Berlin, 2011.

Wilson, C., T. Wilson, P. G. Johnston, D. B. Longley, and D. J. J. Waugh. Interleukin-8 signaling attenuates TRAIL-and chemotherapy-induced apoptosis through transcriptional regulation of c-FLIP in prostate cancer cells. *Molecular Cancer Therapeutics* 7, no. 9 (2008): 2649–2661.

Winterer, M., and Hahn, H. Nanoceramics by chemical vapour synthesis. *Zeitschriftfur Metallkunde* 94 (2003): 1084–1090.

Wong, S. K., I. Zainol, M. P. Ng, C. H. Ng, and I. H. Ooi. Dendrimer-like AB 2-type star polymers as nano-carriers for doxorubicin delivery to breast cancer cells: synthesis, characterization, in-vitro release and cytotoxicity studies. *Journal of Polymer Research* 27 (2020): 1–19.

Yasmin, A., Ramesh, K., and Rajeshkumar, S. Optimization and stabilization of gold nanoparticles by using herbal plant extract with microwave heating. *Nano Convergence* 1 (2014): 1–7.

Yazdi, M. H., Z. Sepehrizadeh, M. Mahdavi, A. R. Shahverdi, and M. A. Faramarzi. Metal, metalloid, and oxide nanoparticles for therapeutic and diagnostic oncology. *Nano Biomedicine & Engineering* 8, no. 4 (2016a): 246–267.

Yazdi, M. T., S. van Riet, A. van Schadewijk, M. Fiocco, T. van Hall, C. Taube, and P. S. Hiemstra. The positive prognostic effect of stromal CD^{8+} tumor-infiltrating T cells is restrained by the expression of HLA-E in non-small cell lung carcinoma. *Oncotarget* 7, no. 3 (2016b): 3477.

Zhao, C., X. Song, Y. Liu, Y. Fu, L. Ye, N. Wang, F. Wang, et al. Synthesis of graphene quantum dots and their applications in drug delivery. *Journal of Nanobiotechnology* 18 (2020): 1–32.

Zhu, P., S. Wang, Y. Zhang, Y. Li, Y. Liu, W. Li, Y. Wang, X. Yan, and D. Luo. Carbon dots in biomedicine: a review. *ACS Applied Bio Materials* 5, no. 5 (2022): 2031–2045.

Zhu, Y., Y. Lin, Y. Z. Zhu, J. Lu, J. A. Maguire, and N. S. Hosmane. Boron drug delivery via encapsulated magnetic nanocomposites: a new approach for BNCT in cancer treatment. *Journal of Nanomaterials* 2010 (2010): 1–8.

Ziylan Yavas, A., Y. Mizukoshi, Y. Maeda, and N. H. Ince. Supporting of pristine TiO_2 with noble metals to enhance the oxidation and mineralization of paracetamol by sonolysis and sonophotolysis. *Applied Catalysis B* 172 (2015): 7–17.

Index

active transport 33, 39, 67, 68, 92, 212, 223
agrochemicals 1, 197, 211
alkaloids 145, 147
amorphous silica 2, 39, 58
anthocyanins 47, 56, 147, 149
anthropogenic sources 1, 4
anthroposphere 1
antibacterial 19, 58, 280, 287
apoplastic pathway 2, 73, 92
aquaglyceroporins 32, 33, 43, 121, 136, 223, 225, 231
aquaporins 2, 33, 35, 37, 47, 92, 118, 135, 223, 231, 272
aquatic ecosystems 3, 4, 222, 225, 236
assimilation 19, 39, 102, 133, 135, 198, 202, 273
astatine (At) 1
ATP 33, 37, 74, 99, 114, 120, 132, 135, 181, 182, 225, 255
auxins 181, 204

bacterial siderophore 251, 252
beneficial roles of metalloids 18
bioaccumulation 4, 97, 100, 112, 113, 118, 226, 228, 231, 247
bioaugmentation 196
biochemical adaptations 73
biocompatibility 279, 286, 287, 290, 291
biodegradation 196, 199, 200, 211, 214
biofertilizer 160, 169, 205, 215
biofiltration 196
biogeochemical cycling 18, 212, 217, 222
biomagnification 4, 196, 221
biosorption 197, 213, 227, 228, 247
biosurfactants 197, 234
biotransformation 198
bismuth 182, 285, 287, 288
borates 92, 188
boric acid 2, 19, 33, 34, 56, 91, 92, 135, 169
boron deficiencies 93
boron toxicity 121, 183
boron transport 21

cadmium 27, 37, 57, 72, 110, 121, 148, 165, 184, 196, 211, 224, 272, 287, 288
calcium metasilicate 69
callose 90, 121, 168, 212
cancer risk 17
carbohydrate metabolism 27, 88, 89
carbohydrates 20, 23, 27, 58, 65, 71, 73, 88–90, 117, 134, 145, 150, 168, 250
carboxylates 247, 261, 292
carcinogenic 19, 23, 36, 98, 164
cardiovascular diseases 167
carotenoids 47, 56, 71, 90, 117, 132, 135, 136, 180, 185
casparian strip 35
catalysis 18, 211, 283, 285
catecholates 247, 261
cell development 102
cell division 20, 22, 24, 27, 46, 89, 90, 99, 161, 168, 178, 181, 184, 187, 188, 212
cell elongation 5, 53, 178, 181, 189
cell wall formation 56, 89, 168, 270
cellular homeostasis 54, 66, 144, 271
cellulose 74, 75, 90, 100, 121, 147, 184, 185, 234
chelation 43, 45, 46, 48, 57, 58, 137–139, 149, 168, 212, 248, 272, 273
chelators 46
chloroplast 20, 27, 48, 73, 102, 110, 132, 133, 135, 144, 150, 161, 165, 178, 182, 255, 271
chlorosis 6, 7, 24, 38, 72, 88, 99, 134, 135, 182
CO_2 absorption 88
CO_2 assimilation 102, 133, 135
competitive interactions 3
CRISPR 108, 109, 111, 118, 274, 275

dehydrogenase enzymes 90
dimethylarsinic acid (DMA) 181
DNA damage 23, 25, 167
DNA synthesis 19, 90

earthworms 26, 27, 203
ecosystems 2–4, 72, 97, 99, 103, 144, 221–223, 229, 233, 234, 236, 288, 291
efflux channels 44
electron transport chain 55, 66, 132
encapsulation 286
endophytic bacteria 26
environmental pollution 144, 272
ex situ bioremediation 197–199, 226

FBPase 90
flavanones 147
flavones 147, 149
flavonoids 54, 56, 71, 75, 117, 145–147, 149, 150
flavonols 147
fluorescence 165
foliar application 174
food chain 1, 2, 4, 9, 15, 23, 28, 45, 97, 98, 144, 149, 156, 163, 181, 196, 205, 216, 230, 259
food safety 23, 28, 76, 166, 178, 211, 272
fossil fuels 1, 2
fungal siderophore 251, 254, 255, 269

genetic engineering 133, 169, 205, 206, 222, 228
genetic modification 28
genomics 39, 99, 108, 111, 112, 116, 119
genotoxicity 16, 98
glutathione 19, 43, 56, 73, 100, 102, 113, 120, 121, 137, 138, 185, 201, 214, 259, 273, 288
glycolipids 89
glycolysis 23, 90, 145
glycoproteins 75, 90, 234

Haber-Weiss 145
health risks 15, 16, 28, 144, 181
hexose transport 38, 48
human health 99, 144, 156, 270, 272, 289
hydroxamates 247, 261
hyperaccumulation 48, 114, 212, 230, 275
hyperaccumulator plants 48, 205, 206, 273

immune system enhancement 21
immune system imbalance 23
in situ bioremediation 199, 200, 225
industrial effluents 25
industrialization 39, 144, 196, 211, 233, 246
integrated pest management 27
ionomics 108, 110, 111, 119
isoflavone 147, 149
isoprenoids 147, 149

lipidome 116
liver diseases 21

membrane proteins 33, 67, 221, 223, 257
membrane transport 24, 93, 135, 224
metabolic pathways 72, 103, 109, 110, 115, 122, 132, 144, 145, 149, 161, 181, 272
metabolic profiling 112
metabolomics 108–112, 119
metal homeostasis 179, 221, 224, 229, 234, 246
metal transporters 101, 118, 134, 177, 201
metalloid complexation 101
metalloid hyperaccumulators 26
metalloid sequestration 48, 149
metalloid translocation 272
metalloid transport 1, 32, 37, 39, 41, 43, 47, 48, 138
metalloid transporters 23, 48, 138
metalloid uptake 25, 27, 45, 48, 103, 138, 185, 189, 212, 223, 271
metallomics 101, 119, 126
metallothionein 43, 73, 102, 118, 137, 149, 186, 204, 229, 257
methylerythritol 145
mevalonic acid 145, 150
microalgae 27, 228, 231, 234, 236
microbial activity 58, 199, 201, 280, 286
microbial remediation 28, 211, 212
micronutrient 3, 4, 21, 57, 97, 100, 135, 156, 159, 160, 169, 211, 273, 292
mineral deficiency 162
monomethyl arsenic acid (MMA) 2
municipal and industrial wastes 1

NADPH 132, 256
nanoparticles 213, 231, 279, 287, 289, 293
nanotechnology 67, 78, 279, 285, 286, 289, 291–293
necrosis 6, 21, 24, 38, 72, 88, 99, 134, 168
nematodes 74
nitric oxide 17, 121, 187
nitrogen metabolism 180
nucleic acid metabolism 93
nutrient cycling 4, 169, 236
nutrient transporters 48
nutritional security 97

oligotrophic 217
osmolytes 54, 55, 59
oxidase 150
oxidative damage 27, 56, 71, 74, 100, 102, 133, 134, 137, 162, 168, 185
oxidative stress 99, 102, 103, 121, 135–137, 147, 182, 184, 201, 225, 253, 259, 271, 286

passive diffusion 2, 33, 34, 92, 135, 136, 270
passive transport system 68
pathogenesis 150
pectin molecule 89
pentose 145
peroxidase (POD) 19, 27, 38, 55, 76, 78, 137, 150, 201, 216, 228, 259, 288
phenol metabolism 5, 89
phenomics 111, 112
phenylalanine 73, 147, 150
phenylpropanoid 146, 150
phosphoinositides 89
photoinhibition 133
photophosphorylation 135
photosystem II 55, 56, 74, 100, 134, 182, 225
physical and chemical characteristics 1, 2, 15
physicochemical properties 3, 271, 286
phytoalexins 66, 75, 150
phytochelatins 11, 25, 43, 102, 117, 118, 120, 137, 149, 204, 272
phytodegradation 138, 202, 213, 232, 233, 274
phytoextraction 138, 185, 201–204, 213, 232, 234, 272, 274
phytoremediation techniques 4, 115, 231
phytosiderophores 212, 252, 254
phytostabilization 138, 201–203, 232, 233
phytotoxicity 16, 47, 76, 101, 185
phytotransformation 201, 202, 274
phytovolatilization 201, 202, 213, 214, 232, 233
plant health 9, 22, 56, 93, 108, 157, 159, 223, 224, 272
plant hormones 74, 75, 90, 178–180, 188, 205
plasma membrane permeability 89
pollen tube wall 90
polonium 15, 156, 196
polyols 92
polyphenol 71, 75, 125, 147, 234
proline 53–55, 75, 119, 124, 137, 179, 188
protein families 32, 43, 44, 118
proteomics 39, 101, 108, 110, 119

quantum 134, 136, 137, 279, 285

radionuclides 118, 197, 200, 228, 232
reactive oxygen species 4, 17, 45, 54, 113, 145, 215, 247, 288
redox potential 3, 16, 17, 19, 55, 97, 162, 201, 224
rhizodegradation 213–215
rhizofiltration 201, 202, 213, 214
RNA sequencing 101, 108
root development 88, 90, 181, 188
root exudates 25, 118, 202, 234
ROS generation 99, 135
rubisco 55, 90, 102, 136

salicylic 17, 25, 74, 145, 173, 177, 204, 247, 248
sediment enrichment 1
selenoproteins 19, 162
semiconductor 183, 211, 270, 290
sequestration 1, 26, 32, 43, 47, 48, 120, 139, 149, 200, 201, 212, 215, 229, 257, 259, 274
serotonin 145
shikimic 145
shoot necrosis 88
Si transporters 67
siderophores 7, 203, 212, 234, 247–260
signal transduction 179
silica gel 65, 69
silicon 69, 70, 73, 75, 77, 78, 118, 134, 144, 164, 175, 181, 183

sodium metasilicate 69
soil amendments 158, 169, 178, 202, 215, 243
soil contamination 1, 120, 270
soil degradation 196
soil erosion 205
soil microbes 197
soil organic matter 1, 97, 203
soil pH 57, 158, 203, 211, 234
soil properties 16, 158
soil remediation 205
soil solution 2, 18, 57, 67, 76, 92, 93, 134, 161
stomatal conductance 7, 73, 74, 90, 102, 133, 188, 270
stress resilience 47, 108, 110
stress responses 178, 179, 222, 230
stress tolerance 17, 53–55, 69, 70, 73, 74, 110, 115, 150, 186, 190
sugar translocation 56, 90
sulfate transport 19, 32, 48, 58
sulfur 19, 25, 26, 58, 101, 121, 136, 183, 185–187
sustainable agriculture 53, 59, 76, 157, 221–223, 236
symbiotic partnerships 221
symplastic pathway 92, 201
synthesis 4–7, 18–20, 24–28, 46, 56, 57, 65, 66, 71, 146, 149, 166, 168, 185–190, 255, 259, 261

tannins 145, 146
tellurium 1, 15, 32, 149, 156, 181
temperature 22, 35, 53, 55, 56, 58, 67, 147
terpenes 145, 147, 149, 150
terpenoids 145–147
thiols 11, 45, 46, 137, 273
thylakoid 74, 134–136
toxicity symptoms 6, 16, 99
transcriptomics 108–112, 119, 228
transgenic plants 274–276
translocation 2, 6, 22, 27, 37, 44, 56, 57, 73, 90, 92, 93, 135, 161, 168, 169, 201–204
transpiration 2, 7, 35, 39, 44, 55, 65–67, 69, 71–74, 92, 93, 112, 165, 185, 214, 270, 274
transpiration stream 2, 39, 44, 69, 92, 93
transporter proteins 36, 37, 43, 271
transporters 1, 2, 19, 20, 22–25, 32, 35, 37, 39, 43, 67–69, 92, 101, 113, 114, 170, 177, 204, 221, 223–225, 258–260, 272, 273

vacuolar sequestration 101, 137–139, 274
vacuole 20, 27, 37, 43, 45–48, 118, 121, 137, 149, 182, 201, 204, 225, 231, 272, 273
vascular 21, 23, 44, 93, 120, 138, 164, 167, 168, 215, 257
volatile 27, 120, 138, 198, 199, 202, 213, 233, 272–274

water contamination 229
water remediation 199, 209
water stress 53, 70, 88, 292

xylem loading 34, 35
xylem parenchyma 93

yield 93, 99, 102, 103, 112, 113, 118, 134, 136, 137, 159, 160, 164, 168, 169, 177, 181, 223, 252, 273, 288, 292, 293

FSC
www.fsc.org
MIX
Papier | Fördert
gute Waldnutzung
FSC® C083411

Kolibri 360 GmbH
Ferdinand-Jühlke-Straße 7
99095 Erfurt, Deutschland
produktsicherheit@kolibri360.de